T0189139

Sustainable Production, Life Cycle Engineering and Management

Series Editors

Christoph Herrmann, Braunschweig, Germany

Sami Kara, Sydney, Australia

SPLCEM publishes authored conference proceedings, contributed volumes and authored monographs that present cutting-edge research information as well as new perspectives on classical fields, while maintaining Springer's high standards of excellence, the content is peer reviewed. This series focuses on the issues and latest developments towards sustainability in production based on life cycle thinking. Modern production enables a high standard of living worldwide through products and services. Global responsibility requires a comprehensive integration of sustainable development fostered by new paradigms, innovative technologies, methods and tools as well as business models. Minimizing material and energy usage, adapting material and energy flows to better fit natural process capacities, and changing consumption behaviour are important aspects of future production. A life cycle perspective and an integrated economic, ecological and social evaluation are essential requirements in management and engineering.

Indexed in Scopus

To submit a proposal or request further information, please use the PDF Proposal Form or contact directly: Petra Jantzen, Applied Sciences Editorial, email:petra.jantzen@springer.com

More information about this series at http://www.springer.com/series/10615

Yusuke Kishita · Mitsutaka Matsumoto ·
Masato Inoue · Shinichi Fukushige
Editors

EcoDesign and Sustainability I

Products, Services, and Business Models

Editors
Yusuke Kishita
Department of Precision Engineering
The University of Tokyo
Bunkyo, Tokyo, Japan

Mitsutaka Matsumoto
National Institute of Advanced Industrial Science and Technology (AIST)
Tsukuba, Ibaraki, Japan

Masato Inoue
Department of Mechanical Engineering Informatics
Meiji University
Kawasaki, Kanagawa, Japan

Shinichi Fukushige
Department of Mechanical Engineering
Osaka University
Suita, Osaka, Japan

ISSN 2194-0541 ISSN 2194-055X (electronic)
Sustainable Production, Life Cycle Engineering and Management
ISBN 978-981-15-6781-0 ISBN 978-981-15-6779-7 (eBook)
https://doi.org/10.1007/978-981-15-6779-7

This Springer imprint is published by the registered company Springer Nature Singapore Pte Ltd.
The registered company address is: 152 Beach Road, #21-01/04 Gateway East, Singapore 189721, Singapore

Preface

EcoDesign has been a core concept for the manufacturing industry in its efforts to transform the mass production, mass consumption, and mass disposal paradigm toward achieving sustainability. In the last two decades, the ecosystem of the manufacturing industry has been rapidly changing, especially when we look at Sustainable Development Goals (SDGs), circular economy, and digitalization. In addition, the COVID-19 pandemic has drastically changed our lifestyles to the "New Normal," where a typical example is to shift from working in office to working from home using a teleconference system. In response to such emerging needs and enablers, though the impact of COVID-19 has not yet been considered, the scope of EcoDesign has been expanding to cover more diversified areas, such as environmentally conscious design of products, services, manufacturing systems, supply chain, consumption, economies, and society.

This book collates 79 papers out of 205 papers presented at EcoDesign 2019—the 11th International Symposium on Environmentally Conscious Design and Inverse Manufacturing, which was held in Yokohama, Japan from November 25 to 27, 2019. All the 79 papers were peer-reviewed by the EcoDesign 2019 Executive committee. Celebrating the 20 years anniversary of the symposium since its first occurrence in 1999, EcoDesign 2019 provided the excellent platform to share the state-of-the-art research and practices in the field of EcoDesign. The total of 278 researchers and practitioners from 28 countries participated in EcoDesign 2019.

The book consists of two volumes, i.e., the first volume focuses on "Products, Services, and Business Models" and the second volume focuses on "Social Perspectives and Sustainability Assessment." Reflecting the expansion of the symposium scope, the book chapters cover broad areas—product and service design, business models and policies, circular production and life cycle management, green technologies, sustainable manufacturing, sustainable design and user behavior, sustainable consumption and production, ecodesign of social infrastructure, sustainability education, sustainability assessment and indicators, and energy system design. We believe that the methods, tools, and practices described in the chapters are useful for readers to facilitate value creation for sustainability.

Last but not least, we would like to express our sincere appreciation to all the contributors, supporters, and participants of EcoDesign 2019. This book cannot be published without the help of the executive committee members who cooperated in the peer review of the papers.

Bunkyo, Japan
Yusuke Kishita
EcoDesign 2019 Program Chair

Tsukuba, Japan
Mitsutaka Matsumoto
EcoDesign 2019 Executive Committee Chair

Kawasaki, Japan
Masato Inoue
EcoDesign 2019 Executive Committee Co-chair

Suita, Japan
Shinichi Fukushige
EcoDesign 2019 Executive Committee Co-chair

Contents

Part IV Green Technologies

Part V Sustainable Manufacturing

Part I
Product and Service Design

Chapter 1
Industrial Designers Towards Design Concepts Based on the Water and Fire Themes—A Review and Comparison of Sustainability Considerations

Ueda Edilson

Abstract The preliminary research contributes to an initial discussion in the field of industrial design regarding the natural resources of water and fire for sustainability. It reviews and compares the aptitudes of industrial designers from around the world and submitted to the respected International Design Competition, Osaka organized by the Japanese Design Foundation and supported by Japanese Ministry of Economy, Trade, and Industry. It contributes to efforts to support and promote the importance of natural resources of water and fire sustainability from an industrial design perspective. According to the results, the large number of design concepts in differents categories suggests that industrial designers had the most flexibility to express their design concepts in this manner. The respective themes of water and fire had a relationship with each product design fields. The products design from water theme were characterized by common elements of evolutions, like methods of utilizing the nature, non consumption energy, clean use and so on. While in the fire theme was characterized by products design that produce energy through sunlight and products design that previne and protection from accidents caused by fire.

Keywords Water · Fire · Product design concepts · Industrial design · Sustainability · Sustainable design · Ecodesig

1.1 Introduction

This preliminary research examines the dimensions of sustainability and sustainable design strategies considered by industrial designers. The water and fire design are explored in its conceptual phase, the stage at which industrial designers have the greatest freedom to generate ideas and design solutions with the fewest constraints. The lack of information and concrete recommendations to address the sustainability of natural resources has kept industrial designers from demonstrating their skill and creativity in environmental product design.

U. Edilson (✉)
Department of Industrial Design, Faculty of Engineering, Chiba University, Chiba, Japan
e-mail: edilsonsueda@faculty.chiba-u.jp

Y. Kishita et al. (eds.), *EcoDesign and Sustainability I*, Sustainable Production, Life Cycle Engineering and Management, https://doi.org/10.1007/978-981-15-6779-7_1

1.1.1 Industrial Design and Sustainability

During sixty year's history of industrial design field in Japan, the promotion of environmental concern and ecology in design field was focused initially on the activities related to the design competitions. Only in the early 90 s through the symposiums and seminars, the terms like ecodesign, green design, design for environment and more recently sustainable design came to be considered in the design academic fields (Japan Design Foundation 1989; Green Designing in Yamagata 1992; Ecodesign 1999).

The term ecological design in the Dictionary of Today's Design (Katsui et al. 1992) was introduced in the year 1992, after many ecological design events like: the second and third International Design Competition Osaka (IDCO) with the themes water and fire. IDCO was a biannual international design competitions that covered all areas of design and served to redefine the role played by design while presenting a concrete vision of better future for manking (Japan Design Foundation (JDF) 1987, 1989).

1.2 The Objective of Research

This research explores the sustainability aspects of design concepts related to the theme of the element of nature water and fire, submitted by industrial designers from around the world to the respected International Design Competition, Osaka (IDCO), which is supported by the Japanese Ministry of Economy, Trade, and Industry (METI). These design concepts are studied and compared.

The Japan Design Foundation (JDF) coordinated this unique and original theme of water and fire (Japan Design Foundation (JDF) 1987, 1989). Analysis of these design concepts offers a point of study and understanding for the relationship between the theme of water and fire and their respective design categories and ecodesign strategies, as well as their social, cultural and environmental dimensions.

1.2.1 Justification of Choice Water and Fire Themes

The choice to examine design concepts of the water and fire themes from the IDCO are justified because these themes were a unique and original reference for the global history of design. Their importance are recognized in international dictionaries of design (the Oxford Dictionary of Modern Design (Woodham 2006) and Japanese Dictionary of Today's Design (Katsui et al. 1992)).

The global history of industrial design has no similar event or activity. Its original "Design for Every Being" concept, its international scale, the number of participating designers from around the world, and its historical importance to the Japanese design and global design justify detailed review and study of submitted design concepts for the benefit of current and future professional industrial designers.

From an academic perspective, no similar or specific research has examined the elements of nature water and fire competition from an industrial design perspective, even in the papers published from 1999 to 2013 by the International Symposium on EcoDesign (Ecodesign Symposium 1999) and in the special, environmentally focused issues from Japanese Society for Science of Design—JSSD (Japanese Society for Science of Design—JSSD 1997; JSSD 2011).

1.3 Design of Research

Given the facts, the sustainability aspects of design concepts are examined from the industrial design perspectives. An inventory of sustainable design strategies from viewpoints of industrial designers are considered in their concepts. They are formulated based on the three sustainability dimensions: social, cultural and environmental. To do so, the tools available for industrial designers are extensively reviewed, with the most relevant found to be the ecodesign strategies. Special reports (Japan Design Foundation (JDF) 1987, 1989) of IDCO contribute to the analysis of the submitted design concepts.

1.3.1 Ecodesign Strategies

Concept design works are analyzed based on the Ecodesign Strategy by Brezet (Brezet and val Hemel 1997), which is useful at both the early and late stages of the product-development process, including for re-evaluation after improvements. The strategy, which helps to identify on which areas to prioritize, is based on the lifecycle, which is a technical approach.

New factors and elements of evolution are adapted in this research to analyze the design concepts in order to explore the new characteristics that the early stage of design development commonly explores.

These factors and elements were found through the review of each design work by the author and also by two Phd experts in Ecodesign.

Each design works were reviewed based on the description as the title and subtitle and also the contents summarized with 300 words by the design author. The design categories of each design work were also defined by the design author through choice options of IDCO application form, e.g.: (A) poster, (B) Transportation, lighting and others (C) Furniture, toys and others (D), Urban Planning, Landscape and others (E) Design research and others. These design categories were based on the recommended by the ICOGRADA (International Council of Graphic Design Association), the ICSID (International Council of Societies of Industrial Design) and the IFI

(International Federation of Interior Architects/Interior Designers) (Japan Design Foundation 1989).

Through the analysis of each design work, we found the frequency of contents characteristics that varied between two or more times. These characteristics are: (1) innovative use natural resources (water and fire), (2) sharing and leasing (new way of consumption of natural resources), (3) natural resources information (environmental concern, learning with nature), (4) natural resources education (discovering natural resources), (5) emotionally design (user's internal feeling, re-think the nature) and (6) multi-functional (mixing natural resources). Each design work was judged with values 0 and 1. To validate the analysis, three experts in ecodesign, four researchers and two lectures in the industrial design carefully reviewed the environmental aspects of each submission. Table 1.1 shows the basic principles and strategies adopted in this research.

Table 1.1 Ecodesign strategies related with water and fire themes

Principles	(N)	Strategies approaches to sustainability
(1) New concept development	1	Innovative use of natural resources
	2	Sharing and leasing
	3	Natural resources information (essence of life, knowing)
	4	Natural resources education (discovering natural resources)
	5	Emotionally design
	6	Multi-functional
(2) Selection of material phase	7	Cleaner material
	8	Renewable materials
	9	Minimizing materials
	10	Recycled materials
(3) Production phase	11	New methods of production
	12	Cleaner production
(4) Use phase	13	Easy to maintain
	14	Easy to repair
	15	Low energy consumption
	16	Cleaner energy consumption
(5) End-of-life phase	17	Easy to disposal
	18	Cleaner disposal
	19	Reuse of product
	20	Recycling of material
	21	Not emit waste and toxic substance

Adapted from ecodesign strategy (Brezet and val Hemel 1997)

Table 1.2 The basic sustainable dimensions found in the product design concepts based on the water theme

	Water theme
(A)	Social dimension
1	Saving natural resources
2	Local-global
3	Pollution concerns
(B)	Cultural dimension
1	Symbolic culture
2	Knowing the essence of water
3	Emotionally design
4	Leisure
(C)	Environmental dimension
1	Sustainable use natural resources
2	Discovering natural resources
3	Sharing use natural resources
4	Minimizing materials
5	Reduce garbage weight
6	Clean energy consumption

1.3.2 Sustainable Dimensions of Water and Fire Themes

Tables 1.2 and 1.3 summarize the basic sustainability dimensions; (a) social dimension; (b) cultural dimension and (c) environmental dimension of water and fire themes. The main keywords of each dimension came from the analysis (quality and quantity) of description each design concept selected by ten international jurors that represented the IDCO. The description of each design works included: 300-word comments by authors, dimensional drawing, and photos of prototypes.

1.3.3 Terms

To enable consensus around the keywords used in this research, the following terms are described.

(a) *Design concept*

The early phase of a product during the process of its development, involving functions created to fulfill customer needs and forms and behaviors created to realize those functions. Industrial designers have the freedom to generate and explore ideas without constraint by the parameters of the later stages of design (Takala et al. 2006).

Table 1.3 The basic sustainable dimensions found in the product design concepts based on the fire theme

	Fire theme
(A)	Social dimension
1	Fire safety
2	Local-global
3	Functional
(B)	Cultural dimension
1	Symbolic culture
2	Lifestyle
3	Warm feeling
4	Leisure
(C)	Environmental dimension
1	Active solar energy
2	Passive solar energy
3	Carbon dioxide removal
4	Recyclable material
5	Reduce garbage weight
6	Modularity

(b) *Sustainability Aspects*

It refers to the concept of sustainable design (also called environmentally sustainable or environmentally conscious design) is the philosophy of designing physical objects, the built environment, and services to comply with the principles of social, economic, and ecological sustainability. The term is broader and more complex than traditional ecodesign, being concerned with minimizing the full lifecycle impacts of products and services (A Review of Sustainable Development Principles 2016).

(c) *Active and Passive Product Design*

Two categories of products with different energy concerns, environmental impacts during use, and complexities of lifecycle, which might considerably influence the industrial designer. For example, passive products, like furniture, tableware, garden tools, or bicycles, have minimal environmental impact during use (energy consumption, air pollution, water pollution, solid waste, and water consumption) compared to manufacturing and end-of-life. Active products, like appliances, office machinery, and vehicles, all have the greater part of their lifetime environmental impacts during use (Akemark 1999; Ueda et al. 2000).

1.4 Preliminary Analysis of Product Design Concepts with Water and Fire Themes

1.4.1 Water Theme

According to the information of IDCO (Japan Design Foundation (JDF) 1987), in Table 1.4, 1144 design concepts were submitted in total to the first phase of judging by industrial designers from 47 countries, showing the interest of designers worldwide towards the water theme. A limited number of these design concepts: 44 (3.8%), respectively—were approved for final selection phases. Only works from 17 countries were approved for the final selection phase: the most from the United States and Germany, with several entries from Japan, and one or two works from 14 other countries.

365 submissions in visual communication (A), 256 works in product design (B), 196 works in product design (C), 298 works in environmental design (D), and 29 works in research and other categories (E).

In the first phase, design categories comprised (A) 31.9%, (B) 22.4%, (C) 17.1%, (D) 26.0%, and (E) 2.5% of entries, while the final selection phase comprised entries in categories (A) 22.7%, (B) 22.1%, (D) 9.4%, and (C) 8.3%, with no submissions in category (E) reaching the final phase.

Table 1.4 Categories of design concepts based on water and fire themes

Categories		Water theme				Fire theme			
		Entry works		Selected works		Entry works		Selected works	
		n/s	(%)	n/s	(%)	n/s	(%)	n/s	(%)
A.	Poster, small printed materials, photography, etc.	401	39.7	35	36.5	365	31.9	20	22.7
B.	Transportation, lighting, machinery, etc.	313	31	38	41.8	256	22.4	15	22.1
C.	Furniture, toy, miscellaneous, textile, etc.	153	15.2	6	6.6	196	17.1	4	8.3
D.	Urban planning, landscape, architecture, etc.	111	11	12	13.2	298	26	5	9.4
E.	Design research, others	22	2.2	0	0	29	2.5	0	0
Total		1009	100	91	100	1144	100	44	100

1.4.2 Fire Theme

According to the information of IDCO (Japan Design Foundation (JDF) 1989), in Table 1.4, 1009 design concepts were submitted in total to the first phase of judging by industrial designers from 52 countries, showing the interest of designers worldwide towards the fire theme. A limited number of these design concepts: 91 (8.91%), respectively—were approved for final selection phases. Only design concepts from 25 countries were approved for the final selection phase: the most from the Japan, United States and Germany, and one or two works from 22 other countries.

(A) 401 visual communication, (B) 313 product design (I), (C) 153 product design (II), (D) 111 environmental design, and (E) 22 research and other categories.

In the first phase, design categories comprised A. (39.7%), B. (31.0%), C. (15.2%), D. (11.0%), and E. (2.2%) of entries. The 91 entries selected in the final selection were reviewed and recategorized with the following results: categories A. (36.5%), B. (41.8%), C. (6.6%) and D. (13.2%), with no submissions in category E. reaching the final phase.

1.5 Results

1.5.1 Ecodesign Strategies and Sustainable Dimensions of Product Designs Based on the Water and Fire Themes

Product Design based on Water Theme

Table 1.5 shows the general approaches of ecodesign strategies of 24 product design concepts based on the water theme.

The sub-categories of product design concepts with water theme in Table 1.4 are: (B): equipment, camera, device, mobile, appliances and lighting; (C): flask, furniture, partition and plastic sheet; and (D): arquitecture and urban development).

Table 1.5 Basic ecodesign strategies of product design concepts based on the water and fire themes

Ecodesign strategies		Water (%)	Fire (%)
1. New concept	New solutions (needs)	75.2	29.4
2. Selection material	Cleaner material, recycled material	9.1	7.8
3. Production phase	New methods of production	2.2	7.8
4. Use phase	Low energy consumption, modular	6.8	45.1
5. End-of-life phase	Reuse, recycling of materials, safe	6.7	9.8
Total	3000	100	100

Table 1.6 Basic sustainable dimensions of product design concepts based on the water and fire themes

Sustainability dimensions	Water (%)	Fire (%)
1.Social	8.9	37.0
2.Cultural	10.1	37.0
3.Environmental	81.1	25.9
Total	100	100

Analyzing basic phases of ecodesign strategies in relation to the total percent item (T%) of water theme in Table 1.5, industrial designers gave attention on the new concepts, in order to explore new solutions in relation with the theme water (75.2%), the selection-of-material phase by minimizing materials (9.1%) was designers' second approach, in which they simplified their designs to reduce the number of parts and facilitate their assembly and disassembly.

Regarding the use phase, the item ease of maintenance and cleaner energy consumption were the most important aspects designers considered (details in Table 1.1).

In the end-of-life phase, designers considered only the element not emit waste and toxic substances, demonstrating basic knowledge and concern about the life-cycle principle. However, industrial designers did not pay attention to production phase (2.2%), that might have been integrated into their design concepts, such as renewable materials, recycled materials, ease of repair, low energy consumption, ease of disposal, cleaner disposal, reuse of product, or recycling materials.

The analysis of 24 product design concepts, in Table 1.6 shows that the environmental dimensions (81.1%) was first-most common, the cultural dimension (10.1%) was second-most common, and social dimension (8.9%) was third-most common.

The environmental dimension based on the design for education and care (Table 1.2) was represented by products that promote rethinking the importance of water, products with messages related to the environmental protection of water. Its main element of evolution was represented by products that expressed the natural essence of water, and it was represented by products related to issues of water pollution. A concrete example was a device that analyzes the condition of drinking water in certain regions that have been affected by environmental pollution.

From social and cultural dimensions were identified as products with new, innovative features, such as the integration of sharing and leasing, multi-functional and emotional aspects of design, and innovative and unique approaches of products based on the water theme. A concrete example was the integration of diving mask with water camera that captures the aquatic images in three dimensions and highlighting natural details of the sea floor. Other example was a floating light that explored the traditional ceremony of floating paper lanterns on a river, with the water surface becoming a gigantic mirror and its reflection of lighting through the wind eliciting visual and mental comfort.

Product Design based on the Fire Theme

In the same Table 1.5 shows the general approaches of ecodesign strategies of 56 product design concepts based on the fire theme.

The sub-categories of product design concepts with fire theme in Table 1.4 are: (B): equipment, lighting and appliances; (C) pot, cooking utensils and toys: and (D) public art and urban planning.

Analyzing basic phases of ecodesign strategies in relation to the total percent item (T%) of fire theme in Table 1.5, industrial designers gave attention on the use phase (45.1%) and new concept (29.4%) stages were the most important aspects considered by industrial designers. However, others strategies of lifecycle principles, such selection material (7.8%), production phase (7.8%), end-of-life phase (9.8%), had low consideration by industrial designers (details in Table 1.3).

Reviewing in details the 44 product design concepts of categories B and C, the majority of concepts (86.4%) were characterised as active products. As such, in their lifecyle during the use phase, they used non-renewable energy sources that will run out or will not be repleshided, such as the fossil fuels, natural gas and nuclear energy. Only a small percent (13.6%) did not use any kind of energy during the use phase.

In Table 1.6, the sustainability dimensions of the 44 samples, the design concepts with a symbolical culture of fire and social aspect of fire (37.0%) were the most significant approach explored by industrial designers in their proposal designs. Industrial designers had the intention to reintroduce fire to ceremony and domestic utility. Other approaches such as lifestyle, warm feeling, and leisure aspects were also explored by designers in their product design concepts.

From the social dimension, the product design concepts with fire safety was the most important aspects explored by industrial designers in their product design concepts. The most significant example were a public life guard mask and proposal of a fire fighting and emergency guidance system using a sensor robot.

Local and global approaches were also explored by industrial designers in their product design concepts, which they were related with energy or power production, such as a biogas and a the offshore solar hydrogen farm, with the latter characterized as global and the former as local approaches to energy solutions.

From the environmental dimension, the majority of product design concepts exploited the natural resources of sunlight, which industrial designers viewpoints they used as method of energy production through the active solar energy and passive solar energy.

The product design concept that used passive solar energy was the designbased on solar heat-storage cooker, which used the energy of direct for sunlight for heating and cooking.

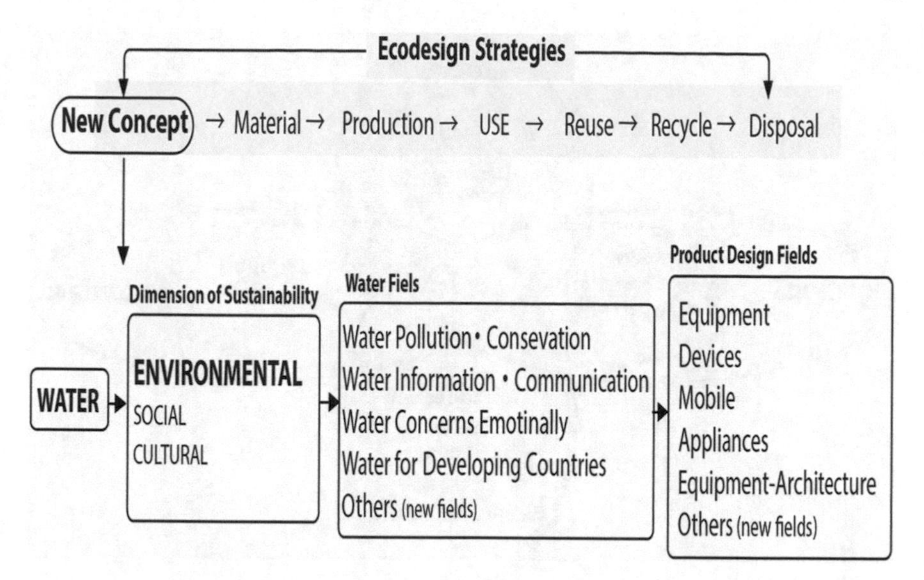

Fig. 1.1 The characteristics of product design concepts based on the water theme

1.5.2 The Characteristics of Product Design Concepts Based on the Water and Fire Themes

Summarizing the characteristics of the product design concepts based on the themes water and fire in Figs. 1.1 and 1.2, we can note that industrial designers had more possibilities and flexibility to propose innovative or new concept design in their respective fields of acting.

According to the Sect. 5.2, in both themes industrial designers focused on equipments field, which the reduction of environmental impact during the use of product was characterized by the following elements: products that not emit toxic substance and waste, any energy consumption and cleaner use. Methods of the utilization the nature and reuse of product from optimization of the end-of-life systems were also noted by designers in their works.

The products design from water theme were formed by common elements of evolution, like methods of utilizing the nature, non consumption energy, cleaner use, not emit toxic substances and wastes and so on.

Industrial designers showed familiarity with exploration of resources of water, e.g.: production of food using seawater and the sun.

While in the fire theme was characterized by equipments that produce and consume energy. In other words, the theme fire in the majority of entry works was interpreted by industrial designers as protection and prevention from accidents caused by fire and energy production through sunlight.

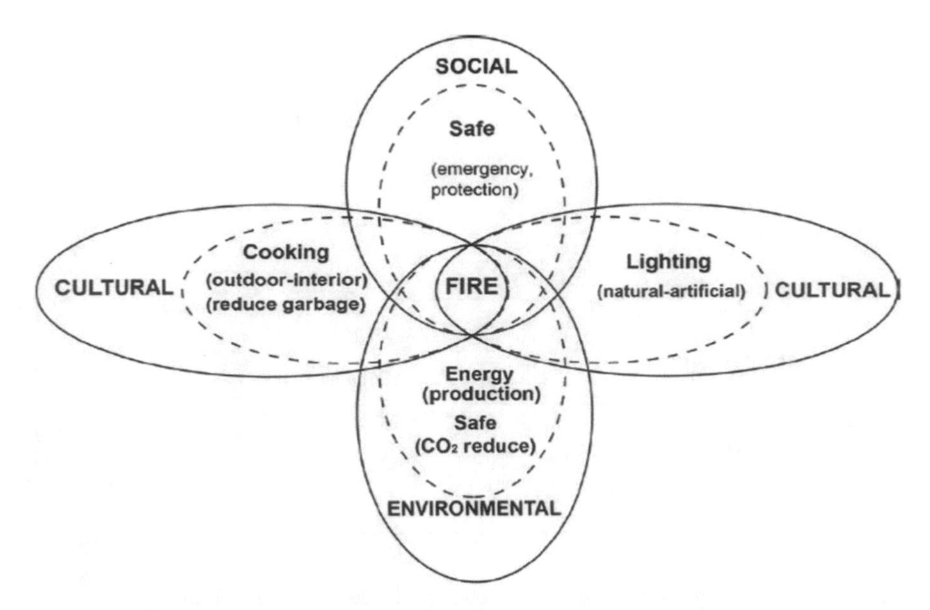

Fig. 1.2 The characteristics of product design concepts based on the fire theme

1.6 Conclusions

This research is a preliminary study of characteristics of water and theme from industrial designers perspective. The large number of design concepts in differents categories suggests that industrial designers had the most flexibility to express their concepts in this manner.

The respective themes of water and fire had a relationship with each product design field.

From the general view of the water theme, 24 product design concepts were characterized with the following dimensions: environmental approach (81.1%), cultural approach (10.1%) and social approach (8.9%).

In the water theme, industrial designers showed interest in the solve environmental problems (81.1%) through the new concepts (75.2%) and sustainable solutions. These new concepts were characterized as integration of new features, such as sharing, multi-functionality, and emotional aspects. One example are the products with sustainable approaches through unique and original ideas based on the use of natural resources of water, such as safe drinking water solutions, use of hydrogen energy for mobility, and production of food using seawater and the sun. From selection of material (9.1%) were represented by minimizing the use of material, in which respect industrial designers explored in their proposals simplicity in form and shape, thus reducing the number of component parts (Fig. 1.1).

Regarding the use phase, industrial designers showed interest in the item ease of maintenance and cleaner energy consumption. In the end-of-life phase, industrial designers considered only the element not emit waste and toxic substances, showing basic knowledge and concern about the lifecycle principle. The production phase was few considered by industrial designers.

From the general view of 44 product design concepts with fire theme, the majority of design concepts implemented the ecodesign strategies as new concept development (29.4%) and use phase (45.1%), which they were represented by the category of active products.

Other ecodesign strategies as selection of materials (7.8%), production phase (7.8%) and end-of-life (9.8%) stages had low consideration by industrial designers. During the use phase of these products, the non-renewable energy consumption were most common. The fire design concepts with a cultural dimensions were characterised as lighting and cooking design fields, while the social dimension were characterised as safe emergency and protection design fields. Finally, the environmental dimension concepts were characterised as new method of energy production through sunlight.

1.7 Limitations of Research

Unfortunately the JDF has closed its activities in 2008, which the theme "Water" was not realized for second time, and also a specific main theme with "Water" approach has not been proposed by others national and international design competitions from 1989 until at the present moment. These facts did not permit up-to-date of design works with water theme by industrial designers (Woodham 2006; Ecodesign Symposium 1999; Japanese Society for Science of Design—JSSD 1997; JSSD 2011).

1.8 Example of Strategies on the Water and Fire Themes

Based on the main findings of the previous sections, examples of main strategies design concept based on the water and fire themes are illustrated in Table 1.7. It shows the examples of key characteristics of products design with water and fire themes from social, cultural and environmental viewpoints. These examples could to stimulate industrial designers to explore new way of generation ideas to intigate new sustainable design solution. This proposal will be tested in the practical modules of the industrial design course.

Table 1.7 Designing strategies of product design concepts based on the water and fire themes

	Categories	Water theme			Fire theme		
	Product design	(a) Social	(b) Cultural	(c) Environmental	(a) Social	(b) Cultural	(c) Environmental
1	Furniture, toy, miscellaneous, textile, etc.	Information	Leisure	Natural resources	Safety	Cooking (outdoor-interior)	Passive/active
2	Transportation, lighting, machinery, etc.	Communication	Conservation	Pollution	Global-local	(reduce garbage)	Energy production
3	Urban planning, landscape, architecture, etc.	Sharing	Emotional	Global-Local	Safe emergency (protection)	Natural-artificial	Safe (CO_2 reduce)
4	Design research, etc.	Concern	Knowing	Sustainable use	Functional	Symbolic	Recycled material
5	Others	Others	Others	Others	Others	Others	Others

1.9 Future Research

In future research, a more detailed and specific inventory of sustainable design strategies in each theme and category of products is needed to identify new features for more innovative sustainable product design.

This inventory could support and facilitates industrial designers to demonstrate their potential creativity and innovation skill during the conceptualization or idea generation process stage of a sustainable product development.

References

A Review of Sustainable Development Principles (2016) http://mri.scnatweb.ch/en/afromontcontent/afromont-discussion-documents/2602-sustainable-development-review-2016/file. Accessed 10 Dec 2017

Akemark A (1999) Design for environment from the industrial designers perspective. In: Ecodesign symposium first international symposium on environmentally conscious design and inverse manufacturing. Ecodesign 1999: 47–50

Brezet HC, val Hemel C (1997) Ecodesign: a promising approach to sustainable production and consumption, United Nations Environment Programme, Industry and Environment, Cleaner Production, France, 52, 140–142, 346, 1997

Ecodesign '99: First International Symposium on Environmentally Conscious Design and Inverse Manufacturing. Tokyo

Ecodesign Symposium (1999–2013) First~eighth international symposium on environmentally conscious design and inverse manufacturing, ecodesign. IEEE Comput Soc, (1999) 10–859, (2001) 10–948, (2003) 1–838, (2005) 25–973, (2007) 29–978, (2009) 43–1246, (2011) 42–1272, (2013) 43–931

Green Designing in Yamagata (1992–1997) Search in various resorts of 1st, 2nd, 3rd, 4th, 5th and 6th

JSSD, Ecodesign, Special Issue (2011) 18–3 (71), 1–74

Japan Design Foundation (1989–1995) Search in various reports of 2nd, 3rd, 4th and 5th International design competition, Osaka

Japan Design Foundation (JDF) (1987) Special report of international design competition, Osaka (IDCO), 3–70

Japan Design Foundation (JDF) (1989) Special report of international design competition, Osaka (IDCO), 3–49

Japanese Society for Science of Design—JSSD (1997) The consideration of the relationship between design and ecology, Recycle PL-Law, Special Issue 1997, 4(3):1–51

Katsui M, Tanaka K, Mukai S (1992) Dictionary of today's design. Heibon Publishing, Japan, p 244

Takala R, Keinonen T, Mantere J (2006) Processes of product concepting. In: Product concept design: a review of the conceptual design of products in industry, pp 57–90. Springer

Ueda ES, Shimizu T, Kiminobu S (2000) The role of passive and active product designers involved in the environmental policy of the Japanese companies. Toward Eco-Des JSSD 47(2):57–66

Woodham JM (2006) Oxford dictionary of modern design, Oxford University Press, pp 224–225

Chapter 2
Time Axis Design as an EcoDesign Method

Kentaro Watanabe, Fumiya Sakamoto, Yusuke Kishita, and Yasushi Umeda

Abstract Time axis design is a design method with which a designer considers temporal changes of social situation, surrounding environment of design objects, users (including their values), and design objects. Although some products were designed with considering temporal changes, the concept of time axis design is not well organized until now. For clarifying the basic concepts of time axis design, this paper first illustrates cases of time axis design and extracts indispensable elements of time axis design. Second, this paper proposes a time axis design support method. Third, as a case study, this paper illustrates time axis design of umbrella. As a result, this paper shows that the proposed design method is effective in stimulating a designer for deriving design ideas.

Keywords Time axis design · Design methodology · Scenario

2.1 Introduction

Since the current manufacturing paradigm cannot get out of the mass production and mass consumption in correspondence with diversification and expansion of customer needs, today's manufacturing industry causes many problems such as sustainability issues typified by SDGs (United Nations 2015). The conventional design theory used in today's manufacturing is not sufficient for these problem because it does not concern about our values to products. For solving this problem, we study "time axis design," which explicitly deals with temporal changes in design (Matsuoka 2012).

K. Watanabe (✉) · Y. Kishita
Department of Precision Engineering, School of Engineering, The University of Tokyo, Tokyo, Japan
e-mail: watanabe@susdesign.t.u-tokyo.ac.jp

F. Sakamoto
Department of Precision Engineering, School of Engineering, The University of Tokyo, Nissan Motor Co., Ltd., Tokyo, Japan

Y. Umeda
Research into Artifacts, Center for Engineering, School of Engineering, the University of Tokyo, Tokyo, Japan

Y. Kishita et al. (eds.), *EcoDesign and Sustainability I*, Sustainable Production, Life Cycle Engineering and Management, https://doi.org/10.1007/978-981-15-6779-7_2

Design objects designed with the time axis design can respond to temporal changes. By introducing the view of time axis in design object, the designer can correspond with the changes of value we do not expect now. However, at present the concept of time axis design is not organized well.

The objective of this paper is to clarify the concept of time axis design and propose a method of time axis design applicable to various products. In this paper, we describe a method to find out time axis design solutions based on collected cases of time axis design.

The rest of this paper is structured as follows. Section 2.2 describes the role of time axis design for eco-design. Section 2.3 summarizes research efforts on time axis design. Section 2.4 describes cases of time axis design cases. Section 2.5 proposes time axis design method based on the collected cases. Section 2.6 shows a case study of time axis design. Section 2.7 discusses the effectiveness and challenges of the proposed method. Section 2.8 concludes the paper.

2.2 The Role of Time Axis Design for Sustainability

As discussed in SDGs and Circular Economy (CE) (European Commission 2015). One of the most important issues for sustainability is the design of value supply system considering future effects such as resource depletion and global warming. This requires product design for longer use in order to avoid resource depletion. However, the conventional product design was to design a product so as to perform the best at the beginning of use. Therefore, (Takahashi et al. 2011) proposed the concept of time axis design to cope with sustainability issues; for realizing time axis design, he proposed multi-space design model representing relationship between a design object and the places of use, and multi-timescale model expressing temporal changes of products with multiple temporal magnitudes. By introducing "time axis" to product design including support systems that enable the product to address temporal changes, we can design products that work longer and cope with sustainability issues typified by SDGs and CE.

In the context of eco-design, we considered about the extension of product life. Daimon et al. (2003) proposed decision support method for life cycle strategy by considering product physical lifetime and value lifetime. This discussion is a typical example of time axis design, while we were not aware of the concept of time axis design at that time. Moreover, if we discuss extremely, all sustainability issues are related to temporal changes, and the essential conflict resides within different time scales; for example, resource depletion and global warming in long time scale and pursuing quality of life in relatively short time scale.

So, by introducing the concept of time axis design, we can discuss such temporal changes in the context of design in an integrated and systematic manner.

2.3 Time Axis Design

In this paper, we describe the framework of time axis design as shown in Fig. 2.1. The designer designs a product service system so as to provide service to a user and the user feels the value from the service. In the modern context of product design, as service engineering Tukker (2004) says, wc should understand that a product more or less consists of hardware and service provided by human. In this paper, we take product service systems as design objects. In the places of use, various artifacts and natural objects exist and interact each other in addition to the designed product. In this situation, we can find out various temporal changes including temporal changes of the design object, the changes of the user, the changes of the place of use, and the changes of social situation. The changes of the design object (i.e., a product service system) include the temporal changes of hardware, such as degradation and failure, and the changes of service such as the improvement of the service. The changes of the user include physical changes such as growing tall and mental changes such as changes in preferences. In addition, we consider that the places of use, in other words the environment surrounding the product service system, also changes. Moreover, social situations such as legislation, fashion, and market environment also change.

The concept of time axis design is to design a product service system considering all of these kinds of temporal changes by introducing "time axis" to design space (Matsuoka 2012). In other words, from the viewpoint of time axis design, a product service system can provide value for users only when the state of these changing

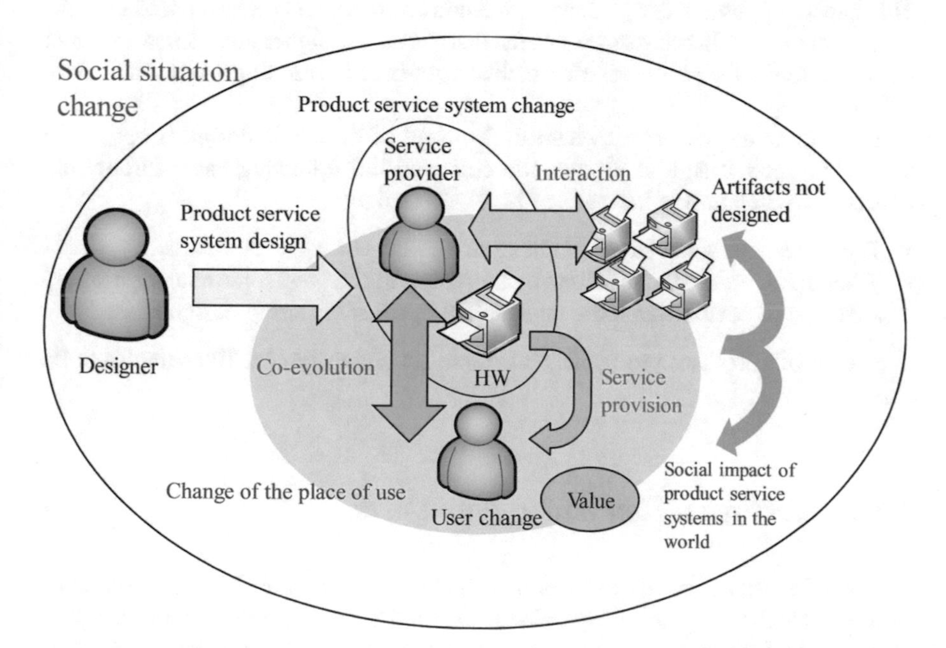

Fig. 2.1 Concept of time axis design

elements (the product service system, users, places of use, and social conditions) fulfill a certain set of conditions. We call such a set of conditions "value provision conditions." While the traditional design is to design a hardware product without considering all these types of changes, the time axis design is to design a product service system so as to satisfy the value provision conditions for a certain period of time with assuming all these types of changes.

For example, we have been studying upgrade design (Umemori et al. 2001), which extends the life of a product by upgrading some functions of the product at the use stage in response to changes of users' needs. We have also proposed self-maintenance machines (Umeda et al. 1995) that can recover indispensable functions by themselves when failure or degradation occur. They have built-in systems that can deal with problems that arise over time. We have developed these design methods without knowing the concept of time axis design. In this way, the concept of time axis design has not been discussed explicitly in the design community, although it is simple, clear, and very powerful tool for solving the sustainability issues.

2.4 Cases of Time Axis Design

2.4.1 Approach

This paper proposes a design support method for time axis design, based on cases we collected. This is because we assume that, when a designer faces a new problem, he or she looks for past similar cases like case-based reasoning (Agnar and Enric 1995).

Since there are few cases where the concept of time axis design is apparently utilized in their design at present, we collected the following cases through the Internet and literature as the cases of time axis design:

- The cases have some temporal changes.
- When some problems are caused by the temporal changes, some countermeasures are taken or countermeasures are installed in advance during design phase.

A total of 130 cases were collected. Table 2.1 shows typical 10 examples of the cases.

2.4.2 Classifying Time Axis Design Cases

We classified the collected cases based on the types of temporal changes and countermeasures for the temporal changes; Tables 2.2 and 2.3 indicate the types of temporal changes and the types of countermeasures, respectively. In this paper, we focus on temporal changes of users, design object, places of use. We do not discuss social

Table 2.1 Examples of time axis design cases

No.	Case name	Description
1	Message soap	A letter is embedded in the soap and when the user washes his/her body with the soap, the letter appears at a certain time
2	BABUBU furniture (http://shinse-i.jp/babubu/)	If the bed size does not fit a baby because of his/her growth, it is customized to a desk by applying additional parts
3	*Bonsai*	*Bonsai* refers to potted plants grown with special care for appreciation. A user can tailor the *bonsai* to his/her taste by taking care of it as the plant grows and the environment changes. (https://www.bonsaimyo.com/bonsai/Japanese/)
4	Copying machine	A copying machine is provided with maintenance service in case of trouble
5	LEGO (https://www.lego.com/en-us)	By preparing different sets of LEGO according to age, LEGO can respond to the changes in children's preferences
6	Wine	Wine is matured by storing it until it is ready to drink after purchasing it
7	Skype translation (https://www.skype.com/ja/)	As the AI learning, it leads to improve the performance of translation technology
8	Baggage made of bamboo	Bamboo used as material of a baggage is getting attractive as the user uses
9	Beaker pudding container (https://www.harioscience.com/)	By adding a measuring function in pudding container, the container can be used as a beaker after eating the pudding
10	Moist hair dryer: Areti (https://areti.jp/)	The dryer has the function to adjust air volume in response to the changes in the user's age and season

changes because it is ambiguous. By collecting and classifying the cases of time axis design, we made a matrix between the changes and the countermeasures as shown in Table 2.4. This Table 2.4 shows the number of the cases corresponding to each change and each countermeasure. Since the red cells in this Table 2.4 have enough number of cases, we consider that the countermeasure in this cell is effective to the change in this cell. And, we use these cells to find out design ideas in Sect. 2.6.

Table 2.2 Types of temporal changes (A-H)

Type of temporal change	Subtype of temporal changes	Example
Change of a user	(A) Change of the outside of a user	Growing tall
	(B) Change of the inside of a user	Change of user's taste preference
	(C) Switch of users	Used by different users
	(D) Change of behavioral pattern of a user	Change contents to match user needs
Change of a design object	(E) Change the structure/attribute of design object	Change of color of the design object
	(F) Change the information held by design object	Accumulating usage history in the design object
Change of the place of use	(G) Change the environment in the place of use	Change of temperature according to the change of season
	(H) Change the places of use by the user	Use the design object in different places

2.5 Design Support Method for Time Axis Design

2.5.1 Approach

In order to solve the problem mentioned in Sect. 2.3, we construct a support method for time axis design as shown in Fig. 2.2. In this method, a designer identifies temporal changes of a design object by examining the use case and derives countermeasures for the temporal changes. First, a designer defines a persona that uses the product service system to design. Then, the designer draws use case scenarios in which the persona uses the design object (see Sect. 2.5.2). The use case scenarios allow the designer to identify plausible temporal changes of the design object, users, and the place of use. Next, the designer identifies a type of temporal change for each temporal change with referring Table 2.2. At this time, the designer classifies the temporal change into two types of long timescale (year/month) and short timescale (one cycle of use of the design object) based on the concept of multi-time scale (Matsuoka 2012). Then, the designer selects countermeasures corresponding to the identified types of the temporal changes with referring Table 2.4. For supporting these steps, we constructed a database of the cases of time axis design we collected. The designer can refer the cases in Table 2.4 (see Sect. 2.5.3). Next, the designer finds out design ideas realizing the selected countermeasures. Finally, by selecting the design ideas and integrating them, the designer derives a design solution (see Sect. 2.5.4).

Table 2.3 Types of countermeasures (a–i)

How to respond	Type of countermeasure	Description	Example
Embed countermeasure in the design object	(a) Add a function to adjust	The design object has adjustable structure to cope with temporal changes	Height adjustable structure
	(b) Change of configuration of design object	The design object has structure in which multiple parts are combined or recombined	Structure that can be freely recombined to the size user needed
	(c) Make countermeasure service easier	The design object has a mechanism that allows users and service providers to execute service easily	Applying design for maintenance
	(d) Use the structure that have a function in self-maintenance	The design object has a structure that can return to its normal state after a failure happen	Structure that repairs scratches on the surface
	(e) Prepare multiple usages	The design object has potential functions	Pudding containers also has a function as measuring cup
	(f) Have a function to learn	The design object has a mechanism to change the behavior based on the accumulated data	Function to learn selection history of the user in the Internet shopping
	(g) Have a function to sense changes	The design object has a mechanism that senses temporal changes and notifies the user	Function to sense changes in the places of use with a change brightness
Create a system to respond as service	(h) Prepare countermeasure service	During use of the design object, services are provided to cope with temporal changes	Repair service is provided
Encourage the user to respond	(i) Provide the user with the opportunities to respond	Facilities and equipment are provided for users to cope with temporal changes as needed	Spare parts are prepared and the user replaces them with faulty ones

Table 2.4 The number of collected time axis design cases

		Types of countermeasures								
		(a)	(b)	(c)	(d)	(e)	(f)	(g)	(h)	(i)
Types of temporal changes	(A)	12	9	0	0	3	0	1	0	0
	(B)	11	7	0	0	4	20	15	6	6
	(C)	5	2	0	0	5	13	12	3	3
	(D)	12	12	0	0	6	13	11	6	4
	(E)	6	1	12	3	3	0	8	12	2
	(F)	7	0	1	1	5	50	14	7	3
	(G)	5	0	3	1	1	4	9	3	0
	(H)	4	3	1	1	3	0	3	4	0

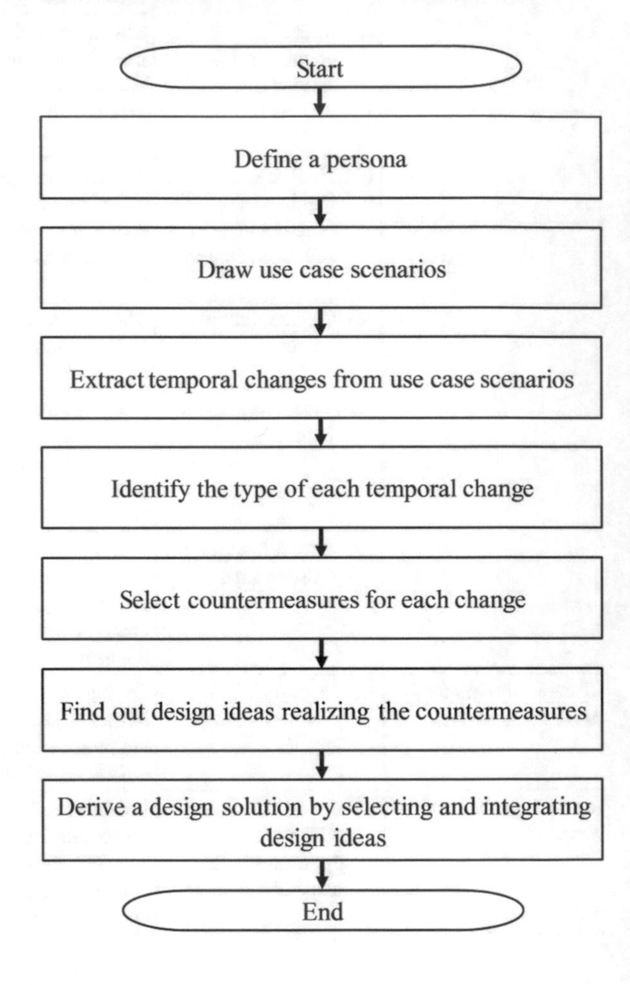

Fig. 2.2 Flow of time axis design support method

2.5.2 *Define a Persona and Draw Use Case Scenarios*

When considering temporal changes of a design object, it is necessary to consider the user who uses the product and the place of use of the product. In this method, the designer clarifies them by describing persona of a user and his/her use case scenarios.

2.5.3 *Extract Temporal Changes from Use Case Scenarios, Identify Key, and Select Countermeasures*

From the described use case scenario, the designer extracts all plausible temporal changes of the design object, the user, and the places of use. Second, the designer identifies a type for each extracted temporal change with referring Table 2.2 and classifies the temporal changes into two types of timescale.

Third, the designer selects countermeasures corresponding to the selected types of the changes.

By using the uniform expression about the type of the temporal changes summarized in Tables 2.2, 2.3 and 2.4 and the database of the cases, we think the designer can create design ideas efficiently.

2.5.4 *Find Out Design Ideas and Derive a Design Solution*

The designer finds out design ideas that can realize the countermeasures selected in Sect. 2.5.3. Then, by selecting appropriate countermeasures and integrating them, the designer derives a design solution.

2.6 Case Study

2.6.1 *Problem Setting*

We executed time axis design based on the proposed method in Sect. 2.5. In this paper, we take an umbrella as a design object. The reason why we choose an umbrella is that it is one of the most popular products daily used by everyone. Therefore, when proposing design ideas about an umbrella, we easily assume the shape, the size, and use case. In this case study, the product service system at the beginning is not clear.

2.6.2 Define a Persona and Draw Use Case Scenarios

We described the use case scenario as follows:

Persona

A Man, 30 years old, office worker

He often goes out for work with his umbrella in rainy days.

Use Case Scenario

On a sunny day, he does not use his umbrella. When it rains or snows all day, he uses his umbrella from his house to his office. He sometimes loses his umbrella on the public transportation. When the wind is strong, he closes the umbrella and stops using it. If he dares to use it, the umbrella might be flapped by the wind, breaking the bones and the surface cloth. When the same umbrella is used for long time, the bones of the umbrella become rusty and it becomes difficult to open and close. Furthermore, the bones are distorted according to time and the cloth also breaks, and the water-repellent function is deteriorated. When the umbrella can no longer prevent rain or he gets bored with the design of the umbrella, he replaces it with a new umbrella of desired color and stops using the old one.

We described this persona and use case scenario since we often see in daily life.

2.6.3 Extract Temporal Changes from Use Case Scenarios, Identify Key, and Select Countermeasures

Next, we extracted plausible temporal changes from the use case scenario and related them to types of temporal changes in Table 2.2. Then, we classified the temporal changes into two timescale. In this case study, the short time scale is one cycle for using an umbrella. Table 2.5 shows the results of this step.

For each type of temporal changes, we selected countermeasures by using Table 2.4 and the database.

2.6.4 Find Out Design Ideas and Derive a Design Solution

We found out design ideas for the countermeasures. Table 2.6 summarizes the design ideas.

As a result of integrating the design ideas, the result of devised design solution is described as below.

Design solution

The Umbrella is made of cloth that is more water-repellent when we use. The number of the bone can be changed and even if several bones are broken, it could be dealt with by the rearrangement. The umbrella can be changed to a raincoat according to the user's mood. If the bones get rusty or the cloth deteriorates, the

Table 2.5 Extracted temporal changes in the case study and their time scales

Patterns		Type of the time scale	Design ideas
Temporal changes	Countermeasures		
(B)	(e)	Short	Change raincoat depending on user's emotion
(D)	(a)	Short	Add an attachment that supports loads
(D)	(i)	Long	Provide a service that allows users to change to durable bones based on frequency of use
(E)	(d)	Long	Use water-repellent cloth
(E)	(i)	Long	Provide a service that allows users to repairs cloth
(G)	(h)	Short	Provide umbrellas rental service where users can borrow and return their umbrella when it's sunny
(H)	(b)	Short	Change the number of bones to cope with snow
(H)	(g)	Short	Add a function to notify users when the rain stops

Table 2.6 Examples of design ideas

Temporal changes from use case scenario		Type of the temporal change	Type of the time scale
User	Preferences change (get tired of umbrella design)	(B) Change of the inside of a user	Long Short
	The luggage changes (large/small)	(D) Change of behavioral pattern of a user	Long Short
Design object	In long time scale, umbrella broken (bones breaking and cloth breaks) and the condition of umbrella has changed (rusting and water repellent function deteriorated)	(E) Change the structure/attribute of design object	Long
	In short time scale, umbrella changes states (open to close)		Short
Places of use	Climate changes	(G) Change the environment in the place of use	Long Short
	Location changes due to the user's movement	(H) Change the places of use by the user	Long Short

provider prepare a product that allows users to repair. Provider also prepares a rental service that users can rent anywhere and return whenever the rain stops.

2.7 Discussions

The results of the case study show that attention to the types of temporal changes in a design object, a user, and the place of use, and their countermeasures contribute to the derivation of new time axis design ideas. In particular, we found that the use of a case-based design approach using the cases of time axis design was effective in deriving design ideas.

There are some remaining problems to be addressed. One is to validate our design support method. For example, we need to investigate our design solution really works. In response to these questions, we need to consider a validation method about our design support method.

Another problem is that the proposed design support method is case based and concrete, so it might be more abstract in order to increase generality.

2.8 Conclusion

In this paper, we clarified the concept of time axis design and proposed a time axis design support method, which aims to respond to changes in people's lives and sense of values as the temporal changes by introducing time axis concept into the design theory.

The design support method we proposed is based on 130 cases of time axis design we collected. In the proposed design method, plausible temporal changes are derived by drawing a use case scenario and design ideas are derived by utilizing the mapping between temporal changes and their countermeasures. As the case study, we illustrated the time axis design of an umbrella. By using the proposed method, several time axis design ideas were derived.

Future work will include validation of the derived design ideas and generalization of design support methods.

Acknowledgements This work was supported by the Grant-in-Aid for Scientific Research (No. 16K12667) from the Japan Society for the Promotion of Science.

References

Agnar A, Enric P (1995) Case-based reasoning: foundational issues, methodological variations, and system approaches. AI Commun 7(1):39–59

Daimon T, Kondoh S, Umeda Y (2003) Proposal of decision support method for life cycle strategy by estimating value and physical lifetimes. In: proceedings of EcoDesign 2003, pp 109–116, IEEE

European Commission (2015) Closing the loop—an EU action plan for the circular economy, 2015.12

Matsuoka Y (2012) Dawn of timeaxis design, Maruzwn (In Japanese)

Takahashi S, Sato K, Matsuoka Y (2011) Proposal of timeaxis design model and for a value growth design. Japanese Soc Sci Des (In Japanese)

Tukker A (2004) Eight types of product service system: Eight way to sustainability? Experiences Sus-ProNet, Business Strategy Environ 13(4):246

Umeda Y, Tomiyama T, Yoshikawa H (1995) A design methodology for self-maintenance machines. J Mech Des ASME 117(3):355–362

Umemori Y, Kondoh S, Umeda Y, Shimomura Y, Yoshioka M (2001) Design for upgradable products considering future uncertainty. In: Proceedings of EcoDesign 2001, pp 87–92, IEEE

United Nations (2015) Sustainable development http://www.un.org/sustainabledevelopment/

Chapter 3
Design of Household Appliances Considering Remanufacturing: A Case Study

Hong-Yoon Kang, Yong-Sung Jun, Ji-Hyoung Park, and Eun-Hyeok Yang

Abstract Remanufacturing is a manufacturing technology that incorporates cost-effective energy savings and sustainability principles to improve the effective reuse of materials at the end-of-life. Leading global remanufacturing countries are promoting the remanufacturing market through the diversification of remanufactured items and the improvement of consumer reliability. Remanufacturing business can be value-added combining with the rental service which is so called product-service system (PSS). In particular, remanufactured home appliances can easily be applied as product-service business model. Initial important condition for successful remanufacturing is how the product is well designed for disassembly at its end of life. In general, the product is characterized at the initial stage of its development. So, it is necessary to apply the design considering remanufacture at the beginning of the product development if the product is planned to be remanufactured at its end of life. This paper thus presents a method to diagnose the structure of the product and extract the major improvement factors for easy disassembly. Also, the application of these factors to small household appliances in the product design stage is highlighted. The validity of the design methodology is also analyzed through the demonstration of the water purifier in terms of assemblability/ disassemblability, cleaning efficiency and processing time.

Keywords Design for remanufacturing · Disassembly · Home appliances · Product-Service business

---

H.-Y. Kang · Y.-S. Jun (✉)
Korea National Cleaner Production Center, KITECH, Seoul, Korea
e-mail: yxjasp@kitech.re.kr

J.-H. Park
School of Environmental Technology & Safety Technology Convergence, Inha University, Incheon, Korea

E.-H. Yang
Production Operation Department, Cowey Co. Ltd., Gyeonggi-Do, Korea

Y. Kishita et al. (eds.), *EcoDesign and Sustainability I*, Sustainable Production, Life Cycle Engineering and Management, https://doi.org/10.1007/978-981-15-6779-7_3

3.1 Introduction

Remanufacturing is a manufacturing technology that incorporates cost-effective energy savings and sustainability principles to improve the effective reuse of materials at the end-of-life. Leading global remanufacturing countries are promoting the remanufacturing market through the diversification of remanufactured items and the improvement of consumer reliability. Korea is also trying to maximize the economic and environmental effects of resource circulation by promoting the addition of remanufacturing to existing 3Rs (Stead and Stead 2004). The automotive parts sector has been the most active area in remanufacturing over the world so far. However, various remanufacturing industries such as aerospace, heavy duty & off-road (HDOR), machinery, IT products and medical devices are formed in the US and Europe (Jun et al. 2019).

According to the latest survey results, Korea's remanufacturing industry consists of automobile parts, toner cartridges, construction machine parts, and household appliances (Kang et al. 2018). Japan's remanufacturing involves automobile parts, printer cartridges, As a manufacturing sector, this article highlights the role and responsibilities of OEMs in the End-of-Life Vehicle Recycling Law (Matsumoto and Umeda 2011). As a result of analysis on automobile parts and remanufactured toner cartridges, it is known that the effect of reducing environmental burden due to remanufacturing is effective (Hj et al. 2011). On the other hand, research on product design considering remanufacturing has been conducted in some fields. A study on the design considering the remanufacturing of the product has been carried out to develop a method for evaluating the degradability considering the ease of disassembly and the decomposition time using the external case of computer as a sample (Yi and Chang 2007). As the amount of electronic waste generated increases, there has been a case in which design for remanufacturing (DfRem) is emphasized so that original equipment manufacturers (OEMs) can prepare for remanufacturing in advance (Hatcher et al. 2013).

Remanufacturing can be associated with a product service system (PSS) to provide OEMs with new business areas. The PSS is a business that can maintain some ownership of the products sold by OEMs and provides opportunities to further optimize OEM remanufacturing activities. There is a case for the maintenance, repair and overhaul (MRO) of a drilling machine used in a construction site (Sakao and Mizuyama 2014; Korea Institute of Industrial Technology 2018). In the case of eco-friendly products, DfE (Design for Environment) design is applied considering reduction of raw materials, energy use and ease of recycling, but it is not well known that new products are produced by applying product design considering remanufacturing. Damaged parts of the product used in the remanufacturing process is to be replaced with a new part in the re-manufacturing process, but some products are difficult to dismantle and re-assemble because of a design applied to fastening methods such as thermal bonding. Systematic management over the entire life cycle of the product is needed to understand the quality level of recovered core for remanufacturing and eliminate uncertainty, and OEMs need to consider design for remanufacturing aspects for better remanufacturing (Sundin 2004). Initial important condition for successful remanufacturing is how the product is well designed for disassembly at its end of life. In general, the product is characterized at the initial stage of its development. So, it is necessary to apply the design considering remanufacture at the beginning of the product development if the product is planned to be remanufactured at its end of life.

This paper thus presents a method to diagnose the structure of the product and extract the major improvement factors for easy disassembly. Also, the application of these factors to small household appliances in the product design stage is highlighted. The validity of the design methodology is also analyzed through the demonstration of the water purifier in terms of assemblability/disassemblability, cleaning efficiency and processing time.

3.2 Research Method

3.2.1 Evaluation Method of Remanufacturing Convenience

In this study, we selected one representative model of remanufacturing water purifier that C Company sells on rental basis, and conducted research to improve the convenience of remanufacturing for this product. When the remanufacturing business of household appliances is combined with a rental service, it is easy to acquire a history-managed core for manufacturing and it can create more added values.

Table 3.1 describes the specifications of the water purifier used as a sample. The remanufacturing convenience evaluation of the product was conducted in the order of selection of parts to improve remanufacturing convenience, establishment of improvement strategies for selected parts, detailed design and evaluation of improved remanufactured products. Finally, the improved results of the remanufacturing method are compared.

To improve the convenience of remanufacturing, the target product was selected through remanufacturing convenience evaluation. As shown in Table 3.2, the convenience items of remanufacturing were composed of six factors affecting the convenience of each remanufacturing process. The evaluation of the number of parts is based on Boothroyd's concept of the minimum number of parts (Bettles 2002), which quantifies the remanufacturing convenience results to the theoretical minimum

Table 3.1 Specifications of water purifier

Items	Purifier Spec.
Size (W * L * H)	181 * 420 * 385 mm
Release date	2013.09
Weight	9.0 kg
Voltage	220 V/60 Hz

Table 3.2 Evaluation of remanufacturing category

<table>
<tr><th rowspan="2">Remanufacturing process</th><th colspan="6">Category</th></tr>
<tr><th>Items</th><th>Influencing factors</th><th colspan="3">Measure</th><th>Score</th></tr>
<tr><td rowspan="5">Assembly/disassembly</td><td rowspan="2">Reduction of the number of part</td><td rowspan="2">Number of parts</td><td colspan="2">Theoretical minimum number of part</td><td rowspan="2">× 20</td><td rowspan="2">20</td></tr>
<tr><td colspan="2">The number of past</td></tr>
<tr><td>Shape of the part</td><td>Rotational symmetry, size, ease of handing</td><td colspan="3">Score card (ease of handing, rotational symmetry)</td><td>20</td></tr>
<tr><td rowspan="2">Fastening</td><td>Fastening method, secondary process, accessibility</td><td colspan="3">Scorecard(Fastening method, secondary assembly process, accessibility)</td><td>20</td></tr>
<tr><td>Number of elements to fasten</td><td>20</td><td>–</td><td>(Quantity of fastening elements—1)
The number of part that need to replace</td><td>20</td></tr>
<tr><td>Cleaning</td><td>Convenience of cleaning</td><td>Cleaning method, accessibility, shape of the part</td><td colspan="3">Scorecard(cleaning method, cleaning tool accessibility, shape of the part)</td><td>20</td></tr>
<tr><td rowspan="2">Testing</td><td rowspan="2">Minimize the number of parts</td><td rowspan="2">Number of parts to be tested</td><td colspan="2">Theoretical minimum number of part—the number of part that need to replace</td><td rowspan="2">× 20</td><td rowspan="2">20</td></tr>
<tr><td colspan="2">The number of parts—the number of part that need to replace</td></tr>
</table>

number of parts per product. For the evaluation items of parts shape, fastening method and washing convenience, the evaluation results for each factor were quantified by scoring using the scoring table. The criterion of remanufacturing convenience score for each factor was derived from AHP questionnaire for expert interview and remanufactured workers.

Each evaluation was performed at the component level that constitutes the product. In order to derive the parts to be improved, the parts that are equal to or less than the average score of each evaluation item were selected as the first improvement target parts for convenience of remanufacturing. After that, the parts to be improved were finally compiled by combining the improvement potential of each part and the expected effect of improvement.

3.2.2 Strategy and Design to Improve Ease of Remanufacturing

The first step to improve remanufacturing convenience is to derive a conceptual improvement strategy based on the failure factors of the parts to be improved. In this study, a conceptual improvement strategy was developed by applying the remanufacturing convenience design technique. As shown in Table 3.3, the remanufacturing convenience design method suggests criteria and solutions for remediation strategies for each failure cause in remanufacturing process. These were used to find basic solutions to improve remanufacture convenience for each problem. The second step is the detailed step of the conceptual improvement strategy, which are the detailed design of the improvement strategy and the development of the prototype. At this stage, the detailed design and prototype development were repeatedly carried out to formulate the improvement strategy, to identify the cause of the failure by each product development field, and to find a solution.

3.3 Results

3.3.1 Improvement of Remanufacturing Convenience in Decomposition Process and Assembly Process

(1) *Derivation of parts to be improved by evaluating ease of remanufacturing of products*

Component assembly in the production process of the product is closely related to the convenience of disassembly and assembly for the remanufacturing of the recovered product after use. In particular, if a non-destructive fastening method is selected, re-manufacturing itself is not possible, or a relatively large amount of time and resources are required for remanufacturing. Easy access way to the fastening area is a prerequisite for disassembly and reassembly, the fastening area should be noticeable, and the tool must be easily accessible at the same time.

As a result of evaluation of the fastening method of the target product, the average score of the product was 18.6/ 20.0. For the purpose of improving the fastening

Table 3.3 Strategies for improving remanufacture convenience: example of disassembly and assembly steps

Influence factor	Failure factor	Criteria for determining improvement strategy	Improvement strategy
Number of parts	High number of parts	Parts of the same material	• Unification of parts – Integrate two or more parts of the same material into one part
		Parts of the same lifetime	• Modularization of parts
		If parts can be reduced by changing the fastening method	• Reduce parts by changing the fastening method – Reduced parts by injection in shelf edge latch type instead of shelf support parts
Volume of parts	Bulky	There is an unnecessary injection site.	• Eliminate unnecessary injection
Weight of parts	Heavy parts	If you can change the part to a material of low weight	• Reduced product weight – Change to low weight material
Material of parts	Difficult to disassemble	Using parts of laminated materials	• Change to a single material
Type of fastening	Many types of fasteners	If the dimensions of the fasteners* and other parts used for each part are different * bolts, nuts, screws, etc.	• Unification of fastener specifications to be used – Head shape, driver type must be considered
		If bolts and nuts can be replaced with screws (+)	• Replace bolts and nuts with screws (+)
		Separate screws and washers	• Changing the screws and washers into one piece
Number of fasteners	High number of fastenings	When the number of fastening can be reduced by improving the fixed structure	• Reduce the number of fasteners – Slide coupling instead of screw

(continued)

Table 3.3 (continued)

Influence factor	Failure factor	Criteria for determining improvement strategy	Improvement strategy
Convenience of fastening and disassembling	Difficulty in fastening and disassembling	In case of non-destructive cracking method using adhesive or welding, it is possible to change by bolt or snap fit method	• Changed to fastening method that can be used for non-destructive disassembly
Extraction and Insert ability	Access to coupling elements is inefficient	If it is difficult to visually identify the fastening position	• Valid marks
Tool accessibility	Inefficient accessibility	If the depth of the screw hole is too deep	• Injection structure change such as embossing of joint part

method, the parts below the contracted average score are derived as the primary improvement parts. The parts subject to the primary improvement are shown in Fig. 3.1, three kinds of stickers and one kind of deco-front assembly, all of which are fastened by an adhesive method. In the remanufacturing convenience evaluation, the disassembly score is 7.4 points out of 20 points, and the reassembly score is 8.3 points out of 20 points, which requires relatively more time and resource input during disassembly and reassembly.

Figure 3.2 shows a remanufacturing process for improving the scratch of the front frame cover. In the case of stickers (Certification Label, etc.) with low engagement points, it is difficult to change the fastening system due to the characteristics of external certification, etc., and the deco-front assembly parts excluding the stickers are derived as improvement parts. Deco-cover front is an external part fastened with frame and adhesive method, so it is not possible to replace only the parts when

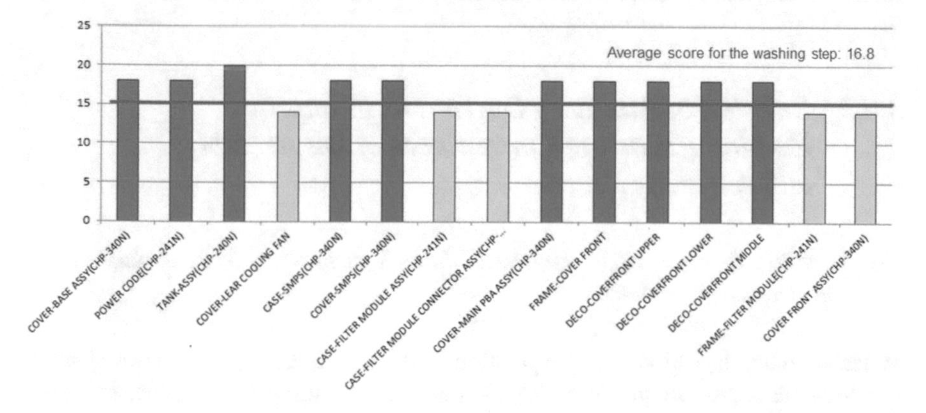

Fig. 3.1 Evaluation of assembling method result

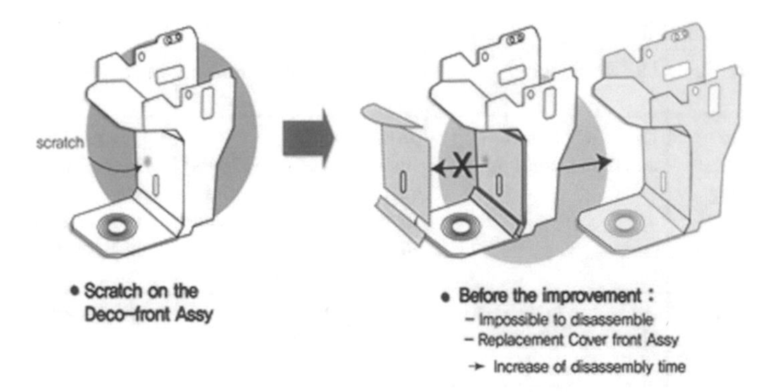
scratch
• Scratch on the Deco-front Assy
• Before the improvement :
- Impossible to disassemble
- Replacement Cover front Assy
→ Increase of disassembly time

Fig. 3.2 Remanufacturing process for improving the scratch of the front frame cover

scratches occur on the cover-front part, so it is necessary to replace the entire cover front assembly. As a result, it takes 225 s of the total remanufacturing process time of 1500 s. If this part is improved, the shortening time of the entire re-manufacturing process will be considerable.

(2) *Development of improvement strategies and prototypes*

In order to improve the fastening method of the deco-cover front part, the possibilities of changing the fastening method were examined with priority to the fastening method with high remanufacturing convenience score in the remanufacturing convenience designing and evaluation stage. Figure 3.3 shows a comparison of before and after improvement of the front frame-cover. After the final improvement strategy and detailed design process, the prototype was developed. As a result of the improvement of the fastening method, the guide ribs of the existing deco-cover front were changed to the hook shape and changed to the snap fit fastening method. The time required for the assembly process in remanufacturing was analyzed to be 225 s before the improvement, and about 185 s after the improvement.

3.3.2 Revision of Auxiliary Operations to Improve Re-Manufacturing Convenience of Disassembly and Assembly Process

(1) *Selection of parts to be improved through evaluation of re-manufacturing convenience of products*

When auxiliary operations such as position adjustment, heating, and the use of auxiliary tools are required at the time of extracting and inserting parts in the disassembling and assembling process, it is necessary to add additional working time and resources to the existing remanufacturing process. Therefore, the case filter module assembly,

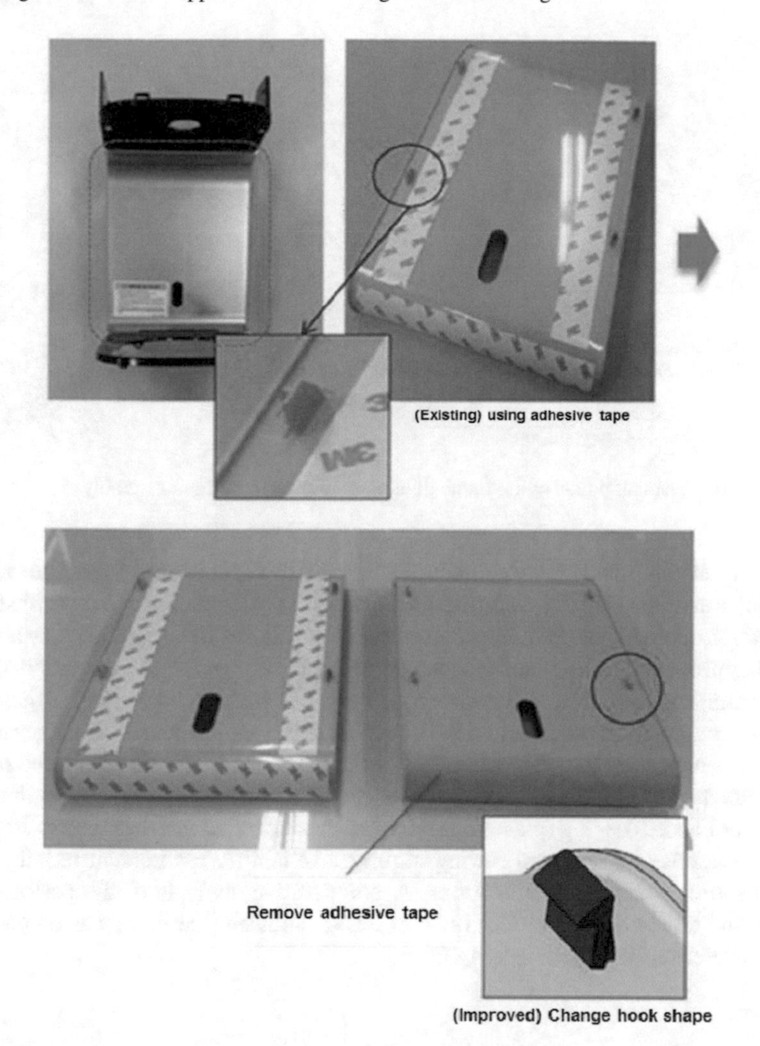

Fig. 3.3 Before and after the improvement of frame-cover front

which is a part that requires auxiliary work in the disassembly process, was selected as a component for improvement in remanufacturing convenience. If the connector (filter connection) is inserted into the case filter module assembly as shown in Fig. 3.4, the filter module and filter must be disassembled together to clean the relevant parts. In this case, the auxiliary equipment is inserted due to the increase in the weight of the parts to be disassembled, and it is analyzed that the disassembling process time of the case filter module is longer.

(2) *Development of improvement strategies and prototypes*

The connector, which is a part connected to the filter, is detached from the case filter assembly to solve the problem that the auxiliary operation occurs due to the

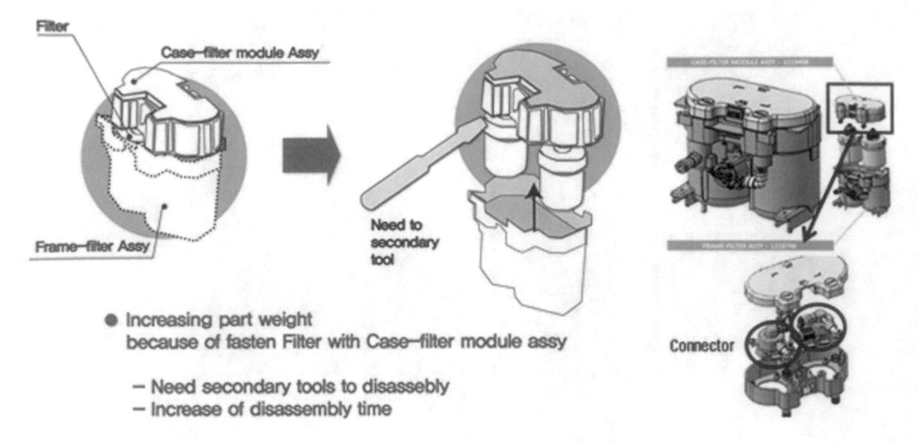

Fig. 3.4 Improvement points of the frame filter assembly and case filter assembly

connector inserted into the case filter assembly. In the separation process, the existing connector and filter O-ring sealing connection structure is changed to a simple fitting connection method, thereby improving the convenience of separation. As a result of the improvement, disassembly became possible sequentially in the order of case filter, connector, and filter. Therefore, the weight of parts to be worked is dispersed in the decomposition process, and the use of auxiliary equipment is no longer necessary, thus improving the convenience of remanufacturing. The time required for the remanufacturing assembly process was shortened by about 7 s in the conventional 10 s (7 s (tool access) + 3 s (placement)) to 2 s. Products that are designed to improve the convenience of remanufacturing require a reliability assessment based on the manufacturer's quality specifications. A comparative analysis of the performance before and after improvement of the part design showed that there was no problem such as deterioration of the quality level.

3.4 Summary

In this study, we performed a series of processes from product planning to design, including remanufacturing convenience evaluation, selection of key improvement items, improvement of derived items, and detailed design. Through these processes, a product development case considering the convenience of remanufacturing that can be applied in the new product production process is presented. The results of this study are as follows.

A series of remanufacturing processes such as disassembly, reassembly, cleaning and inspection were carried out in order to identify the parts to improve remanufacturing convenience in existing products. As a result of the evaluation, the deco-cover front assembly with the adhesive fastening method which cannot be disassembled and the case filter module assembly (with the fastening point of 17.0/20 points), which requires disassembly through the auxiliary equipment, were derived as the target parts requiring improvement. In the case of the deco-cover front assembly, which was fastened by the adhesive method and could not be disassembled, it improved from 7.9 points to 20.0 points after improving the convenience of remanufacturing by changing the fastening method by snap fit method. The time required for the work process was reduced from 225 to 40 s, which is about 185 s shorter. The case filter module assembly, which was hampered by the convenience of remanufacturing due to the insertion of auxiliary equipment in the disassembly stage, improved the convenience by changing the connection method between the case filter assembly and the filter. As a result of improving the convenience of remanufacturing of the water purifier model selected as the sample, the working time was reduced by 192 s in total.

References

Bettles (2002) Design for manufacture & assembly (DFMA)-the boothroyd & dewhurst approach. In: Competitive performance through advanced technology

Hatcher GD, Ijomah WL, Windmill JF (2013) Design for remanufacturing in China: a case study of electrical and electronic equipment. J Remanuf 3:3

Hj JO, Hwang YW, Park JH, Kang HY (2011) Environmental impact evaluation for the automotive remanufacturing parts and remanufacturing toner cartridge using the LCA methodology. Korea Solid Wastes Eng Soc 28(7):770–777

Jun YS, Kang HY, Jo HJ, Baek CY, Kim YC (2019) Evalustion of environmental impact and benefits for remanufactured construction equipment parts using life cycle assessment. Proc Manuf 33:288–295

Kang HY, Jun YS, Jo HJ, Baek CY, Kim YC (2018) Korea's remanufacturing industry in comparison with its global status: a case study. J Remanuf 8(1–2):81–91

Korea Institute of Industrial Technology (2018) Ministry of Trade, Industry and Energy, Korea Institute of Energy Technology Evaluation and Planning. Resource Circulation Technology Roadmap (In Korean)

Matsumoto M, Umeda Y (2011) An analysis of remanufacturing practices in Japan. J Remanuf, 1:2

Sakao T, Mizuyama H (2014) Understanding of a product/service system design: a holistic approach to support design for remanufacturing. J Remanuf 4 1:1 24

Stead JG, Stead WE (2004) Sustainable strategic management. In: A greenleaf publishing book, pp 20–21

Sundin E (2004) Product and process design for successful remanufacturing. Linköpings Universitet, Linköping, Sweden, pp 1–3

Yi HC, Chang YS (2007) Development of the disassemblability evaluation methods of the products for remanufacturing. Clean Technol 13(2):134–142

Chapter 4
Persuasive Design for Improving Battery Swap Service Systems of Electric Scooters

Li-Hsing Shih and Yi-Tzu Chien

Abstract Automatic battery swap stations have been recently set widely in Taiwan while the system operators have found significant differences among the battery utilization rates of the stations. To reduce the battery idle time in the less visited stations, this study looks for effective persuasive design strategies that persuade users to choose the less visited stations. After useful persuasive strategies were collected from the related literature, eighteen feasible design strategies were proposed by considering the problem characteristics. A questionnaire survey was conducted to estimate the persuasion effects of the eighteen design strategies by using the storyboard method. The persuasive design strategies with higher persuasion effect are identified and recommended for reducing the gap of battery utilization rates. Furthermore, by using the statistical analysis like ANOVA to analyze the persuasion effect with respect to demographical variables such as gender and age, the results could help choose effective persuasive strategies for different target customer groups.

Keywords Persuasive design · Battery swap · Service design · Electric scooters

4.1 Introduction

Since battery charging time is still longer than expected, battery swap service with significant shorter time has been welcome to users of electric scooters. Recently, automatic battery swap stations have been set in various sites such as convenient stores, parking lots, and sidewalks to provide better services in Taiwan by two competing private companies. The battery swap service systems are major drivers for the fast growth of eclectic scooter sales in Taiwan. However, system operators have found that there are significant gaps among the battery utilization rates of the stations. The phenomenon implies several disadvantages. For those less visited stations, the battery utilization rates are low, incurring idle cost that should be avoided. For those more visited stations, users often complain about no fully charged battery available,

L.-H. Shih (✉) · Y.-T. Chien
Department of Resources Engineering, National Cheng Kung University, Tainan, Taiwan, Republic of China
e-mail: lhshih@mail.ncku.edu.tw

Y. Kishita et al. (eds.), *EcoDesign and Sustainability I*, Sustainable Production, Life Cycle Engineering and Management, https://doi.org/10.1007/978-981-15-6779-7_4

incurring low customer satisfaction. From the sustainability point of view, too many idle batteries also mean that the society does not get the environmental benefit of the battery sharing systems, since the benefit of a PSS is increasing the utilization rate of products.

This study tries to solve the problem by looking for effective persuasive design strategy that persuade users to use the less visited stations. The work was conducted in three stages:

(1) Candidate persuasive design strategies are collected from practical cases and related literature on persuasive technology and persuasive design. By considering the problem characteristics of the current servive systems, eighteen feasible design strategies were screened and suggested.
(2) The eighteen feasible design strategies are presented in a questionnaire that aims to estimate the persuasion effects of the eighteen design strategies. In the questionnaire, each strategy is presented by using a storyboard that illustrates the expected use experience assuming the design strategy is implemented. Questions measuring to what extent the respondents are persuaded by each design strategy were presented soon after each storyboard.
(3) By going through statistical analysis, the persuasive design strategies with higher persuasion effect are identified and recommended for reducing the gap of battery utilization rates and raising the customer satisfaction. Furthermore, by using the statistical analysis like ANOVA to analyze the persuasion effects with respect to demographical variables such as gender and age, significant difference of persuasion effects among different groups of users can be identified.

4.2 Literature Review

To find candidate persuasive design strategies before conducting the questionnaire design, related literature are reviewed. Major searching keywords include persuasive technology, persuasive design, and persuasive system design. To explain how a persuasion is successful, Fogg (2009) proposed a behavior model that includes three major elements: motive, ability and trigger (prompt). He mentioned three motives including sensation, anticipation and belonging. Six types of abilities are often required for behavior changes including time, money, physical effort, brain cycle, social deviance, and non-routine while three types of trigger can be utilized for persuasion, including spark, facilitator, and signal. The model has been widely used in many areas for persuading people for behavior change. For example, Lee et al. (2014) used the model to raise the rate of women doing health examination. Ackermann et al. (2018) studied consumers' purchase behavior and overconsumption based on this behavior model. This study uses this model to check whether design strategies are feasible for persuading people for behavior change.

Oinas-Kukkonen and Harjumaa (2009) proposed a complete model/procedure for persuasive design and called the model "persuasive system design (PSD)". The procedure include three stages: (1) understanding key issues behind persuasive systems, (2) analyzing the persuasion context, and (3) proposing design strategies for use. The procedure is basically adopted in this study. They also suggested twenty eight useful design strategies that are divided into four categories: primary task support, dialogue support, system credibility support and social support. Each category contains seven useful strategies for use. Many literature adopted the PSD model for finding appropriate persuasive design. Chou (2010) used the model to persuade women to ride bicycles. Lehto and Kukkonen (2011) used the 28 strategies to find the best strategies for quitting smoking and alcohol. Kluchner et al. (2013) studied the persuasive strategy for home energy saving. Karppinen et al. (2016) developed a "health behavior change support system" for more healthy behavior based on the PSD model. Since many of the example literature found the twenty eight design strategies useful in persuasive design, they are considered as candidate strategies in this study for further screening.

To find more candidate design strategies other than the 28 strategies suggested in PSD, more literature were examined. For example, Dayan et al. (2011) suggested a "nudges" strategy for changing dining pattern against obesity. Lu et al. (2014) tried to find useful design for persuading college students to conduct more exercise and sports instead of being addicted to the internet. Coombes and Jones (2016) studied how to persuade children to walk and ride bicycles instead of taking automobiles. Oh et al. (2018) used persuasive design on website interface to persuade people to take action against obesity. Eisenbeiss et al. (2015) studied the real effects of using strategies like discount and time constraint in marketing. Whillans et al. (2018) proposed useful persuasive strategies using reminder strategy for persuading busy people to make purchase plan in advance. Payne et al. (2018) suggested some useful persuasive strategies for buying daily necessities. Chen et al. (2012) suggested a virtual object like a virtual aquarium to provide feedback on users' energy use. Lu et al. (2016) suggested several feedback strategies like using different colors to reduce energy use. Zhong and Huang (2016) suggested a "point reward system" to encourage users' recycling behavior. All the persuasive strategies in these literature were considered for screening out feasible design strategies in this study.

4.3 Selecting Feasible Persuasive Design Strategies

In this section, feasible persuasive design strategies are selected in two steps: (1) identifying candidate design strategies mentioned in the literature and (2) screening out feasible ones considering the problem characteristics of behavior changes for users taking the battery swap service. The second step is similar to the suggestion of Oinas-Kukkonnen and Harjumaa (2009) about analyzing the persuasion context before determining the design strategy. The candidate design strategies are adopted from the literature mentioned in Sect. 2 while the 28 design strategies in PSD model are adopted as a basis. Some strategies are deleted and some from other literature are added in based on the consideration of whether the strategies are feasible to be applied in the battery swap service system. In Sect. 3.1, the current user behavior/experience for swapping batteries are discussed and divided into five stages. The analysis result is used to help the authors see the feasibility of each candidate design strategy. In Sect. 3.2, eighteen feasible design strategies are selected based on the feasibility and then ready for use in the questionnaire design.

4.3.1 Behavior Analysis for Battery Swap

The current battery swap service systems in Taiwan contain several components: (1) battery, (2) automatic charging station that provides charged batteries ready for swap, (3) cloud system providing data storage, information processing and computing, and (4) scooter. Users use their cell phones to connect with the four. To find useful and feasible persuasive design strategies, this study analyzes the user experience and related behaviors in the current swap service systems. The results can help identify whether a candidate strategy could be implemented and current users' behavior could be changed and lead to the less visited stations. If the answer of the question is positive, the candidate design strategy is called a 'feasible' one. A typical battery swap process contains five stages, including:

(a) Being aware of battery status: Users check the battery status by inspecting the control panel of the scooters. The panel shows how much electricity power is left by using ten blocks where each block represents one tenth of electricity left.
(b) Finding out nearby charging stations: Users find nearby charging stations from the map using the app provided by service providers. The locations of the charging stations are highlighted on the map noting how many fully charged batteries are available in each station.
(c) Selecting stations: Users select an intended charging station among the ones shown on the map based on users' need and other personal consideration.
(d) Checking detail information of the selected station: Users look into the detail of the selected station including the route and additional information of the surroundings of that station.
(e) Swapping battery on the station: Users interact with the unmanned charging station and physically swap the batteries inside the scooter and the station. If the station is very popular and users take some time to reach the spot, users may find there is a waiting line or no fully charged battery available and then need to find another station.

The goal is to persuade users to choose and use the charging stations that are less visited. The persuasive design could be implemented in the five stages as mentioned above. Although those popular charging stations have the advantages like near popular spots, easy to reach, on the road of good traffic flow, or near important public facilities, users may face the risk of finding no battery available. The authors keep the three elements of the behavioral model (Fogg 2009) in mind and consider the characteristics of the current swapping behavior as presented above to find the feasible design strategies from the candidate design strategies suggested in the literature.

Whether the design strategies obtained from literature are suitable for changing the behavior of choosing charging stations was checked one by one by figuring out if the strategy can be implemented in the five stages of swapping battery behaviors. Eighteen feasible design strategies were screened out based on the feasibility and the potential of persuading users.

4.3.2 Feasible Persuasive Strategies for Users Choosing Less Visited Stations

Whether the design strategies are feasible ones also depends on if there are ideas of how the strategies can be implemented in the swap service systems. This is important because a storyboard for each design strategy has to be drawn illustrating the hypothetical use experience assuming the design strategy is implemented. In other words, how the design strategy can actually persuade users will be presented in the story board so that the persuasion effect of each design strategy can be measured in the survey. The description of the eighteen feasible design strategies are listed in the followings, where the first 11 strategies are adopted from PSD model (Oinas-Kukkonen and Harjumaa 2009) and the last seven strategies are from other literature.

1. Rewards: Systems that reward user's target behaviors may have persuasive powers. (adopted from PSD) For example, users get gifts in the less visited stations.
2. Reminders: If a system reminds users of their target behavior, the users will more likely achieve their goals. (adopted from PSD) For example, messages are sent to users to swap battery earlier.
3. Cooperation: A system can motivate users to adopt a target attitude or behavior by leveraging human beings' natural drive to cooperate. (adopted from PSD) For example, users are encouraged to team up to achieve target behavior.
4. Simulation: Systems that provide simulations can persuade by enabling users to observe immediately the link between cause and effect. (adopted from PSD) For example, systems simulate the situation users will face for changing swap behavior.
5. Tunneling: Using the system to guide users through a process or experience provides opportunities to persuade along the way. (adopted from PSD) For example, systems guide users to swap battery based on the users' past record and daily schedule.
6. Authority: A system that leverages roles of authority will have enhanced powers of persuasion. (adopted from PSD) For example, systems provide some experts advice for utilizing less used batteries for sustainability purpose.

7. Personalization: A system that offers personalized content or services has a greater capability for persuasion. (adopted from PSD) For example, systems suggest where and when to swap battery to save users' time based on users' daily plan.
8. Social comparison: System users will have a greater motivation to perform the target behavior if they can compare their performance with the performance of others. (adopted from PSD) For example, systems present the comparison result between users and others.
9. Praise: By offering praise, a system can make users more open to persuasion. (adopted from PSD) For example, the screen of the less visited station will praise users' behavior.
10. Self-monitoring: A system that keeps track of one's own performance or status supports the user in achieving goals. (adopted from PSD) For example, systems provide user's own records on battery swap and let user decide when and where to swap battery.
11. Liking: A system that is visually attractive to its users is likely to be more persuasive. (adopted from PSD) For example, more lovely or attractive design for the less visited stations should be applied.
12. Diversified function: The system provides more services will have a greater capability for persuasion. (adopted from (Fogg 2009)) For example, more functions could be added on the less visited stations like washing scooters for free.
13. Discount: The discount level will enhance the appeal of an offer and increase purchase likelihood. (adopted from Eisenbeiss et al. (2015)) For example, some discount may be applied for users visiting the less visited stations.
14. Data visualization and interactive narratives: The use of different website features, such as sliders, drags, or mouse-overs, for enabling content engagement, and further, increasing its persuasive intent. (adopted from Oh et al. (2018)) For example, systems allow users to know more information about the swap frequency and available batteries with additional features.
15. Media richness and interactivity: Richer media contain more communication modes and social visual cues and interactivity is defined as the extent to which media would let the designer exert an influence on the content/form. (adopted from Lu et al. (2014)) For example, systems provide more pictures or using AR (augmented reality) to attract users.
16. Color associations: Colors associations could help users to easily understand the feedback messages, and thereby, increase the persuasive effectiveness. (adopted from Lu et al. (2016)) For instance, different colors are used to show the charging status of batteries in different stations.
17. Virtual object: Users interact with the virtual object, resulting in the promotion of behavior attuned towards energy conservation by influencing emotional or rational thinking. (adopted from Chen et al. (2012)) For example, systems may illustarte a virtual vegetable growing as users visit the less visited stations.
18. Threat and coping appraisal: Presenting threat (ex: Don't speed. Don't kill your mates) and coping appraisal (ex: Better arrive late than not at all) messages

sequentially might increase the likelihood that they will be committed to memory. (adopted from Cathcart and Glendon (2016)) For example, systems send messages to users about the risk of visiting the crowded stations.

4.4 Questionnaire Design and the Survey

With the eighteen feasible persuasive strategies, a questionnaire is designed for estimating the persuasion effect of each strategy. The questionnaire survey was conducted to collect responses from the potential users in Taiwan.

4.4.1 Questionnaire Design

To estimate the persuasive effects of the feasible design strategies, eighteen storyboards were drawn to let respondents understand how each design strategy works and what a user can interact with the system and obtain service during the battery swapping process. Recent examples of using storyboards for studying persuasive effect include Orji et al. (2014) and Shih and Jheng (2017). According to the suggestion of Truong et al. (2006), a good storyboard design should contain presentation to a certain level of detail, verbal or graphic expression, calling the as-real feelings for the respondents, and showing the time or sequence of the use experience. In this questionnaire, three cartoons were designed to show how and what a user would experience assuming the design strategy is implemented in the battery swap service system. For instance, a storyboard designed for the strategy "rewards" is shown in the Fig. 4.1 where a gift is given to the user who goes to the less visited station.

Fig. 4.1 An example storyboard for strategy "rewards"

Two questions are put under each storyboard to collect the responses about persuasion effect. The 7-point Likert scale is used to let respondents express their agreement on:

(1) I have the same feelings as that of the person in the storyboard and
(2) I will react/do the same as the person in the storyboard.

The answers to the two questions are treated as persuasion scores (estimate of persuasion effect) for the design strategy. There are eighteen storyboards as the one shown in Fig. 4.1 in the questionnaire. In other words, a respondent will read eighteen storyboards and answer 36 questions in addition to the questions collecting the respondent's demographic information.

4.4.2 A Questionnaire Survey

The questionnaire survey was conducted via internet in January, 2019 and had 330 effective responses out of 368 responses. The background information of the sample are as follow.

(a) Gender: female: 51.8%, male: 48.2%
(b) Age: 21–30: 51.5%, 31–40: 32.1%, 40 and above 16.4%.
(c) Education: College 56.1%, graduate school 35.8%, others 8.1%.
(d) Vocation: students 24.8%, manufacturing 18.5%, commerce 13.6%, service industry 12.4%, government employee 10.3%, others 20.4%.

Since the survey was conducted via internet, social network platform, and portal websites, the respondents tend to be younger. This should be noted in the result interpretation.

4.5 Statistical Results & Recommended Design Strategies

In this section, statistical results of the questionnaire survey are presented. Design strategies are ranked according to the persuaion scores (persuasion effects) that collected from the responses. The strategies with higher pesuasion scores are recommended for use. Section 5.1 presents the results from overall respondents while Sect. 5.2 presents the ANOVA results that show whether the persuasion scores are affected by demographic variables like gender and age of the respondents.

4.5.1 Design Strategies Recommended

Table 4.1 shows the recommended design strategies with the rankings based on persuasion scores. Figure 4.2 contains an error bar chart of the results of the first ten design strategies. The 'color association' strategy that uses different colors to show the availability of batteries in the charging stations has the highest persuasion score. The users can know not only the number of fully charged batteries but also the

Table 4.1 Persuasion effect of 18 design strategies

Strategy	Average	Stand dev	Rank
Color	6.02	0.96	1
Threat	5.82	1.02	2
Diversified functions	5.77	1.07	3
Reminder	5.77	1.22	4
Tunnel	5.73	1.08	5
Simulation	5.72	1.09	6
Discount	5.65	1.15	7
Data visual	5.62	1.05	8
Self-monitor	5.53	1.09	9
Personalize	5.48	1.18	10
Authority	5.21	1.20	11
Media richness	5.16	1.28	12
Reward	5.12	1.37	13
Cooperation	4.63	1.56	14
Praise	4.46	1.55	15
Virtual object	4.36	1.49	16
Social comparison	4.28	1.50	17
Liking	4.08	1.49	18

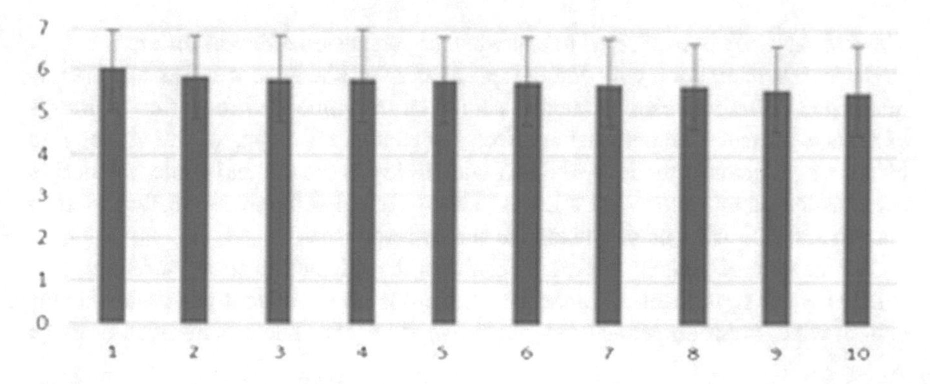

Fig. 4.2 Average and standard deviation of persuasion effect of the first ten strategies

Table 4.2 Recommended strategies in five stages

Behavioral stages	Strategy (rank)
I. Being aware of battery status	Threat (2)
	Reminder (4)
II. Finding information of charging stations	Tunnel (5)
	Simulation (6)
	Data Visualization (8)
	Self monitor (9)
III. Selecting stations	Color (1)
	Diversified function (3)
	Discount (7)
IV. Checking detail information of the selected station	Media richness and interactivity (12)
V. Swapping battery	Color (1)

number of partially charged batteries so that users could be aware of the likelihood of getting a partially charged battery in the crowded station. The 'threat and coping appraisal' strategy has the second highest persuasion score. The users consider the type of message like 'you should swap the battery earlier to avoid the risk of not finding available charged battery' is persuasive.

Table 4.2 puts the design strategies with higher persuasion scores (rank 1 to rank 9) in the five behavior stages of battery swap process. For example, The strategies of 'threat and coping appraisal (rank 2)' and 'reminder (rank 4)' could be used in the first stage 'being aware of battery status' in order to let users start the process of battery swapping. The implementation of a strategy (e.g. color) is not limited in one behavioral stage.

4.5.2 Design Strategies for Different Users

The RM-ANOVA is conducted to test whether persuasion effect of different design strategies has interaction with demographic variables like gender and age. If there is interaction, different design strategies should be recommended for different groups of users to increase the persuasion effect. For example, the 'persuasive strategy' is taken as one factor in the RM-ANOVA with 18 levels (strategies) while 'gender' is taken as the other factor with 2 levels. The statistical software SPSS was used to conduct the ANOVA analysis using the survey responses.

For the case of gender, the F-test results of the interaction ($F = 1.563$ and $p = 0.121$) are not significant. In other words, there is no significant persuasion (interaction) effect between gender and persuasive strategies. There is no need to make

Table 4.3 Recommended strategies for two groups

	Age under 40		Age above 40	
Rank	Strategy	Average	Strategy	Average
1	Color	6.00	Color	6.18
2	Threat	5.82	Media richness*	5.84
3	Remind	5.80	Threat	5.84
4	Diversified	5.79	Tunnel	5.79
5	Simulation	5.75	Data visual	5.78
6	Tunnel	5.73	Diversified	5.63
7	Discount	5.68	Self monitor	5.54
8	Data visual	5.60	Simulation	5.51
9	Self monitor	5.52	Remind	5.51
10	Personalize	5.48	Personalize	5.44
11	Authority	5.21	Discount	5.41
12	Reward*	5.18	Authority	5.24
13	Media richness*	5.08	Cooperate	4.71
14	Cooperate	4.62	Reward*	4.59
15	Praise	4.49	Virtual object	4.35

different persuasion design for female or male customers. Other demographical variables such as education level and vocation were also tested and no significant effect was found.

For the case of age, the ANOVA results ($F = 1.866$, $p = 0.001$) show a significant interaction effect. In other words, the persuasion effects of persuasive strategies are significantly different for users in different ages. In order to make the recommendation on design strategy more practical, two groups of users (under and above the age of 40) are taken for further analysis. Table 4.3 shows the averages and standard deviations of persuasion scores for different strategies for the two groups of users. For example, the strategy "color associations" ranks the first in both groups. The strategy "threat and coping appraisal" ranks the second in the group of age under 40 while "media richness and interactivity" ranks the second for the group of age above 40. However, if strict statistical testing like ANOVA is taken, there are only two strategies: 'rewards' and 'media richness and interactivity' have significantly different persuasion effects on the two groups. This should be considered when the designers want to use the results in Table 4.3 for practical implementation.

4.6 Summary

This study focuses on choosing appropriate persuasive design strategies for improving battery swap service systems. To reduce the idle cost of battery and raise the customers' satisfaction, the goal is to persuade users not to go to the more visited charging stations and go to the less visited stations instead. Feasible persuasive design strategies were carefully selected from literature review and considering the problem characteristics. Eighteen feasible design strategies were adopted in the questionnaire survey and the storyboard method was used to estimate the persuasion effect of each design strategy. The statistical results of persuasion effects for eighteen strategies are presented and the results recommend the strategies with higher persuasion scores like color associations, threat and coping appraisal, diversified functions, reminders, and tunnel. In addition, ANOVA method is used to find whether different strategies should be recommended for persuading specific types of users. The results show that there is no significant interaction effect on gender and persuasive strategy while there is significant interaction effect on age and persuasive strategy. By conducting further statistical analysis, appropriate persuasive strategies are recommended respectively for the groups of users under age of 40 and above 40.

References

Ackermann L, Mugge R, Schoormans J (2018) Consumers' perspective on product care: An exploratory study of motivators, ability factors, and triggers. J Cleaner Production 183:380–391

Cathcart RA, Glendon A (2016) Judged effectiveness of threat and coping appraisal anti-speeding messages. Accid Anal Prev 96:237–248

Chen H, Lin C, Hsieh S, Chao H, Chen C, Shiu R, Ye S, Deng Y (2012) Persuasive feedback model for inducing energy conservation behaviors of building users based on interaction with a virtual object. Energy Build 45:106–115

Chou (2010) Investigating persuasion in female cycling design. Master's thesis of the Inst of Applied Arts, National Chiao Tung University, Taiwan

Coombes E, Jones A (2016) Gamification of active travel to school: A pilot evaluation of the Beat the Street physical activity intervention. Health & Place 39:62–69

Dayan E, Bar-Hillel M (2011) Nudge to obesity II: menu positions influence food orders. Judgment Decis Making 6:333–342

Eisenbeiss M, Wilken R, Skiera B, Cornelissen M (2015) What makes deal-of-the-day promotions really effective? The interplay of discount and time constraint with product type. Int J of Res Mark 32:387–397

Fogg BJ (2009) A behavior model for persuasive design. Proceedings of the 4th international conference on persuasive technology. ACM, 1–7, Clarement, California

Karppinen P, Oinas-Kukkonen H, Alahäivälä T, Jokelainen T, Keränen AM, Salonurmi T, Savolainen M (2016) Persuasive user experiences of a health behavior change support system: a 12-month study for prevention of metabolic syndrome. Int J Med Inf 96:51–61

Kluckner PM, Weiss A, Sundström P, Tschelig M (2013) Two actors: providers and consumers inform the design of an ambient energy saving display with persuasive strategies. In: First Int Conf on behavior change support systems (BCSS), Adjunct PERSUASIVE Position Paper, p 33–44

Lee H, Koopmeiners JS, Rhee T, Raveis VH, Ahluwalia JS (2014) Mobile phone text messaging intervention for cervical cancer screening: Changes in Knowledge and Behavior Pre-Post Intervention. J of Med Internet Res 16:8

Lehto T, Oinas-Kukkonen H (2011) Persuasive features in Web-based alcohol and smoking interventions: a systematic review of the literature. J Med Internet Res 13:3

Lu Y, Kim Y, Dou X, Kumar S (2014) Promote physical activity among college students: Using media richness and interactivity in web design. Comput Hum Behav 41:40–50

Lu S, Ham J, Midden C (2016) The influence of color association strength and consistency on ease of processing of ambient lighting feedback. J Environ Psychol 47:204–212

Oh J, Lim H, Chadraba E (2018) Harnessing the persuasive potential of data: The combinatory effects of data visualization and interactive narratives on obesity perceptions and policy attitudes. Telematics Inform 35:1755–1769

Oinas-Kukkonen H, Harjumaa M (2009) Persuasive systems design: key issues, process model, and system features. Communications of the Association for Information Systems 24:485–500

Orji R, Vassileva J, Mandryk RL (2014) Modeling the efficacy of persuasive strategies for different gamer types in serious games for health. User Model User-Adap Inter 24(5):453–498

Payne C, Niculescu M (2018) Can healthy checkout end-caps improve targeted fruit and vegetable purchases? evidence from grocery and SNAP participant purchases. Food Policy 79:318–323

Shih LH, Jheng YC (2017) Selecting persuasive strategies and game design elements for encouraging energy saving behavior. Sustainability 9(7):1281

Truong KN, Hayes GR, Abowd GD (2006) Storyboarding: an empirical determination of best practices and effective guidelines. Proc DIS 2006:12–21

Whillans AV, Dunn EW, Norton MI (2018) Overcoming barriers to time-saving: reminders of future busyness encourage consumers to buy time. Soc Influence 13:117–124

Zhong H, Huang L (2016) The Empirical research on the consumers' willingness to participate in e-waste recycling with a points reward System. Energy Procedia 104:475–480

Chapter 5
Service Design of Rehabilitative Exoskeleton for Sustainable Value Creation: A Case Study of Stroke Rehabilitation in China

Jing Tao and Suiran Yu

Abstract Stroke is characterized by high morbidity and high disability rate. Physical rehabilitation can effectively alleviate functional disability of stroke patients, and thus reduce potential nursing costs and save social resources. Given the prominent conflicts between limited resources provision-especially the qualified rehabilitative personnel-and the huge demands of rehabilitation, China is now actively encouraging R&D as well as popularization and application of rehabilitation robots. Aside from the design of rehabilitative exoskeleton itself, which is the focus of most studies in this era, the service design is also critical if rehabilitative exoskeletons are to realize the declared social-economic benefits. Service design may provide solutions to wider and fairer access to exoskeleton, better affordability and satisfaction of exoskeleton-based treatments, while creates monetary or non-monetary benefits for manufacturers, medical facilitates and other possible stakeholders. This study is focused on rehabilitative exoskeleton value creation. The service design of rehabilitative exoskeleton for stroke patients, which is considered the most promising application era of exoskeleton technology in China, are exploited based on the national conditions. First, sustainable value was conceptualized based on the understanding of value from the sustainable view point that includes economic, social and environmental perspectives. Then, the application scenarios of stroke patients' exoskeleton-based rehabilitation were developed based on China's multi-level rehabilitation service delivery system. In accordance to a modular product architecture, the service design of rehabilitative exoskeleton driven by sustainable value requirements with life cycle considerations was presented. The proposed sustainable value-driven service design may provide insights to rehabilitative exoskeleton developers and manufacturers in servitization of the emerging technology.

Keywords Rehabilitative exoskeleton · Service design · Sustainable value · Life cycle

---

J. Tao · S. Yu (✉)
School of Mechanical Engineering, Shanghai Jiao Tong University, Shanghai, People's Republic of China
e-mail: sryu@sjtu.edu.cn

Y. Kishita et al. (eds.), *EcoDesign and Sustainability I*, Sustainable Production, Life Cycle Engineering and Management, https://doi.org/10.1007/978-981-15-6779-7_5

5.1 Introduction

With the advent of aging society, China is suffering form the high incidence, mortality and disability rates of stroke. The population over 40 years old in China who have suffered from and have suffered from stroke is estimated to be 12.42 million and 70% of the survivors of stroke have different degrees of disability (Wang et al. 2019). Stroke and disability caused by stroke are the major contributors to "poverty due to illness" in China.

Modern rehabilitation theories and practice have proved that rehabilitation training can alleviate functional disability of patients, accelerate the rehabilitation process of stroke, reduce potential nursing costs, and save social resources (Ostwald et al. 2008). Despite of the government's efforts on developing the nation-wide rehabilitation delivery system, China is still struggling with the conflicts between limited resource provision and the huge demands of rehabilitation. According to the National Survey on Resources of Rehabilitative Medicine in 2009, there was a personnel gap of 15,000 rehabilitative specialists and 28,000 therapists based on the requirements of personnel quota listed in the Guide on Establishing and Managing Rehabilitative Units in General Hospitals (Chinese Association of Rehabilitation Medicine 2009). In terms of treatment method, therapists carry out most rehabilitation treatments manually. Therefore, therapists are often overworking and it is difficult to ensure the recommended amount and consistent quality of rehabilitation training.

Exoskeletons are basically wearable machines providing power and/or structural supports to human body. Given the nature of rehabilitation exoskeleton being a kind of advanced robotic technology, it has the advantages of automation, programmability, controllability and accuracy. Studies showed that the exoskeleton-based rehabilitation is a promising alternative to conventional manual therapy for improving motor function of stroke patients (Norouzi-Gheidari et al. 2012), because the device is designed to have multi-DOF to mimic various limb movements to accommodate all types of exercises (Rahman et al. 2015). It is also suggested that exoskeletons can be a potential solution to high quality rehabilitation training and a possible replacement (or supplement) of therapists. Given the huge patient population and the increasing desire for better life quality, there is huge potential for rehabilitative exoskeleton to prevail in China. However, currently there are very few rehabilitative exoskeletons certified for medical use and most of them are too expensive to popularize in China.

Given the encouraging policy of domestication of advanced medical devices and technologies as well as the emergence of a Blue Sea market, rehabilitative exoskeleton has been drawing attentions and interests of Chinese high-tech companies, universities, and research institutions [e.g. Chen et al. (2019, 2018), Zhang et al. (2019)]. However, aside from the design of rehabilitative exoskeleton hardware and software, which is the focus of most studies in this era, the service design is also critical if rehabilitative exoskeletons are to realize the declared social-economic benefits. Service design may provide solutions to wider and fairer access to exoskeleton, better affordability and satisfaction of exoskeleton-based treatments, while create monetary or non-monetary profits for manufacturers, medical facilitates and other possible stakeholder in the business.

To design effective service, a number of approaches such as product and service system (PSS) (Tukker and Tischner 2006), customer-oriented design (Kimita et al. 2009), are suggested. However, service design is still challenging with the increasing environmental pressure and demand of customization. With growing demands for sustainable products, there are clear implications for developing engineering capabilities of holistic application of sustainability concepts to products, processes and services design. Tao et al. (2018) proposed a QFD-based approach of life cycle scheming driven by sustainable value requirements. The proposed approach help bring experts in fields of product and process engineering, industrial management and ecological assessments to a common vision, and therefore accelerate design convergence for more sustainable products, processes and business. There are also evident needs of the systematic methodical support for the entire design process from requirements identification to concept configuration for product and service customization. To address the needs, Song et al. (2016) proposed a customization-oriented framework for design of sustainable product/service system, while Fargnoli et al. (2019) proposed a product/service design process with synergic use of QFD for PSS, Axiomatic Design (AD) and the service blueprint tools.

Rehabilitative exoskeleton is considered advanced medical instrument in China. According to the authors knowledge, service design of such exoskeleton is still new in practice. The purpose of this study is to exploit service design of rehabilitative exoskeleton for stroke patients based on the Chinese national conditions. Taking advantages of some existing methods, in this study, sustainable value is conceptualized from engineering perspectives and introduced as the anchor of stakeholder interests and motive power to drive the service design process. This study may shed lights on servitization of the emerging technology.

5.2 Conceptualization of Sustainable Value

With the three pillars of sustainability in mind, "sustainable value", in this study, is conceptualized as the total satisfaction of value requirements from perspectives of economic, social and environment. Also, under the umbrella of life engineering, the total sustainable value of a product and/or service include values delivered to customers and businesses partner throughout the life cycle. Thus, sustainable value requirements (SVRs) answer to the critical question of WHAT is valuable to different shareholders from sustainable perspectives. Without loss of generality, the system of SVRs is proposed as shown in Fig. 5.1. Therefore, the ultimate goal of sustainable value engineering is to maximize the overall value perceived by different stakeholders by increasing level of satisfaction of value demands.

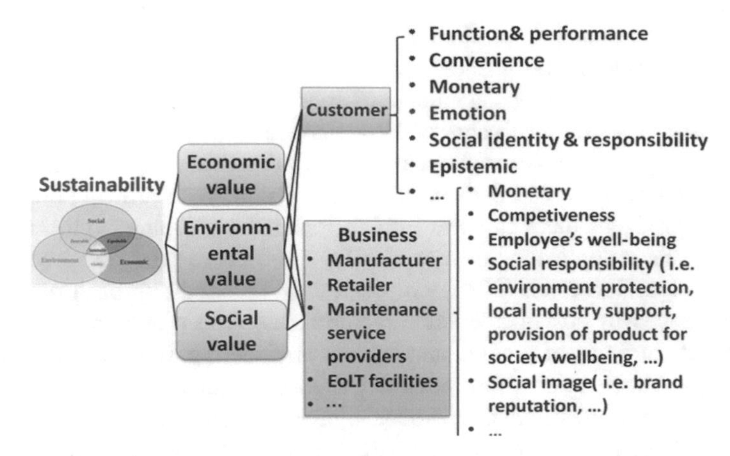

Fig. 5.1 General sustainable value requirements (Tao and Yu 2018)

5.3 Simplified QFD for Sustainable Value Driven Service Design

In this study, the service design is carried out as the structured and strategized process of translation of sustainable value requirements into product and service design solutions (see Fig. 5.2). The concept of "domain" from Axiomatic Design is employed to characterize stages of the translation, while the simplified QFD is employed for organized mapping between design semantics in different domains.

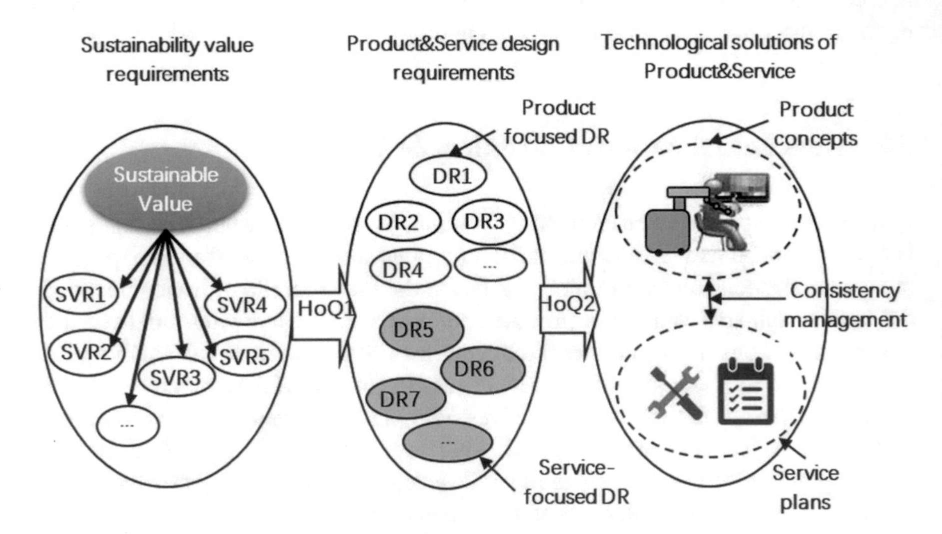

Fig. 5.2 Domain mapping for sustainable value-driven product and service design. Adapted from Tao and Yu (2018)

With the proposed method, three domains are defined, which are the sustainable value requirement domain, the product and service design requirement domain, and product service solution domain. Design requirements (DRs) also known as HOWs of engineering capabilities to fulfil the 'value' requirements. In this study, DRs are categorized into product-focused and service-focused DRs. Product-focused DRs are those on product itself including function and performance, quality, structure, appearance, material, etc., while service-focused DRs are those on various service activities throughout product life cycle, such as efficiency and effectiveness, availability and affordability, technicality of service. Capability of value delivery is enhanced by translation of sustainable value requirements into product and/or service focused DRs.

The total design solution is characterized by the combination of product concepts and service plans. The product concept (PC) and service plan (SP) tell the major engineering characteristics of the physical product entity and service activities. In this study, the service options for technology-intensive products (i.e. rehabilitative exoskeleton) include pre-sale services such as information revealing and (or) inquiry, pre-sale consultation, customization, after sale and in-use services such as delivery and installation, training and qualification of operators, in-use consultation and supervisions, maintenance and upgrade, and end-of-life services such as collection, refurbishment and circulate, recycling and safe disposal of retired product and components. It should be notice the product concept should always be consistent with the service plan. For instance, if upgrade service is provided on certain component of the product, the component should be designed to be easy to assemble and dissemble and functionally upgradable. As shown in Fig. 5.3, the simplified QFD creates two interlinked mappings. The correlation matrix in the second HoQ is divided into three matrices: two self-correlation matrices of PCs and SPs, and one matrix of correlation between each PC and SP. The PC-SP matrix is used for product-service consistency management which produce the detailed service plans on module and component levels.

5.4 Sustainable Value Driven Service Design of Exoskeleton for Stroke Rehabilitation in China

As shown in Fig. 5.4, The Chinese rehabilitation delivery network is characterized by a "3+3" architecture (Tao and Yu 2019 ; Xiao et al. 2017). The three level network is consistent with stroke rehabilitation progress, including the acute stroke patient stage, the mild stroke patients or midway in recovery process stage and the full recovery stage (Zhang 2011).

In this study, the exoskeleton use scenario is characterized by 'who the patient is', 'training with whom', 'where the training taking place' and 'what training delivered'. Thus, the considered customers include stroke patients who are the end user of the exoskeleton, the therapist and healthcare personnel who are the operators and in

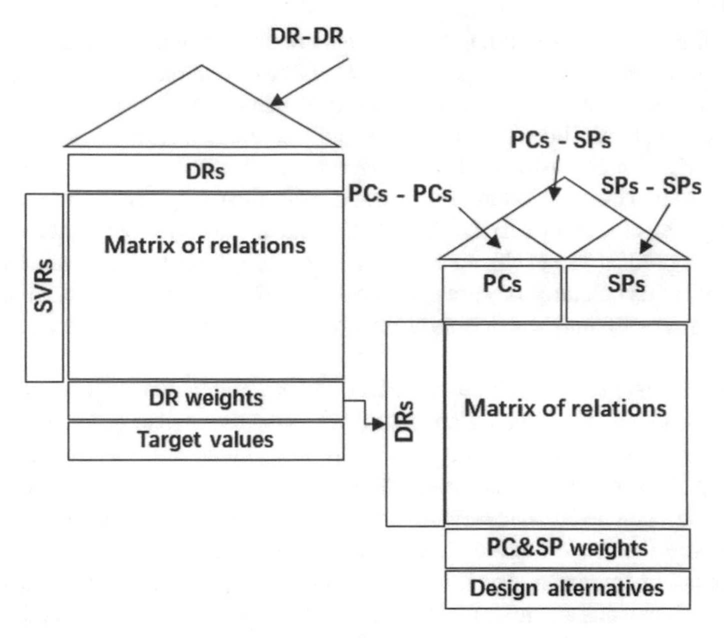

Fig. 5.3 Simplified QFD for sustainable value-driven product and service design

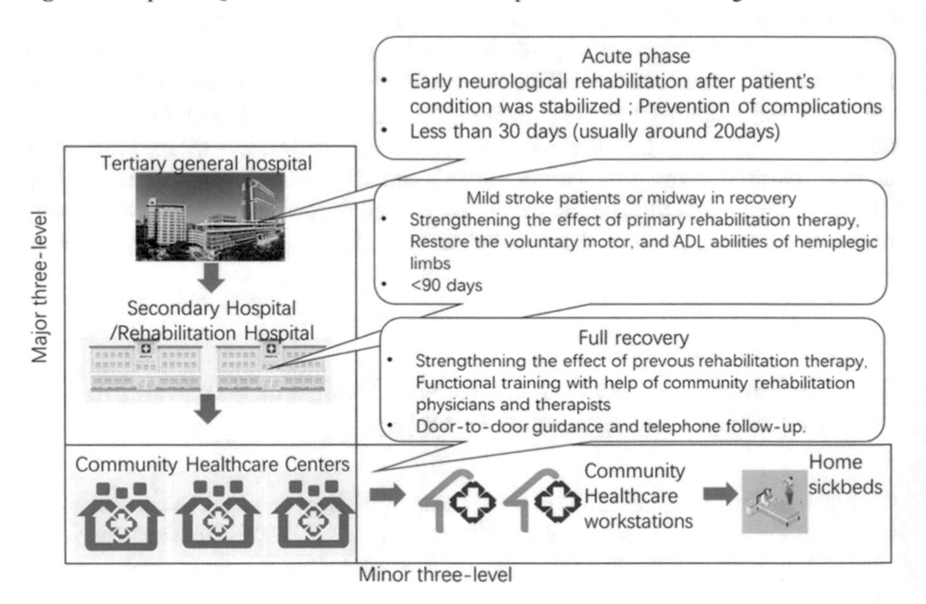

Fig. 5.4 Stroke rehabilitation in China

Table 5.1 Critical sustainable value requirements of customers and business

General customers	*Sustainable value requirements*
Patients; Therapist&Health care personnel; Medical facilities	• Function&Performance: Adaptivity to different patients; Multi-functional; Long useful lifetime & few malfunction • Convenience: Ease to put-up&off; Compact&Space-saving; Easy to operate; Visualization&Digitalization • Emotion: Safety&Security; Fun&Encouraging; Assitance&Care • Availability: Availability of treatment to patients; Availability of exoskeleton to healthcare facilities • Monetary: Affordability of treatment; Affordability of exoskeleton • Social Identity&Responsibility: Creditability of technology; Domestication; Warranty
Business	*Sustainable value requirements*
Manufacturing; Distribution; Maintenance; End-of-Life Treatment	• Monetary: Business profitability • Competitiveness: IP&Patents; Creditability of Technology; Brand awareness & Market share • Social responsibility: Qualification & Certification; Standardization; Legitimacy

charge of management of the exoskeleton, and the healthcare facilities who are decision maker of exoskeleton purchase. The critical sustainable value requirements of general customers and the business are presented in Table 5.1.

It is found that the functional requirements of exoskeleton vary in different stages of stroke rehabilitation. According to the training principles, the exoskeleton is required of fully powered and controlled movements of patient's affected joints for acute rehabilitation, while 'assist-as-needed' and resistive movements, respectively, for mid-way and full recovery phases. Thus, the proposed exoskeleton and services should be able to adapt different training needs of different patients in different rehabilitation stages.

According to the interview with patients and therapists, as the period of rehabilitation gets longer, the patients usually become more worried about treatment costs, as the medical insurance coverage on rehabilitation is still limited. Patients value availability of rehabilitation. This is typically due to the insufficient medical devices and healthcare personnel to carry out treatment, especially in base level facilities.

Also, it is found that the healthcare personnel's specialty in rehabilitation may decrease as patients are transferred from top to base of the rehabilitation network, and so does healthcare facilities' the affordability of expensive medical devices. Thus, demands for easy-to-operate, automated and intelligent, cost-effective exoskeletons and just-in-time technology support from the manufacturers or service providers may increase.

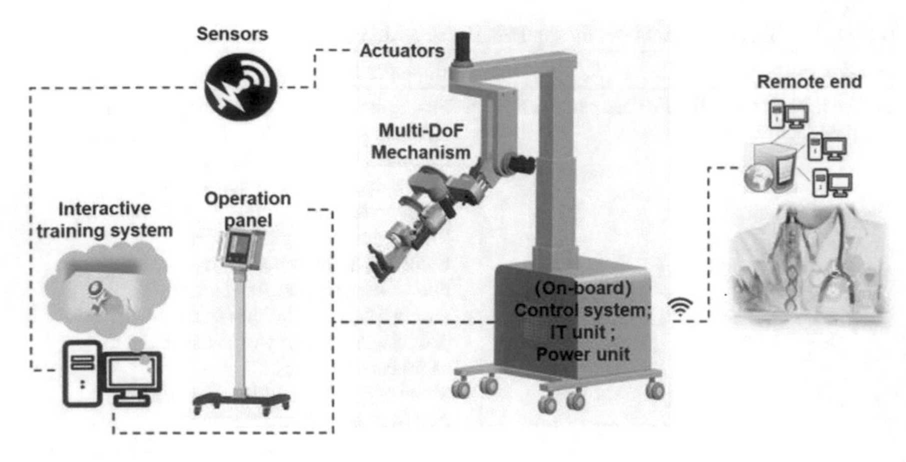

Fig. 5.5 Modular product concept

Technology creditability is identified as the major influencer on purchase decision, as it is closely related to the value of social responsibility and reputation of healthcare facilities. Technology credibility is also important to business, as it is directly linked to the competiveness of the exoskeleton in the market, and therefore the profitability of the business.

The simplified QFD process is carried out for generation of product and service solutions. The full QFD matrix is not be presented in this paper due to length limit. The product concepts are generated based on reference product, such as Armeo and Lokomat of Hocoma (see Fig. 5.5). Given the emphasis of this study being service design, more explanation of the service plan generation is presented as followed (Fig. 5.6).

The pre-sale, after-sale/in-use, and end-of-life service solutions are proposed to maximize the value delivery to stakeholders. The certification inquiry and pre-sale consultation services are propose to address customers' concerns on rehabilitative exoskeleton technology. Such services are also beneficial to brand building and marketing. The customized configuration service is proposed for better adaptability to exercise needs in different stages, as well as expansion of device price range. The financial leasing service is proposed to address the economic concerns of base level healthcare facilities (see Fig. 5.4) of owning an exoskeleton. In that case, the manufacturer or an exoskeleton service provider, rather than the healthcare facility, owns the device and is responsible for the maintenance of it, while the healthcare facility pays for functionality of the exoskeleton. Therefore, instead of a one-off payment for ownership, there will be a much smaller cost for the healthcare facilities.

With the development of telecom and internet technologies (e.g. IoT), it is also possible to provide in-use tele-monitoring and supervision services, which in combination with intelligent exoskeleton, may enable self-rehabilitation at community or by home sickbeds. In that case, fairer access to exoskeleton-based rehabilitation technology, which contributes to social well-being, may be realized. Cloud-based

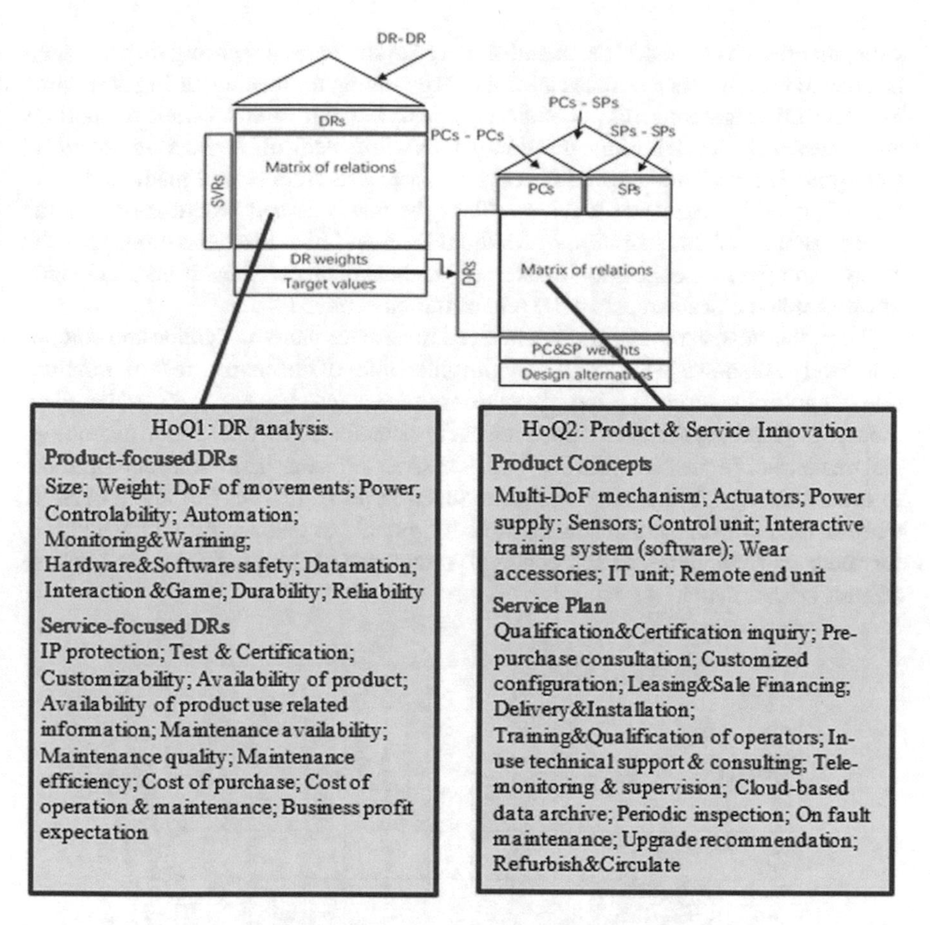

Fig. 5.6 QFD-based product and service design of rehabilitative exoskeleton

data archive service is proposed for value creation for both customers and the business. Exoskeleton can easily adapt to patients' treatment needs based on training profiles and schedule downloaded (with proper authority) from the cloud. Therefore, patients can conveniently go to any nearby healthcare facility with an exoskeleton for training, while healthcare facilities can attract more patients with the flexible and efficient treatment with the exoskeleton. At same time, the business can benefit from the improved utility of their exoskeletons.

Some end-of-life service options are also proposed. As the exoskeletons are expensive and technology-intensive products, the collection of retired devices or parts and return to manufacturer is beneficial to intellectual property protection. The circulation of refurbished exoskeleton (or components) can help reduce manufacturing and ownership costs, and consequently increase the customers' access to exoskeleton and exoskeleton-based treatments. Especially, it is technically feasible to reuse the mechanical parts of the device, as they have much longer useful lifetime than the valuable lifetime of the device. In addition, reuse strategies may bring forward environmental benefits, such as material and energy saving for manufacturing new parts. To address the legal concerns of circulating exoskeletons, related laws and regulations are then investigated. It is found that companies with "Medical Device Manufacturing Enterprise License" are allowed to recycle spare parts from retired medical device, and sell qualified renovated devices. Also, the newly issued "Regulations on the Supervision and Administration of Medical Devices" liberalizes the transfer of the safety and effectiveness of the transferred medical devices. Thus, it opens channel of exoskeleton cascade reuse in the rehabilitation service system.

Then, the consistency matrix is generated for service plans on module and component levels. As shown in Fig. 5.7, the highlighted lines of the matrix are the configuration of applicable service options for major modules, which are determined based on module engineering characteristics such as functionality, cost, market and technological trends, etc. For instance, the DoFs of the exoskeleton mechanism are configurable to meet training needs of single joint, reduced DoFs of limb or full DoFs of limb, while actuators with the mechanism are configurable to realized fully powered (i.e. for acute phase training), partial powered or unpowered (i.e. for full recovery phase training) exoskeleton.

Product concept / Service Plan	Multi-DoF mechanism	Actuators	Power supply	Sensors	(on borad) Control	Operation panel	Interactive training system (software)	Wear accessories	IT unit	Remote ends
Qualification& Certification inquiry										
Pre-purchase Consultation										
Customized configuration	√	√	√			√	√	√	√	√
Leasing&Sale Financing										
Delivery&Installation										
Training&Qualification of operators										
In-use technical support & consultation										
Tele-monitoring&supervison				√			√		√	√
Cloud-based data achive									√	√
Periodic inspection	√	√	√	√				√		
On fault maintenance	√	√	√	√	√	√	√	√	√	√
Upgrade recommendation	√			√	√		√		√	√
Refurbish&Circulate	√	√								

Fig. 5.7 Consistency management of product and service

5.5 Conclusion

This paper first presents the methods of service design driven by sustainable value goals. The sustainable value was conceptualized based on the understanding of value from a sustainable point of view that includes economic, social and environmental perspectives and then characterized by general sustainable requirements. Thus, the total sustainable value of product and/or service can be evaluated as the level of total satisfaction of value requirements. The sustainable value driven product and service design is proposed as the strategic process of translating sustainable value requirements into product and service technological solutions. The concept of "domain" from Axiomatic Design is employed to characterize stages of the translation, while the simplified QFD is employed for organized mapping between design semantics among domains.

Then, based on the architecture of the Chinese rehabilitation service delivery system, the three-stage stroke rehabilitation scenario is proposed. The critical sustainable value requirements of general customers and business are identified. By the proposed QFD-based method, service options are generated in consistency with product design concepts. In addition to conventional services such as pre-sale consultation, in-use technical support, maintenance and upgrade, the customized configuration, leasing and sale financing, tele-monitoring and supervision, cloud-based data service are proposed to address requirements from boarder value perspectives. Thus, the proposed sustainable value-driven service design may provide insights to rehabilitative exoskeleton developers and manufacturers in servitization of the emerging technology.

Acknowledgements This study is financially supported by National Natural Science Foundation of China (NSFC), Project No. 51875358.

References

Chen S, Li J, Shuai M et al (2018) First multicenter clinical trial of China's domestically designed powered exoskeleton-assisted walking in patients with paraplegia. Ann Phys Rehabil Med 61(S):e495

Chen B, Zi B, Wang Z et al (2019) Knee exoskeletons for gait rehabilitation and human performance augmentation: a state-of-the-art. Mech Mach Theory 134:499–511

Chinese Association of Rehabilitation Medicine (2009) The National Survey on Medical Rehabilitation Resources [in Chinese], pp 73–191

Fargnoli M, Haber N, Sakao T (2019) PSS modularisation: a customer-driven integrated approach. Int J Prod Res 57(13):4061–4077

Kimita K, Shimomura Y, Arai T (2009) A customer value model for sustainable service design. CIRP J Manuf Sci Technol 1(4):254–261

Norouzi-Gheidari N, Archambault PS, Fung J (2012) Effects of robot-assisted therapy on stroke rehabilitation in upper limbs: systematic review and meta-analysis of the literature. J Rehabil Res Dev 49:479

Ostwald SK, Davis S, Hersch G et al (2008) Evidence-based educational guidelines for stroke survivors after discharge home. J Neurosci Nurs 40(3):1

Rahman MH, Rahman MJ, Cristobal O et al (2015) Development of a whole arm wearable robotic exoskeleton for rehabilitation and to assist upper limb movements. Robotica 33:19–39

Song W, Sakao T (2016) A customization-oriented framework for design of sustainable product/service system. J Cleaner Prod 140(3):1672–1685

Tao J, Yu S (2018) Product life cycle design for sustainable value creation: methods of sustainable product development in the context of high value engineering. Procedia CIRP 69:25–30

Tao J, Yu S (2019) Developing conceptual PSS models of upper limb exoskeleton based post-stroke rehabilitation in China. Procedia CIRP 80:750–755

Tukker A, Tischner U (2006) Product-services as a research field: past, present and future. Reflections from a decade of research. J Cleaner Prod 14(17):1552–1556

Wang L, Liu J, Yang Y et al (2019) The prevention and treatment of stroke still face huge challenges-brief report on stroke prevention and treatment in China [in Chinese]. Chin Circ J 34(2):105–119

Xiao Y, Zhao K, Ma Z X et al (2017) Integrated medical rehabilitation delivery in China. Chronic Dis Trans Med 3(2):75–81

Zhang T (2011) Guideline for rehabilitation of stroke in China (in Chinese). People's Health Press

Zhang L, Li J, Su P et al (2019) Improvement of human–machine compatibility of upper-limb rehabilitation exoskeleton using passive joints. Robot Auton Syst 112:22–31

Chapter 6
State-of-the-art on Product-Service Systems and Digital Technologies

Clarissa A. González Chávez, Mélanie Despeisse, and Björn Johansson

Abstract Digitalization has undoubtedly revolutionized the way businesses think, plan and operate. This transition finds origin in the dramatic increase in demand of digital solutions, from those that target every day activities to highly-specific manufacturing processes. The last decades have been characterized for having both researchers and practitioners join efforts to innovate through solutions that are smarter, more productive and more efficient. This constant effort has accompanied the appearance of other impactful phenomena, often considered as a new industrial revolution. In a historically parallel line, companies made leaps towards exploring alternative ways to organize their business structure and relate to their customers. So is the case of Product-Service Systems (PSS), which have called for increased attention in the last years due to the extensive opportunities they offer. Recent literature raises the question of which and how sustainability advantages can be derived from PSS implementation. This study aims to understand the interactions of digital technologies and PSS through a state-of-the-art review. The authors have identified that academic literature encounters the challenge of finding digital technologies under a large cloud of different terminologies, which complicates systematization. Therefore, increased efforts will be applied to clarify the area of study and provide novel insights and results. This process will include the dissemination of enablers, constrains and possible effects of integrating digital technologies with PSS. The authors aim to contribute to the on-going discussion regarding the relationship between PSS and sustainability, specifically in applications that have high impacts, such as digital technologies.

Keywords Product-service systems · Digital technologies · Sustainability

C. A. González Chávez (✉) · M. Despeisse · B. Johansson
Department of Production Systems, Chalmers University of Technology, Gothenburg, Sweden
e-mail: clarissa.gonzalez@chalmers.se

Y. Kishita et al. (eds.), *EcoDesign and Sustainability I*, Sustainable Production, Life Cycle Engineering and Management, https://doi.org/10.1007/978-981-15-6779-7_6

6.1 Introduction

The last decades have successfully demonstrated the world's potential for exploitation of human and industrial capabilities. Megatrends in manufacturing such as digitalization, automation, and connectivity are changing the way organizations and their customers perceive value. The manufacturing sector finds itself in a fierce competition for differentiators that provide a competitive advantage to achieve growth within the firm and the possibility of developing permanence within highly-innovative markets (Gebauer et al. 2011). Efforts towards the development of more efficient, flexible and automated manufacturing environments have benefitted from the revolutionary Internet-of-Things (IoT) and Internet-of-Services which enable smart manufacturing with vertically and horizontally integrated production systems (Thoben et al. 2018). Additionally, manufacturing companies have invested in the development of personalized products with value-added services, servitised value proposition which is better known as Product-Service Systems (PSS) and finds value with lifecycle concerns through cooperation and interaction of a range of stakeholders (Zheng et al. 2018).

The imminent intersection of the digital and material world could provide the infrastructure required for implementing feedback-rich systems throughout the product lifetime, facilitating information transparency and process circularity (Alcayaga et al. 2019), generating additional possibilities of achieving more sustainable processes. The convergence of these research trends has contributed to strong PSS development which although is supported by more than 20 years of research, it still requires further exploration to maximize the utilization of its capabilities.

This paper aims to understand the interactions of digital technologies and PSS by identifying enablers, challenges, and effects these two domains have on one another. A literature review is performed, justified by existing academic work which has generated a large cloud of terminologies and made systematization of this knowledge a highly complex process. Additionally, this study aims to contribute to the on-going discussion regarding the relationship between PSS and sustainability.

6.2 Theoretical Background

6.2.1 Digital Technologies

Digitalization refers to the digital representation of a product or service, which at the same time, allows easier delivery and manipulation of the mentioned assets (Bitner 2010). Consequently, digital technology has emerged as an umbrella term that aims to englobe the tools used to achieve digitalization means. According to Cambridge University's dictionary, it is directly related to applications related to the usage of connectivity and internet.

The concept of digital technologies has englobed many terms, one of them being the Internet of Things (IoT). IoT was first coined by Ashton (2009) to describe the interconnection of physical objects through added sensors. IoT has impacted companies' processes in an unusual way, and can be considered a key element of the fourth industrial revolution (Suppatvech et al. 2019). IoT specifically comprises hardware and software to enable objects to interact and to communicate with each other, supporting the development of new services (Senzi et al. 2015).

In a recent publication, Evangelista et al. (2014) explains that the move towards a digital society does not consist in achieving that people use technology, but on the actual impact of its use and how it can transform people's lives. On one hand, dealing with digitalization and assessing its socioeconomic impact, requires more comprehensive indicators on the actual use of Information and Communications Technology (ICT) in the economy and society at larger scales. On the other hand, the intelligence capability of organizations, such as the ability to configure hardware components to sense and capture information represents a first step case companies take towards digitization (Lenka et al. 2017).

Today, both academia and industry attribute great opportunities to the emergence of "big data", a term that not just addresses the volume of information, but also refers to its variability, variety, velocity, veracity, and value (Chen and Zhang 2014). In the existing literature, several terms are used to referrer to digital technologies which vary according to context. Through this literature review, the authors aim to systemize and identify those that have a bigger impact when integrating them with a PSS perspective.

6.2.2 Product-Service Systems

In a recent definition, Annarelli et al. (2016) integrates two of the pillar contributions of PSS (Baines 2007; Mont 2001), to suggest PSS as a market proposition focused on final user's needs rather than on the production process, which allows a need-fulfilment system with radically lower impacts and enhanced environmental and social benefits.

Leaping back to 2010s, it is visible that developments such as the mobile revolution and social media, specifically led to the digital transformation of businesses giving place to anything-as-a-service business models where platforms with business networks and ecosystems were promoted. This has intensified the development of PSS and its interaction with digital technologies (Rachinger et al. 2018).

PSS is by definition not a circular business model, but it can drive organizations towards reaching targets that are intrinsic of circular business models (Antikainen et al. 2018). Sustainability concerns have created a pull towards digitalized solutions that maximize the use of tangible resources through services. This suggestion is supported by some of the original conceptualizations of PSS which included dematerialization and reinforce sustainability and competitiveness goals (Annarelli et al. 2017).

Several references suggest PSS as one of the schools of thoughts rooted in CE. However, some research studies assume positive sustainability impacts from PSS and do not consider that they can have negative environmental impacts if they are not implemented carefully (Barquet et al. 2016). The findings from this study could be considered as potentially beneficial for the design of PSS.

6.2.3 Conceptual Integration

Several contributions in literature have explored alternative terms to PSS which induce intrinsically the digitalized conditions brought by new technologies. In this way, the value creation process needs to cope with conditions of high complexity, dynamics and ambiguity. A clear example is the proposal of Smart PSS as a concept, driven by embedded electronics and software, multisensory systems and integrated actuators (Kuhlenkötter et al. 2017). Smart PSS is a result of the digitalization of products and services as the digital connectivity between components allows autonomous interaction and enables opportunities for further development (Kuhlenkötter et al. 2017). Zheng et al. (2018) argues that its innovation possibilities are enabled by a platform-based approach and it is generated in a data-driven manner.

From one end, servitisation has been demonstrated to be a process with new distributed sources of unstructured and structured data with a high level of variety, while ensuring relative veracity and needed velocity. Servitisation can be thought of as a data-intensive process (Opresnik and Taisch 2015). From another perspective, data analytics is expected to drive the next wave of servitisation and, therefore, has the potential to become a new source of competitive advantage (Schüritz et al. 2017).

6.3 Methodology

The literature review was developed through the following three main stages suggested by Tranfield et al. (2003):

1. Planning the review
2. Conducting the review
3. Reporting and dissemination.

This methodology was followed because its argument embeds the strict development of unbiased outcomes and reliable knowledge on context-sensitive research topics.

In the first step, where the review was planned, the authors chose academic databases (Google Scholar and Scopus) to explore a representative set of keywords such as "Product-Service Systems", "Industry 4.0", "Digital Technology", "digitalisation" along with their linguistic variations. Both conference and journal papers were considered and publications from 2009 onwards were selected. In the second step, the authors followed the steps represented in Fig. 6.1 to achieve reliable content

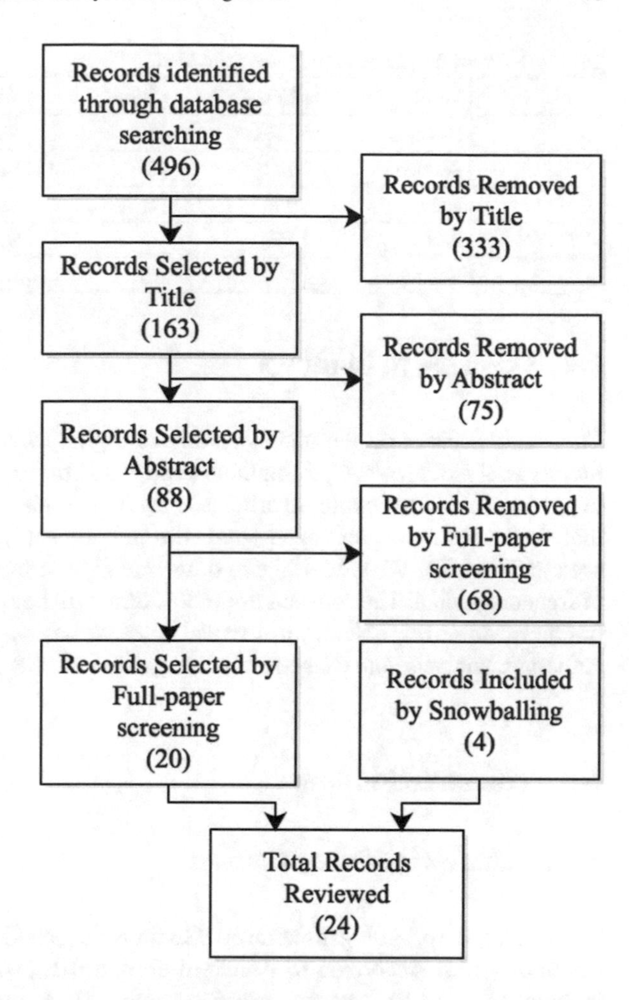

Fig. 6.1 Representation of record selection process

extraction from the records retrieved. The records which used the term "servitisation" only (i.e. not mentioning "PSS") were excluded. This decision not only helped managing the sample size, but also avoided confusion regarding the inclusion of certain study objects as services or PSS (Baines et al. 2009); e.g. controversy around the description of streaming services as part of a sharing economy (Curtis and Lehner 2019).

For the descriptive findings, the authors chose to use NVIVO as a tool for qualitative data analysis which allows to import and code textual data (Opresnik and Taisch 2015). The results from the thematic findings are summarized in Table 6.2, which was filled after an intensive analysis of records content.

Table 6.1 Word frequency in fully reviewed files

Rank	Word	Count	Rank	Word	Count
1	Product	1131	6	Value	662
2	Service	1121	7	Services	631
3	Data	981	8	Smart	597
4	PSS	755	9	Systems	569
5	Business	688	10	Digital	553

6.4 Descriptive Findings

This study focuses on the understanding of digital technologies and PSS through literary analysis. However, the authors believe that the explanation of some constitutional facts about the explored articles is useful for this study. Some of the findings include the ten most mentioned words through the analysed records, which can be seen in Table 6.1. We would like to draw our sight to the third word in the ranking of frequency: data. The constant use of this term will be further explored in thematic findings. Moreover, it shows that recent research has emphasized the need for data gathering, analysis, and usage for the integration of PSS and digital technologies.

6.5 Thematic Findings

6.5.1 Systemisation of Concepts

In literature, the on-going research on PSS finds a big challenge in the lack of systemisation of terms. According to Annarelli et al. (2016), some of the most discussed include industrial PSS, product-service combinations, product-to service, servicification (i.e. increased reliance from the manufacturing sector on services), post-mass production paradigm, functional product, total care product, integrated solutions, hybrid product, hybrid value bundles, hybrid value creation.

However, in the last years, some terms have enriched this list, such as Smart PSS, Circular PSS, and Digital PSS. Other authors refer to service additions in products as "intelligent digital systems" which allow human intervention and machine communication (Lerch and Gotsch 2015). Additionally, there are some further inconsistencies allocated to PSS and servitisation which frequently leads to semantic confusion that has negative impacts on the perception of the sharing economy (Curtis and Lehner 2019).

Likewise, the usage of diversified concepts for digitalisation and technologies can create complications for researchers and practitioners on their attempt to identify relevant literature. Some of the concepts located in (Antikainen et al. 2018) include "CPS", "big data", "Data mining", "Data analytics", "Internet of Things

(IoT)", "Mobile internet" and "Cloud computing". The diffusion of digitalization and the enhancement of digital technologies present potentialities to support the transformation towards more service-oriented strategies (Ardolino et al. 2013).

Also, the term of ICT and the growing academic discussion on how their exploitation is integral to a growing number of services (Kowalkowski et al. 2013) exemplifies through quotes such as: "the service revolution and the information revolution are two sides of the same coin" (Ardolino et al. 2016).

6.5.2 Enablers

This subsection describes and explores each of the seven enablers listed in Fig. 6.2 and summarized in Table 6.2. Note that big data and business analytics are strongly connected and thus discussed together.

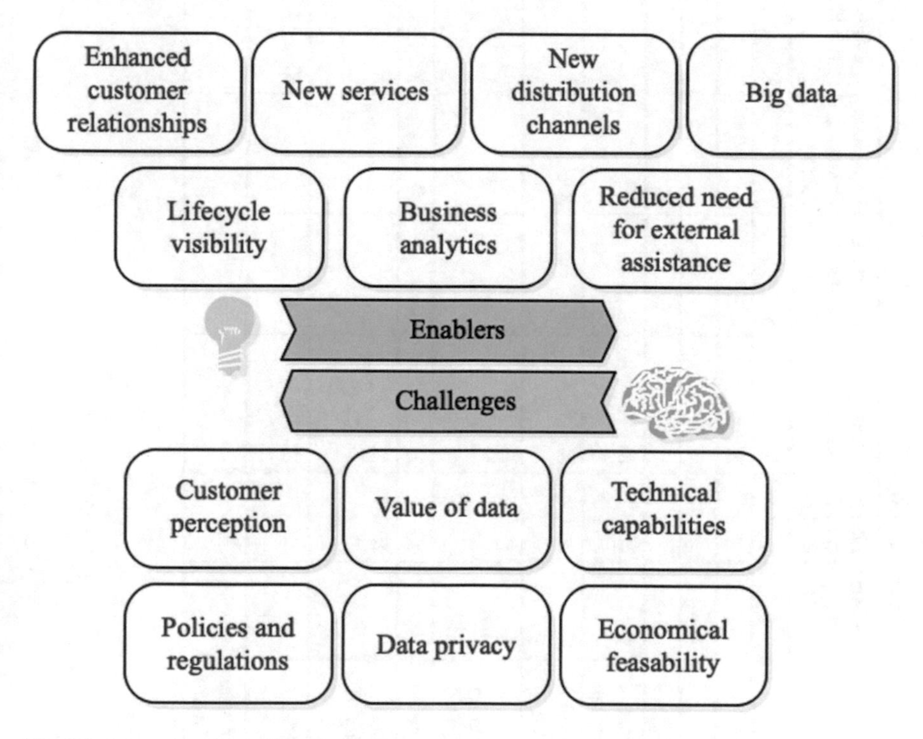

Fig. 6.2 Representation of key findings

Table 6.2 Summary of literature review findings

Reference		Thoben et al. (2018)	Zheng et al. (2018)	Alcayaga et al. (2019)	Curtis and Lehner (2019)	Lerch and Gotsch (2015)	Antikainen et al. (2018)	Ardolino et al. (2013)	Stark (2014)	Cimini et al. (2018)	Mourtzis et al. (2018)
Concept	PSS			X	X		X		X		X
	Smart		X	X						X	
	PSS	X									
	CPSS					X					
Enablers	Customer relationships	X	X	X	X	X		X	X	X	X
	New service provision	X	X	X	X	X	X	X	X	X	
	New Distribution channels	X	X					X	X	X	
	Big Data		X	X	X	X	X		X	X	X
	Increased visibility of LC		X	X			X	X	X	X	X
	Business Analytics			X			X	X	X	X	X
	Reduce need for ext. assist.										

(continued)

Table 6.2 (continued)

Reference		Thoben et al. (2018)	Zheng et al. (2018)	Alcayaga et al. (2019)	Curtis and Lehner (2019)	Lerch and Gotsch (2015)	Antikainen et al. (2018)	Ardolino et al. (2013)	Stark (2014)	Cimini et al. (2018)	Mourtzis et al. (2018)
Constraints	Value of Data	X				X	X	X	X		
	Customer perception	X		X	X			X		X	
	Tech. capabilties of customer			X		X	X		X	X	X
	Policies and regulations				X						
	Privacy of Data						X			X	
	Economical feasibility			X		X	X				

Reference		Marini and Bianchini (2016)	Süße et al. (2018)	Kans and Ingwald (2016)	Pagoropoulos et al. (2017b)	Coreynen et al. (2017)	Roy et al. (2016)	Kuhlenkötter et al. (2017)	Rymaszewska et al. (2017)	Bressanelli et al. (2018)	West et al. (2018)
Concept	PSS	X	X	X	X	X			X	X	X
	Smart							X			
	PSS										
	CPSS										
Enablers	Customer relationships		X	X		X		X	X	X	
	New service provision		X	X	X		X		X		

(continued)

Table 6.2 (continued)

Reference		Marini and Bianchini (2016)	Süße et al. (2018)	Kans and Ingwald (2016)	Pagoropoulos et al. (2017b)	Coreynen et al. (2017)	Roy et al. (2016)	Kuhlenkötter et al. (2017)	Rymaszewska et al. (2017)	Bressanelli et al. (2018)	West et al. (2018)
	New Distribution channels		X	X	X		X		X		
	Big Data	X	X	X	X	X	X	X	X	X	X
	Increased visibility of LC	X		X			X	X	X	X	
	Business Analytics	X	X	X	X		X		X	X	X
	Reduce need for ext. assist.				X				X	X	X
Constraints	Value of Data		X			X	X		X		X
	Customer perception			X		X		X	X		X
	Tech. capabilties of customer			X		X	X	X	X	X	X
	Policies and regulations				X				X		
	Privacy of Data		X	X		X	X	X	X		X
	Economical feasibility			X		X	X		X	X	

6.5.2.1 Enhanced Customer Relationships

The literature reviewed mentions repeatedly the exploitation of additional services as a potential enabler for customer loyalty and business growth (Ardolino et al. 2016). For instance, the concept of digital intelligence was found to refer to an enabler for distributed knowledge, structure, ownership and customization (Antikainen et al. 2018).

6.5.2.2 New Distribution Channels

The concepts of delivery and distribution are approached from a perspective that indicates the delivery of services and solutions through digital challenges (Ardolino et al. 2013). Additional to new distribution channels, digital technologies can enable diversified value configuration and foster partner networks (Cimini et al. 2018). These ideas are often linked to the concept of dematerialisation. Besides, the adoption of technologies such as additive manufacturing could contribute with benefits to supply chains by providing new options on the development of spare parts which increases the decentralization of production (Ardolino et al. 2013) and would mean relevant improvements in total operating and downtime costs for customers.

6.5.2.3 Big Data and Business Analytics

As mentioned in the previous subsection, literature shows repeated use of the concept of big data and data analytics, some articles even merge these two concepts into big data analytics (Wang et al. 2016; Mourtzis et al. 2018). The frequency of these concepts in the selected records can also be located in Table 6.2. The utilisation of big data is a complex decision which requires that organizations possess abilities and advanced sensing technologies (Marini and Bianchini 2016) to make use of it in an efficient, responsible and sustainable way (Süße et al. 2018). When organizations possess the abilities to make proper usage of digital technologies, they can strive to improve their maintenance activities (Kans and Ingwald 2016). For instance, it is possible to capture data from sensors located on critical components of the tangible product, transmit data gathered, (e.g. temperature or pressure), keep a record of fault codes (e.g. overheating or direct requirement of scheduled maintenance), send it directly to the manufacturer, store and analyse data. These actions enable the possibility to transform data into useful information about the product and develop responsive functions and actions such as repair, inform the customer or arrange maintenance (Ardolino et al. 2013; Lightfoot et al. 2013).

6.5.2.4 New Services

New service development was repeatedly discussed as a consequence of the usage of big data (Paschou et al. 2018). Some authors (Marini and Bianchini 2016; Ostrom et al. 2015; Lim et al. 2018) argue that big data impacts on manufacturing competitiveness not only by uncovering opportunities for new service offerings but also by capturing and analysing service-related information for effective and real-time decision making.

6.5.2.5 Lifecycle Visibility

The integration of data and information is a major enabler for the visualization of the materials cycle which can allow capturing sustainable value (Antikainen et al. 2018).

Several studies (Ardolino et al. 2016; Bressanelli et al. 2019; Pagoropoulos et al. 2017a) have studied how IoT, big data and analytics facilitate a transition towards CE. Eight main functionalities have been identified as enabled by these technologies: improving product design, attracting target customers, monitoring and tracking products, providing technical support, providing maintenance, optimizing the product usage, upgrading the product, enhancing renovation and end-of-life activities. Additionally, if organizations have data feeding their sustainability-oriented decision-making process big data and analytics can positively support lifecycle management advance towards CE (Bressanelli et al. 2019).

6.5.2.6 Reduced Need for External Assistance

Digital competence is defined as the ability to act in a digitized knowledge-based society. Additionally, this concept underlines stakeholders' capability to utilise the advantage of symmetric information as it becomes digitally spread and collectively provided (Süße et al. 2018). The understanding of this concept and its dimensions helps organizations on their path towards actively developing digitally-enabled PSS and further research is suggested in the in-depth analysis of competences within organizations (Süße et al. 2018; Paschou et al. 2018).

Also, reference is made to how data sharing can improve supplier/customer interaction (West et al. 2018). In this way (West et al. 2018) indicates that proactive data sharing can prepare the OEM to be ready to support with troubleshooting or provision of spare parts. This point is directly co-related to the reduced external assistance required by the customer and increments the responsibility of the OEM over the performance of their PSS. Augmented Reality (AR) is mentioned as a potential tool to support guidance, diagnostics, and training (Roy et al. 2016).

6.5.3 *Challenges and Constraints*

This subsection describes and explores each of the six challenges and constraints listed in Fig. 6.2 and summarized in Table 6.2.

6.5.3.1 Customer Perception

Regarding the identified challenges, literature expresses that on the demand side, some customers have found to have "service-for-free attitudes" where they become reluctant to pay additional fees for services added to their tangible goods (Coreynen et al. 2017; Ulaga and Loveland 2014). This situation can make the service pricing process highly difficult and poses an issue that has not been solved as of today.

6.5.3.2 Value of Data

Similarly, digital capabilities are one of the most recurrent identified challenges faced by firms who are working towards their migration to PSS offerings. In this way, we do not refer solely to the collection of data, but to the correct visualization and usage of the information that it represents (Stark 2014). The visualisation of data supports human analytical thinking and decision making. Some visualization tools, also known as visual analytics synthetize multi-dimensional knowledge from complex and dynamic data which can help support assessment, planning, and forecasting within firms (Roy et al. 2016).

6.5.3.3 Technical Capabilities

The strong requirement for companies to increase their capabilities in managing products and service variants, being agile and being ready to develop "ad-hoc solutions for customers through the so-called "co-design" process" (Alghisi and Saccani 2015) promotes value co-creation (Lenka et al. 2017). The development of suitable knowledge management approach has been expressed as essential for the successful management of all the variants (Cimini et al. 2018). Knowledge management is mentioned as one of three major areas of changes which demands that all the personnel involved remain consistent with one another to ensure long-term agreements (Cimini et al. 2018).

Kuhlenkötter et al. (2017) refers to six points of transformation through which organizations should go in order to achieve the major changes in the socio-technical system which can allow the proper development of digitalized PSS; some of these include high degrees of autonomy, strong human centration, variability and solutions along lifecycle, among others.

The capabilities of firms translate into both digital infrastructure and the availability of qualified employees who can develop and provide services with increased complexity, abstraction and the required problem-solving mindset (Lerch and Gotsch 2015). In this way, the development of abilities that allow dealing with technological devices and applications efficiently and effectively is a critical component.

Literature suggests assessment methods to ensure that companies are ready to perform in this regard (Kans and Ingwald 2016). Hence, it is commonly recognised that there is an increasing need for technical and social competencies of the employees (Cimini et al. 2018; Romero et al. 2016).

6.5.3.4 Policies and Regulations

The sharing economy is widely promoted by policy makers due to its sustainability potential, although the specific impact and possible effects are still unexplored. Some authors suggest that policy-makers who desire to institutionalise sharing as a consumption practice need to draw their sight to the possibility of negative perceptions that may exist among customers (Curtis and Lehner 2019).

In recent work, Pagoropoulos makes reference to the potential challenges that regulators and policy makers might face when integrating physical assets with service developments through digital technologies (Pagoropoulos et al. 2017b). Additionally, authors are concerned about the policy difficulties related to the use of IT (Rymaszewska et al. 2017) which is discussed further in the subsection on Data privacy.

6.5.3.5 Economic Feasibility

Another major concern is the economic feasibility of digitally-integrated PSS. It is unlikely that through vast computing power sensors with simultaneous sending and reception of data is achieved without simplicity and autonomy of end devices (Rymaszewska et al. 2015).

Besides, companies that create value offerings through PSS BMs, is that some financial risks are transferred from users to providers. More specifically, providers are often financially exposed to the risk of an early suspension of the contract by customers, when they have financed in advance the entire solution. Moreover, some cases present a time mismatch between revenue and cost streams, where providers who convert their offerings into PSS provisions find themselves having to finance in advance the capital costs of the solution and facing high-risk contracts (Bressanelli et al. 2018).

6.5.3.6 Data Privacy

Further, from a managerial perspective, the trend of digitalization i.e., the extensive use of digital technologies has enabled major business improvements such as the development of business eco-systems that are constantly reshaped by highly interdependent value co-creating actors (Süße et al. 2018). While this might be true, this collaborative ecosystem presents two of the major challenges, which is to first, ensure a safe data exchange (West et al. 2018) and protect the intellectual property from hackers' attacks (Cimini et al. 2018; Khan and Turowski 2016) and second, the monetising of product data through data analytics with the dilemma whether data should be available to entities that have no connection to the tangible products (Rymaszewska et al. 2015). Cybersecurity, then, becomes a technological matter that has cross importance for managing a range of elements in the transformation towards digitalization and servitisation (Kuhlenkötter et al. 2017; Cimini et al. 2018; Roy et al. 2016).

6.6 Discussion

The adoption of PSS, particularly if intended to interact with digital technologies requires a long-term perspective and in-depth understanding and involvement of users. One of the most important issues in PSS research relies on customers' requirements. The findings of this literature review have stressed on the need to further investigate the customer's perception of value associated with a PSS to further address the manufacturer's strategies (Fargnoli et al. 2018).

Some authors argue that if technological capabilities remain unchanged, the capabilities of PSS for new service development cannot be explored to their maximum potential (Rymaszewska et al. 2017; Pagoropoulos et al. 2017b). This has led recent articles to seek for interactions between digital technologies and PSS, and the role of an enabler is directed towards digitalization (Suppatvech et al. 2019; Lenka et al. 2017; Kowalkowski et al. 2013; Cimini et al. 2018; Süße et al. 2018). From this perspective, technologics such as ICT do not only enable servitisation through the provision of effective diagnostics services on the product, but it also has the potential to reduce costs, improve internal efficiency and aids increase the service business orientation of the company (Ardolino et al. 2013; Kowalkowski et al. 2013). We believe that the role and interaction of these concepts can be subjective to individual perspective and vary according to the unit of study.

An interesting point is only one of the records explored referred to organisational culture as an enabler of the servitisation process and explained the particular role of leadership, vision, and marketing as part of the effectiveness directional of servitisation strategies (Ardolino et al. 2013). A big majority of authors referred to this element of analysis from a negative perspective considering it a challenge. This study finds a limitation on the usage of terms which have absorbed a broad range of concepts and on the consideration of case studies from a wide set of industries. To overcome these concerns, the authors suggest focalizing companies that share as many characteristics as possible.

6.7 Conclusions

In this paper, the authors aimed to characterise the interactions between PSS and digital technologies, and to position further research which can contribute to the implementation of more successful and sustainable PSS.

Through the findings, a list of enablers and constraints were identified to understand the current scenario. This way a set of seven enablers and six constraints are presented and explained thoroughly. Through their analysis authors conclude that implementation and operation of efficient PSS which comply with sustainability concerns depend highly on the appropriate and in-depth consideration of other organizations and stakeholders involved in the process. Consequently, some of the most discussed results include the relevance of data monitoring, collection, and analysis and it is emphasized as an enabler for better-informed decisions which can support more sustainable and efficient operations through a PSS lifecycle.

The organizational capabilities in both cultural and technological aspects have shown potential for a strong impact in PSS transition. Also, the possibility of new distribution channels is highlighted in the literature as an unexplored area where the usage of manufacturing technologies could represent revolutionary changes in the costs of logistics and downtimes.

References

Alcayaga A, Wiener M, Hansen EG (2019) Towards a framework of smart-circular systems: an integrative literature review. J Clean Prod 221:622–634

Alghisi A, Saccani N (2015) Internal and external alignment in the servitization journey—overcoming the challenges. Prod Plan Control 26:1219–1232

Annarelli A, Battistella C, Nonino F (2016) Product service system: a conceptual framework from a systematic review. J Clean Prod 139:1011–1032

Annarelli A, Battistella C, Borgianni Y, Nonino F (2017) Predicting the value of product service-systems for potential future implementers: result for multiple industrial case studies. Procedia CIRP 47:436–441

Antikainen M, Uusitalo T, Kivikytö-Reponen P (2018) Digitalisation as an enabler for circular economy. Procedia CIRP 73:45–49

Ardolino M, Saccani N, Perona M (2013) Preliminary analysis on the enabling role of digital technologies and platform strategy for the servitization of companies. Ind Syst Eng Prelim 73–78

Ardolino M, Saccani N, Gaiardelli P, Rapaccini M (2016) Exploring the key enabling role of digital technologies for PSS offerings. Procedia CIRP 47:561–566

Ashton K (2009) That "Internet of Things" thing: in real world things matter more than ideas. RFID J 22:97–114

Baines TS (2007) State-of-the-art in product-service systems. J Eng Manuf 221:1543–1552

Baines TS, Lightfoot HW, Benedetti O, Kay JM (2009) The servitization of manufacturing: a review of literature and reflection on future challenges. J Manuf Technol Manag 20:547–567

Barquet AP, Seiden J, Seliger G, Kohl H (2016) Sustainability factors for PSS business models. Procedia CIRP 47:436–441

Bitner MJ (2010) Handbook of service science. Boston MA, pp 199–218

Bressanelli G, Adrodegari F, Perona M, Saccani N (2018) The role of digital technologies to overcome circular economy challenges in PSS business models: an exploratory case study. Procedia CIRP 73: 216–221

Bressanelli G, Adrodegari F, Perona M, Saccani N (2019) Exploring how usage-focused business models enable circular economy through digital technologies. Sustainability 10:639

Chen CLP, Zhang CY (2014) Data-intensive applications, challenges, techniques and technologies: a survey on big data. 275:312–347

Cimini C, Rondini A, Pezzotta G, Pinto R (2018) Smart Manufacturing as an enabler of servitization: a framework for the business transformation towards a smart service ecosystem. Proc Summer Sch Fr Turco 341–347

Coreynen W, Matthyssens P, Van Bockhaven W (2017) Boosting servitization through digitalization: pathways and dynamic resource configurations for manufacturers. Ind Mark Manag 60:42–53

Curtis SK, Lehner M (2019) Defining the sharing economy for sustainability. Sustainability 11:567

De Senzi E, Takey SM, Pezerra Barquet AP, Heiji Kuwabara L (2015) Business process support for IoT based product-service systems (PSS). Eur J Mark 49:1484–1504

Evangelista R, Guerrieri P, Meliciani V (2014) The economic impact of digital technologies in Europe. Econ Innov New Technol 23:802–824

Fargnoli M, Constantino F, Di Gravio G, Tronci M (2018) Product Service-Systems implementation: a customized framework to enhance sustainability and customer satisfaction. J Clean Prod 188:387–401

Gebauer H, Gustafsson A, Witell L (2011) Competitive advantage through service differentiation by manufacturing companies. J Bus Res 64:1270–1280

Kans M, Ingwald A (2016) Business model development towards service management 4.0. Procedia CIRP 47:489–494

Khan A, Turowski K (2016) A survey of current challenges in manufacturing industry and preparation for Industry 4.0. Proceed Int Sci Conf 15–26

Kowalkowski C, Kindström D, Gebauer H (2013) ICT as a catalyst for service business orientation. J Bus Ind Mark 28:506–513

Kuhlenkötter B, Wilkens U, Bender B, Abromovici M (2017) New perspectives for generating smart PSS solutions—life cycle, methodologies and transformation. Procedia CIRP 64:217–222

Lenka S, Parida V, Wincent J (2017) Digitalization capabilities as enablers of value co-creation in servitizinfg firms. Psychol Mark 34:92–100

Lerch C, Gotsch M (2015) Digitalized product-service systems in manufacturing firms: a case study analysis. Res Manag 58:45–52

Lightfoot H, Baines T, Smart P (2013) The servitization of manufacturing: a systematic literature review of interdependent trends. Ing JOper Prod Manag 33:1408–1434

Lim CH, Kim MJ, Heo JY, Kim KJ (2018) Design of informatics-based services in manufacturing industries: case studies using large vehicle-related databases. J Intell Manuf 29:497–508

Marini A, Bianchini D (2016) Big data as a service for monitoring cyber-physical production systems. In: Proceeding of the for 30th European conference on modelling and simulation. pp 579–586

Mont O (2001) Introducing and developing a Product-Service System (PSS) concept in Sweden. Manug Eng 80:133–138

Mourtzis D, Fotia S, Boli N (2018) Metrics definition for the product-service system complexity within mass customization and Industry 4.0 environment. In: International conference engineering technology innovations engineering technology innovations. pp 1166–1172
Opresnik D, Taisch M (2015) The value of big data in servitization. Int J Prod Econ 165:174–184
Ostrom AL, Parasuraman A, Bowen DE, Patricio I, Voss CA (2015) Service research priorities in a rapidly changing context. 18:127–159
Pagoropoulos A, Pigosso DCA, McAloone TC (2017a) The emergent role of digital technologies in the circular economy: a review. Procedia CIRP 64:19–24
Pagoropoulos A, Maier A, McAloone TC (2017b) Assessing transformational change from institutionalizing digital capabilities on implementation and development of product-service systems: learnings from the maritime industry. J Clean Prod 166:369–380
Paschou T, Adrodegari F, Rapaccini M, Saccani N, Perona M (2018) Towards service 4.0 a new framework and research priorities. Procedia CIRP 73:148–154
Rachinger M, Rauter R, Mûller C, Vorraber W, Schirgi E (2018) Digitalization and its influence on business model innovation. J Manuf Technol 20:341–362
Romero D, Stahre J, Wuest T, Noran O (2016) Towards an operator 4.0 typology: a human-centric perspective on the fourth industrial revolution technologies. In: Proceedings of the international conference on computers and industrial engineering (CIE46). pp 29–31
Roy R, Stark R, Tracht K, Takata S, Mori M (2016) Continuous maintenance and the future-Foundations and technological challenges. CIRP Ann Manuf Technol 65:667–688
Rymaszewska A, Helo P, Gunasekaran A (2015) IoT powered servitization of manufacturing—an exploratory case study. Int J Prod Econ 192:92–105
Rymaszewska A, Helo P, Gunasekaran A (2017) IoT powered servitization of manufacturing—an exploratory case study. Int J Prod Econ 192:92–105
Sala R, Zanetti V, Pezzota G, Cavalieri S (2017) The role of technology in designing and delivering Product-service Systems. In: 2017 International conference on engineering technology and innovations. pp 1255–1261
Schüritz R, Satzger G, Seebacher S, Schwarz L (2017) Datatization as the next frontier of Servitization- understanding the challenges for transforming organizations. In: Proceeding of the 38th international conference on information systems. pp 1098–1118
Stark R (2014) Advanced technologies in life cycle engineering. Procedia CIRP 22:3–14
Suppatvech C, Godsell J, Day S (2019) The roles of internet of things technology in enabling servitised business models: a systematic literature review. Ind Mark Manag 20:547–567
Süße T, Wilkens U, Hohagen S, Artinger F (2018) Digital competence of stakeholders in Product-Service Systems (PSS): conceptualization and empirical exploration. Procedia CIRP 73:197–202
Thoben KD, Wiesner A, Wuest T (2018) “Industrie 4.0” and smart manufacturing—a review of research issues and application examples. J Autom Technol 11:4–16
Tranfield D, Denyer D, Smart P (2003) Towards a methodology for developing evidence-informed management knowledge by means of systematic review. 14:207–222
Ulaga W, Loveland JM (2014) Transitioning from product to service-led growth in manufacturing firms: emergent challenges in selecting and managing the industrial sales force. Ind Mark Manag 43:113–125
Wang S, Wan J, Zhang D, Li D, Zhang C (2016) Towards smart factory for Industry 4.0: A self-organized multi-agent system with big data based feedback and coordination. Comput Netw 101:158–168
West S, Gaiardelli P, Resta NB, Kujawski D (2018) Co-creation of value in product-service systems through transforming data into knowledge. IFAC Pap Online 51:1323–1328
Zheng P, Lin TJ, Chen CH, Xu X (2018) A systematic design approach for service innovation of smart product-service systems. J Clean Prod 201:657–667

Chapter 7
Material-service Systems for Sustainable Resource Management

Marco Aurisicchio, Anouk Zeeuw Van Der Laan, and Mike Tennant

Abstract In current supply chains, material suppliers sell raw material resources to producers who sub-sequentially sell produced resourced to consumers. Ownership of resources, therefore, shifts from a few organisations to many consumers who are responsible to deal with them at the end of life. Product-service systems are business models where producers retaining ownership of produced resources have increased control on obsolete resources. Motivated by the need to facilitate an unlimited use of materials and eliminate waste, this research has introduced the concept of material-service systems, which are business models where material suppliers offer *materials as a service* to product producers. These systems offer the advantage that material suppliers are in control of resources and are incentivized to revalorise them. A scenario is explored in which a material-service system operates in conjunction with a product-service system and one in which it functions on its own. Finally, the benefits and incentives of the proposed service systems are discussed along with potential enablers and challenges.

Keywords Material-service system · Business model · Resource efficiency · Circular economy · Sustainability

7.1 Introduction

For a long time, our economy has been linear. We have taken material resources from our planet; used them to make products; and disposed of them as waste. Within this system, consumerism and population growth have propelled a use of resources that has outpaced the sustainable capacity of the ecosystem and led to the accumulation of waste (Stahel 2010). The depletion of material resources is making their prices

M. Aurisicchio (✉) · A. Z. Van Der Laan
Dyson School of Design Engineering, Imperial College, London, UK
e-mail: m.aurisicchio@imperial.ac.uk

M. Tennant
Centre for Environmental Policy, Imperial College, London, UK

Y. Kishita et al. (eds.), *EcoDesign and Sustainability I*, Sustainable Production, Life Cycle Engineering and Management, https://doi.org/10.1007/978-981-15-6779-7_7

volatile and creating uncertainties in resource markets (Ellen MacArthur Foundation 2015). Importantly it is increasingly acknowledged that if we do not intervene now to preserve our material resources, we risk compromising the ability of future generations to source and use materials to address their manufacturing needs.

At present, consumption is predominantly based on one-off payments made by consumers to own products and dispose of them at the end of use. Despite the present recycling infrastructure aiming to close resource loops, acquisition and recovery of products and materials is poor. Disposed products largely end up in landfills because either the infrastructure for recovery does not exist or is not suitable for many types of products; and because not all consumers are willing or able to access the infrastructure (Steg and Vlek 2008). The products and materials that are intercepted for recovery are generally cascaded to lower level applications as the collected streams remain contaminated. This is due both to the way products are designed (Mestre and Cooper 2017; Bocken et al. 2016) and the ability to revalorise material resources, which is limited by sorting and separating technologies, material properties and material recovery technologies.

To transition to an economy that is circular, products can be designed in conjunction with use and result oriented services to offer value to consumers, while efficiently using fewer resources. These product-service systems (PSSs) offer consumers access to products, while ownership of the products remains with or returns to producers (Ellen MacArthur Foundation 2015). Rather than paying to obtain ownership of products, consumers are charged for the time spent with products as they 'pay per use'. A notable application is the 'Pay-per-lux' intelligent lighting service introduced by Philips (Rau and Oberhuber 2016). In this product-service system, Philips retain control over the products that they produce enabling better maintenance, reconditioning and recovery. The producer, therefore, is incentivised to increase its control over resources that are consumed to deliver the performance of resources and dematerialise the performance where possible. Increased control over the flow of resources offers two main opportunities for the circular management of materials. First, producers would be expected to design products according to circular design principles (Mestre and Cooper 2017), e.g. easier to disassemble, sort, identify and separate. This improves the ability to produce pure or purifiable resource flows. Second, centralised ownership of products encourages to develop product-specific recycling infrastructure and activities. This permits companies to reduce the risk of contamination of the collected resources and optimise the recovery and reuse of products, components, and possibly materials. Despite the opportunities to flow materials continuously using this model, the resources in these models will eventually become obsolete and producers still have to dispose of them (Zink and Geyer 2018). The issue of whether resources are revalorised and returned to materials suppliers to function as a continuous feedstock remains dependent on the response of producers to obsolescence; and the roles to be fulfilled by collectors, recovers and recyclers.

This paper introduces preliminary research to investigate the concept of offering *materials,* not products*, as a service.* Material-service systems (MSSs) are positioned as new business models for the circular economy where material suppliers shift from the selling of material resources to the provision of material services.

With these systems, material suppliers are in control of resources and therefore have significant incentive to revalorise them in collaboration with collectors, recoverers and recyclers. After introducing the concept of MSS, the paper explores the expected benefits, enablers and challenges of the proposed service systems. In this research we take a resource-centric view of materials and the outcomes of their transformations over the production, use and end of life phases. We pose that resources have multiple states including raw resources, produced resources, wholesale resources, operative resources, obsolete resources, recoverable resources, recovered resources and revalorised resources. We define materials as raw resources and products as produced resources.

7.2 Literature

7.2.1 Materials as a Service

This research on *materials as a service* has been influenced by two main concepts: Material Matters (Rau and Oberhuber 2016); and Chemical Looping (Stoughton and Votta 2003). Materials Matter is the title of a book by Thomas Rau and Sabine Oberhuber capturing their vision that it is service, not ownership, the answer to facilitate an unlimited use of materials and the elimination of waste in the construction sector (Rau and Oberhuber 2016). To enable this shift, they have proposed an online registry of material passports, which allows to know where materials are located, and preserve and reuse materials while saving costs. With this information, obsolete buildings become a mine of materials.

Research on chemical leasing (Stoughton and Votta 2003), referring to the selling of the function performed by a chemical, is also related to offering *materials as a service*. However, this work has mainly focussed on effective use of chemicals for cleaning and coating purpose in manufacturing (e.g. charging customers by m2 of coated surface rather than Kg of paint). Hence, it does not address the proposed vision to manage material resources throughout the entire supply chain.

7.2.2 Ownership and Business Models

Ownership is the state or fact of exclusive rights and control over property including objects such materials and products, and land, real estate and intellectual property. In traditional business models, the property of objects and the rights over them are exchanged as a result of financial transactions between sellers and buyers. Consumers, for example, buy goods from producers or retailers and as a result own new products. Use and result oriented product-service systems are business models where producers grant consumers access to products or offer consumers the performance of products as experiences. Ownership of products in these business models is retained by producers. Futures have been proposed where business models based on access, not ownership, will become the dominant market offering and the idea of ownership is perceived as old fashion (Rifkin 2000). In a market where consumers are granted access to goods that they feel theirs though they do not legally own them, research has emerged to understand how to design to satisfy psychological ownership using a human centred approach (Baxter et al. 2018, 2015).

In recent reviews of business models for the circular economy and sustainability (Pieroni et al. 2019; Lüdeke-Freund et al. 2019; Bocken et al. 2014), models are reported where access to products is granted by paying-per-use instead of paying per-ownership (i.e. by delivering functionality rather than ownership), but there is no reference to a model where materials are offered as a service.

7.3 Approach

This research is based on literature review and the experience of the authors in the fields of ownership, design, product-service systems, the circular economy and business along with insights gathered during research projects with industrial partners. The paper presents a theoretical exploration of the concept of *materials as a service*. Two scenarios are used to explain how a material-service system would work. In the first scenario we explore its interaction with a product-service system. In the second scenario we review its independent use in the resource lifecycle system.

7.4 The Concept of Material Service System

The concept of MSSs revolves around the fact that the principle to 'pay per use' can be moved up the supply chain. In a MSS, suppliers of materials offer producers the use of material resources through services. MSSs can be thought of as marketable sets of materials and services capable of jointly fulfilling the needs of producers. They offer producers access to the performance of materials, while ownership of resources and corresponding responsibility of material management remains with suppliers. A MSS, therefore, implies a new role for producers, as they shift from consuming, to using material resources, which remain owned by material suppliers. Producers enter a business relationship with suppliers based on a 'pay per use' model rather than traditional purchasing of material resources. In Table 7.1 a MSS is contrasted to a PSS. As it can be seen the fundamental difference between the two models is that in a MSS material suppliers retain ownership of material resources and sell the function of materials, whereas in a PSS producers retain ownerships of produced resources and sell the function of products. It is noteworthy that material suppliers retain ownership of resources, not of the intellectual property resulting from the

Table 7.1 Main differences between a MSS and a PSS

Material service system	Product service system
Material suppliers offer producers the use of materials through services	Producers offer consumers the use of products through services
Marketable set of materials and services capable of jointly fulfilling the needs of producers	Marketable set of products and services capable of jointly fulfilling the needs of consumers
Selling the function of materials	Selling the function of products
Producers enter a business relationship with suppliers based on a 'pay per use' model	Consumers enter a business relationship with producers based on a 'pay per use' model
Material suppliers retain ownership of the resource	Producers retain ownership of the resource
Ownership of the IP for materials, manufacturing processes and products remains with the inventors	Ownership of the IP for materials, manufacturing processes and products remains with the inventors

downstream transformations made to resources by producers or other stakeholders. Two scenarios are envisaged to deploy a MSS, namely a MSS used in conjunction with a PSS and a MSS used on its own. The scenarios are explored in the next two sections.

7.4.1 *Use of a MSS with a PSS*

A MSS can be operated in conjunction with a PSS, see Fig. 7.1a. In this scenario a material supplier markets *materials as a service* to a producer, and in turn the producer markets products as a service to consumers. Consumers have an obligation to return obsolete products to the producer, while the producer has an obligation to return them

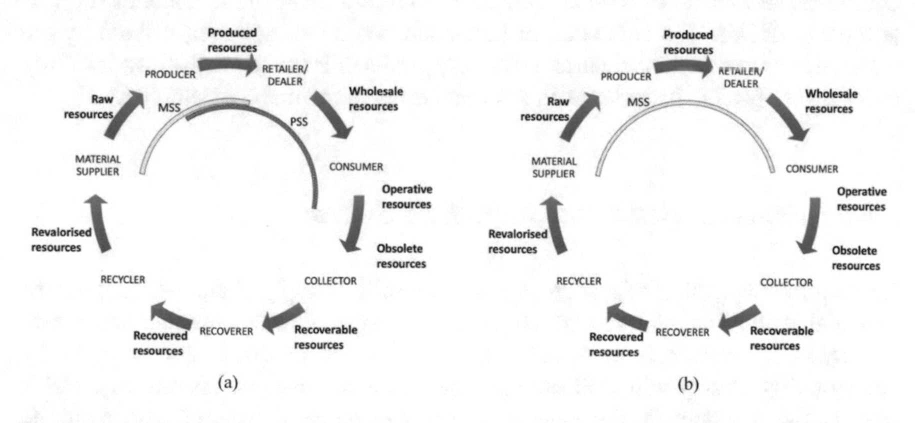

Fig. 7.1 **a** MSS and PSS versus **b** MSS only over the resource lifecycle and stakeholders

to the material supplier. This scenario is now explored using the case of energy storage through lithium-ion batteries. A material supplier of lithium cobalt oxide (i.e. cathode material) or lithium titanium oxide (i.e. anode material) offers to the producer of e-vehicle batteries the possibility to use its *materials as a service*. The material supplier requests that its materials are used on standardised battery designs, which follow predetermined design for disassembly rules. The material supplier also requests that the materials are returned at the end of the contract stipulated with the producer. After entering this contractual agreement, the producer manufactures battery cells, modules and packs for e-vehicle applications thinking carefully about the life of its battery products. The battery producer, conscious of the requirement to return the batteries to the material supplier, offers the batteries to manufacturers and consumers of e-vehicles as a product-service system. This means that consumers lease batteries from the producer and have a requirement to return them to the producer at the end of life as they have to be returned to the material supplier. Depending on the contractual agreement with the material supplier, the producer could decide to keep the batteries longer before returning them to the material supplier for recycling. For instance, the producer could repurpose the batteries for less energy intensive applications or operate a refurbishment service and offer them for appropriate applications.

7.4.2 Independent Use of a MSS

A MSS can also be operated independently of a PSS, see Fig. 7.1b. In this scenario the material supplier markets *materials as a service* to a producer, and the producer markets products to consumers. Consumers, however, have an obligation to return obsolete products to the material supplier. This scenario is now explored using the case of energy storage through lithium-ion batteries. A material supplier sells lithium cobalt oxide or lithium titanium oxide to a producer to manufacture batteries for e-vehicles. The battery producer then markets batteries to manufacturers and consumers of e-vehicles with an obligation to return them to the material supplier at the end of life. The collection or take-back service is, however, offered by the material supplier. In this scenario the battery producer is free from the responsibility to recollect obsolete batteries as this is transferred to the material supplier.

7.4.3 Expected Benefits to the Whole System

Leaving the ownership of resources at the top of the supply chain, MSSs have the potential to allow suppliers to exert more control on the flow of resources and consequently to use resources more efficiently (Allwood et al. 2011), for example, by collaborating closely with collectors, recoverers and recyclers. Specifically, MSSs would allow to reduce the dependency on virgin resources and permit to bring to life the concept of pure and closed resource flows, i.e. less contamination.

7.4.4 Expected Benefits for Material Suppliers

The following benefits and incentives have been identified for material suppliers, see Table 7.2.

Table 7.2 Expected benefits for material suppliers

Benefits and incentives	Description
Ability to *develop long term business relationships with producers*	Retaining customers through new value propositions
Ability to *sustain feedstock*	Sourcing obsolete returned resources rather than finite resources
Ability to *sustain business*	Building business that is based on services and partnerships (McAloone and Pigosso 2018) rather than the single transaction of finite resources
Ability to *exert control on material resources prior and after use*	Controlling flows of material resources similarly to how PSSs take control over product flows
Incentive to ensure that *resources flow uncontaminated* and in high volume to function as feedstock	Motivation and encouragement to exert control on systems across the resource lifecycle, as returning resource flows are seen as creating value (Krikke et al. 2013) and suppliers have a demand for their own used product (Zeeuw van der Laan and Aurisicchio 2019a)
Incentive to *develop a portfolio of material resources that are recyclable* and retain high quality after multiple cycles	Motivation and encouragement to develop materials that perform across multiple successive resource lifecycles
Ability to *react to emergent legislation*	Anticipating emerging legislation which could set tighter targets on the uptake of cycled materials e.g. recycled plastics (Ellen MacArthur Foundation 2019)
Ability to *react to new competition*	Reacting to the emergence of suppliers of recycled materials and suppliers of biodegradable materials
Ability to *lower material development pressure*	Lowering the pressure to develop alternative materials, e.g. biodegradable or compostable materials
Ability to *react to volatile raw material costs*	Reducing the need for and dependency on costly finite resources
Ability to *protect IP*	Protecting IP by retaining ownership of resources

Table 7.3 Expected benefits for producers

Benefits and incentives	Description
Ability to benefit from *material services offered by the supplier*	Repairing, maintaining, replacing or upgrading materials as part of contracted services, and avoiding material disposal taxes
Incentive to *set more ambitious sustainability targets*	Motivation, encouragement and opportunity to set and achieve realistic sustainability targets (Ellen MacArthur Foundation 2019)
Incentive to *design circular products that can be easily disassembled and in which materials remain identifiable*	Motivation and encouragement to design for disassembly to suit the demand of suppliers (Mestre and Cooper 2017)
Incentive to *optimise the time spent with resources*	Motivation and encouragement for lean consumption (Womack and Jones 2005) e.g. timely delivery and collection of obsolete resources from consumers, operating lean disassembly processes and returning obsolete resources to materials suppliers
Ability to lower social pressure	Mitigating the transfer of social pressure from producers to material suppliers

7.4.5 Expected Benefits for Producers

The following benefits and incentives have been identified for producers, see Table 7.3.

7.4.6 Expected Benefits for Consumers

If the time spent with material resources is charged, producers or consumers respectively will have incentives to optimise the time spent with them (Stahel 1982). The incentives include using only what they really need, for the time they need it. Similar to the effect of slowing loops of PSSs (Bocken et al. 2016), resources will consequently be flowing slower as they will become available quicker to consumers e.g. no hibernation of resources.

7.4.7 Expected Enablers of MSSs

A concept such as MMSs would benefit from the following enablers to come to life, see Table 7.4.

Table 7.4 Expected enablers

Enablers	Description
Material identification	Technology and systems to identify materials (Corbin et al. 2018)
Material monitoring	Technology to track and trace materials (Corbin et al. 2018) such as IoT, sensors, blockchain, RFID, etc.
Material processing at the end of life	Technology for sorting, separating, and further purifying materials, for example through chemical recycling (Rahimi and García 2017)
Product design guidelines	Rules on design for disassembly and design for material identification (Mestre and Cooper 2017)
Standardised material portfolio	Standardised materials to optimise revalorisation processing and increase recovered volumes (Prendeville et al. 2014)

7.4.8 Expected Challenges for the Adoption and Implementation of MSSs

The following challenges have been identified for MSSs, see Table 7.5.

7.5 Discussion

At present, material suppliers make business by sourcing, processing and selling virgin material resources extracted from our planet. Producers buy these material resources and manufacture products that are sold to consumers. At the end of product life only a small proportion of these resources are recovered and recycled. PSSs are business models through which producers sell the performance of products. PSSs can help control the flow of material resources as product ownership is shifted from consumers to producers. MSSs depict a future where resource usage is transformed and where business models, design and manufacturing processes, and product value and ownership are redefined compared to both traditional business and PSS models. MSSs have the potential to disrupt the current linear economy by shifting us faster towards a circular economy in which resources are more effectively managed and reliance on new resources is reduced. In particular, MSSs can extend the circular economy as a regenerative system to the material level (as opposed to the component level) contributing to increasing the recycling of technical materials. MSSs can transform the future manufacturing landscape by:

- Introducing new business models and performance offerings. Material suppliers will sell the performance of material resources as a service to producers, who in turn will sell access to products.

Table 7.5 Challenges

Challenges	Description
Suitability of materials	Not all materials will be suitable for this model due to their qualitative properties (Zink and Geyer 2017)
Volume of materials	Available volumes will influence the value of materials in flows
Degradation of materials	Some materials are subject to degradation when recovered. Technology is needed to retain quality. There is a risk that 'secondary' materials come to exist next to 'primary' materials and instead of replacing the need for virgin, and reducing the demand, they create a new market (Zink and Geyer 2017)
Environmental impacts of material revalorisation	Many recycling or recovery processes are energy intensive. There will be critique around choosing material recovery over component recovery (Allwood et al. 2011)
Competing with resource flows at higher utilities	The reuse of products and components typically seems to have a higher economic value than the recovery of materials (Schenkel et al. 2015)
Decontamination of resource flows	The purification of contaminated waste streams of material resources is dependent on technology (Rahimi and García 2017)
Material identity	The current market is producer-led. Materials are anonymous i.e. it is rarely possible to identify which company has supplied a material. The anonymity has advantages for the suppliers, as they cannot easily be blamed. Instead, the blame goes to the producers who put their name and identity on the products that the materials embody. To make MSS work, material suppliers may have to come out of the anonymity and materials be branded
Consumer behaviour	MSSs involve a systematic change with new roles for consumers (Zeeuw van der Laan and Aurisicchio 2019b). Eliciting the required behaviour has been found to be one of the main barriers to adopt PSSs
Dependence on product-service systems	In general, the adoption of PSSs by industry is low due to corporate, cultural and regulatory challenges that require structural systemic changes (Ceschin 2013; Vezzoli et al. 2012)
Circularity performance metrics	Until there is agreement on and adoption of the indicators of material flow and circularity, it will be hard to prove the success of circular business models (Moriguchi 2007; Saidani et al. 2019)

(continued)

Table 7.5 (continued)

Challenges	Description
Viability of business model	There is a need for research to estimate the revenue streams and expenses of the proposed business model. There is also a need for empirical research on the business model and for methods to validate and implement them (Pieroni et al. 2019)
Collaboration and supply chains	There is a need to understand the implications on collaboration and supply chains of the proposed business model. It is in fact possible that MSSs will disrupt existing inter-firm relationships and power dynamics resulting in new collaborative structures or vertical integration
Maintaining the status quo	Politics around material markets that aim to retain current practice for governmental and industrial benefits (Gregson et al. 2015)

- Helping secure material resources. Material suppliers will retain ownership of the materials embodied in products contributing to treating resources as banks; this will be useful to guarantee that material suppliers have their own future supply, and facilitate resource management between material suppliers, producers and consumers.
- Supporting new partnerships. Collaborative production of products and services between suppliers, producers, consumers, collectors, recoverers and recyclers will emerge with the aim to achieve pro-environmental outcomes.
- Propelling the development of a new industrial sector. Robust reverse supply chains have to emerge to recover value from obsolete material resources.
- Incentivising new approaches to material development and product design. Material suppliers have to introduce better recyclable materials and producers have to design products that perform effectively in the whole lifecycle.

7.5.1 Limitations and Future Work

This research has shown how repositioning ownership of material resources has allowed to introduce and explore a novel concept such as material-service systems. The work presented in this paper is based on a theoretical exploration of the concept and its relation to a construct such as product-service systems. Empirical work is necessary to deepen current understanding of material-service systems and shed light on their feasibility and viability.

To advance the concept of MSSs it would be beneficial to work in collaboration with industrial organisations including material suppliers, producers, collectors, recoverers and recyclers. Specifically, there is a need to identify what material and product types are more likely to benefit from MSSs; demonstrate the feasibility of

MSSs as new business models for materials that need circular management; map and address the technical, social and business challenges posed by resources in supply chains based on MSSs; and understand how MSSs would disrupt current business models and supply chains.

7.6 Conclusions

MSSs introduce new business models, in which material suppliers sell material performance as a service to producers, who, in turn, sell access to products to consumers. MSSs can help secure resources as material suppliers retain ownership of the materials embodied in products contributing to treating resources as banks and guaranteeing a future supply. MSSs have the potential to contribute to a paradigm shift in existing thinking about material resource management, leading to new and disruptive business models and accelerating the shift towards a structural change in consumption systems for a circular economy. Research on MSSs is ambitious and risky, but it has high potential to produce findings that could be transformative for industrial ecology and our society.

References

Allwood JM, Ashby MF, Gutowski TG, Worrell E (2011) Material efficiency: a white paper. Resour Conserv Recycl 55:362–381

Baxter WL, Aurisicchio M (2018) Ownership by design. In: Peck J, Shu S (eds) Psychological ownership and consumer behavior. Springer, Cham, pp 119–134

Baxter WL, Aurisicchio M, Childs PRN (2015) A psychological ownership approach to designing object attachment. J Eng Des 26:140–156

Bocken NMP, Short SW, Rana P, Evans S (2014) A literature and practice review to develop sustainable business model archetypes. J Clean Prod 65:42–56

Bocken NMP, de Pauw I, Bakker C, van der Grinten B (2016) Product design and business model strategies for a circular economy. J Ind Prod Eng 33:308–320

Ceschin F (2013) Critical factors for implementing and diffusing sustainable product-Service systems: insights from innovation studies and companies' experiences. J Clean Prod 45:74–88

Corbin L, Gladek E, Tooze J (2018) Materials Demcoracy: an action plan for realising a redistributed materials economy. Mak Futur 5:1–30

Ellen MacArthur Foundation (2015) Towards a circular economy: business rationale for an accelerated transition

Ellen MacArthur Foundation (2019) New plastics economy global commitment

Gregson N, Crang M, Fuller S, Holmes H (2015) Interrogating the circular economy: the moral economy of resource recovery in the EU. Econ Soc 44:218–243

Krikke H, Hofenk D, Wang Y (2013) Revealing an invisible giant: a comprehensive survey into return practices within original (closed-loop) supply chains. Resour Conserv Recycl 73:239–250

Lüdeke-Freund F, Gold S, Bocken NMP (2019) A review and typology of circular economy business model patterns. J Ind Ecol 23:36–61

McAloone TC, Pigosso DCA (2018) Designing product service systems for a circular economy. In: Charter M (ed) Designing for the circular economy. Routledge, pp 102–112

Mestre A, Cooper T (2017) Circular product design. A multiple loops life cycle design approach for the circular economy. Des J 20:S1620-1635

Moriguchi Y (2007) Material flow indicators to measure progress toward a sound material-cycle society. J Mater Cycles Waste Manag 9:112–120

Pieroni MPP, McAloone TC, Pigosso DCA (2019) Business model innovation for circular economy and sustainability: a review of approaches. J Clean Prod 215:198–216

Prendeville S, Sanders C, Sherry J, Costa F (2014) Circular economy: is it enough?

Rahimi A, García JM (2017) Chemical recycling of waste plastics for new materials production. Nat Rev Chem 1:0046

Rau T, Oberhuber S (2016) Material Matters—Het alternatief voor onze roofbouwmaatschappij, 5th ed. Bertram + de Leeuw Uitgevers

Rifkin J (2000) The age of access. Penguin, Harmondsworth

Saidani M, Yannou B, Leroy Y et al (2019) A taxonomy of circular economy indicators. J Clean Prod 207:542–559

Schenkel M, Caniëls MCJ, Krikke H, van der Laan E (2015) Understanding value creation in closed loop supply chains—Past findings and future directions. J Manuf Syst 37:729–745

Stahel WR (1982) The product life factor. In (Orr, G.S. (ed), An inquiry into the nature of sustainable societies: the role of the private sector. Houst Area Res Cent, pp 72–105

Stahel WR (2010) Sustainability and the performance economy. In: The performance economy. Palgrave Macmillan, UK, pp 269–270

Steg L, Vlek C (2008) Encouraging pro-environmental behaviour: an integrative review and research agenda. J Environ Psychol 29:309–317

Stoughton M, Votta T (2003) Implementing service-based chemical procurement: lessons and results. J Clean Prod 11:839–849

Vezzoli C, Ceschin F, Diehl J, Kohtala C (2012) Why have "Sustainable Product-Service Systems" not been widely implemented? Meeting new design challenges to achieve societal sustainability. J Clean Prod, pp 288–290

Womack JP, Jones DT (2005) Lean consumption. Harv Bus Rev

Zeeuw van der Laan A, Aurisicchio M (2019) Designing product-service systems to close resource loops: circular design guidelines. Procedia CIRP 80:631–636

Zeeuw van der Laan A, Aurisicchio M (2019) Archetypical consumer roles in closing the loops of resource flows for fast-moving consumer goods. J Clean Prod 236

Zink T, Geyer R (2017) Circular Economy Rebound. J Ind Ecol 21:593–602

Zink T, Geyer R (2018) Material recycling and the myth of landfill diversion. J Ind Ecol 23:541–548

Chapter 8
Designing for Vehicle Recyclability from the Perspectives of Material and Joining Choices

Vi Kie Soo, Paul Compston, and Matthew Doolan

Abstract The increasing use of lightweight materials and multi-material vehicle designs has improved the environmental impacts during the use phase through reduced vehicle mass. However, the growing complexity of vehicle designs has led to challenges in the choice of joining techniques. Previous ecodesign guidelines encourage less use of joints to reduce potential weak spots. However, minimizing the number of joints contradict with the increasing variety of multi-material combinations and their associated joining techniques. Consequently, the use of multi-material designs and their associated joining choices further reduces the effectiveness of current shredding and sorting recycling processes. This paper proposes a vehicle design framework for material and joining preferences to assist in material recyclability. Observations from previous case studies in Australia and Europe supplemented by literature data are used to assess the correlation between different material types, material separability through commonly used recycling practices, and material compatibility during the metallurgical processing. The method used to evaluate the design preferences for material separation during recycling are discussed. This work provides critical insights into the linkage between vehicle design and recycling phases to promote environmentally conscious design of products.

Keywords Design for recycling · Joining choices · Multi-material designs · Vehicle recycling · End-of-life vehicles

8.1 Introduction

Lightweight vehicle concept is used extensively in the vehicle industry to reduce fuel consumption towards achieving the strict CO_2 emission legislations. Vehicle manufacturers have focussed on reducing the vehicle mass through the use of multi-material designs which led to the growing complexity in vehicle designs and manufacturing over time. Vehicle structural parts are increasingly replaced with different

V. K. Soo (✉) · P. Compston · M. Doolan
Research School of Electrical, Energy and Materials Engineering, College of Engineering and Computer Science, The Australian National University, Canberra, Australia
e-mail: vikie.soo@anu.edu.au

Y. Kishita et al. (eds.), *EcoDesign and Sustainability I*, Sustainable Production, Life Cycle Engineering and Management, https://doi.org/10.1007/978-981-15-6779-7_8

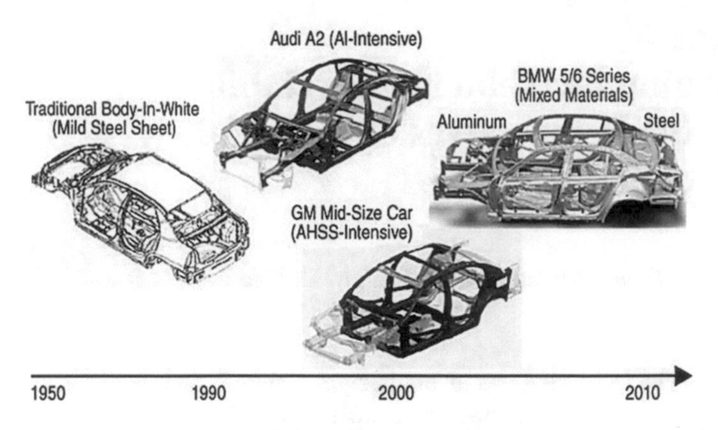

Fig. 8.1 The evolution of automotive BIW structure from 1950 to 2010 Taub et al. (2007)

materials with a higher strength-to-weight ratio. The body-in-white (BIW) structure, for example, is one of the core body structures showing significant material changes, as shown in Fig. 8.1.

The mass-optimised design approach to combine a variety of lightweight materials (e.g. advanced high strength steel (AHSS), aluminium (Al), magnesium (Mg), polymers, and carbon fibre reinforced polymers (CFRP)) also led to the increasing use of different joining techniques (Davies 2012). Therefore, the feasibility of intricate multi-material vehicle designs is affected by the choice of material types and also the viability of the joining techniques used to connect the different materials.

The choice of joining techniques to satisfy the design requirements for the complex vehicle structures is becoming challenging (Martinsen et al. 2015). Manufacturers are limited by the types of joining techniques that are feasible for multi-material combinations while remaining cost-effective for large-scale production. Previous work has shown the increasing use of non-welding techniques to join multi-material combinations, particularly for metal to non-metal structures (Groche et al. 2014). There is a growing trend of mechanical fasteners, adhesive bonding or combination of these joining techniques used for multi-material combinations, as shown in Table 8.1 (Soo et al. 2014, 2007a). The narrow joining process selection is mainly due to the need for lower manufacturing cost (Davies 2012) and the flexible adaptability to changing material designs (Martinsen et al. 2015).

The commonly used shredding and sorting recycling processes no longer catering well for a high material recycling efficiency of complex multi-material vehicle designs. Frequently used non-welding joining techniques (e.g. mechanical fasteners, adhesive bonding or a combination of these joining methods) are widely applicable for joining the same or different material types while remaining cost-effective (Meschut et al. 2014). In newer vehicle designs, there is a growing use of mechanical fasteners, such as screws, to combine plastic materials (Kah et al. 2014). Consequently, perfect material separation is becoming more difficult when the joints are not liberated during the recycling process. Moreover, the commonly used joining

Table 8.1 Multi-material joining matrix Soo et al. (2014)

Material combination		Light metal			Non-metal	
		AHSS	Aluminium	Magnesium	Plastic	CFRP
Light metal	AHSS	a b c d e* f g*	a b c d* e* f* g	a b c d* e* f* g*	b c e*	b c e*
	Aluminium		a b c d e* f g	a b c d* e* f* g*	b c e*	b c e*
	Magnesium			a b c d* e* f g*	b c e*	b c e*
Non-metal	Plastic				b c e*	b c e*
	CFRP					b c e *

[a]TIG, MIG welding
[b]Adhesive bonding
[c]Mechanical fastening
[d]Resistance welding
[e]Ultrasonic spot welding
[f]Laser welding
[g]Friction stir spot welding
*Not in large production

methods, such as mechanical fastening and adhesive bonding, introduce additional materials that are often the source of contaminants. In most cases, the unliberated joint designs also contributed to the impurities in recovered output streams or the material losses in the automotive shredder residues. For instance, steel screws used to join Al materials that are not liberated during the shredding process are likely to end up in the recovered Al output stream (Soo et al. 2015).

The choice of joining methods is critical to assist in high material recycling efficiency through the shredder-based end-of-life vehicle (ELV) recycling processes. Vehicle manufacturers often focus on the choice of material types during the design phase to improve the ELV recyclability. However, the choice of joining techniques used to combine the different material types can have a significant influence on the efficiency of material recovery at the end-of-life (EoL) phase. Therefore, it is critical to better understand the linkage between vehicle designs and their impacts on current ELV recyclability. This can ensure that the general guidelines remain relevant for changing vehicle designs.

8.2 Design Framework

8.2.1 Design Framework for Sustainable ELV Recycling

Design for Recyclability (DfR) is one of the aspects of sustainability framework tightly-linked to EoL strategies for products. Designing a vehicle for recyclability needs to consider both the material choice and their associated joining techniques, and their impacts on the shredder-based ELV recycling approach (Bogue 2007). The gap between vehicle manufacturers (designers) and the recyclers need to be addressed and is becoming more critical (Bras 1997). Although manufacturers have used DfR concept for many years, the effect of changing vehicle designs on the efficiency of material recycling processes is not well understood. This can be attributed by the contradicting design requirements, cost and technical limitations.

8.2.2 Design for Vehicle Recycling Guidelines

Vehicle manufacturers have used different decision-making approaches and design guidelines to address conflicting requirements during the design phase (Girubha and Vinodh 2012). One of the criteria that is often given more importance is the selection of materials to achieve lightweight structures. The variety of material combinations used for vehicle components are shown in Table 8.2. Toward producing more lightweight vehicles, the choice of material replacement while remaining the vehicle structural integrity is critical to achieving substantial mass reduction with the highest potential (Fuchs et al. 2008). The selection of material is based on different criteria such as material supply risk (Knoeri et al. 2013); mass reduction potential based on functional equivalence (Li et al. 2017); and material production efficiency (Ashby 2012). The most basic material selection guidelines for manufacturers, such as Volvo (Luttropp and Lagerstedt 2006), are the white, grey, and black material lists. The white list shows the materials that are encouraged to be used in the product. On the other hand, the grey and black lists identify materials that should be considered for material substitution and avoided respectively.

The current ecodesign strategies are limited and do not cater well for the improvement in material recyclability (Worrell and Reuter 2014). This is caused by the increasing challenges to link the changing product design phase and their impact on current industrial recycling practices. The use of complex multi-material designs has led to lower material separation efficiency and contaminations that prevents the reused of material for the same application.

The choice of materials will influence the options for feasible joining methods. Although there are available guidelines specific to joining choices, they often address the selection based on the ease for disassembly, maintenance and repair (mainly non-destructive recycling approach) (Argument et al. 1998; Edwards et al. 2006; Ghazilla et al. 2014; Shu and Flowers 1999,1996). Hence, these guidelines may not apply to the destructive ELV recycling approach (shredder-based recycling processes) (Newell 1965). One of the most comprehensive design guidelines from the joining perspective is the German recycling rating, VDI 2243 guidelines (VDI 2243 1993), as seen in Fig. 8.2. The guidelines take into consideration the characteristics of fastening and their implication on recycling to suggest preferred joining methods.

Vehicle manufacturers often face conflicting ecodesign guidelines (Luttropp et al. 2001). For example, the 'Ten Golden Rules'—a set of ecodesign guidelines used

Table 8.2 Multi-material combinations for the different vehicle components Elgowainy et al. (2015)

Category	Component	Steel/cast iron	Stainless steel	Aluminium (wrought)	Aluminium (cast)	Copper	Magnesium	Lead	Glass	Plastic	Rubber	CFRP
Glider	Body	■							▒	▒		
	Exterior	▒				▒				■	▒	
	Interior	■		▒		▒				■		
	Chassis	■		▒		▒				▒	▒	
	Weld blanks and fasteners	■								■		
Powertrain	Engine	■			■	▒	▒			▒	▒	
	Engine fuel storage system	■										
	Powertrain thermal	■								■		
	Fuel cell stack		■							▒		
	Fuel cell auxiliaries	■		▒		▒				▒	▒	▒
	Exhaust	■						▒				
	Powertrain electrical					■				■		
	Emission controls					■				■		
	Weld blanks and fasteners	■										
Transmission	ICEV	■		■						▒	▒	
	HEV, FCEV and PHEV	■		▒		▒						
	Traction motor	■			■	▒						
Wheels component	Wheels	■										
	Tires	■									■	
Battery	Lead-acid battery (ICEV)							■		▒		
	Li-ion battery (HEV, FCEV, PHEV, BEV)	▒		▒		▒				▒		

■ Major material (at least 30wt.% of the component) ▒ Minor material (less than 30wt.% of the component)

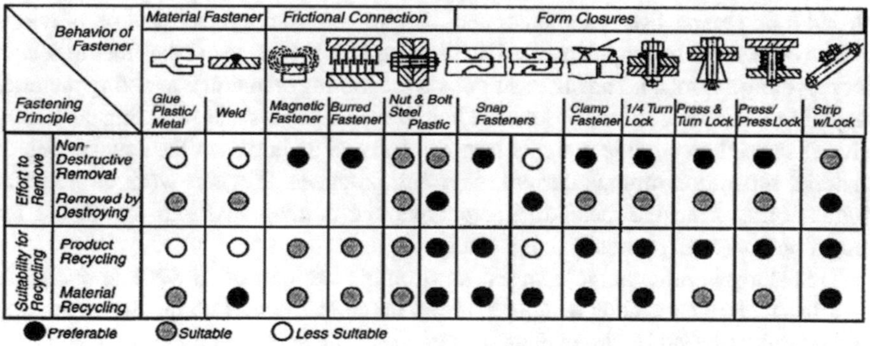

Behavior of Fastener / Fastening Principle		Material Fastener		Frictional Connection				Form Closures						
		Glue Plastic/ Metal	Weld	Magnetic Fastener	Burred Fastener	Nut & Bolt Steel	Nut & Bolt Plastic	Snap Fasteners (1)	Snap Fasteners (2)	Clamp Fastener	1/4 Turn Lock	Press & Turn Lock	Press/ PressLock	Strip w/Lock
Effort to Remove	Non-Destructive Removal	○	○	●	●	◍	◍	●	○	●	●	●	●	◍
	Removed by Destroying	◍	◍			◍	●		●	◍	◍	◍	◍	●
Suitability for Recycling	Product Recycling	○	○	◍	◍	◍	●	●	○	●	●	●	●	●
	Material Recycling	◍	●	◍	◍	◍	●	●	●	●	●	◍	◍	●

● Preferable ◍ Suitable ○ Less Suitable

Fig. 8.2 Joint selection for recycling based on the VDI 2243 guidelines Rosen et al. (1996)

by companies and researchers (Luttropp and Lagerstedt 2006) encourages vehicle designs with efficient resource consumption during the use phase. This has led to multi-material designs to reduce vehicle mass that contradicts with the guideline to use fewer joints. Manufacturers often resolve such conflict based on priorities. For instance, manufacturers are obliged to achieve the vehicle emission standards and therefore prioritise efforts to improve fuel efficiency. However, from an EoL perspective, the design for ELV scrap material recycling should be prioritised to reduce the effort for material separation and to optimise the quality of recycled material recovered through the shredding process (Shu and Flowers 1999).

To cater for the conflicting design guidelines for recycling, a more effective approach is required. In this paper, the design framework to assess the material and joining preferences for vehicle recyclability is presented. The application of the design approach is demonstrated using the observations from case studies and literature data based on current recycling practices.

8.3 Method

A framework for mapping design for recycling knowledge from the material and joining perspectives is proposed. The risk assessment approach is adopted since the recycling impacts of increasingly complex vehicle designs are dynamic. Risk assessment analysis has been used as a tool to evaluate the uncertainties of interdependent variables to assist in decision-making in various industries (Fletcher 2005; Zeng et al. 2007; Theoharidou et al. 2011). This method is highly adaptable and allows ongoing monitoring of new joining advancements for multi-material combinations, and changing recycling technologies in the vehicle industry.

In this study, the framework to map the design for recycling approach from a closed-loop perspective is demonstrated. Empirical observations based on previous industrial case studies supplemented by literature data are used to interpret the design for recycling approach. The observations on the joining techniques assisting material recycling are based on industrial shredder trials in large-scale recycling facilities. This is crucial to provide insights into the linkage of joints on the efficiencies of material separation through current recycling practices. Previous work on material compatibility from the metallurgical perspective is integrated into the design for recycling approach proposed in this study.

This study highlights the material and joining preferences in vehicle design to facilitate material recycling without material or quality losses through the commonly used shredder-based ELV recycling approaches. The assessment matrix used in the design approach of this study only focuses on the perspective of material recycling without degradation. Nevertheless, the proposed design framework for recycling can be implemented for changing requirements or recycling technologies (e.g. non-destructive disassembly recycling process) by reviewing the assessment matrix.

Table 8.3 Observations on the input joining audit and the unliberated joining types used for different or similar material combinations in the various output streams

Output stream	Input joining audit	Unliberated joining types
Ferrous (Fe)	Steel fasteners, steel rivets, steel clips plastic rivets, plastic clips, spot welding, MIG welding, adhesive bonding, brazing	Steel fasteners, plastic rivets, adhesive bonding
Non-ferrous (NF)	Steel fastener, steel rivets, adhesive bonding	Steel fasteners, steel rivets, adhesive bonding
Automotive shredder residue (ASR)—valuable material loss	Steel fasteners, steel rivets, plastic rivets, steel clips, adhesive bonding, plastic clips	Steel fasteners, plastic rivets

8.4 Observations from Case Studies and Literature Data

The case studies presented in Sects. 8.4.1 and 8.4.2 focus on the observations for unliberated shredded output samples due to the choice of joining methods and material combinations. The material and joining audit for the Australian car door shredder trials is used to identify design preferences for material liberation during recycling (Soo et al. 2017a). The joining choices causing unliberated joints are further supported by the recovered Al sampling from ELV through a Belgian recycling facility (Soo et al. 2018). Previous work on material compatibility of different material combinations from the metallurgical processing perspective is then presented in Sect. 8.4.3.

8.4.1 Australian Car Door Case Study

The observations on the input material and joining audits for car doors, and the unliberated joints in the collected output streams are shown in Table 8.3. Some of the unliberated samples observed from the case study are shown in Figs. 8.3, 8.4 and 8.5.

8.4.2 European Case Study—Recovered Al Sampling

Observations from the European case study are used to support the unliberated joint types in the Australian case study through a more rigorous recycling facility. The observations shown in this study provide insights on unliberated joint characteristics that are contradicting the generic eco-design guidelines in Fig. 8.2. Based on the Al shredded output samples collected through a more advanced recycling approach used

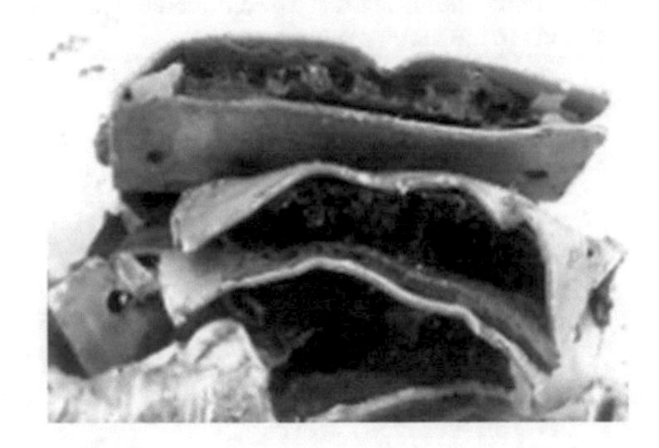

Fig. 8.3 Sandwich layer adhesion for different material types (Al-glass)

Fig. 8.4 Unliberated steel rivet from the car door mechanism and car door handle (aluminium)

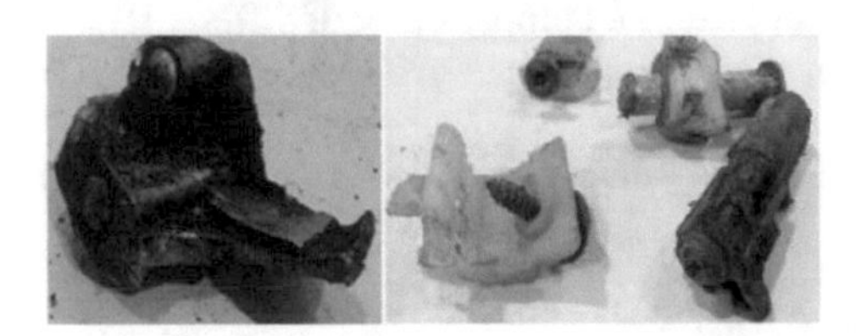

Fig. 8.5 Unliberated steel fasteners for same and different material combinations (Fe–Al, Fe-Plastic)

in the Belgian recycling facility (Soo et al. 2017b), some of the observed samples with unliberated joint are shown in Fig. 8.6. The types of unliberated joints and their characteristics are presented in Table 8.4.

8.4.3 Material Compatibility Matrix

Castro et al. (2004) studied the compatibility of material from the metallurgical aspect. They have suggested the use of the thermodynamic evaluation of material combinations (THEMA) model (Castro et al. 2004). The THEMA model assist

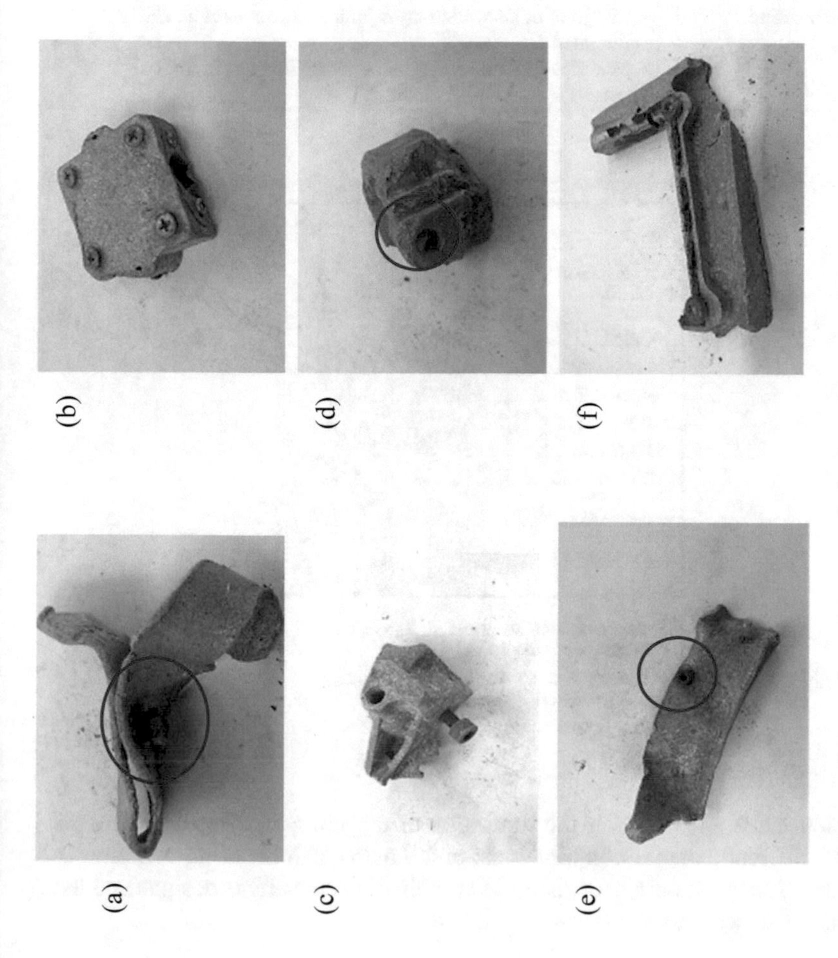

Fig. 8.6 Fe impurities in the recovered Al output stream due to joints **a** steel clip; **b** machine screw; **c** socket screw; **d** bolt screw—missing head; **e** steel rivet; **f** hybrid joining- adhesive bonding and machine screw Soo et al. (2019)

Table 8.4 Joining types causing impurities in the Al output stream and the observed characteristics

Unliberated joint type	Joint characteristics
Steel fasteners	• Partial liberation for threaded fasteners • Fasteners with protrusion (e.g. socket screw) assist in liberation • Corroded steel fasteners cannot be easily liberated
Steel rivets, steel clips, pins	• Smooth surface ease liberation if shredded into smaller particle sizes
Adhesive bonding	• Lap joint or sandwich layer joint design cannot be easily liberated • Material combination with larger differences in material densities assist in liberation

Table 8.5 Material compatibility from the metallurgical processing perspective Castro et al. (2004)

Output stream (impurities/ similar material)	Output recycled stream: Steel/cast iron	Stainless steel	Al (wrought)	Al (cast)	Magnesium	Plastic	CFRP
Steel/cast iron	Good	Average	Poor	Poor	Poor	Poor	Poor
Stainless steel	Average	Good	Poor	Poor	Poor	Poor	Poor
Al (wrought)	Average	Average	Good	Average	Poor	Poor	Poor
Al (cast)	Average	Average	Poor	Good	Poor	Poor	Poor
Magnesium	Average	Average	Average	Average	Good	Poor	Poor
Plastic (same type, colour)	Average	Average	Average	Average	Poor	Good	Poor
Plastic (different type)	Average	Average	Average	Average	Poor	Poor	Poor
CFRP	Average	Average	Average	Average	Poor	Poor	Good

Poor compatibility - avoid mixing; can lead to material quality loss

Average compatibility - avoid if possible; problem can arise

Good combination during processing

in decision-making to consider the limitation of recycling processes; the compatibility of different material combinations; and the constraints during metallurgical processing. The commonly used material combinations in vehicle designs and their metallurgical constraints are shown in Table 8.5.

8.5 Design to Facilitate ELV Recycling

Based on the observations from previous case studies, the joining choices for multi-material combinations have a significant influence on the material separation during the commonly used shredder-based recycling approach. The likelihood of material separation using different joining techniques through the current recycling

Table 8.6 Likelihood of material separation using different joining techniques through the current shredder-based recycling processes Soo et al. (2017a, 2018)

Main material/ Joining	Ultrasonic spot welding	Other welding	Adhesive bonding	Mechanical fastener - Steel	Mechanical fastener - plastic
Steel/cast iron	Low	Low	Average	Low	High
Stainless steel	Low	Low	Average	Low	High
Al (wrought)	Low	Low	Average	Average	High
Al (cast)	Low	Low	Average	Average	High
Magnesium	Low	Low	Average	Average	High
Plastic	Average	Not applicable	Average	Average	High
CFRP	Average	Not applicable	Average	Average	High

Low material separation
Average material separation
High material separation
Not applicable

approaches is shown Table 8.6. It is important to note that the preliminary results of joining preferences for material recycling observed from the case studies may change in accordance with the development in recycling technologies, such as the adoption of non-destructive material disassembly process.

To identify the design preferences for material separation, both the joining choices and material combinations need to be assessed. The observations from Tables 8.5 and 8.6 are used to identify vehicle designs assisting in material recycling without material quality degradation through the current shredder-based recycling processes. The relationship matrix considering the material and joining choices shown in Table 8.7 provides a general ecodesign guideline for material recycling without loss of quality.

8.6 Discussion and Future Work

Observations from case studies have shown that the generic ecodesign guidelines may not cater well for the shredder-based recycling practices largely used for ELV. The changing material designs and the associated joining choices have an implication on the vehicle recyclability. The evolution of joining technologies to cater for changing material designs has led to the uncertainty of material separability during the ELV recycling. This paper highlights the importance of better design approach from the perspective of material and joining choices to assist in material separation for non-compatible material during the metallurgical processes.

Table 8.7 The relationship matrix to design for optimised material separation without quality loss during the metallurgical processing of recycled materials

Likelihood of Separation	Good	Average	Poor
Low	Material combination with low separability but good compatibility leading to no material quality losses during metallurgical processing.	Material combination with low separability and average compatibility. There is a possibility of impurities causing material or quality losses.	Material combination with low separability and compatibility leading to recycling with material or quality losses.
Average	Material combination with average separability but good compatibility leading to no material quality losses during metallurgical processing.	Material combination with average separability and compatibility. There is a possibility of impurities causing the material quality loss.	Material combination with average separability and poor compatibility. The poor material compatibility is the major cause of recycling with material or quality losses.
High	Material combination can be easily separated with good compatibility leading to no material quality losses during metallurgical processing.	Material combination can be easily separated with average compatibility. The ease of separation is the key to material recycling without the presence of impurities.	Material combination can be easily separated with poor compatibility. The ease of separation is the key to material recycling without the presence of impurities.

Material Compatibility

Material quality loss | Possible material quality loss | No material quality loss

The proposed design approach serves as a generic guideline to identify design preferences based on the influence of joining choices on current recycling practices; and the material compatibility for optimised recycling. It is important to highlight that the relationship matrix to assist in DfR shown in Table 8.7 can be used to infer changing requirements and objectives to identify the range of conflicting design guidelines. In this study, the observations from industrial case studies and literature data are used as an initial assessment to identify material and joining choices that can influence material recycling at the ELV phase. There is a potential for further experimental work to investigate the material separability for new joining technologies through a controlled experiment.

The design for recycling framework is based on risk assessment that may subject to different interpretations. This may lead to a lack of differentiation within risk categories. There is a need to clearly state the objective of the assessment supported by robust likelihood and consequences to improve clarity in conducting the risk-assessment framework.

8.7 Conclusions

The design for vehicle recycling guideline needs to consider the joining and material choices, as well as the material compatibility from the thermodynamic perspective. Observations from case studies have shown the contradicting ecodesign guidelines for joining preferences due to the lack of insights into the current recycling practices.

Therefore, there is a need for a generic design approach to account for the linkage between vehicle design and recycling to assist in improving material recyclability.

A generic design approach considering the material compatibility and likelihood of material separation based on the material and joining choices can assist in providing informed decision targeting DfR. This is particularly the case for optimising material recyclability of multi-material vehicle designs with increasing complexity. Although certain material combinations are not compatible from the thermodynamic perspective, the choice of joining techniques used can determine the material separability through the current shredder-based recycling practices. The fundamental design approach proposed in this study can be used for understanding the changing design preferences for recycling.

Acknowledgements This study is supported by the ARC Training Centre in Lightweight Automotive Structure (project number IC160100032), the Australian National University, and is funded by the Australian Government.

References

Argument L, Lettice F, Bhamra T (1998) Environmentally conscious design: matching industry requirements with academic research. Des Stud 19:63–80. https://doi.org/10.1016/S0142-694X(97)00017-3

Ashby MF (2012) Materials and the environment: eco-informed material choice. Elsevier

Bogue R (2007) Design for disassembly: a critical twenty-first century discipline. Assem Autom 27:285–289. https://doi.org/10.1108/01445150710827069

Bras B (1997) Incorporating environmental issues in product design and realization. Ind Environ 20:7–13

Castro MBG, Remmerswaal JAM, Reuter MA, Boin UJM (2004) A thermodynamic approach to the compatibility of materials combinations for recycling. Resour Conserv Recycl 43:1–19. https://doi.org/10.1016/j.resconrec.2004.04.011

Davies G (2012) Materials for automobile bodies. Butterworth-Heinemann

Edwards C, Bhamra T, Rahimifard S (2006) A design framework for end-of-life vehicle recovery. In: Proceedings of the 13th CIRP international conference life cycle engineering. Leuven, Belgium, pp 360–365

Elgowainy A, Han J, Ward J, Joseck F, Gohlke D, Lindauer A et al (2016) Cradle-to-grave lifecycle analysis of U.S. light-duty vehicle-fuel pathways: a greenhouse gas emissions and economic assessment of current (2015) and future (2025–2030) technologies. Argonne National Laboratory

Fletcher WJ (2005) The application of qualitative risk assessment methodology to prioritize issues for fisheries management. ICES J Mar Sci 62:1576–1587. https://doi.org/10.1016/j.icesjms.2005.06.005

Fuchs ERH, Field FR, Roth R, Kirchain RE (2008) Strategic materials selection in the automobile body: economic opportunities for polymer composite design. Compos Sci Technol 68:1989–2002. https://doi.org/10.1016/j.compscitech.2008.01.015

Ghazilla RAR, Taha Z, Yusoff S, Rashid SHA, Sakundarini N (2014) Development of decision support system for fastener selection in product recovery oriented design. Int J Adv Manuf Technol 70:1403–1413. https://doi.org/10.1007/s00170-013-5373-3

Girubha RJ, Vinodh S (2012) Application of fuzzy VIKOR and environmental impact analysis for material selection of an automotive component. Mater Des 37:478–486. https://doi.org/10.1016/j.matdes.2012.01.022

Groche P, Wohletz S, Brenneis M, Pabst C, Resch F (2014) Joining by forming—a review on joint mechanisms, applications and future trends. J Mater Process Technol 214:1972–1994. https://doi.org/10.1016/j.jmatprotec.2013.12.022

Kah P, Suoranta R, Martikainen J, Magnus C (2014) Techniques for joining dissimilar materials: metals and polymers. Rev Adv Mater Sci 36:152–164

Knoeri C, Wäger PA, Stamp A, Althaus HJ, Weil M (2013) Towards a dynamic assessment of raw materials criticality: linking agent-based demand—With material flow supply modelling approaches. Sci Total Environ 461:808–812. https://doi.org/10.1016/j.scitotenv.2013.02.001

Li Z, Yu Q, Zhao X, Yu M, Shi P, Yan C (2017) Crashworthiness and lightweight optimization to applied multiple materials and foam-filled front end structure of auto-body. Adv Mech Eng 9:1687814017702806. https://doi.org/10.1177/1687814017702806

Luttropp C, Karlsson R (2001) The conflict of contradictory environmental targets. In: Proceding of EcoDesign 2001 second international symptoms environmental conscious design inverse manufacturing 2001, pp. 43–48. 10.1109/.2001.992312

Luttropp C, Lagerstedt J (2006) EcoDesign and the ten golden rules: generic advice for merging environmental aspects into product development. J Clean Prod 14:1396–1408. https://doi.org/10.1016/j.jclepro.2005.11.022

Martinsen K, Hu SJ, Carlson BE (2015) Joining of dissimilar materials. CIRP Ann Manuf Technol 64:679–699. https://doi.org/10.1016/j.cirp.2015.05.006

Meschut G, Janzen V, Olfermann T (2014) Innovative and highly productive joining technologies for multi-material lightweight car body structures. J Mater Eng Perform 23:1515–1523. https://doi.org/10.1007/s11665-014-0962-3

Newell AS (1965) Hammer mills. US3482788A

Rosen DW, Bras B, Hassenzahl SL, Newcomb PJ, Yu T (1996) Towards computer-aided configuration design for the life cycle. J Intell Manuf 7:145–160. https://doi.org/10.1007/BF00177070

Shu LH, Flowers WC (1996) Towards life-cycle fastening and joining cost optimisation using genetic algorithms. In: Proceeding 1996 ASME design engineering technical conferences & computers engineering conferences 1996

Shu LH, Flowers WC (1999) Application of a design-for-remanufacture framework to the selection of product life-cycle fastening and joining methods. Robot Comput-Integr Manuf 15:179–190. https://doi.org/10.1016/S0736-5845(98)00032-5

Soo VK, Compston P, Subic A, Doolan M (2014) The impact of different joining decisions for lightweight materials on LCA. Driv. Automot. Innov, Melbourne, Australia, p 5

Soo VK, Compston P, Doolan M (2015) Interaction between new car design and recycling impact on life cycle assessment. Procedia CIRP 29:426–431. https://doi.org/10.1016/j.procir.2015.02.055

Soo VK, Compston P, Doolan M (2017) The influence of joint technologies on ELV recyclability. Waste Manag 68:421–433. https://doi.org/10.1016/j.wasman.2017.07.020

Soo VK, Peeters J, Compston P, Doolan M, Duflou JR (2017) Comparative study of end-of-life vehicle recycling in Australia and Belgium. Procedia CIRP 61:269–274. https://doi.org/10.1016/j.procir.2016.11.222

Soo VK, Peeters J, Paraskevas D, Compston P, Doolan M, Duflou JR (2018) Sustainable aluminium recycling of end-of-life products: a joining techniques perspective. J Clean Prod 178:119–132. https://doi.org/10.1016/j.jclepro.2017.12.235

Soo VK, Peeters JR, Compston P, Doolan M, Duflou JR (2019) Economic and environmental evaluation of aluminium recycling based on a Belgian case study. Procedia Manuf 33:639–646

Taub AI, Krajewski PE, Luo AA, Owens JN (2007) The evolution of technology for materials processing over the last 50 years: the automotive example. JOM 59:48–57. https://doi.org/10.1007/s11837-007-0022-7

Theoharidou M, Kotzanikolaou P, Gritzalis D (2011) Risk assessment methodology for interdependent critical infrastructures. Int J Risk Assess Manag 15:128–148. https://doi.org/10.1504/IJRAM.2011.042113

VDI 2243 (1993) Konstruieren recyclinggerechter technischer produkte (Designing Technical Products for ease of Recycling). Germany

Worrell E, Reuter M (2014) Handbook of recycling: state-of-the-art for practitioners, analysts, and scientists. Newnes

Zeng J, An M, Smith NJ (2007) Application of a fuzzy based decision making methodology to construction project risk assessment. Int J Proj Manag 25:589–600. https://doi.org/10.1016/j.ijproman.2007.02.006

Chapter 9
Eco-innovation by Integrating Emerging Technologies with ARIZ Method

Jahau Lewis Chen and Chi-Yu Chou

Abstract This paper integrates emerging technologies with ARIZ method and presents a process of eco-innovative design. The process is based on emerging technology database and ARIZ-85C (Fey) innovation algorithm. With the connection between the database and ARIZ-85C (Fey), designers can analyze problems step by step and take advantage of the traits of suitable emerging technologies to solve problem and create innovative design. Furthermore, this process adds the concept of functional assessment and eco-assessment to confirm that designers develop an innovative design that not only can effectively solve the problem, but also more friendly to the environment.

Keywords ARIZ · Emerging technologies · Eco-innovation · TRIZ

9.1 Introduction

New emerging and fast growing technologies, such as IoT, 3D printing, self-driving cars, intelligent robots/machines, has attracted lots of attention recently. Those technologies for the future will play a key role in helping businesses design much more sustainable products and move towards sustainability in the future. However, the influence of emerging technologies to the environmental impacts is not clearly realized by inventors and designers. It is therefore necessary to develop eco-innovation method to help designers integrating emerging technologies into their product innovation activities to innovate new product with sustainability concepts.

The definition of emerging technologies has been very vague. Halaweh (2013) points out that emerging technology is not necessary new technology. If any technology is not spread into all technology domains or does not produce great change in business and industry model, then it can be treat as emerging technologies. Rotolo et al. (2015) defined emerging technologies by five attributes: (1) radical novelty, (2) fast growth, (3) coherence, (4) prominent impact, and (5) uncertainty and ambiguity.

J. L. Chen (✉) · C.-Y. Chou
Department of Mechanical Engineering, National Cheng Kung University, Tainan, Taiwan
e-mail: jlchen@mail.ncku.edu.tu

Y. Kishita et al. (eds.), *EcoDesign and Sustainability I*, Sustainable Production, Life Cycle Engineering and Management, https://doi.org/10.1007/978-981-15-6779-7_9

Chen and Chen (2017) combine emerging technologies with TRIZ to propose an eco-innovative design method.

ARIZ combined with different field have emerged. An extensive search of the literature about ARIZ shows that only a few works have been devoted to its study and application. In 1994, Fey et al. (1994) used ARIZ to solve some real life engineering problems for obtaining non-obvious solutions. In 2005, Krasnoslobodtsev and Langevin (2005) used ARIZ to develop a robot that can clean, finish and diagnose any oriented surfaces in space, for instance shop windows and some dangerous tank surfaces, and over 20 patents have been granted to this invention. In 2005, Krasnoslobodtsev et al. (2005) used ARIZ to help Samsung Electronic Company solve a problem of air conditioner and the destruction problem of a driving pin and cam-bush. In 2006, Krasnoslobodtsev and Langevin (2006) researched by using the ARIZ to solve some of Samsung's high-tech problems, such as those related to printer ink cartridges and vacuum cleaners. Chen and Chen proposed to use ARIZ as an innovation tool for eco-innovation problems (Chen and Chen 2009, 2013) and patent design around problems (Chen and Chen 2011, 2014a). Chen and Chen (2014b) integrated biomimetic with ARIZ-85C (Fey) and proposed Bio-ARIZ to make the designers can get the corresponding new idea from biology. Chen and Lin (2018) integrated ARIZ-85C (Fey) and case-based reasoning method on product innovative design for the Bottom of the pyramid.

This paper presents an eco-innovative method by integrating concepts of technologies for the future with ARIZ. An eco-innovation process for products by technologies for the future and ARIZ tools is proposed in this paper.

9.2 Technologies for the Future

9.2.1 New Emerging Technologies and Environment

The potential effect of emerging technologies to environment has caused wide attention. Shaikh et al. (2017) study the necessary key technologies to establish green internet of things (IoT). Gebler et al. (2014) present the qualitative and quantitative assessment of 3D printing process for evaluating the change in product life cycle cost, energy consumption, and CO_2 emissions. Ford and Despeisse (2016) show the advantages and challenges of 3D printing in concerning sustainability. Wadud et al. (2016) present the relationship between self-driving vehicles and amount of energy consumption and CO_2 emissions.

Table 9.1 Domain classification of emerging technology

Domain	Emerging technologies
Energy	Horizontal axis wind turbine, Vertical l axis wind turbine, Challenergy, Vortex bladeless, Altaeros energies, Wireless energy transfer, Perovskite solar cells, Airborne wind turbine
Information technology	Artificial intelligence (AI), Machine learning, Machine vision, Biometric—fingerprint recognition, Augmented reality (AR), Virtual reality (VR), Cloud computing, Radio-frequency identification (RFID), Internet of things (IoT), Swarm robotics, Big data analysis, The blockchain
Manufacturing	3D printing—Plaster-based 3D printing, • Fused deposition modeling (FDM), • Digital light processing (DLP), • Stereo lithography appearance (SLA), • Selective laser sintering (SLS), Module, Distributed manufacturing
Transportation	Battery electric vehicle, Autonomous vehicle, Unmanned aerial vehicle (UAV), Personal ropid transit (PRT), Hyperloop, New type Maglev elevator (MULTI)
Material	2D-materials, Self-healing materials, Recyclable-thermoset plastics, Nanostructured carbon composites
Biological technology	Organs-on-chips, Systems metabolic engineering, Implantable drug-making cells, Lab-grown meat, Liquid biopsy

9.2.2 Case Database of Emerging Technologies

This research collects different emerging technology cases and finds associated TRIZ innovative principles to build innovative database. Designer can use keywords to search related innovations and get new product concepts by innovations and TRIZ innovative principles. Each emerging technologies case is recorded in the database with information about this emerging technology, such as the case number, the name of case, a brief introduction of case, the characteristic of this emerging technology and related TRIZ inventive principles and effect of environmental impact. Currently, we collected 42 emerging technologies cases in the case database, as shown in Table 9.1.

9.3 ARIZ Method

9.3.1 The Development of ARIZ

Russian scientist Altshuller (2000) developed the serial methods of problem-solving which are called TRIZ theory by analyzing more than 40 million patents. ARIZ is one of most important method in TRIZ theory. The goal of ARIZ is to analyze the engineering problems through a series of logical steps and processes, so that problems can be gradually simplified to be solved. There are many versions of ARIZ including ARIZ-56, ARIZ-59, ARIZ-61, ARIZ-68, ARIZ-71, ARIZ-77, ARIZ-82, ARIZ-85, ARIZ-85C and so on (Zlotin and Zusman 1999; Savransky 2000). In the version of ARIZ-85C, the whole framework is complete but has too much parts and steps which is more complitICATED than other versions. Thus, Fey and Rivin (2006) developed ARIZ-85C (Fey) that simplifies the processes of ARIZ-85C. ARIZ method after several years of use becoming more and more complete and systematic.

9.3.2 ARIZ-85C (Fey) Version

ARIZ-85C (Fey) has four parts, such as Formulation of system conflicts, Analysis of the system conflicts and formulation of a mini-problem, Analysis of the available resources, and Development of conceptual solutions. The flow chart is shown as Fig. 9.1.

There are two paths that can generate innovative design through the problem-solving processes of ARIZ-85C (Fey) (Fig. 9.2). One is to convert the engineering problem into Su-Field model (Altshuller 2000; Savransky 2000) and apply standard solutions (Savransky 2000) to find the most suitable suggested solution. Su-field analysis is a modelling approach in TRIZ for the analysis and innovation of physical phenomena in product systems. The other is the main path of ARIZ-85C (Fey) operated from the first part to the fourth part.

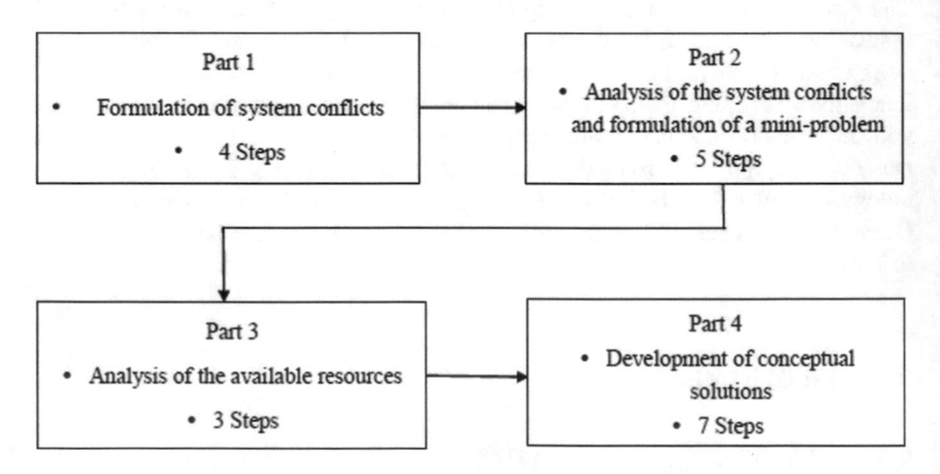

Fig. 9.1 Flow chart of ARIZ-85C (Fey)

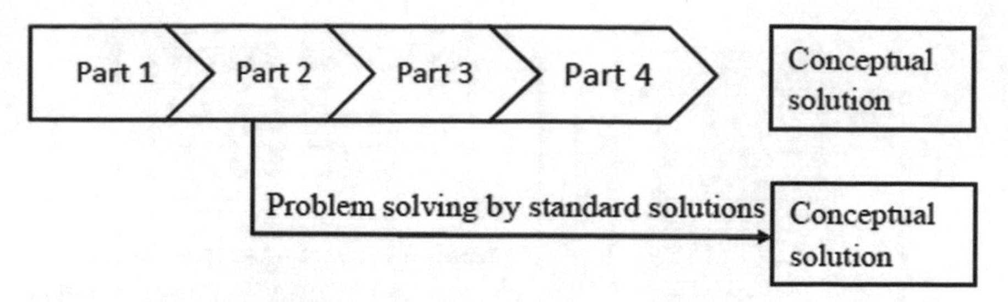

Fig. 9.2 Two problem solving paths in ARIZ-85C (Fey)

The emerging technologies database serves as a reference to make designers understand the emerging technologies with the help of traits of emerging technologies. The traits of emerging technologies include their concept, feature, function and application.

To make the connection between ARIZ-85C (Fey) and the database, this study makes two classifications for each technology in the database. The operation of the first classification is to add Substance term and Field term on the original database according to traits of each technology. With the first classification, the emerging technologies and Su-Field table is found. Emerging technologies and Su-Field table can guide designers to select suitable technology to solve their engineering problem in accordance with Su-Field model and standard solutions during part 2 in ARIZ-85C (Fey).

The operation of second classification is to make the category of applied field for each emerging technology. The emerging technologies and applied field table is established by second classification. Designers need to determine the field of the engineering problem they face, and then select the corresponding technologies as the resources from the emerging technologies and applied field table during part 3 in ARIZ-85C (Fey). These two tables are the link between ARIZ-85C (Fey) and emerging technologies and play an important role in the design process this study presented.

9.4 ECO-innovation Method

9.4.1 Part 1: Formulation of System Conflicts

Designers are able to deal with engineering problem through the innovative eco-design process presented in Fig. 9.3. The innovative eco-design process is based on ARIZ-85C (Fey), emerging technologies database, emerging technologies and Su-Field table, emerging technologies with applied field table, and green assessment.

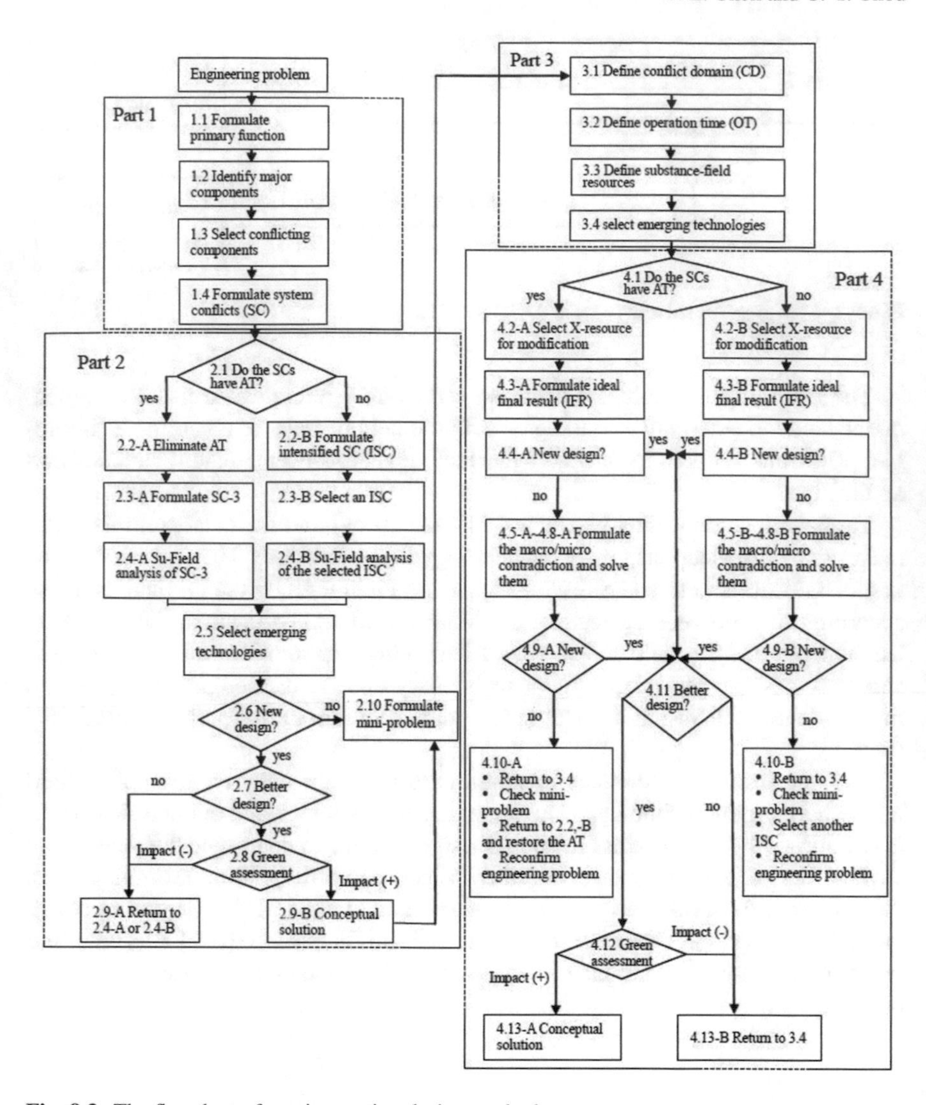

Fig. 9.3 The flowchart of eco-innovation design method

The whole process has four parts. In the part 1, designers can identify the primary function, major components (tools, objects, environmental elements) and system conflicts (SC) of the given engineering problem.

9.4.2 Part 2: Analysis of the System Conflicts and Formulation of a Mini-Problem

Part 2 begins with the checkpoint: do the system conflicts involve an auxiliary tool? If the answer to this question is yes, then the auxiliary tool is eliminated and a new system conflict is formulated (2.1–2.3-A). If the system conflicts do not contain an auxiliary, they are intensified. Designers need to select one intensified system conflicts which better emphasizes the primary function (2.1–2.3-B). At the next step, Su-field analysis helps to determine whether the selected system conflict can be resolved by the standard solutions.

After finding the suggested standard solutions, designers can use the emerging technologies and Su-Field table and the database to select suitable technologies and apply the traits of selected technologies to the given system conflict (2.5).

If designers develop new design in step 2.5, the new design should be evaluated to confirm if the functionality is better than that of the original design. During the evaluation of functionality, it is suggested that the previously defined improving function be taken as the criterion to compare the old and the new design. Additionally, the impact of the new design on the environment should also be considered. The evaluation could be performed with the method of weighing with the help of some of the potential pros and cons of the emerging technologies list in the database (2.6–2.9-B). If the new design is not a better design or has negative impact on the environment, designers have to select the more suitable technologies (2.9-A). If the new design has better functionality and positive impact on environment, then designers can consider the new design the conceptual solution (2.9-B). The final step in part 2 is the formulation of a mini-problem (2.10). The mini-problem for SC with auxiliary tool (2.2-A) is to find a X-resource that can provide useful action without hindering the performance of the primary function. The mini-problem for SC without auxiliary tool (2.2-B) is to find a X-resource that can eliminate the harmful action while preserving the useful one.

9.4.3 Part 3: Analysis of the Available Resources

In the part 3, designers identify the conflict domain, operation time and the substance-field resources within the system conflict (3.1–3.3). In the step 3.4, designers need to determine the field of the engineering problem they face, and then select the corresponding technologies as the resources from the emerging technologies and applied field table. With the operation of step 3.4, available resources can be extended to the area of emerging technologies.

9.4.4 *Part 4: Development of Conceptual Solutions*

In the part 4, designers select the resources from step 3.3 and step 3.4 as the X-resource for modification after checking whether the system conflict have auxiliary tool. Then, an ideal final result (IFR), which is an ideal functioning of the selected resource in the conflict domain is formulated. In essence, the selected resource has to perform the actions specified in the formulation of the mini-problem.

If the IFR can be realized, then the problem is solved (4.1–4.4-A/4.4-B). After realizing the IFR, designers need to check the functionality and the impact on the environment of the new design such as the procedures in part 2. If the new design is not a better design or has negative impact on the environment, designers have to select the more suitable technologies (4.13-B). If the new design has better functionality and positive impact on environment, then designers can regard the new design as the conceptual solution (4.13-A).

If the IFR can't be realized, the separation principles are used to resolve the macro and micro physical contradictions (4.5-A–4.8-A/4.5-B–.8-B). If all of above steps have not resulted in a good conceptual solution, there are four suggestions to guide designers to continue the analysis. The first suggestion is to select other technologies from the emerging technologies and applied field table (3.4). The second suggestion is to check the definition of mini-problem (2.10). The third suggestion is to revive the auxiliary tool eliminated in part 2 or select another intensified system conflict to analyze according to whether the SC involve auxiliary tool (2.2-B/2.3-B). The last suggestion is to redefine or reconfirm the engineering problem.

9.5 Example

Electrical motorcycle problem is selected as example to demonstrate the capability of proposed eco-innovation method. In Taiwan, the gas-powered motorcycle is one of source of air pollution problem in city. The electrical motorcycle is the solution of air pollution problem. However, the low battery power capacity and long charging time for battery are the chief drawback of current most electrical motorcycle. Furthermore, the carbon footprint of electrical motorcycle using electricity from current fossil fuels power plant is still high. Eco-innovation idea is required to solve above problems. Gogoro in Taiwan created a city-based network of battery swapping stations called the Gogoro Energy Network and designed a high performance electric scooter called the "Gogoro Smartscooter". However, if the density of battery swapping stations is low in some area, then the increasing the traveling range of the electrical motorcycle will be a problem.

9.5.1 Part 1: Formulation of System Conflicts

Step 1.1 Formulate Primary Function:

Primary function of electrical motorcycle is no exhaust emissions in the driving area. Therefore, it can reduces air pollution and environmental impacts.

Step 1.2 Identify Major Components (MCs) of system:

The MCs of the system are battery swapping station and electrical motorcycle. The MCs of the surrounding environment are air in the driving area of electrical motorcycle, driving road, and solar light.

Step 1.3 Choose components of system conflicts and associate them with useful and harmful actions:

Table 9.2 matches conflicting components with real components. It lists only the real MCs that are related to each conflict and matches them with the conflicting components.

Step 1.4 Draw diagrams of models of pairs of system conflicts (SC-1 and SC-2):

Figure 9.4 presents SC-1. The more electrical motorcycles driving in the road can reduce exhaust emissions in the driving area, and the need of too much power, resulting in reduced battery life. Figure 9.5 presents SC-2, which is opposite to SC-1. With less electrical motorcycles, the power required is stable, but it will increase exhaust emissions in the driving area.

Table 9.2 Conflicting components matched with real components

Conflicting components	Real components
Objects (O)	Battery swapping station
Main Tools (MT)	Electrical motorcycle
Auxiliary Tools (AT)	None
Environmental Elements (EE)	Air in the driving area of electrical motorcycle, Driving road, Solar light
Useful Actions (UA)	No exhaust emissions in the driving area
Harmful Actions (HA)	Need too much power, resulting in reduced battery life

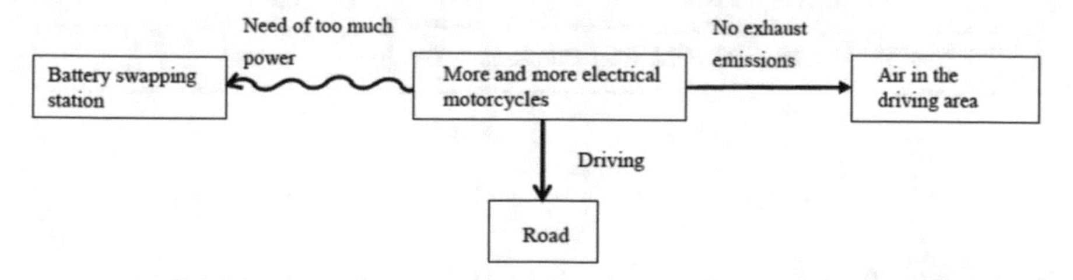

Fig. 9.4 System conflict associated with more electrical motorcycles (SC1)

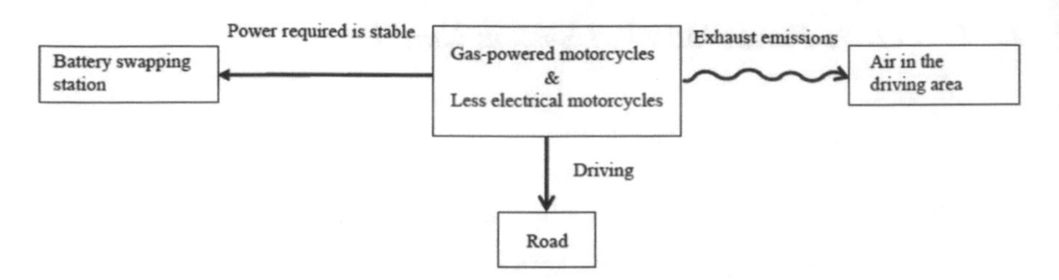

Fig. 9.5 System conflict associated with less electrical motorcycles (SC-2)

9.5.2 Part 2: Analysis of the System Conflicts and Formulation of a Mini-problem

Step 2.1 Check whether the system conflict have auxiliary tool (AT):

From Table 9.2, the system conflict of this problem does not have auxiliary tool (AT).

Step 2.2B Formulate intensified system conflicts (ISC):

The intensified system conflicts of SC-1 (ISC-1) is that all driving motorcycle on road are all electrical motorcycles, as shown in Fig. 9.6. Therefore, this situation will produce super large amount of power requirement for battery swapping station. Furthermor, It will be resulting in reducing battery life and the traveling range of the electrical motorcycle.

The intensified system conflicts of SC-2 (ISC-2) is that all driving motorcycle on road are all gas-powered motorcycles, as shown in Fig. 9.7. The gas-powered motorcycles have no power requirement for battery swapping stations. However, it will produce large amount of exhaust emissions in the driving area.

Step 2.3B Select an intensified system conflicts:

The intensified system conflicts ISC-1 is selected to reach the primary fuction (PF) of this problem. If all driving motorcycle on road are all electrical motorcycles, then it will be no exhaust emissions in the driving area. However, this situation will produce super large amount of power requirement for battery swapping station. Furthermor,

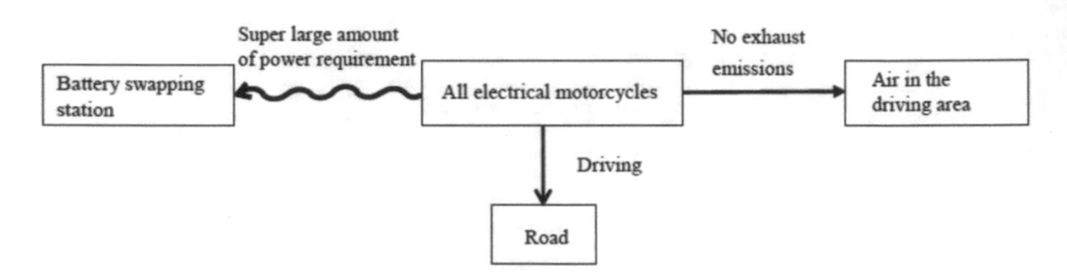

Fig. 9.6 Intensified system conflict associated with totally electrical motorcycles (ISC-1)

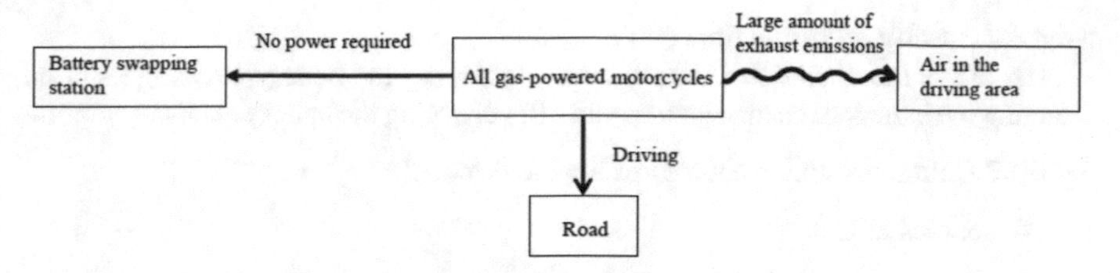

Fig. 9.7 Intensified system conflict associated with only gas-powered moyorcycles (ISC-2)

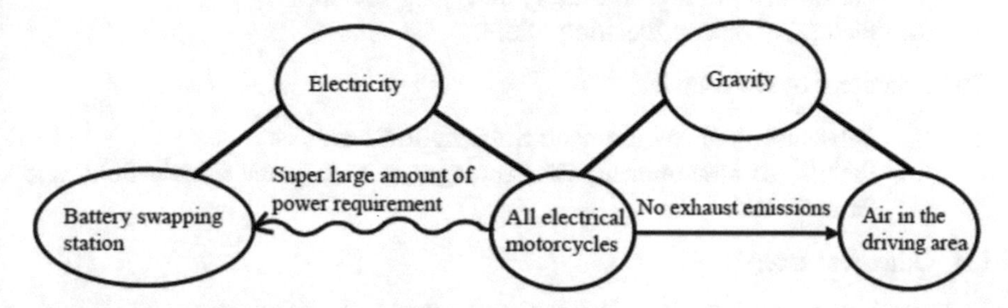

Fig. 9.8 Su-field model

It will be resulting in reducing battery life and the traveling range of the electrical motorcycle.

Step 2.4B Perform an Su-Field analysis of selected ISC:

The su-fielf model of ISC-1 is illustrated in Fig. 9.8.

Step 2.5 Select emerging technologies:

The standard solution from Su-field model can't find suitable emerging technologies to produce ideas.

Step 2.6 New design:

No any new design concept generated in this step.

Step 2.10 Formulate mini-problem:

The mini-problem is "Must find an X-resource to solve the contradictory problems of the system. This resource can maintain the system's useful action UA (no exhaust emissions) while reducing or eliminating harmful actions HA (reduced endurance)". The X-resources represent innovative solutions.

9.5.3 Part 3: Analysis of the Available Resources

Step 3.1 Specify conflict domain (CD):

The CD is the path for electrical motorcycle to reach battery swapping station.

Step 3.2 Specify operation time (OT):

The OT is the operational period before charging the battery exchange station, charging the battery exchange station and after charging the battery exchange station.

Step 3.3 Define possible resources in Su-Field model:

(1) Resources of CD:

(a) Substance(s) of main tool: Electrical motorcycle.
(b) Field(s) of main tool: Electricity field.
(c) Substance(s) of object: Battery swapping station.
(d) Field(s) of object: Electricity field.

(2) Resources of environment:

(a) Substance(s) of environment: Solar light, Road.
(b) Field(s) of environment: Electromagnetic field, Atmospheric field, and Gravity field.

(3) Other resource(s):

(a) Substance(s) of other: Drivers, other electrical motorcycles.
(b) Field(s) of other: Electromagnetic field.

Step 3.4 Select emerging technologies:

Since the system problem in this case is mainly focused on the provision of electric energy, it is recommended that designers choose emerging technology resources from the "energy technology" category in the classification table of emerging technologies, as shown in Table 9.1. This case is based on the "wireless energy transfer" that clearly provides the energy concept in the energy technology category and the energy technology "perovskite solar cells" with the same environmental protection concept as the electrical motorcycles.

The wireless energy transfer technology can transmit power energy to the power receiving device through magnetic induction or magnetic resonance, and no wire connection is needed in the whole process. The perovskite solar cells can be known that the technology is a new type of solar cell, which has higher conversion efficiency than the traditional solar cells, and the energy source is a renewable resource. As a new design direction, designers hope to use these two technologies to improve system problems.

9.5.4 Part 4: Development of Conceptual Solutions

Step 4.1 Check whether the system conflict have auxiliary tool (AT):

The system conflict of this problem does not have auxiliary tool (AT).

Step 4.2B Select one X-resource from Step 3.3 for modification:

Choose solar light, road, electromagnetic field, wireless energy transfer, and perovskite solar cells, as a combination of new X-resource and modify it.

Step 4.3B Formulate an Ideal Final Result (IFR) concerning the selected X-resource:

IFR: Within an appropriate period and space and with a totally electrical motorcycles system, the X-resource can produce no exhaust emissions in the driving area (UA) while the electrical motorcycles can reduce air pollution and environmental impacts (PF). Furthermore, it can improve the situation of need too much power, resulting in reduced battery life (HA).

Step 4.4B New design:

This study proposes a conceptual solution to the characteristics of the X-resource (wireless energy transfer, perovskite solar cells, solar light, road, electromagnetic field) selected in step 4.2-B for the electrical motorcycles and battery charging station system. First of all, since the wireless energy transmission can complete the power transmission without the connection of the wires, the concept of the innovtion is to build the power supply device with wireless energy transmission function under the road surface to form an electrical motorcycle driving lane with wireless charging function. When the electrical motorcycle is driving, the electrical motorcycle can be charged by the electric power source of the road surface, and the battery or induction coil of the electrical motorcyclee is designed at the bottom of the locomotive, and the distance from the ground is relatively close, so as to obtain the maximum receiving power for power supply and transportation. In addition, this concept also allows the electrical motorcycles to perform static wireless charging while waiting for the traffic light, and can receive the power source through the parking and cycling process, increasing the traveling range of the electrical motorcycles. Furthermore, for providing wireless charging power, it can be derived from renewable energy by trying to build a solar cell with a higher conversion efficiency of perovskite solar cells on a separate island or road surface.

Step 4.11 Better design:

In the case of a situation, riding an electrical motorcycle to the battery exchange station to replace the battery for charging, but found that the station's battery is completely dead or is still charging low battery state, can't provide enough batteries to drive. The next battery exchange station is charged. At this time, the wireless power supply lane can play its role. When driving to the next exchange station, the electric motor vehicle can still obtain power at any time to maintain a long battery life and traveling distance. Otherwise, the driver can only wait for the assistance of the operator or waiting for the battery to be charged at the battery exchange station. Therefore, the new design concept can be used as an auxiliary tool for current exchanging battery station. The endurance of the electrical motorcycles is enhanced to avoid above driving situations. If the wireless charging efficiency is improved to be the same with the wired charging, then the demand for the battery exchange station will gradually decrease. Therefore, it shows that the new design concept has better functionality.

Step 4.12 Green assessment:

The positive impact for environment are as follows: (1) The charger can be invisible and the equipment wear rate is low. (2) A charger can charge a plurality of electrical motorcycles, and in the case of multiple electrical motorcycles, multiple chargers can be omitted without occupying multiple power outlets. (3) Renewable energy replaces fossil fuels to reduce carbon emissions. (4) Compared with the production mode of samarium-based solar panels, perovskite solar cells can reduce greenhouse gas emissions.

The negative impact for environment is the farther the distance of wireless transmission, the greater the loss of useless work.

Step 4.13A Conceptual solution:

After functional evaluation and green assessment, it is determined that the new design concept has good functionality and the impact on the environmental level is positive, so the new design concept is taken as an innovative solution.

9.6 Summary

This study integrates different ways including ARIZ-85C (Fey), other TRIZ auxiliary tools, and emerging techonologies database to guide designers to solve their engineering problem step by step and take advantage of the traits of emerging technologies to develop innovative and eco-friendly designs. During designing products, designers can think and break through the perspective of emerging technologies, and provide reference for guiding the design process through case exercises. Electrical motorcycle in Taiwan is used as example to demonstrate the capability and usefulness of proposed eco-innovation method.

Acknowledgements This research is sponsored by the Ministry of Science and Technology, Taiwan, under Grant number MOST 106-2621-M-006-004 and MOST 107-2621-M-006-005.

References

Altshuller G (2000) The innovation algorithm: TRIZ, systematic innovation and technology creativity. Technical Innovation Center, Inc., Worcester

Chen JL, Lin GH (2018) Product innovation design for the bottom of the pyramid by integrating ARIZ and case-based reasoning. In: *Proceedings of the IEEE 2018 International Conference on Applied System Innovation (IEEE ICASI 2018)*, April 13–17, Chiba, Tokyo, Japan

Chen WC, Chen JL (2009) An eco-innovative method by ARIZ. In: Proceedings of the 6th international symposium on environmentally conscious design and inverse manufacturing, Ecodesign09. Sapporo, Japan, 7–9 Dec

Chen WC, Chen JL (2011) Eco-innovation by applying design-around concepts on eco-ARIZ method. In: Proceedings of the 7th international symposium on environmentally conscious design and inverse manufacturing, Ecodesign11. Kyoto, Japan, November 30-Dec 2.

Chen WC, Chen JL (2013) Eco-innovation design method: eco-ARIZ. J Chin Soc Mech Eng 34(5):379–389
Chen WC, Chen JL (2014a) Innovative method by design-around concepts with integrating the algorithm for inventive problem solving. J Mech Sci Technol 28(1):201–211
Chen WC, Chen JL (2014b) Eco-innovation by integrating biomimetic design and ARIZ. Procedia CIRP 15:401–406
Chen JL, Chen CW (2017) Eco-innovation by integrating technologies for the future with TRIZ, In: 10th international symposium on environmentally conscious design and inverse manufacturing, Ecodesign17, Tainan, Taiwan, November 29-Dec 1.
Fey VR, Rivin EI, Vertkin IM (1994) Application of the theory of inventive problem solving to design and manufacturing systems. Annals CIRP 43:107–110
Fey V, Rivin E (2006) Innovation on demand: new product development using TRIZ, 2nd ed., Cambridge, UK
Ford S, Despeisse M (2016) Additive manufacturing and sustainability: an exploratory study of the advantages and challenges. J Cleaner Production 137:1573–1587
Gebler M, Schoot Uiterkamp AJM, Visser C (2014) A global sustainability perspective on 3D printing technologies. Energy Policy 74:158–167
Halaweh M (2013) Emerging technology: what is it? J Technol Manage Innovation 8(3):108–115
Krasnoslobodtsev V, Langevin R (2005) TRIZ application in development of climbing robots. In: TRIZ symposium. Japan, September
Krasnoslobodtsev V, Langevin R (2006) Applied TRIZ in high-tech industry. In: Proceedings of the TRIZCON2006. April. Milwaukee, WI USA
Krasnoslobodtsev V, Lee JY, Lee JB (2005) TRIZ improvement of rotary compressor design. J TRIZ Eng Des 1(1):2–13
Rotolo D, Hicks D, Martin B (2015) What is an emerging technology. Res Policy 44(10):1–40
Savransky SD (2000) Engineering of creativity: introduction to TRIZ methodology of inventive problem solving. CRC Press, USA
Shaikh FK, Zeadally S, Exposito E (2017) Enabling technologies for green internet of things. IEEE Syst J 11(2):983–994
Wadud Z, Mackenzie D, Leiby P (2016) Help or hindrance? The travel, energy and carbon impacts of highly automated vehicles. Transp Res Part a 86:1–18
Zlotin B, Zusman A (1999) ARIZ on the move, TRIZ J

Chapter 10
Video Networks of Sustainable Design: The Doughnut Perspective

Vargas Meza Xanat and Yamanaka Toshimasa

Abstract Economics is going through a radical change to better account for biological and social indicators of health and prosperity, the so-called Doughnut Economics proposed by Kate Raworth. Through this framework, design understood as creating ways of being is an integral component of economic development. On the other hand, some recent studies are proposing to incorporate Doughnut Economics in specific design areas. Previous research by the authors analysed YouTube sustainable design videos in English and Spanish. However, the present study classifies the videos according to Doughnut Economics indicators to provide a more comprehensible description of local design processes and products, the contextual factors tied to them, and communication patterns of sustainable design in YouTube. There was a clear difference on design areas and approaches; and types of communication hubs across country groups. The study has implications for the conceptualization of design as a multidisciplinary and participatory process, where global and local dynamics should account for planetary sustainability and resilience.

Keywords Sustainable design · YouTube · Doughnut economics

10.1 Introduction

Meadows and colleagues reported in 1972 that our current economic system can not continue to grow forever (Meadows et al. 1972). Today, we face financial excesses, cultural colonialism and exploitation of living systems at an unprecedented rate. Based on (Hanington and Egenhoefer 2018), the consequences in the design field

V. M. Xanat (✉)
Department of International Industrial Engineering, National Institute of Technology (KOSEN), Ibaraki College, Hitachinaka, Japan
e-mail: kt_designbox@yahoo.com

Y. Toshimasa
Faculty of Art and Design, University of Tsukuba, Tsukuba, Japan

Y. Kishita et al. (eds.), *EcoDesign and Sustainability I*, Sustainable Production, Life Cycle Engineering and Management, https://doi.org/10.1007/978-981-15-6779-7_10

are failed products saturating land, sea and atmosphere; materials and processes wasted; physically and/or psychologically hurt living beings; and lost jobs. Better accountability and assessment of design processes is needed, which is challenging in our densely interconnected world.

10.1.1 Literature Review

Different systems have been proposed to assess sustainability, like the Millennium Ecosystem Assessment (Alcamo 2003), The Economics of Ecosystems and Biodiversity (TEEB) (Sukhdev 2008), Planetary Boundaries (Rockstrom et al. 2009), and the Sustainable Development Goals (Assembly 2015). However, their level of complexity and/or disconnection to other aspects of sustainability make them hard to understand/apply from an economic point of view. The biggest issue is within the economic mindset, which was influenced by white male Enlightenment and thus views economics as a science with general laws governing decision making processes.

Economist Kate Raworth proposed a model called Doughnut Economics, where the goals should be meeting the needs of all within planetary boundaries, view the economy as embedded, nurture human nature, consider complex systems, design to distribute, design to regenerate and be agnostic about growth (Raworth 2017). We should note that two of her arguments include the word design, which can be understood as creating ways of being. Hence, through the Doughnut Economics framework, the impact of what designers do is more easily grasped. Design is ubiquitous, its success depends on social context, other fields are realizing its vital role in achieving sustainability, and thus has outgrown our discipline (Escobar 2018).

On the other hand, some studies are proposing methods to incorporate the Doughnut Economics paradigm into industrial design (Stopper et al. 2016), fashion design (Rissanen 2018) and design theory (Boehnert 2018). However, given the complexity of current design processes, there is a need to understand which aspects of Doughnut Economics are being achieved within design and where can we locate gaps.

One option to explore complex human systems and their interactions is through Social Networking Sites (Boyd and Ellison 2007). Among them, YouTube is the second largest search engine after Google and the third most visited site after Facebook (Alexa 2019), counting with a video ecosystem that tends to form small world networks. Considering its high usage rate in creative scenarios, YouTube has potential for research in design communities.

10.1.2 Research Objectives

Research questions are as follows:

RQ1. Which types of design are shown in YouTube videos classified according to Doughnut Economics indicators?
RQ2. Which are the characteristics of relevant video hubs across Doughnut Economics indicators?
RQ3. Which contextual factors are tied to such content and communication patterns?

10.2 Methods

10.2.1 Data Collection and Classification

We employed data previously analyzed by the authors in (Vargas Meza and Yamanaka 2017; Vargas Meza 2019), where an extensive description of our methods is included. Because the original intention was to explore the role of academia in diffusion of sustainable design, keywords included "university" and "lecture". Data on YouTube videos in English and Spanish related to sustainable design was collected in 2015 with YouTube Data Tools (Rieder 2015). Videos were watched to discard unrelated content and to classify them on several categories, including contextual factors like type of uploader, type of speaker, uploading date, YouTube channel's country and type of design.

Given the unequal economic and communication structural power, former analysis mostly highlighted countries like US and UK in the English sample, and Spain and Colombia in the Spanish sample. However, their social and environmental issues are different from those in other countries with equal or larger number of netizens. Therefore, the present study also classified the videos according to Doughnut Economics indicators (O' Neill et al. 2018) of their countries of origin, which are based in measurements revised in the literature review section and in Raworth's proposals.

It should be noted that so far, no nation on Earth counts with thriving economy, society and living systems at the same time. Nevertheless, groups of countries were considered according to their biological and social indicators. A high average in biological indicators means that the nation is trespassing planetary boundaries, while a high average in social indicators means human well-being.

Therefore, if the biological indicators average was moderately low and the social one was moderately high (0.52–3.33 and 0.70–1.13 respectively), countries were classified as Bio Socially Compromised (BSC). If both averages were low (0.18–1.99 and 0.29–0.89 respectively), countries were classified as Socially Compromised (SC). Finally, if both averages were high (2.08–7.98 and 0.91–1.20 respectively), countries were classified as Biologically Compromised (BC). It should be noted that most countries (40) detected in the YouTube video samples fell in the BC category, followed by 27 SC countries and 13 BSC countries.

To answer RQ1, statistical tests for comparison of categorical and numerical data were computed with R (Team 2013). For RQ2 basic social network metrics were calculated and graphs visualized with Gephi (Bastian et al. 2009), in order to

Table 10.1 Description of network centralities

Centrality	Definition
Degree	Total number of connections a video has with other videos
Indegree	Number of connections directed to a video
Outdegree	Number of connections a video directs to other videos
Betweenness	Number of shortest paths that connect other videos in the network by passing through a given video
Closeness	Average number of steps a video requires to access all the other videos in a network
Clustering Coefficient	Measure of how close is a video to be part of a group
Local Clustering Coefficient	Measure of how close is a video to be part of its local group

detect relevant videos and communication hubs. Table 10.1 includes a description of network centralities employed in the present study. A connection refers to videos likely to be featured in the recommendation area of YouTube, which are based on overall viewing patterns, user interactions, playlists, etc.

Also, to compare network centralities across country groups, ANOVA was computed with Ucinet (Borgatti et al. 2002). Finally, to answer RQ3, qualitative analysis was performed in top videos in terms of YouTube metrics and network metrics.

10.3 Results

10.3.1 Types of Design

Table 10.2 shows that English videos from BC countries were more numerous across design areas but differences were non-significant; while in Spanish, most videos

Table 10.2 Design across country groups

Design area	English videos			Spanish videos		
	BSC	SC	BC	BSC	SC	BC
Architecture	13	57	915	129	97	157
Industrial D	8	32	180	30	58	46
Graphic D	4	1	59	10	10	17
General	5	7	258	32	20	32
Mixed	4	21	231	38	61	52
Total	34	118	1643	239	246	304
	$x^2 = 46.981$, df = 8, $p > 0.05$			$x^2 = 26.038$, df = 8, $p < 0.005$		

about architecture and graphic design were from BC countries; most videos about industrial design and the mixed area were from SC countries; and most videos about the general area were from BSC and BC countries ($p < 0.005$). The architecture content was related to both cities and rural localities.

10.3.2 Uploaders and Speakers in the Doughnut

Most English videos from BC countries were uploaded by Universities (699, 38.9%), most from SC countries by media channels (34, 1.8%), and most from BSC countries by students (8, 0.4%). These student videos were mostly related to Shell. As for Spanish, most videos from BC and SC countries were uploaded by Universities (100, 12.6%; 64, 8.1%), with content predominantly about architecture and fashion design respectively. Most videos from BSC countries were uploaded by media (63, 7.9%), which included public television channels.

As for speakers, most English videos from BC countries showed professors (419, 23.3%); most from SC countries showed designers (30, 1.6%); and most from BSC countries, students or didn't depict humans at all (9, 0.5% in both cases). Videos by students were once more related to Shell. In contrast, most Spanish videos from BC countries showed a mix of education related individuals with other people (58, 7.35%). Most videos from SC countries showed students (56, 7%), predominantly related to toys. Finally, videos from BSC countries showed designers (56, 7%) principally.

We will now revise the role of women across country groups, as there is little research on women designers outside the architecture and engineering fields. Table 10.3 shows the number of videos where women were identified as uploaders or speakers. Although differences were non-significant across country groups, it is worth mentioning that women speakers were found in nearly half of English BSC countries videos and in all the Spanish groups.

Most of English videos from BC countries depicted architecture students (71, 3.9%), followed by designers (53, 2.9%), citizens (45, 2.5%) and other actors (37, 2%); and professors (33, 1.8%) and researchers (18, 1%) in the general area. This

Table 10.3 Presence of women across country groups

	English videos			Spanish videos		
Uploaders	BSC	SC	BC	BSC	SC	BC
W/Women	5	1	46	17	33	7
W/o Women	29	117	1597	222	213	297
Speakers	BSC	SC	BC	BSC	SC	BC
W/Women	16	36	625	117	113	141
W/o Women	18	82	1018	122	133	163

result reflects the considerable quantity of English architecture videos, and a preference for content related to theory and interdisciplinarity. Women in SC and BSC countries were mostly industrial design students (12, 0.6% and 4, 0.2%; respectively) shown in Shell related videos.

As for Spanish videos, 16 (2%) from BC countries showed diverse women actors in architecture, and 9 (1.1%), women professors in the mixed area. The video content was related to architecture; and engineering and fashion design respectively. 14 (1.7%) videos from SC countries showed women students and 8 (1%), business related women in the mixed area. Most of such content was about fashion design. Finally, 16 (2%) videos from BSC countries showed women architects.

10.3.3 YouTube Metrics and Qualitative Descriptions of Top Videos

Previous literature (Vargas Meza 2019) found that the most reliable metric of YouTube videos is number of views. Table 10.4 shows this measure across country groups. In English, videos from BC countries were significantly the most viewed ($H(2) = 6.6471$, $p < 0.05$), while differences were not significant in Spanish. This result reflects the algorithmic bias in YouTube that favors content from English BC countries.

We will proceed to analyze the most viewed videos in detail. The Top 10 English videos from BC countries were uploaded mostly in the U.S., although videos from U.K, Denmark, Canada and Australia were also among the most viewed. The content was largely related to industrial design and was uploaded by media channels, notably TED Talks. A noted exception was a video from the Story of Stuff educative channel, ran by an NGO. Half of the content showed women. On the other hand, Top 10

Table 10.4 YouTube metrics across country groups

English videos			
Views	BSC	SC	BC
Average	489.82	6349.25	12698.31
Median	305	507.5	556
Std. Dev	626.42	4130.74	8585.911
$H(2) = 6.6471$, n = 1795, $p < 0.05$			
Spanish videos			
Views	BSC	SC	BC
Average	11748.14	3463.74	3735.07
Median	299	219.5	316
Std. Dev	74012.27	18068.33	30231.93
$H(2) = 3.934$, n = 789, $p > 0.05$			

Spanish videos were uploaded in Spain principally, although videos from Chile and the U.S. were also present. The content versed mostly on architecture techniques, was uploaded by a diverse set of actors and half had ties with universities. 3 videos showed women.

Top 10 English videos from SC countries were mostly uploaded in India, although content from Uganda, South Africa, Philippines, and Peru was also found. Design areas were architecture and the mixed area (natural resources and waste management, often related to architecture). Videos were mostly uploaded by educational entities and one included women. As for Spanish, videos were uploaded in Colombia, although one video from Peru was found. The content largely versed on techniques in architecture, industrial design and the mixed area (notably fashion design). Videos were uploaded by education actors, including 2 women. The content showed mostly citizens, although 3 videos included women.

As for Top 10 English videos from BSC countries, they were mostly uploaded in Malaysia, although content from Thailand, Vietnam, Lebanon, Moldova and Brazil was also frequently viewed. Most of the Malaysian content was related to Shell and was education oriented. 3 videos were uploaded by women, 4 depicted women and 4 did not show humans. On the other hand, Spanish videos were mostly uploaded in Mexico, followed by Argentina and Panama. The content largely versed in architecture techniques and concrete design examples; and was uploaded by businesses, 3 including women. The videos showed citizens and 3 included women. It is worth mentioning that only one top video in each language was uploaded in 2015, which was the year of the data collection.

10.3.4 Network Metrics and Qualitative Descriptions of Top Videos

Table 10.5 shows significant ($p < 0.0005$) differences in terms of network metrics for English videos across country groups. We included the whole network sample ($n = 2452$) in this analysis, labelling videos from countries without Doughnut Economics indicators as "Unclear". The centrality that accounted for most of variance ($R^2 = 0.962$) was degree, followed by indegree ($R^2 = 0.699$). Videos from SC countries were the highest across all network indicators, suggesting that they form a community subnetwork. However, betweenness centrality was low for this sample, pointing to an overall disconnected network.

On the other hand, Table 10.6 shows significant ($p < 0.0005$) differences in terms of network metrics for Spanish videos across country groups in the whole network sample ($n = 1075$). Once more, the centrality that accounted for most of variance ($R^2 = 0.918$) was degree, followed by indegree ($R^2 = 0.896$). This suggests that the English network videos are more dependent on outdegree than the Spanish ones. Although averages tended to be more similar across groups, videos from SC countries

Table 10.5 ANOVA in English videos network metrics

Group	BSC	SC	BC	Unclear
Degree	F = 1054.7639, R^2 = 0.962			
Mean	2.4411	**6.0338**	4.6171	3.7488
Std. Dev	3.6696	9.9896	9.0081	7.1277
Indegree	F = 255.8935, R^2 = 0.699			
Mean	0.7352	**3.1440**	2.3791	1.7001
Std. Dev	1.1885	7.2209	6.6648	4.7227
Outdegree	F = 1228.7771, R^2 = 0.334			
Mean	1.7058	**2.8898**	2.2379	2.0487
Std. Dev	2.7472	3.9174	3.3312	3.1000
Closeness	F = 1165.4085, R^2 = 0.322			
Mean	0.1285	**0.2148**	0.1630	0.1775
Std. Dev	0.2438	0.2929	0.2473	0.2757
Clust. Coef	F = 1165.4085, R^2 = 0.322			
Mean	0.1197	**0.2137**	0.1251	0.1351
Std. Dev	0.2469	0.2965	0.2285	0.2373
Loc. Clust. Coef	F 47.8688, R^2 = 0.262			
Mean	0.0431	**0.0827**	0.0490	0.0539
Std. Dev	0.0919	0.1289	0.0974	0.1035
Betweenness	F 491.9396, R^2 = 0.167			
Mean	0.0001	**0.0004**	0.0003	0.0002
Std. Dev	0.0004	0.0022	0.0016	0.0013

tended to be the highest across network indicators. The exceptions were betweenness, which was highest for BC countries; and local clustering coefficient, which was highest for countries without Doughnut Economics indicators. Interestingly, betweenness centrality was as low as in the English network.

In order to visualize indegree and doughnut indicators, network graphs were drawn in Gephi. Figure 10.1 shows the English video network in the left and the Spanish video network in the right. In the case of English, interlinked groups of architecture and academic videos mostly from BC countries tended to be more central, while videos from SC and BSC countries versed more often in eco villages. In contrast, the Spanish video network has more central nodes from SC and BSC countries, particularly from Mexico. The video groups labelled in the figure were usually uploaded by university students.

Next, we will analyze videos with the highest indegree more in depth. Top 11 English videos from BC countries were mostly uploaded in the U.S., followed by the Netherlands and UK. 5 were related to industrial design, uploaded mostly by education related actors including one woman. 7 of the videos were related to universities, 6 included women, 5 showed education related actors, 4 were about Shell and 2 were

Table 10.6 ANOVA in Spanish videos network metrics

Group	BSC	SC	BC	Unclear
Degree	F = 534.2640, $R^2 = 0.918$			
Mean	3.9246	**4.5853**	3.7960	3.7979
Std. Dev	5.8563	6.5750	5.8052	5.9007
Indegree	F = 295.9653, $R^2 = 0.826$			
Mean	2.0669	**2.2317**	1.9309	1.8292
Std. Dev	3.8908	3.6810	3.7896	3.3452
Closeness	F = 1076.9187, $R^2 = 0.501$			
Mean	0.1860	**0.2669**	0.1884	0.2000
Std. Dev	0.2687	0.3638	0.2853	0.2935
Outdegree	F = 1014.1048, $R^2 = 0.486$			
Mean	1.8577	**2.3536**	1.8651	1.9686
Std. Dev	2.6163	3.2702	2.4839	2.9640
Betweenness	F = 475.1337, $R^2 = 0.307$			
Mean	0.0003	0.0002	**0.0004**	0.0002
Std. Dev	0.0012	0.0013	0.0013	0.0011
Loc. Clust. Coef	F 1.5698, $R^2 = 0.030$			
Mean	0.0362	0.0481	0.0394	**0.0520**
Std. Dev	0.0657	0.0898	0.0718	0.1066

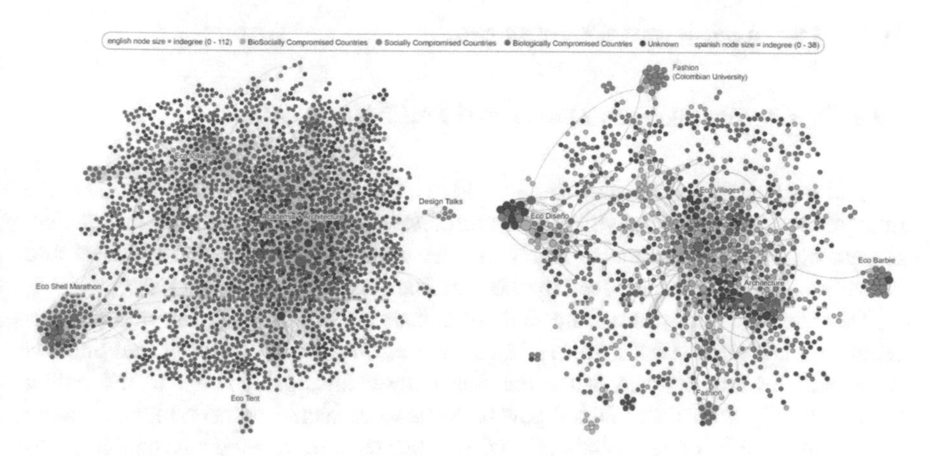

Fig. 10.1 Video networks of sustainable design

uploaded in 2015. Meanwhile, top 11 Spanish videos were from Spain and only one was from France. 8 videos were about architecture, 7 were related to universities, 8 were uploaded by education related actors, 5 showed citizens and 2 included women.

The Top 10 English videos from SC countries were mostly from South Africa, although videos from Philippines, India, China, Indonesia and Pakistan were also found. Content was largely related to industrial design. 8 videos had ties to universities, 6 were about Shell, 5 were uploaded by education related actors, 5 showed students, 4 were uploaded in 2015 and 3 included women. As for Spanish videos, the top 11 were from Colombia and only one was from Peru. 9 videos had ties with universities and 7 versed on industrial design, including 5 about toys. 6 videos were uploaded by students, including women in 2 cases. 6 videos showed education related actors and 3 included women.

Regarding BSC countries, 6 English Top videos were mostly from Malaysia, although videos from Brazil, Lebanon and Thailand were also found. The content was diverse and uploaded by education related actors, including 1 woman. 4 videos were related to universities, and they tended to show a diverse set of speakers, 3 including women. 2 videos had ties to Fuji Xerox and 1 with Shell. Finally, 2 videos were uploaded in 2015.

In contrast, Top 8 Spanish videos were mostly from Argentina and Mexico, although a video from Panama was also present. 6 had ties with universities, 5 videos were about architecture, 5 included women, 4 were uploaded by universities and 4 showed education related actors mixed with citizens. In particular, 3 videos were programs produced by a private university in Mexico and ran by students, explaining sustainable design and presenting products and services in a didactic manner.

10.4 Discussion and Conclusions

10.4.1 Content Across Doughnut Indicators

We will start the discussion with English videos. Urban architecture content was more numerous in this sample, while rural architecture was more frequent in Spanish. Regarding videos uploaded by students, the most visible in terms of views and network metrics were sponsored by the multinational Shell Oil Company.

This enterprise has teamed up with students since 1979, allegedly due to environmental concerns (Wellington 1982). However, projections of fossil fuel impacts were held by the company since the 80s without disclosing them to the public (Franta 2018), while they lobbied governments to curb environmental laws (Laville 2019). Some universities collaborate with enterprises to develop sustainable products (Esslinger 2013), but as it is unclear how much fuel based energy efficiency has improved thanks to the Shell Eco Marathon, this initiative could be considered as a case of greenwashing, understood as marketing used to promote a false perception of an organization as environmentally conscious at the expense of university collaboration.

In contrast, some of the few noted videos related to graphic design in BSC countries (Malaysia in this case) had ties with Fuji Xerox, a Japanese company that produces printers. Their model, developed in 2000, was based on Corporate Social Responsibility and divided sustainability in Strategic Human Sustainability and Strategic Ecological Sustainability to develop a cycle of remanufacturing and material recovering (Benn and Dunphy 2004).

This model has been widespread in similar industries across Japan (Matsumoto and Umeda 2011). Some of the reasons why it was shown in Malaysia might be the importance of Japanese investment in that country (Karim and Majid 2009), the high penetration rate of internet and social media (we Are Social 2019), and the collectivism of Malaysian society. This created a network effect, where online communities add value to platforms (Regalado 2014).

Regarding speakers across country groups, BC content depicted professional designers while non-BC countries showed students. SC and BSC countries videos versed on more interdisciplinary and collective approaches to the design process but tended to limit participation to professors, designers, researchers and students, the "professionals". In other words, non-BC countries students who are exposed to video sources from BC countries are at risk of decontextualizing the design process and of being culturally colonized.

As for Spanish videos, SC countries had popular content generated by public television and actors often mixed with common citizens, communicating less hierarchical societies where governments are active diffusors of sustainable design. The big quantity of videos versed on fashion design and toy production in BSC countries was partly due to a set of laws promoting sustainable development, design and children welfare, particularly in the case of Colombia (Abello Llanos 2010; Cálad Idárraga 2013; Gómez Barrera 2010). However, Latin America has a long tradition of maquila for the fashion industry that should not be ignored as a warning for what accounts as sustainability for brands (Villanueva Ponce et al. 2015) versus working conditions in such industries (Frey 2003; Guadarrama 2006).

BSC countries tended to be more related to businesses and architecture. In the case of Mexico, this is influenced by solid frameworks, financial incentives and government participation (Galvez 2011).

Despite that equality is one of the Sustainable Development Goals and should be included by default in the social dimension of sustainable design, what our data suggests is that women participation in this particular area is more prominent in English speaking BSC countries (particularly in Malaysia) and in Spanish videos. Given that gender equality is one of the 3 less financed goals (Ramos 2019), such results are not surprising. Regardless of the numerous women students participating in sustainable design, systemic barriers still largely difficult their progress towards professional and academic positions.

10.4.2 Communication of Sustainable Design in YouTube

Despite being well connected, English SC countries videos and Spanish videos lack visibility partly due to algorithmic bias influencing exposure to BC countries content.

However, there is also a psychological bias confirmed in (Vargas Meza 2019), a perception of virtual colonialism, where YouTube users from non-BC countries consider BC content as of higher quality and more "professional".

What is problematic is that content generated by media is also preferred and watched more often than content generated by universities. Media, particularly from BC countries, tended to portray sustainable design as a profession for loner visionaries (both men and women). On the other hand, Spanish videos tended to show a more participative approach where, as Manzini mentions, "everyone is a designer" (Manzini 2015).

Another aspect to be mentioned was the small amount of recent content among the most viewed and most connected videos, regardless of YouTube algorithmic priority for fresh content. Based in our data, such content is better connected in English SC countries. However, average videos related to sustainable design in Spanish tended to be from recent years, which was not the case for those with highest number of views or indegree.

Taking on account that Facebook is behind YouTube in terms of web traffic and that its penetration is high in Latin American countries (Kemp 2019), it is likely that traffic for average videos is partly influenced by multiplatform sharing, while a more specific but fragmented audience related to education seeks high quality video content regardless of the year it was uploaded. Therefore, despite the many structural, social and environmental issues provoked by social media platforms, valuable content is still viewed and appreciated in some scenarios, generating micro spaces for sharing economies.

10.4.3 Conclusions

Based in our analysis and discussion, we conclude the following:

RQ1. Videos from BC countries tended to verse on urban architecture and graphic design, while videos from SC countries tended to depict more industrial design and design from the mixed area; and BSC videos were more centred on rural architecture. Multidisciplinarity and participatory design were shown more frequently in Spanish videos.

RQ2. Regarding video hubs:

RQ2.1. In terms of video views, media channels were relevant for English BC countries, while Spanish videos were diverse. Education related actors were relevant for SC countries, and corporations for BSC countries.

RQ2.2 In terms of network centralities, education related actors were relevant in BC countries and SC countries. Actors in BSC countries tended to be diverse.

RQ3. Algorithmic bias, corporative influence and psychological bias played in favor of videos from English BC countries. In contrast, comprehensive political and educational frameworks, and government and citizens participation have contributed to the development and practice of sustainable design in SC and BSC countries.

The present study is a first re-exploration of YouTube data that functions as knowledge source for many without access to formal education, and as electronic repository for education related actors. We could uncover communication patterns and characteristics of groups of countries that are largely absent in Video Social Networks literature. Limitations include video content in other languages used by large numbers of netizens like Chinese and Portuguese. Therefore, future analysis could target such languages. Further, a next step is to analyze public discourse in the YouTube videos to uncover more communication patterns and other factors across country groups related to sustainable design from a Doughnut Economics perspective.

References

Abello Llanos R (2010) Factores claves en las alianzas universidad–industria como soporte de la productividad en la industria local: hacia un modelo de desarrollo económico y social sostenible. Inv Desarrollo 15(1)

Alcamo J (2003) Ecosystems and human well-being: a framework for assessment. Island Press, Washington DC

Alexa (2019) The top 500 sites on the web. www.alexa.com/topsites

Assembly G (2015) Sustainable development goals. SDGs), Transforming our world: the 2030

Bastian M, Heymann S, Jacomy M (2009) Gephi: an open source software for exploring and manipulatin networks. In: International AAI Conference on Weblogs and Social Media, San Jose, California, 17–20 May 2019

Benn S, Dunphy D (2004) A case study in strategic sustainability: Fuji xerox eco manufacturing centre. Innovation 6(2):258–268

Boehnert J (2018) Anthropocene economics and design: heterodox economics for design transitions. She Ji: J Des Econ Innovation 4(4):355–374

Borgatti SP, Everett MG, Freeman LC (2002) Ucinet for windows: software for social network analysis. Analytic Technologies, Harvard, MA

Boyd DM, Ellison NB (2007) Social network sites: definition, history, and scholarship. J Comput Mediated Commun 13(1):210–230

Cálad Idárraga ML (2013) Propuesta de educación para el desarrollo sustentable en el reciclaje y la reutilización de materiales en juegos y juguetes en la educación inicial. REXE, Rev de Estud y Experiencias Educación 12(24)

Escobar A (2018) Designs for the pluriverse: radical interdependence, autonomy, and the making of worlds. Duke University Press, US

Esslinger H (2013) Design forward: creative strategies for sustainable change. Arnoldsche Art Publishers

Franta B (2018) Shell and Exxon's secret 1980s climate change warnings. The Guardian

Frey S (2003) The transfer of core-based hazardous production processes to the export processing zones of the periphery: The maquiladora centers of northern Mexico. J World-Syst Res 9(2):317–354

Galvez DM (2011) Edificacion sustentable en Mexico: Retos y Oportunidades. Mexico City, Mexico, 1 December, 2011

Gómez Barrera YNI (2010) La cultura del diseño, estrategia para la generación de valor e innovación en la PyME del área metropolitana del Centro Occidente, Colombia. Cuad Del Centro De Estud En Diseño Y Comunicación, Ensayos 34:109–209

Guadarrama R (2006) Identidades, resistencia y conflicto en las cadenas globales: Las trabajadoras de la industria maquiladora de la confeccion en Costa Rica. Destacados 21:67–82

Hanington B (2018) Empathy, values and situated action. Sustaining people and planet through human centered design. In: Egenhoefer B (ed) Routledge handbook of sustainable design, 1st edn. Routledge, Oxon, pp 193–205

Karim BA, Majid MSA (2009) International linkages among stock markets of Malaysia and its major trading partners. J Asia-Pac Bus 10(4):326–351

Kemp S (2019) We are social. Digital Report, pp 52–89

Laville, S (2019) Top oil firms spending millions lobbying to block climate change policies, says report. The Guardian.

Manzini E (2015) Design, when everybody designs: an introduction to design for social innovation. MIT Press, U.S.

Matsumoto M, Umeda Y (2011) An analysis of remanufacturing practices in Japan. J Remanufacturing 1(1):2

Meadows DH, Meadows DH, Randers J, Behrens III WW (1972) The limits to growth: a report to the club of Rome, Google Scholar

O' Neill DW, Fanning AL, Lamb WF, Steinberger JK (2018) A good life for all within planetary boundaries. Nat Sustain 1(2):88

Ramos G (2019) Women deliver conference. Vancouver, 5 June 2019

Raworth K (2017) Doughnut economics: seven ways to think like a 21st-century economist. Chelsea Green Publishing, Whiter River Junction, Vermont

Regalado A (2014) The economics of the internet of things. MIT Technology Review

Rieder B (2015) YouTube Data Tools. Comput. Softw

Rissanen T (2018) Possibility in fashion design education—a manifesto. Utopian Stud 28(3):528–546

Rockstrom J, Steffen WL, Noone K, Persson A, Chapin FS III, Lambin E, Lenton TM, Scheffer M, Folke C, Schellnhuber HJ, Nykvist B (2009) Planetary boundaries: exploring the safe operating space for humanity. Ecol Soc 14(2):32

Stopper M, Kossik A, Gastermann, B (2016) Development of a sustainability model for manufacturing SMEs based on the innovative doughnut economics framework. In: Proceedings of the international multi conference of engineers and computer scientists, Hong Kong, 15–17 Mar 2016

Sukhdev P, Pushpam K (2008) The economics of ecosystems and biodiversity (TEEB). In: Wesseling, Germany, European communities, Aug 2008

Team RC (2013) R: A language and environment for statistical computing

Vargas Meza X (2019) Development of an educational intervention to enhance interest on sustainable design. Ph.D. thesis dissertation, the University of Tsukuba, Japan, 25 Mar 2019

Vargas Meza X, Yamanaka T (2017) Sustainable design in YouTube. Int J Affec Eng 17(1):39–48

Villanueva Ponce R, Garcia Alcaraz JL, Cortes Robles G, Romero Gonzalez J, Jimenez Macias E, Blanco Fernandez J (2015) Impact of suppliers' green attributes in corporate image and financial profit: case maquiladora industry. Int J Adv Manuf Technol 80(5–8):1277–1296

We Are Social (2019) Digital 2019. Malaysia, p 15

Wellington RP (1982) The mileage marathon competition. An exercise in realistic engineering education. In: Papers presented at the annual conference of research and development in higher education. vol 5, p 179

Part II
Business Models and Policies

Chapter 11
Barriers for Remanufacturing Business in Southeast Asia: The Role of Governments in Circular Economy

Mitsutaka Matsumoto, Kenichiro Chinen, Khairur Rijal Jamaludin, and Badli Shah Mohd Yusoff

Abstract Remanufacturing is one of the key determinants in enhancing resource efficiency of economies and pursuing circular economy. Facilitating international remanufacturing supply chain enhances the effects of remanufacturing. This study focused on the current scenario in remanufacturing businesses and related policies in Southeast Asia through interviews with 12 companies in 4 Southeast Asian countries, namely Malaysia, Indonesia, Singapore, and Philippines, and 5 Japan-based companies having remanufacturing facilities in Southeast Asia. The study presents the barriers for the remanufacturing businesses and debates on the roles of the governments to promote remanufacturing.

Keywords Sustainable consumption and production · Remanufacturing · Southeast Asia · Global supply chain · Policy

11.1 Introduction

Consumerism is the human desire to satisfy oneself by owing, obtaining, and receiving products, goods, and services above one's basic needs. We are often convinced that we need to repurchase and replace goods that would have lasted for much longer.

There is increasing academic interest across the world on how the circular economy may contribute to the goals of sustainability. More than 35% of respondents

M. Matsumoto
National Institute of Advanced Industrial Science and Technology (AIST), Tsukuba, Japan

K. Chinen (✉)
California State University, Sacramento, CA, USA
e-mail: chinen@csus.edu

K. R. Jamaludin
Universiti Teknologi Malaysia (UTM), Kuala Lumpur, Malaysia

B. S. M. Yusoff
Universiti Kuala Lumpur (UNIKL), Kuala Lumpur, Malaysia

Y. Kishita et al. (eds.), *EcoDesign and Sustainability I*, Sustainable Production, Life Cycle Engineering and Management, https://doi.org/10.1007/978-981-15-6779-7_11

in the report by APICS (2014) stated that remanufacturing is an important method of complying with sustainability policies, goals, and requirements.

Supply chains, among the various sectors, have garnered greater worldwide attention due to the dynamic nature of present-day business environments (e.g., Angelis et al. 2017; Genovesea et al. 2017; Gupta et al. 2011). Global supply chains involve and connect multilevel businesses from various geographical contexts. Because the industry spans across sectors, regulated by different stakeholders across countries, inter-nation coordination is key for the growth of the industry. In other words, raising awareness among government officials regarding remanufacturing business and remanufactured goods is essential to ease trade restriction on cores and remanufactured goods, especially among the members of regional economic integration. While most remanufacturing activities have taken place historically in the United States and Europe, Asia holds excellent potential for the future growth of remanufacturing (Liu et al. 2014).

This study focused on the current scenario in remanufacturing businesses and related policies in Southeast Asia through interviews with 12 companies in 4 Southeast Asian countries, namely Malaysia, Indonesia, Singapore, and the Philippines, and 5 Japan-based companies having remanufacturing facilities in Southeast Asia. The aim of this paper is to seek to explore the following questions:

- What are some of the challenges in inbound and outbound logistics?
- What is the role of governments in mitigating the consumers' perceived risks of remanufactured products?

The present study contributes to the literature on the circular economy on how ASEAN and their respective governments can play a critical role by formulating and implementing policies to manage better supply chains in emerging economies.

11.2 Literature Review

11.2.1 Framework

Nasr et al. (2017) used sensitivity analysis for supply chain remanufacturing processes to classify barriers into four categories: regulatory and access, technical, market, and recovery. Figure 11.1 summarizes four barriers. Technical barriers address challenges of the remanufacturing process. Challenges in the activities composing a remanufacturing process, such as disassembly, inspection, sorting, cleaning, reprocessing, reassembly, and inspection/testing, are well-documented in Steinhilper (1998) and Kurilova-Palisaitiene et al. (2018). Recovery barriers are "typically a reflection of infrastructure and predominant end-of-use behavior in an economy," but the scope is too broad, encompassing the social norms associated with diversion versus disposal, efficiency, cost, and convenience of diversion programs,

Fig. 11.1 Barriers in remanufacturing value chain [Modified from Nasr et al. (2017)]

Technical Barriers
• Access to skilled labor and equipment • Cost of reverse logistics and recovery • Access to required material inputs and cores • Production efficiency and waste generation • Access to distribution channel

Recovery Barriers
• End-of-use behaviors • Quality and efficiency of infrastructure • Diversion to recycling rate • Diversion to secondary market rate • Disposal to environment rate

Regulatory & Access Barriers
• Product/material-level transaction restrictions • Ability to import cores and export reman products • Ability to supply reman products • Permission to produce reman products domestically • Unclear definition of « reman» vs. « used » products

Market Barriers
• Attitudes towards « new » vs. « used » products • Previous experience with reman products • Access to and awareness of reman options

allocation of the cost associated with reverse-logistics, overall diversion rate, and so forth.

The present study focuses on barriers that have "a real impact on the intensity of remanufacturing" (Nasr et al. 2017, p. 15): regulatory and access barriers and market barriers.

11.2.2 Regulatory and Accessibility Concerning Cores and Remanufactured Products in ASEAN

When regional economies agree on integration, trade barriers between the member countries fall, and their economic and political coordination increases. If it is successful, the integration can draw foreign direct investment, reduce tariffs, and enable the rise of a major manufacturing hub in the region.

Association of Southeast Asian Nations (ASEAN) is a regional political and economic organization comprising ten countries in Southeast Asia. ASEAN mainly aims to curb regional measures that lead to inefficiencies in global economic integration (Lloyd and Smith 2004). Until the late-1990s, the increase in economic interdependence in the region took place without any formal framework of economic cooperation. While there has been a strengthening of its institutions in recent years,

Schwartz and Villinger (2004) argue that ASEAN does not have a strong framework. It lacks a formal, detailed, and binding options which enable the institution to prepare, enact, coordinate and execute policies (Hew 2006). It is argued that this weak institutional structure "has been a major reason for the relatively low impact of ASEAN's initiatives to reduce tariffs and eliminate non-tariff barriers" (Hew 2006). Some ASEAN members intentionally restrict imports of goods to protect their domestic industries (Kojima 2017). In the case of trade of remanufacturing, "unintentional trade restrictions on cores and [remanufactured products] exist" (Kojima 2017, p. 642). Whether intentional or not, a government's restrictive trade policy can impact businesses by making it more challenging for inbound and outbound supply chains to trade across international borders. Therefore, it is essential for ASEAN countries to facilitate trade through harmonization of logistics policies and international trade procedures to lower total costs of production (Nguyen et al. 2016).

11.2.3 Market Barriers: Building Trusts in Remanufactured Products

Credible quality certifications can guarantee product quality, and help in mitigating consumers' perceived risks of remanufactured products (Michaud and Llerena 2011; Kissling et al. 2013; Wang et al. 2013). Building trust and confidence in remanufactured products is vital because the remanufacturing started receiving wider recognition since 1970s in the US (Hormozi 1997). Certification from authorized parties provides strict guidelines for remanufactured products or remanufacturing processes and can "reduce the frequency and mitigate consequences of market failures" (Vertinsky and Zhou 2000, p. 231). Credible quality certifications assure the consumers regarding product quality and thus help mitigate the consumers' perceived risks of remanufactured products (Michaud and Llerena 2011, Kissling et al. 2013; Wang et al. 2013).

Matsumoto et al. (2018) found that certifications of remanufactured auto parts influence purchase intentions. They show that, in the Southeast Asian countries, consumers are more willing to purchase certified remanufactured auto parts than uncertified ones. They find that automobile OEMs are trusted the most, followed by those certified by remanufacturing industrial associations, international standard organizations, government or public organizations, and remanufacturing companies. Certifications provide information regarding environmentally friendly features of remanufactured products, including energy saving, extending the lives of landfills, and reducing air pollution.

11.3 Interviews

Based on the literature review, the present study focuses on the two likely barriers: (1) regulations on the trade of cores and remanufactured products and (2) trust on remanufactured products.

The study adopted the interview method with remanufacturing companies that included questions on the aforementioned items. The study is based on the interviews with 12 remanufacturing companies in 4 Southeast Asian countries (Malaysia, Indonesia, Singapore, and the Philippines) and five Japan-based companies having remanufacturing facilities in these countries. A part of the interviews (nine interviews) were commissioned to a survey company in which the question items were prepared by the authors, and the other interviews (eight interviews) were conducted directly by the authors. The product areas included: auto parts, electronics products (mainly personal computers), photocopiers, heavy-duty and off-road (HDOR) equipment components, and ink and toner cartridges. In the interviews, we first asked the basic features of the companies' businesses that include the companies' profiles, their remanufacturing businesses, market properties, and the companies' market shares. Then, we inquired the regulations on trade of cores and remanufactured products. Next, we asked the companies' customers' perceptions of remanufactured products. The interviews were supplemented with desktop surveys on regulations and markets.

11.4 Results

11.4.1 Regulatory and Access Barriers

A desktop survey of the regulations in the four countries was conducted before the interviews. Table 11.1 summarizes the results. This study found that the Indonesian government restricts the imports of used capital goods. The cores for remanufacturing are mostly in the category of used capital goods. A large mining industry thrives in Indonesia, and major global mining machine OEMs (or HDOR OEMs) such as Caterpillar, Komatsu, and Hitachi Construction Machinery have set up their remanufacturing facilities here. A few of these companies import used parts for remanufacturing. The government issues an importer's identification number (API) to a company in order imports goods. Acquiring an API is often difficult, especially for foreign-affiliated companies. The permit to import used goods is even more difficult to obtain. Global HDOR OEMs that import used parts for remanufacturing collaborate with the Indonesian companies having the said permits. The opportunities for such collaborations are also limited. The interview results revealed that the regulation restricts remanufacturing and lacks fairness.

In Indonesia, refurbished photocopiers have a high market share. The local Indonesian companies import used photocopiers and supply refurbished photocopiers to the local market. A photocopier refurbisher stated in the interview that Indonesia restricts

Table 11.1 Regulations on trades of cores and remanufactured products

Country	Regulations or NTM on trades of non-newly manufactured products
Malaysia	None on remanufactured products. Approved Permit (AP) is required to import used cars and construction machines. Restrictions on imports of old electronics equipment
Indonesia	Restrictions on imports of used capital goods. Restrictions on imports of used cars
Singapore	None
Philippines	Restrictions on imports of used cars and tires. Imports of parts of used trucks or buses are allowed only for accredited rebuilding centers

Compiled from the information in the following sites and (Kojima 2017)
https://www.apec.org/Groups/Committee-on-Trade-and-Investment/Market-Access-Group/NTM/
http://www.federalgazette.agc.gov.my/outputp/pua_20110826_perintah%20kastam(nilai-nilai%20kenderaan%20motor%20pasang%20siap%20yang%20diimport)%20-complete.pdf
https://www.jetro.go.jp/world/asia/my/trade_02.html https://www.jetro.go.jp/world/asia/ph/trade_02.html
https://www.jetro.go.jp/world/asia/sg/trade_02.html
https://www.jetro.go.jp/world/qa/04J-101001.html
https://www.jetro.go.jp/world/qa/04J-101103.html
https://www.jetro.go.jp/world/qa/04A-031102.html
https://www.jetro.go.jp/ext_images/jfile/country/idn/trade_02/pdfs/idn2F010_exp_hinmoku.pdf
http://www.regulasi.kemenperin.go.id
http://setkab.go.id/en/minister-of-industry-to-allow-import-of-second-hand-capital-goods/

the imports of used products but permits the imports of used "black and white" photocopiers. However, the imports of used "color" photocopiers are not permitted, and it forms a major obstacle for the company's business today.

In other countries, the interview responses did not find extensive regulatory barriers that strongly affected trade in both remanufactured goods and cores. Among the eight auto-parts remanufacturers or traders interviewed in ASEAN, namely Malaysia, Singapore, and the Philippines, at least five companies are importing cores from countries such as Japan, the United States, China, Germany, the United Kingdom, and other European countries.

Core import can also be affected by global agreements, such as the Basel Convention. The Basel Convention restricts the movements of hazardous waste between nations, and specifically to prevent the transfer of hazardous waste from developed countries to less-developed countries (Kojima 2017). However, the interviews results revealed that some OEMs have concerns regarding the trend of tightening trade restrictions of used products. Currently, it is unclear whether the Basel Convention prohibits the trade of cores for remanufacturing purposes if they contain harmful substances, for example, in some products in the electrical apparatus and IT product sectors. Several countries, companies, and lobby groups have sought clarification on the Basel convention with regard to repair, recondition, and remanufacturing, contending that imported/exported goods for these purposes (and particularly if the restored products are destined to be exported back to the original market) should not be classified as waste. Malaysia, one of the major remanufacturing countries in ASEAN, supports this issue to open up trade of cores for refurbishment and remanufacturing. Moreover, the 3R Action Plan adopted at the G8 summit in 2004 stipulated that it seeks "to reduce barriers to the international flow of goods and materials for recycling and remanufacturing, recycled and remanufactured products, and cleaner, more efficient technologies, consistent with existing environmental and trade obligations and frameworks."

Our interview revealed a host of technical and administrative complications in ASEAN, including lack of clarity in the understanding and defining the nature of remanufacturing, tariff and non-tariff measures, or intentional and unintentional trade barriers. So-called 'spaghetti bowl effect' (Hew 2006) which may be a stumbling block to ASEAN (Schwartz and Villinger 2004; Kojima 2017). According to an interviewee, the unclear definition of used versus remanufactured goods results in customs authorities treat remanufactured goods sometimes as used products and as remanufactured products in other cases. Therefore, a common definition of remanufactured goods in regional integration is urgently required.

11.4.2 Building Trusts in Remanufactured Products

Customers' acceptance of remanufactured products is another key factor for remanufacturing to proliferate in the market (Matsumoto et al. 2018a, b). Several of the companies interviewed, in particular, those remanufacturing auto parts and consumer products such as ink toners and home electronics products, mentioned that customers' unawareness about the quality of remanufactured products is obstacles for their businesses.

In remanufacturing of OEMs, the remanufactured products are generally inspected similarly as new products, and they are provided the same warranties as new products. The major customers are aware of the inspection processes. On the other hand, in the case of non-OEM remanufacturing, customers generally are larger concerned about product quality.

Some non-OEM auto-parts remanufacturers stressed the importance of warranty. For example, an auto-parts remanufacturer interviewed in Malaysia claimed to provide 2-year warranties for products.

Many of the interviewed non-OEM remanufacturers rely on ISO 9001 accreditation to show that they have robust quality management systems as a fundamental requirement for producing high-quality products. Other relevant standards included the IEC series that outlines the safety and performance requirements for electrical apparatus. Malaysian government (MATRADE) supports remanufacturers to understand the relevant import requirements that may affect their ability to trade.

Some interviewees stated that a few certifications by the government for remanufactured products are effective because credible quality certifications help in guaranteeing the product quality and help in mitigating customers' perceived risks of remanufactured products. Quality seals applicable to remanufacturing also exist. For example, the EC mark and NOM mark certifying that a new product is compliant with the import requirements of the EU and Mexico, respectively, are equally applicable for remanufactured products that wish to enter those markets. The effectiveness of the government-issued quality certificates concerning that of industry-led systems depends on the extent of the market development of remanufactured products. In the nascent phase of the market, the government-led certificate systems are more likely to be effective. Because credible quality certifications can guarantee product quality, and help in mitigating consumers' perceived risks of remanufactured products, we recommend ASEAN's policymakers to consider measures to educate consumers and improve image of remanufactured products in the region.

11.4.3 Notes on Policy in Malaysia

Malaysia currently has a narrow focus on the remanufacturing industry. Toner and inkjet cartridges are the primary subsectors with majority of the companies (nearly 40% of the identified firms). However, it has the potential to build a strong automobile-related remanufacturing industry with the existing manufacturing base, coupled with the availability of an established and sound automotive recycling industry.

The present-day remanufacturers believe that the Malaysian government can play an important role in developing the industry by influencing policy levers. For example, offering investment tax incentives, amending or clarifying import policies, and creating a certification program to instill consumer confidence in high-quality remanufactured products are all considered to be beneficial to further develop the remanufacturing industry. For the policies on quality control and customer awareness, the emphasis should be on introducing a remanufactured product quality seal for remanufactured parts and supporting awareness campaigns on remanufacturing. The existing quality standards for new goods can be applied to remanufactured goods as well. Economic incentives should be granted to firms certified to meet the required standards for high-value remanufacturing.

The Malaysian government is currently in the process of stipulating quality standards for critical automotive parts, which will also include a quality seal. A few countries, including the United States, China, and Korea use a remanufactured product quality seal, although each country administers it differently. Quality seals in the United States are designed and administered by industry associations and in China by the government. Given the nascent market in Malaysia and the cross-cutting nature of the industry (motor vehicle parts, machinery, electrical, and more sectors), quality control for remanufactured products in Malaysia is currently more suited to a government-led model than industry-led. Relevant bodies in the Malaysian government, including MITI, Standards Malaysia, DOSH, DOE, SIRIM, and MAI, are well-positioned to deliver the key elements of quality control.

Malaysia needs to attract major global OEMs to set up remanufacturing facilities and significantly launch and grow the industry. The existing remanufacturers are still in a nascent stage, and they may find it challenging to access the export market. OEMs and OEM-authorized remanufacturers have a greater competitive advantage compared with independent players along each step of the remanufacturing value chain. OEMs not only have access to the detailed technical specifications for each part to guide the remanufacturing process, but they can also leverage the existing capabilities from their manufacturing facilities, such as testing and machining equipment and technical know-how. At the sales and distribution stage, OEMs can lend their brand to the remanufactured product, thus providing better support for the product. Malaysian government led by MITI can play a role in promoting awareness of remanufactured products. This can be done through awareness campaigns, mainly targeted at businesses.

For increasing ease of trade, the Malaysian government should maintain the existing trade policies on cores, including the prohibition on import of items considered as e-waste, but continue to be open to case-by-case consideration of exemptions, as well as continue to allow imports of cores to be remanufactured under warranty.

Furthermore, to increase the ease of trade and promote remanufacturing in the country, the industry expects the government to coordinate and reduce the existing double taxation for trades, namely tax to import cores and to export remanufactured goods.

Malaysian government should enhance waste management (e.g., by establishing extended producer's responsibility, such as a take-back or deposit-refund system, or requiring to use a minimum amount of recycled material), enhance end-of-life vehicle (ELV) regulations in the medium term (e.g., by mandating a maximum percentage of an end-of-life vehicle that may be scrapped), and incorporate "consider remanufactured product first" guideline into the government procurement policy. These policies will help enhance both the supply of cores and the demand for remanufactured goods while creating a more sustainable economy and achieving cost savings for the government.

Malaysia has announced several policies such as the National Policy on the Environment that include sustainable development. The 10th Malaysia Plan also promotes sustainable development. The government should make a special mention in environmental regulations and policies that remanufacturing is an attractive means of reuse and sustainable development. This would build greater public awareness regarding the environmental benefits of choosing remanufactured products. Consumer awareness campaigns in this regard would also be useful. This would ensure that consumers have remanufactured products at the "top of mind" when making an environmentally friendly choice and increase customers' understanding of the quality and value proposition of remanufacturing to promote a sustainable economy, as well as increase the demand for remanufactured goods.

11.5 Summary

Remanufacturing is key to increasing the resource efficiency of economies and pursuing circular economy. Furthermore, enabling international remanufacturing supply chain is likely to enhance the effects of remanufacturing in increasing resource efficiencies.

This study found that restrictions on the trade of cores were a constraint on developing the remanufacturing supply chain. Policymakers must prioritize reduction of the barriers to the international flow of cores for remanufacturing while preventing the risks of inappropriate flow of e-wastes that can cause environmental pollution. The study also found that customers are unaware of remanufactured products and their value. Overcoming the unfavorable perception through marketing campaign and education is another priority. Developing credible quality standards or certifications is effective because it helps in mitigating customers' perceived risks of remanufactured products. The move toward international standards is effective, and it should include the perspectives of industry, government, and market stakeholders.

Remanufacturing provides a viable approach to enabling circular economies. It is important that industries are provided the market opportunities to create a strong supply chain and value chain in remanufacturing.

For our future research, it is interesting to examine the remanufacturing industry in ASEAN from historical perspectives. It is generally difficult to follow the process because precise market data of remanufactured products usually do not exist. This historical review may provide an insight into the growth of the remanufacturing market in ASEAN.

Acknowledgements This study was partially supported by the Environmental Research and Technology Development Fund (Project S-16) of the Environmental Restoration and Conservation Agency of Japan.

References

Angelis RD, Howard M, Miemczyk J (2017) Supply chain management and the circular economy: towards the circular supply chain. Prod Plann Control 29(6):425–437

APIC (2014) Examining remanufacturing in supply chain and operations management. http://www.apics.org/docs/default-source/scc-non-research/apics_research_reverse_short_1214.pdf?sfvrsn=2

Genovesea A, Acquayeb AA, Figueroaa A, Koha SCL (2017) Sustainable supply chain management and the transition towards a circular economy: evidence and some applications. Omega 66(Part B):344–357

Gupta S, Goh M, Desouza R, Garg M (2011) Assessing trade friendliness of logistics services in ASEAN. Asia Pac J Market Logistics 23(5):773–792

Hew D (2006) Economic integration in East Asia: an ASEAN perspective. UNISCI Discussion Papers. 11(May 2006)

Hormozi AM (1997) Parts remanufacturing in the automotive industry. Prod Inventory Manag J 38(1):26–31

Kissling R, Coughlan D, Fitzpatrick C, Boeni H, Luepschen C, Andrew S, Dickenson J (2013) Success factors and barriers in re-use of electrical and electronic equipment. Resour Conserv Recycl 80:21–31

Kojima M (2017) Remanufacturing and trade regulation. Procedia CIRP 61:641–644

Kurilova-Palisaitiene J, Sundin E, Poksinska B (2018) Remanufacturing challenges and possible lean improvements. J Clean Prod 172:3225–3236

Liu Q, Goh M, Grag M, Souza RD (2014) Remanufacturing in Asia: location choice and outsourcing. Int J Logistics Manag 25(1):20–34

Lloyd P, Smith P (2004) Global economic challenges to ASEAN Integration and competitiveness: a prospective look. REPSF Project 03/006a (September). http://www.aadcp-repsf.org/docs/03-006a-FinalReport.pdf

Matsumoto M, Chinen K, Endo H (2018a) Paving the way for sustainable remanufacturing in Southeast Asia: an analysis of auto parts markets. J Clean Prod 205:1029–2014

Matsumoto M, Chinen K, Endo H (2018b) Remanufactured auto parts market in Japan: historical review and factors affecting green purchasing behavior. J Clean Prod 172:4494–4505

Michaud C, Llerena D (2011) Green consumer behavior: an experimental analysis of willingness to pay for remanufactured products. Bus Strateg Environ 20(6):408–420

Nasr N, Kreiss C, Russell J (2017) Barriers to advancing remanufacturing, refurbishment, repair, & direct reuse: insights from the International Resource Panel. Rochester Institute of Technology and IRP. http://ec.europa.eu/environment/international_issues/pdf/7_8_february_2017/Nabil_Nasr_sec2.pdf

Nguyen AT, Nguyen T, Hoang GT (2016) Trade facilitation in ASEAN countries: harmonisation of logistics policies. Asian Pac Econ Lit 30(1):120–134

Schwartz A, Villinger R (2004) Integrating Southeast Asian economies. The McKinsey Quarterly (November 1, 2004)

Steinhilper R (1998) Remanufacturing: the ultimate form of recycling. Frauenhofer IRB Verlag, Stuttgart

Vertinsky I, Zhou D (2000) Product and process certification—systems, regulations and international marketing strategies. Int Mark Rev 17(3):231–252

Wang Y, Wiegerinck V, Krikke H, Zhang H (2013) Understanding the purchase intention towards remanufactured product in closed-loop supply chains: an empirical study in China. Int J Phys Distrib Logistics Manag 43(10):866–888

Chapter 12
Efforts to Reduce CO_2 Emissions in the Japanese Automobile Recycling Industry

Sosho Kitajima and Hiroshi Onoda

Abstract A CO_2 reduction program "Green Point Club", researched and developed by Waseda University and Japan Automobile Parts Recyclers Association (JAPRA), was announced in 2007 as a groundbreaking effort that has been developed to add a new "Environmental Contribution Index" to recycling parts. With the momentum of CO_2 reduction such as abnormal temperature increasing, we promoted the use of recycled parts of vehicles will help to reduce the CO_2 reduction effect. In this study, we summarize the CO_2 reduction figures from using recycled parts in the past 10 years in Japan and verified it. We would like to the spread the method such as cooperation with the administration, information exchange with foreign countries, point return etc. and aim at further usage expansion.

Keywords CO_2 · Automobile · Used auto parts

12.1 Introduction

It will be 15 years in 2020 since the enactment of the "Act on Recycling, etc. of End-of-Life Vehicles" (Automobile Recycling Law). This effort is evaluated as "generally going well" by report on the evaluation and examination of the enforcement status of the automobile recycling system of the Industrial Structure Council of the Ministry of Economy, Trade and Industry in Japan on September 2015. However, as a result of the China's movement of import restrictions on waste materials including waste plastics since Autumn 2018, industrial wastes that had been exported overseas other than automobiles have been concentrated in domestic shredding and sorting yards, causing delays in the collection of dismantled end-of-life vehicles for their final disposal. Apart from scrap business, which has a major impact on economic trends and social responses, the factor that contribute to the profit stabilization in car dismantling businesses is selling of recycled auto parts. In Japan, since Bigwave's predecessor was launched in 1979, the sales of recycled parts in the automobile aftermarket has been

S. Kitajima (✉) · H. Onoda
Graduate School of Environment and Energy Engineering, Waseda University, Tokyo, Japan
e-mail: kitajima@jara.co.jp

Y. Kishita et al. (eds.), *EcoDesign and Sustainability I*, Sustainable Production, Life Cycle Engineering and Management, https://doi.org/10.1007/978-981-15-6779-7_12

increased. In the 1980s, grouping was promoted to improve "distribution effects", which include an increase in the delivery rate of recycled parts and an increase in shipment volume. Amid this movement, the business has grown steadily. However, since enactment of the Automobile Recycling Law, it has become impossible for automobile dismantlers to emphasize "low price" that is the biggest benefactor to expand distribution of recycled parts, due to increased costs for facilities and other factors. In addition, while "recycled parts are good for the environment", exactly what is good was unclear, and it was impossible to clearly differentiate recycled parts from new parts or external parts.

Under such circumstances, in cooperation with Nagata Laboratory in the Faculty of Science and Engineering, Waseda University, LCA (Life Cycle Assessment) method, which quantitatively evaluates the level of "environmental friendliness" of recycled parts (especially reuse parts), was developed in 2001. This method presents a new evaluation axis that is not just the amount of money. Also, it has become possible to send easily understandable messages to society by developing and operating an "environmental load evaluation system" that is applied to the use of auto recycled parts, through industry-academia collaborations.

In nine years after the deployment of this "environmental load evaluation system", CO_2 has been reduced by about 1,273,000 t-CO_2 or more. In this report, the history from the development to the current trend of a CO_2 reduction program named as "Green Point System" is described. Its environmental load evaluation system has been used and worked on by many auto recycled parts sales networks in Japan, such as JARA and JAPRA, which are deemed to be representing these networks.

12.2 Automobile Recycling Law

The past situation of recycling of end-of-life vehicles (as of 2002).

Approximately 4 million end-of-life vehicles are generated in Japan every year (approximately 5 million when exported used cars are included), Since these cars were valuable as resources as they are made from useful metals and parts, recycling was carried out in the course of distribution (buying & selling) by dismantling, shredding and sorting operators. While the recycling rate of end-of-life vehicles was not low with about 80%, the landfill sites for treating ASR (Automobile Shredder Residue) that is left behind after metals and parts are recycled became tight. Also, due to the price drop of iron scrap and application of reverse charge (the situation where car users are required to bear the processing cost when handing over their end-of-life vehicles to the operator), concerns over illegal dumping and improper disposal of end-of-life vehicles arose. In addition, no real progress had been seen in the proper destruction of air conditioner refrigerants (fluorocarbons), which affect global warming, and appropriate disposal of airbags, which requires specialized techniques (Japan Automobile Recycling Promotion Center 2019a). Also, there were the following problems in the recycling of end-of-life vehicles

- Contamination of soil, groundwater, etc. due to outflow of harmful substances from illegally dumped end-of-life vehicles
- Destruction of the ozone layer and change of the ecosystem caused by fluorocarbons that have been released into the air due to improper treatment
- Specialized techniques are required for safe processing of airbags and seat belt pretensioners (collectively, referred to as "airbags") for the protection of passengers from the impact of a collision.
- Most of the shredder dust is landfilled as waste.

To address these problems, as the fifth individual law based on the "Basic Law for Establishing a Recycling-based Society", the Act on Recycling, etc. of End-of-Life Vehicles ("Automobile Recycling Law") was enacted on July 12, 2002 and enforced on January 1, 2005 (Japan Automobile Recycling Promotion Center 2019b).

12.3 Development of Auto Recycled Parts

In Japan, selling of auto recycled (used) parts started in the 1960s and this business has developed to be a very historic industry. At that time, local dealers and repairers found the same model of the vehicle they were repairing among the cars piled up at demolition shops, took the necessary parts off from those cars and bought them from the demolition shop at a low price. Started like this, this business has grown to be a market with annual sales of 200 billion yen. Such demolition shops were so-called "junkyard for stripping".

Later, in the 1980s, started from calling fellow shops to check the stock to fulfill the request of those customers who visited the yard to find the parts that were not stored there, a fax and computer system was constructed in order to collect more information and make more profit. As a result, this business saw a rapid growth, with each company increased the sales by double digit (Japan Automobile Recycling Promotion Center 2010a).

12.4 Launch of Japan Automotive Parts Recyclers Association (JAPRA)

In November 1995, Japan Automotive Parts Recyclers Association was launched after the preparation of the foundation started from 1992 by the persons related to the industry group, for the purpose of broadly contributing to society by responding to the national request for the effective use of resources, environmental protection and continuous supply of high-quality, inexpensive recycled parts to the users through the widespread use and sound transactions of recycled parts for auto repairs based on the "Law concerning the Promotion of Reclaimed Resources Utilization" and the "Environmental Basic Law".

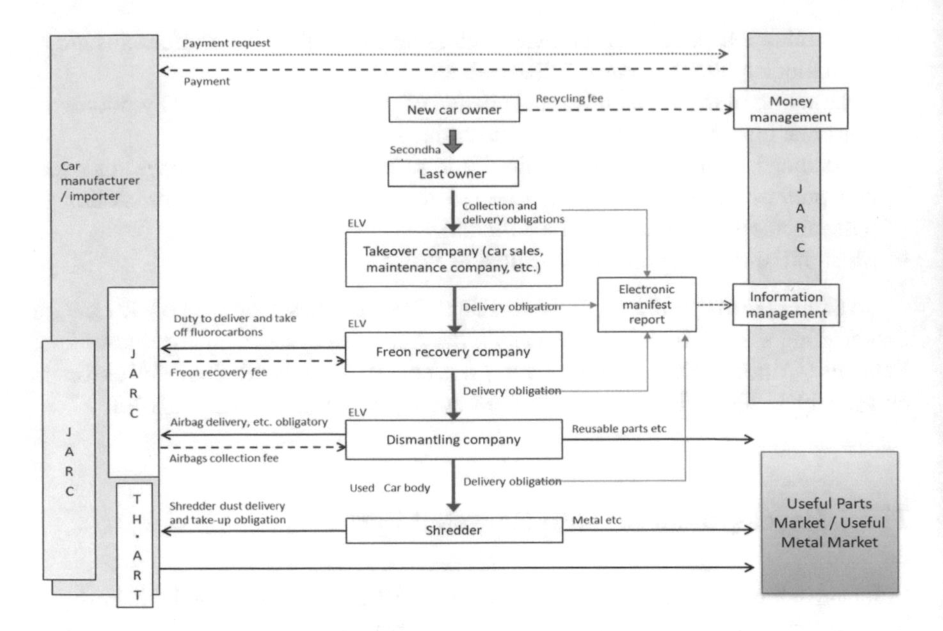

Fig. 12.1 Concept of the automobile recycling law (*Source* Ministry of the Environment HP)

Currently, 12 groups including JAPRA, SAP, NGP and JARA and more than 500 companies are the members of this association, collaborating in the industry (Fig. 12.1).

Since the announcement of the automobile recycling initiative by the Ministry of Economy (currently, the Ministry of Economy, Trade and Industry) in May 1997, these organizations have been advocating that the very promotion of use of "recycled parts" including "reuse parts" is a major factor to increase the automobile recycling rate and acting in line with it. In particular, the industry standardizes quality standards and warranty standards that were scattered throughout the group.

12.5 Background for the Development of Green Point System

It has been long since the dealing of "auto recycled parts" started. With the development of domestic motorization since the 1970s, the number of end-of-life vehicles has increased and social awareness on "environmental problem" and "pollution problem" has risen. In such a situation, the related "Waste Disposal & Public Cleaning Law" has been largely revised, and the "Automobile Recycling Law" was enacted in 2002, implemented in January 2005.

As a result, "dismantling business" has become subject to a permit system, and the "dismantling activities" in the places other than the factory facilities approved by law have been prohibited. Gradually, the name of "environmental industry" has come to be crowned to the business.

However, basically the work process has not been changed, where end-of-life vehicles go through the "liquid removal process" for waste oil & liquids, "flon and air bag disposal process", and finally proceed to the "part removal process". The useful parts removed are "commercialized" after inspection and cleaning, and reused as "recycled parts". This process is "reuse and recycling of products" in terms of "resource saving" and "energy saving", and unmistakably it "eliminates wasteful consumption of resources and energy". That is, "recycled parts are the products friendly to the global environment". Based on this concept, the General Insurance Association of Japan is promoting widespread use of recycled parts.

The concept is understandable somehow as an image, however, there was no ground to answer the question of what truly proves "earth-friendly".

In March 2001, clear expressions of "environmental friendliness" with "numerical figures" was enabled by the LCA method researched by Professor Katsuya NAGATA and Hiroshi ONODA (the author of this report) in Nagata laboratory in the Faculty of Science and Engineering, Waseda University.

In 2003, the implementation plan for the research on "the system to expand the use of reuse & rebuilt parts" was presented by Waseda University. The first-year plan was "to practice environmental load evaluation of reuse parts". After the basic research, (1) investigation and evaluation on highly-demanded parts, (2) sophistication of the evaluation method based on the more detailed data, (3) expansion of target auto models to be evaluated, and (4) effective method of using the results of environmental load evaluation were considered to create a prototype of the current Green Point System.

Under these circumstances, the research based on "disassembly and measurement" was started by the Waseda University Environmental Research Institute in 2005 at U-PARTS for small-car parts, and at Kyoei Jikou (Edogawa, Tokyo) and at Kanazawa Shokai (Saitama Plant) for large-car parts. Indexing of the effect of the using of reuse parts in CO_2 emission reduction completed in 2007. In May 2007, a joint press conference was held by Waseda University and Automotive Parts Recyclers Association (one company) after the demonstration of 90% or more of CO_2 emission reduction effect of reuse parts compared to new parts. "This result could be a guide for promoting the use of reuse parts", they stated in the press conference (Japan Automobile Recycling Promotion Center 2010b).

12.6 Usage Status of Green Point System

12.6.1 Operational Deployment

Since December 2007, named as "Green Point System", and the user group using this system as "Green Point Club", this system has been supervised and operated by Japan Automotive Parts Recyclers Association.

As of 2009, Green Point Club was consisted of 12 groups and 530 companies. The number of the parts sold each month by these members to repair shops is converted to the equivalent of CO_2 reduction amount and published. In fiscal 2008, 3,324,000 parts were sold with CO_2 reduction effect of as much as 160,000 tons.

Also, the number of target items has been increased to 475 from the initial number of 40 at the time of development of the system. The columns to enter CO_2 reduction figure is provided for each product on the invoices for sold parts issued by the members. The figures are automatically recorded to provide the customer with the information of environmental contribution effect (Japan Automobile Recycling Promotion Center 2010c).

12.6.2 System Overview

Green Point System overview is provided in Fig. 12.2. The items in the master data for names and components of those parts used by each auto recycled parts network were unified and standardized. These data are linked to the CO_2 data server on the part of Waseda University for analysis. Also, CO_2 reduction figures can be calculated based on the sales data for each customer. The calculation results can be output into a PDF or CSV file in a fixed format for Green Point Club. The member companies input their respective figures on the invoice for sold parts issued for each customer so that the repair shop (customer) can check CO_2 reduction figures with the invoiced prices of recycled parts purchased from the member.

In addition, each recycled parts network has customized the system in accordance with their respective purpose to increase the convenience (Fig. 12.3).

12.6.3 Evaluation Method for CO_2 Reduction Figure

In Green Point System, the following numerical evaluation method is applied

(1) Evaluation scenario

The difference between environmental load of the new part and environmental load of the reuse part is quantified as the CO_2 reduction effect of the part.

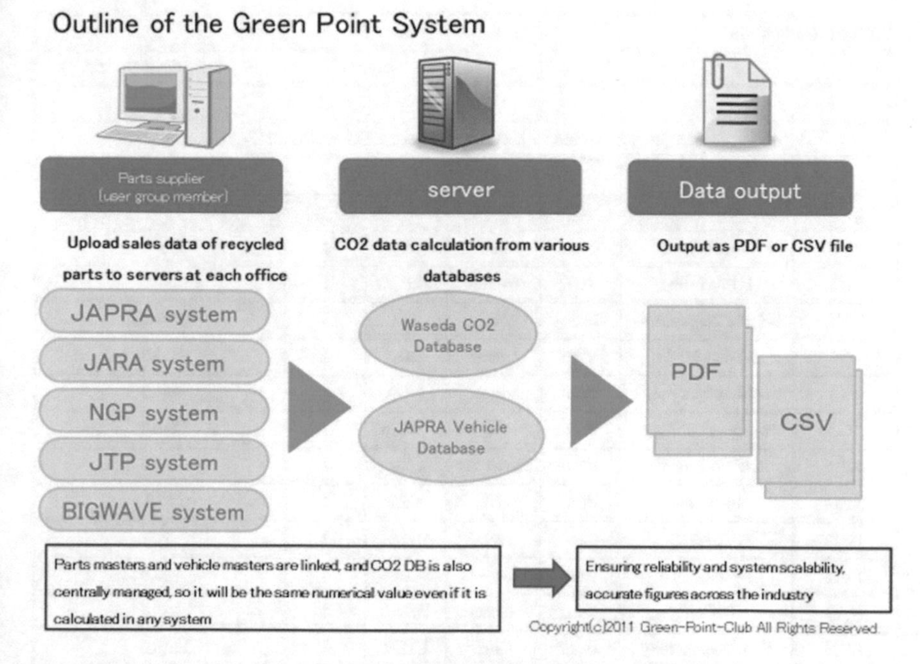

Fig. 12.2 Outline of the green point system (*Source* Greenpoint Club-JAPRA)

(2) Basis of environmental contribution points

CO_2 emissions of the new part − CO_2 emissions of the reuse part = CO_2 emissions that may be reduced by using the reuse part.

12.6.3.1 Evaluation Scenario of Reuse-Parts

Figure 12.4 shows the process of new parts and reuse-parts manufacturing. The ELVs are trucked to Automobile dismantlers to be dismantled. They retrieve trouble-free parts and market after quality checking. In this reuse parts manufacturing process, transportation fuel and energies in dismantling are consumed. Meanwhile, in original system, a new part manufacturing, new resources in new materials and energies in parts manufacturing are consumed.

Therefore, the CO_2 reduction effect of the reuse parts is given by the following equation;

$$R_{X\,reuse} = (Ev + E_M) - (E_T + E_D) \tag{12.1}$$

where CO_2 reduction effect of a reuse parts X is expressed as R_{Xreuse}; CO_2 emissions from producing virgin materials is expressed as E_V, CO_2 emissions from parts

TO: ABC Body shop

Date: 2018-10-31

Automotive Recycled Parts CO2 Reduction Contribution Report

Ver.0908

Date	Parts	Category	Car Name	Model	Qty	CO2 Reduction amount(KgCO2)
10/1	Front bumper	Used	Life	JA4	1	324.0
10/5	L-Head lump	Used	Estima	ACR55W	1	123.0
10/5	L-Front Brake caliper	Used	Acty	HA4	1	137.0
10/6	L-Front Fender	Used	Lancer	CK2A	1	199.0
10/6	L-front Door Assy	Used	Lancer	CK2A	1	139.0
10/8	L-front Door Assy	Used	IS	GSE20	1	139.0
10/10	Bonnet food	Used	Lancer Evo	CT9A	1	159.0
10/10	Rear Gate	Used	Fit	GE6	1	87.0
10/12	AC Compressor	Used	Jeep	HYMX	1	69.0
10/12	R-Head lump	Used	Wagon-R	MC12S	1	122.0
10/15	Radiator	Used	Demio	DE3FS	1	319.0
10/16	Rear Bumper Assy	Used	Move	L900S	1	319.0
10/23	AT transmission	Used	Oddessy	RA5	1	472.0
10/26	R-Head lump	Used	Serena	C25	1	128.0
10/27	R-Front Lowerarm	Used	Serena	C25	1	122.0
	Total				15	2858.0

Member No.12345

Member Name: ABC Body shop

Street address: 0-0-0 Minato-Ku, Tokyo, Japan.

TEL: 03-0000-0000

自動車リサイクル部品の供給で地球温暖化防止に貢献

CO2 Green Point Club

Fig. 12.3 CO_2 reduction introduction sample written on the invoice of a recycling parts sales company that has been output data (*Source* Greenpoint Club-JAPRA)

manufacturing is expressed as E_M; CO_2 emissions from transport is expressed as E_T; CO_2 emissions from dismantling an ELV is expressed as E_D.

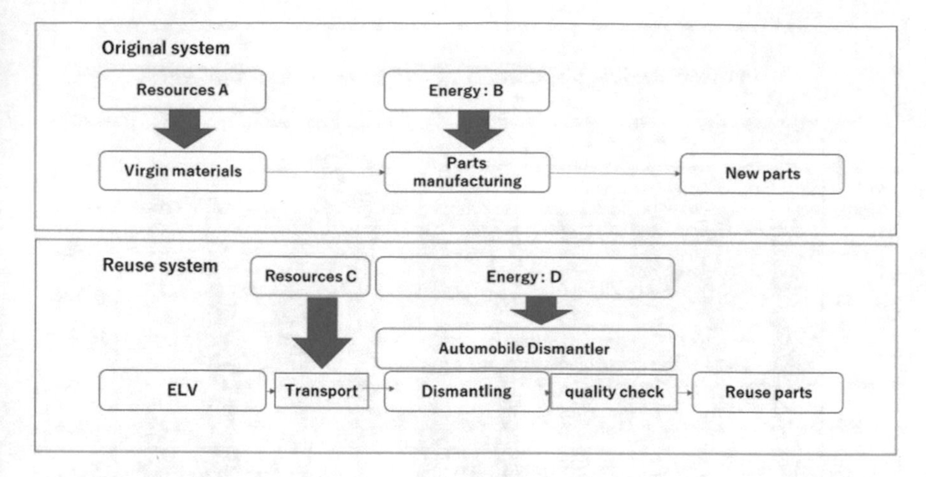

Fig. 12.4 Process of new/reuse-parts manufacturing (*Source* Waseda Environmental Institute Co., Ltd. 2011a)

12.6.3.2 Evaluation Scenario of Rebuilt-Parts

Figure 12.5 shows the process of new parts and rebuilt-parts manufacturing. The ELVs are trucked to Automobile dismantlers and trouble-free are retrieved. The parts needed locally-repaired are torn down, cleaned, replaced to new parts, and assembled once again. The original system is the same as mentioned.

Therefore, the CO_2 reduction effect of the rebuilt-part is given by the following equation;

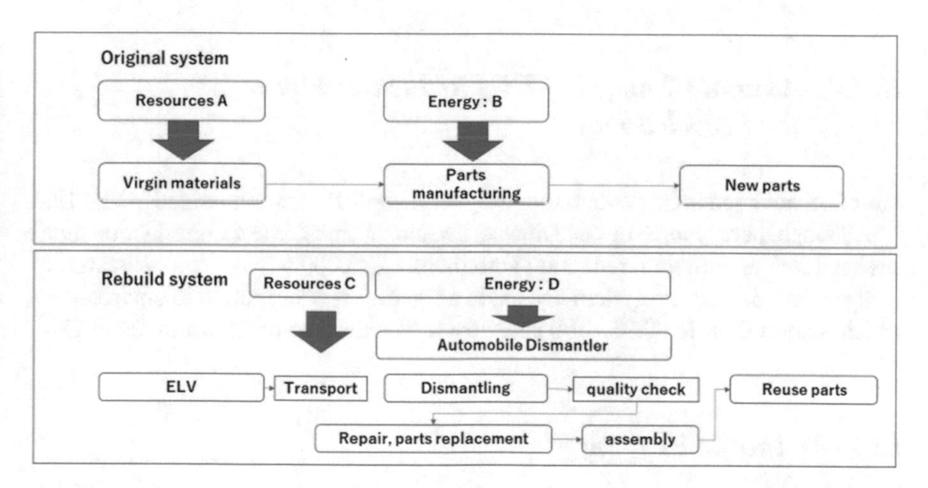

Fig. 12.5 Process of new/rebuilt parts manufacturing (*Source* Waseda Environmental Institute Co., Ltd. 2011b)

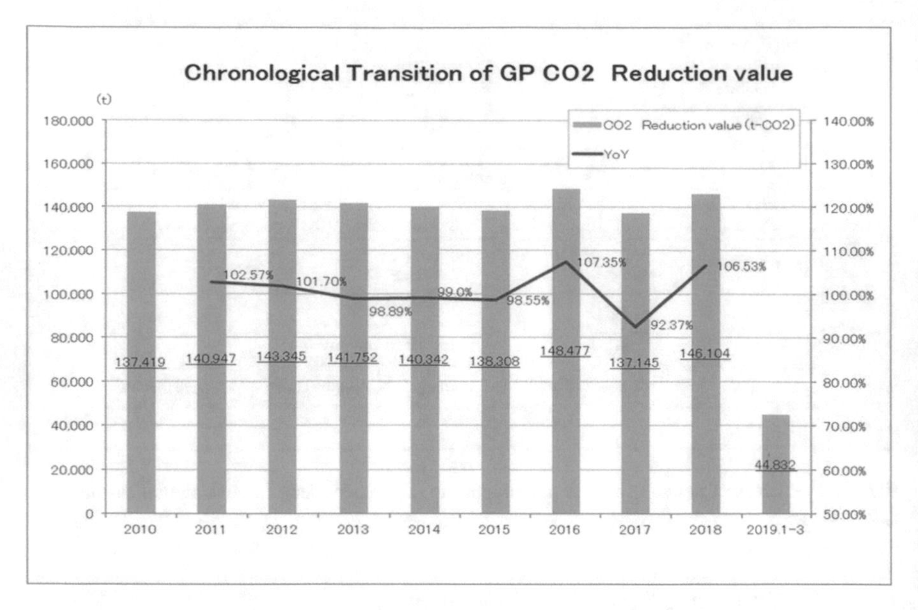

Fig. 12.6 Annual changes of CO_2 reduction figure for recycled parts (*Source* Japan Automotive Parts Recyclers Association & JARA Co.)

$$Rx_{rebuilt} = (E_V + E_M) - (E_T + E_D + E_L) \tag{12.2}$$

where CO_2 reduction effect of a part X is expressed as $R_{Xrebuilt}$; CO_2 emissions from producing a locally-repaired part is expressed as E_L (Nakajima et al. 2012).

12.6.4 Annual Changes of CO_2 Reduction Figure for Recycled Parts

The effect of Green Point System for nine years from 2010 is summarized in Fig. 12.6.

Although there is some ups and downs, the annual changes in CO_2 reduction figure has been stable, with total reduction amount of 1,273,000 t-CO_2 for nine years.

It can be said that the system has made a significant contribution to environment, which is equivalent to 90,928,500 cedar trees. The details are shown in Table 12.1.

12.7 Domestic Response

Domestically, the system has been appreciated by many administrative agencies and companies. Especially, it is highly evaluated by administrative agencies and new collaborations with the organizations using this system have been emerged, which

Table 12.1 CO_2 reduction figures of recycled parts (2010–2018)

	Unit # of sold parts	CO_2 reduction (t)	Year-on-year
2010	3,430,719	137,419	
2011	3,438,252	140,947	102%
2012	3,554,781	143,345	101%
2013	3,573,453	141,752	99%
2014	3,578,607	140,342	99%
2015	3,512,423	138,308	99%
2016	3,444,228	148,477	107%
2017	3,105,215	137,145	92%
2018	2,784,495	146,104	107%
	30,422,173	1,273,839	

Source Japan Automotive Parts Recyclers Association & JARA Co.

include the partnership in the application of reuse parts for the repairing of city official vehicles (Kumagaya City, Saitama Prefecture).

Also, some companies convert the reduction figures into points for collaboration and donation.

Many member companies selling auto recycled parts have created plates and distributed them to their customers (i.e. repair shops) to enable those customers to represent themselves as “a company contributing to CO_2 reduction” by placing it on the reception counter or desk to draw attention to their activities and improve their image as an environmentally friendly company. In addition, as an enlightenment measure for automobile users, we would like to focus on sharing the environmental contribution of automobile recycling parts with non-life insurance companies.

12.8 Response from Overseas

Since its launch, “Green Point System” has been presented as “a new movement in Japan” at international conferences.

When presented at the International Roundtable on Auto Recycling (IRT) held in Liverpool, England in June 2012, the system attracted considerable interests from automobile recycling bodies and companies in Sweden, United States, Canada, Australia, etc., and joint use of the system has been suggested from Australia. Recently, the interest in the system is growing among Chinese companies. They have expressed the desire to use the Japanese system itself.

In the situation where greenhouse gas reduction activities are attracting global attention, the use of this very effective system to grasp the figures related to automobile recycling is expected to be considered with the introduction of automobile recycling scheme in those countries where such scheme has not been introduced, and the needs for the system will increase further.

12.9 Conclusions and Future Outlook

Since the start of Green Point System initiative, inquiries from interested companies have been increased. Previously it was only image-oriented educational activities. Now, the specific figures issued by this system are cited in CSR or other reports of companies and administrative agencies. These organizations have established a new status by the use of auto recycled parts.

The recycled parts on which the Green Point is attached increase its appeal as a product. There is many more potentials in Green Point System, which include;

- Visualization of reduction of the CO_2 emissions caused by the disposal of each end-of-life car
- Visualization of reduction of the CO_2 emissions caused by the processes such as dismantling and removal of parts in the whole factory
- Creating various donation systems using Green Point numbers
- Survey and report on CO_2 figures of used parts exported.

Another potential in this Green Point System is the coordination with other various point systems, which is expected to be an important mean for enlightening activities targeting general users in the future.

References

Japan Automotive Parts Recyclers Association (2010a) 15 years with "recycled parts", pp 55–56 (in Japanese)

Japan Automotive Parts Recyclers Association (2010b) 15 years with "recycled parts", pp 261–264 (in Japanese)

Japan Automotive Parts Recyclers Association (2010c) 15 years with "recycled parts", p 264 (in Japanese)

Japan Automobile Recycling Promotion Center (2019a) Background of car recycling. https://www.jarc.or.jp/automobile/background/. Accessed 25 Aug 2019

Japan Automobile Recycling Promotion Center (2019b) Problems in recycling used cars. https://www.jarc.or.jp/automobile/background/. Accessed 25 Aug 2019

Nakajima T, Onoda H, Nagata K (2012) Development of provision of environmental information system on the method of E2-PA: take an automotive recycle-parts as an example. In: Matsumoto M, Umeda Y, Masui K, Fukushige S (eds) Design for innovative value towards a sustainable society. Springer, Dordrecht, pp 296–301

Waseda Environmental Institute Co., Ltd. (2011a) Comparison of the manufacturing processes for new parts and used parts

Waseda Environmental Institute Co., Ltd. (2011b) Comparison of the manufacturing processes for new parts and rebuilt parts

Chapter 13
Study of Formalization of Informal Collectors Under a Dual-Channel Reverse Logistics: A Game Theoretic Approach

Juntao Wang and Nozomu Mishima

Abstract Proper and appropriate collection and recycling of Waste electrical and electronic equipment (WEEE) are necessary regarding their value and harmfulness. However, the formal collection channel and appropriate recycling of WEEE are still in their infancy in most developing countries. In contrast, informal collection and recycling activities play a big role in end-of-life treatment of WEEE. Informal recycling activities are always associated with environmental pollution, poor working and living conditions for informal recyclers, while informal collection activities are regarded as helpful and beneficial to waste collection as well as jobs opportunity for the low-income population. Considering these, this study is conducted to analyze the possibility of incorporating informal collectors into formal channel and to provide efficient countermeasures for a better formalization extent. Based on game theory, a dual-channel collection and recycling model is established, in which the formal and informal recycling activities are divided. In this dual-channel model, the informal collector has two choices, either sending WEEE to the formal recycler or to the informal recycler. The authors also introduce the bargaining power of the informal recycler in the model. Considering the formal recycler as the Stackelberg leader, the optimal solutions of the dual-channel model are solved. The results indicate that under a certain condition a higher formalization extent can be achieved as the decreased bargaining power of informal collectors and the decreased additional cost of formal recycling. Accordingly, several countermeasures are proposed to promote the formalization of informal collectors.

Keywords WEEE collection · Informal collection · Formalization · Dual-channel · Bargaining power

---

J. Wang (✉)
Graduate School of International Resource Sciences, Akita University, Akita, Japan
e-mail: smartwjt@163.com

N. Mishima
Faculty of Engineering Science, Akita University, Akita, Japan

Y. Kishita et al. (eds.), *EcoDesign and Sustainability I*, Sustainable Production, Life Cycle Engineering and Management, https://doi.org/10.1007/978-981-15-6779-7_13

13.1 Introduction

Waste electrical and electronic equipment (WEEE) is known as components, subassemblies and consumables that are deemed obsolete from users (Bhuie et al. 2004; Ramzy et al. 2008; Babbitt et al. 2011), it is one of the fastest growing waste streams all over the world (Widmer et al. 2005). According to Duygan and Meylan (2015), the world generates 20–50 million tonnes of WEEE every year with an expected 3–5% increase annually. WEEE contains various materials, including many kinds of metals, plastics and rubber. These materials either have high value or limited quantity in earth on one hand, such as gold, silver, copper and palladium (Oguchi et al. 2011; Xue et al. 2012). On the other hand, these materials can lead to environmental pollution and human safety risk if not well treated (Robinson 2009; Tsydenova and Bengtsson 2011).

Proper and appropriate collection and treatment of these WEEE are necessary and important regarding their value and harmfulness. Even though formal collection channel and advanced recycling methodology have been implemented in several developed countries, the formal collection channel and appropriate recycling methodology are still in their infancy in most developing countries. Specific to China, it is reported that more than 80% of WEEE collected in China still depend on informal private traders such as street peddles (Tang and Wang 2014; Chi et al. 2014). Almost all the informal collectors send the collected WEEE to informal recyclers and the WEEE are recycled there with environmentally polluted methodology. At the meantime, the informal recycling activity is always associated with poor working and living conditions for informal recyclers (Medina 2000; Wilson et al. 2006).

Even though informal recycling activities are deemed as polluting environment and damaging human health, the informal collection activities are always regarded as helpful and beneficial to waste collection. Several studies have pointed out the cost reductions to formal waste management systems and the reduced landfill volume (Wilson et al. 2006, 2012). At the meantime, the informal collection activities can also generate some working jobs for low income population (Linzner and Lange 2013).

Considering these benefits of informal collection activities, one constructive proposal is to incorporate these informal collectors into the formal collection and treatment channel. Instead of expelling informal collectors, incorporating these informal collectors into the formal channel can also benefit the safety and health of informal collectors and workers on the one hand. On the other hand, incorporating informal collectors into the formal channel can minimize the overall cost of collection and treatment, which the cost is much lower than expelling informal collectors through launching strict policy on informal collectors as well as through fierce competition between informal and formal collectors (Campos 2014; Li and Tee 2012). Besiou et al. modelled three different scenarios to study the impact of the waste pickers activities in the WEEE recovery system through system dynamics, and revealed that incorporating the informal into the formal waste recovery system (instead of either ignoring or prohibiting their participation) is beneficial for economical, environmental and social sustainability (Besiou et al. 2012).

To analyze the possibility of incorporating informal collectors into formal channel, reveal underlying influence factors on the formalization process, at the meantime, provide efficient countermeasures for a better formalization extent, this study is undertaken. Based on game theory, a dual-channel collection and recycling activities model is established, in which the formal and informal recycling activities are divided. In this dual-channel model, the informal collector has two choices, either sending WEEE to the formal recycler or to the informal recycler. The authors also introduce the bargaining power of the informal recycler to represent the bargaining power of selling the recycled materials to the market. Considering the formal recycler as the Stackelberg leader in this game model, we discuss the formalization possibility of informal collection activities and analyze the influence on formalization extent in following sections.

This paper is organized as follows: the model assumptions and notations are introduced and presented in Sect. 13.2. The detailed game model and results are described in Sect. 13.3. Finally, Sect. 13.4 comes up with the conclusion.

13.2 Model Assumptions and Notations

The goal of this paper is to evaluate the possibility and extent of prompting informal collector to send WEEE to formal recycler so as to reduce the adverse environmental influences along informal recycling process. In this study, an informal collector, a formal recycler, an informal recycler and a market are modelled. We consider a dual-channel model consisting of formal recyclers and informal recyclers simultaneously. The informal collector can send the WEEE either to formal recyclers or informal recyclers. Both the recycled materials will be sold to the same market to gain revenue (Fig. 13.1).

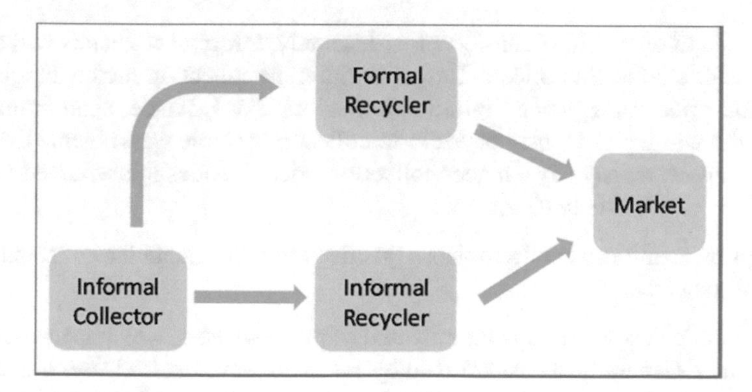

Fig. 13.1 Structure of the considered problem

Table 13.1 Notations

Parameter	Definition
v	Reluctance of sending WEEE to formal recyclers, $0 < v < 1$
β	Bargaining power of the informal recycler, $0 < \beta < 1$
θ	Collection price when collecting the WEEE
c_1	Minimum recycling cost
c_2	Additional cost of the formal recyclers
P	Sell price of recycled materials for the formal recyclers
P_i	Payoff of the informal collector from the informal recycler
P_f	Payoff of the informal collector from the formal recycler
q_i	Recycled amount of WEEE by the informal recycler
q_f	Recycled amount of WEEE by the formal recycler

In the rest of this paper, some assumptions are made for the sake of simplicity.

Assumption 1. Only informal collectors collect WEEE in the market.

In most developing countries, the informal collector is the main player, it is reported that 80% WEEE in China are collected by informal collectors. In some countries, the figure is higher considering the lack of formal treatment system. If these informal collectors can be formalized into the formal treatment channel, formal collection system may be unnecessary. In this study, we mainly focus on the formalization of these informal collectors. And for the sake of simplicity, only informal collectors collecting WEEE in the market is assumed (Table 13.1).

Assumption 2. The informal collector collects a same kind of WEEE with a collection price θ.

Only one kind of WEEE is considered in this study. Informal collectors will collect WEEE with a collection price θ. Based on different collection methodologies, the collection price varies among informal collectors. For instance, apart from labor efforts, the waste pickers may pay zero to collect recyclable waste, while the door-to-door collectors may pay a higher collection price. Besides it is assumed that the collection price is evenly distributed.

Assumption 3. Informal collector has a positive attitude to send the collected waste to formal recyclers.

Informal collectors are always been treated as profit-seekers, which is also adopted in this study. Generally, the payoff paid by informal recyclers is always higher than by formal recyclers. Therefore, under the above condition, the informal collector prefers to send WEEE to the informal recyclers. However, the payoff paid by informal recyclers is highly based on the WEEE quantity and quality, which are uncertain and influenced by the up-front, such as the attitude and behavior of consumers. Hence, the revenue of informal collectors is uncertain and risky if they choose to send WEEE to informal recyclers. In contrast, if the informal collector agreed to be formalized and sends WEEE to formal recyclers, even though the direct payoff paid to each WEEE is less, the risk of revenue can be reduced. For instance, the formal recycler can provide a minimum salary to informal collectors once they choose to be formalized. This undoubtedly can reduce the income risk and uncertainty.

Assume the reluctance of sending WEEE to formal recyclers as v, which v is a discounted rate of the collection price θ. Given the payoff from formal and informal recyclers as P_f and P_i, the utility of informal collectors can be denoted as $U_{f(\theta)} = P_f - v\theta$ for sending WEEE to formal recyclers, and $U_{i(\theta)} = P_i - \theta$ for sending WEEE to informal recyclers.

Undoubtedly, the informal collector will choose to send WEEE to formal recyclers only the utility satisfies $\{\theta | U_{f(\theta)} \geq U_{i(\theta)}, U_{f(\theta)} \geq 0, (P_f - \theta) \geq 0\}$, while choosing to send WEEE to informal recyclers when the utility satisfies $\{\theta | U_{i(\theta)} \geq U_{f(\theta)}, U_{i(\theta)} \geq 0\}$.

Accordingly, when $P_f > \frac{P_i}{2-v}$, the informal collector who offers collection price among $[\frac{P_i - P_f}{1-v}, P_f]$ will choose send WEEE to formal recyclers, while the informal collector who offers collection price among $[0, \frac{P_i - P_f}{1-v}]$ choose informal recyclers as their destination. When $vP_i < P_f \leq \frac{P_i}{2-v}$, and $P_f \leq vP_i$, the informal collector will never consider formal recycler as the destination while the informal collector offers collection price among $[0, \frac{P_i - P_f}{1-v}]$ and $[0, P_i]$ will choose informal recyclers respectively.

Based on these, the amount of WEEE received by formal recyclers and informal recyclers can be summarized as follows:

$$q_f = \begin{cases} \frac{(2-v)P_f - P_i}{1-v} & P_f > \frac{P_i}{2-v} \\ 0 & vP_i < P_f \leq \frac{P_i}{2-v} \\ 0 & P_f \leq vP_i \end{cases}$$

$$q_i = \begin{cases} \frac{(P_i - P_f)}{1-v} & P_f > \frac{P_i}{2-v} \\ \frac{(P_i - P_f)}{1-v} & vP_i < P_f \leq \frac{P_i}{2-v} \\ P_i & P_f \leq vP_i \end{cases}$$

Assumption 4. A low bargaining power is assigned to the informal recycler.

Both the formal recycler and informal recycler will sell the recycled materials to the market to gain profit. The formal recycler is the one who has advanced equipment, they can ensure a higher quality of recycled materials, for instance, pure and with less contaminated materials. Besides, upon the promotion of government on formal recycling, the formal recycler always can gather sufficient cash flow. Contrast to these, the informal recycler is always recognized as backyard working associated with dated technology and small scale. The higher quality of recycled materials and the sufficient cash flow of the formal recycler make the formal recycler a higher bargaining power on selling recycled materials.

Set the sell price of the formal recycler as *P*, the sell price of the informal recycler will be βP given the bargaining power of the informal recycler as β. Therefore, the average benefit of the informal recycler will be $B_i = \beta P - P_i - c_1$. The average benefit of the formal recycler will be $B_f = P - P_f - c_1 - c_2$, where c is the additional cost of the formal recycler associated with environmental treatment and waste management in formal recycling activities.

Combining with the quantity, the profit function of the formal recycler is

$$\pi_f = \begin{cases} \left(P - P_f - c_1 - c_2\right)\frac{(2-v)P_f - P_i}{1-v} & P_f > \frac{P_i}{2-v} \\ 0 & vP_i < P_f \le \frac{P_i}{2-v} \\ 0 & P_f \le vP_i \end{cases}$$

Likewise, the profit function of the informal recycler is

$$\pi_i = \begin{cases} (\beta P - P_i - c_1)\frac{(P_i - P_f)}{1-v} & P_f > \frac{P_i}{2-v} \\ (\beta P - P_i - c_1)\frac{(P_i - P_f)}{1-v} & vP_i < P_f \le \frac{P_i}{2-v} \\ (\beta P - P_i - c_1)P_i & P_f \le vP_i \end{cases}$$

As analyzed previously, no WEEE will be sent to the formal recycler from the informal collector when $vP_i < P_f \le \frac{P_i}{2-v}$, and $P_f \le vP_i$. This means the formalization extent is zero when $P_f \le \frac{P_i}{2-v}$. Since our mainly target is analyzing the formalization extent of informal collection activities, in other words, the amount of WEEE sent to the formal recycler from the informal collector, we will focus on the condition of $P_f > \frac{P_i}{2-v}$ in the following sections.

13.3 Game Theoretic Model

In this section, the quantity change will be analyzed under the optimal price solved through the game theoretic model. The formal recycler Stackelberg game model is considered in following study.

In the formal recycler Stackelberg game model, the formal recycler decides its optimal price P_f first. Then, the informal recycler makes its pricing decision P_i based on the optimal price P_f. Therefore, the model is formulated as follows:

$$\begin{cases} First : \max \pi_f\left(P_f\right) \\ Then : \max \pi_i(P_i) \\ s.t.\ P_f > \frac{P_i}{2-v} \end{cases}$$

Proposition 1: When $0 < v < \frac{3(P-c_1-c_2)-\beta P+c_1}{2(P-c_1-c_2)}$, the optimal price $P_f = \frac{(3-2v)(P-c_1-c_2)+\beta P-c_1}{2(3-2v)}$, and $P_i = \frac{\beta P+P_f-c_1}{2} = \frac{(3-2v)(P-c_1-c_2)+(7-4v)(\beta P-c_1)}{4(3-2v)}$, the received amount of WEEE by the formal recycler and the informal recycler are as follows:

$$q_f = \frac{(3-2v)(P-c_1-c_2)-\beta P+c_1}{4(1-v)}$$

$$q_i = \frac{(5-4v)(\beta P-c_1)-(3-2v)(P-c_1-c_2)}{4(3-2v)(1-v)}.$$

Proof of Proposition 1:

Based on the backward deduction rule, the optimal solutions of the informal recycler are initially analyzed. Given $\frac{\partial \pi_i}{\partial P_i} = 0$, Then, $P_i = \frac{\beta P+P_f-c_1}{2}$.

Substituting P_i in the profit function of the formal recycler, we gain:

$$\max \pi_f(P_f) = (P - P_f - c_1 - c_2)\left[\frac{(3-2v)P_f - \beta P + c_1}{2(1-v)}\right] = f(x)$$

$s.t. P_i = \frac{\beta P+P_f-c_1}{2} < (2-v)P_f$, and set $g(x) = \beta P - c_1 + (2v-3)P_f < 0$.

The Karush-Kuhn-Tucker (KKT) conditions of the above constrained optimization problem is shown as follows:

$$\begin{cases} \nabla f(x) + \lambda \nabla g(x) = 0 \\ \lambda g(x) = 0 \\ \lambda \geq 0 \end{cases}$$

After arrangement,

$$\begin{cases} \frac{(3-2v)(P-c_1-c_2)-2(3-2v)P_f+\beta P-c_1}{2(1-v)} + \lambda(2v-3) = 0 \\ \lambda\left[\beta P - c_1 + (2v-3)P_f\right] = 0 \\ \lambda \geq 0 \end{cases},$$

(1) When $\lambda = 0$,

$$P_f = \frac{(3-2v)(P-c_1-c_2)+\beta P-c_1}{2(3-2v)},$$

and $P_i = \frac{\beta P+P_f-c_1}{2} = \frac{(3-2v)(P-c_1-c_2)+(7-4v)(\beta P-c_1)}{4(3-2v)}$.

Since $g(x) < 0$, substituting P_f with above solution,

$$\beta P - c_1 + (2v - 3)\frac{(3 - 2v)(P - c_1 - c_2) + \beta P - c_1}{2(3 - 2v)} < 0,$$

$$v < \frac{3(P - c_1 - c_2) - \beta P + c_1}{2(P - c_1 - c_2)}$$

When $\frac{3(P-c_1-c_2)-\beta P+c_1}{2(P-c_1-c_2)} \leq 1$, then $0 < v < \frac{3(P-c_1-c_2)-\beta P+c_1}{2(P-c_1-c_2)}$ will be the range of v to satisfy the condition of this case.

(2) When $\lambda \neq 0$,

$P_f = \frac{\beta P-c_1}{3-2v}$, at this point, the informal collector will never send any collected WEEE to the formal recycler. We do not go to the details considering the goal of this study.

In the following study, we will focus on the condition of (1), where $0 < v < \frac{3(P-c_1-c_2)-\beta P+c_1}{2(P-c_1-c_2)}$, under which the following quantity can be derived:

$$q_f = \frac{(3 - 2v)(P - c_1 - c_2) - \beta P + c_1}{4(1 - v)}$$

$$q_i = \frac{(5 - 4v)(\beta P - c_1) - (3 - 2v)(P - c_1 - c_2)}{4(3 - 2v)(1 - v)}$$

Corollary 1. Formalization extent of the informal collector decreases as the increasing bargaining power of the informal collector.

Proof of Corollary 1.

Based on the above q_f, q_i, we obtain $\frac{\partial q_f}{\partial \beta} < 0$, $\frac{\partial q_i}{\partial \beta} > 0$.

The formal recycler will receive a decreased amount of WEEE as the increasing bargaining power of the informal recycler, while more WEEE will be sent to the informal recyclers. The bargaining power of informal recyclers enlarges the sell price of materials recycled by the informal recycler, resulting in an increasing profit of the informal recycler. The informal recycler will increase collection effort to collect more WEEE from informal collectors. According to P_f, P_i, we obtain $\frac{\partial P_f}{\partial \beta} = \frac{P}{2(3-2v)} > 0$, $\frac{\partial P_i}{\partial \beta} = \frac{(7-4v)p}{4(3-2v)} > 0$, and $\frac{\partial P_i}{\partial \beta} > \frac{\partial P_f}{\partial \beta}$. The payoff from the informal recycler will increase more rapidly than the payoff paid by the formal recycler in the optimal condition, which will make more informal collectors choose the informal recycler as their destination.

In order to prompt the formalization of informal collection activities, it is important to restrict the bargaining power of the informal recycler. One possible countermeasure is to set standard on the purchased materials. Instead of standard related to quality of materials, establishing standard on the source of recycled materials. The government can trace and supervise the recycled materials in the market side. Different from regulating the informal recycler directly, the management from the market side may be more easily to achieve for the government.

Corollary 2. Additional cost hinders the formalization extent.

Proof of Corollary 2.

It is easily to derive that $\frac{\partial q_f}{\partial c_2} = \frac{-(3-2v)}{4(1-v)} < 0$, $\frac{\partial q_i}{\partial c_2} = \frac{1}{4(1-v)} > 0$.

Meanwhile, $\frac{\partial P_f}{\partial c_2} = -\frac{1}{2} < 0$, $\frac{\partial P_i}{\partial c_2} = -\frac{1}{4} < 0$ the payoff from the formal recycler decreases more rapidly than from the informal recycler as the increasing additional cost for formal recyclers and increases as the increasing sell price. The informal collector is profit-seeker, they will choose to send the collected WEEE to informal recyclers rather formal recyclers encountering with such payoff change.

The decreasing additional cost is necessary for the formalization extent. The additional cost only consists of environmental protection cost and waste treatment cost. From the technology point of view, more advanced cost-efficient technology should be invented to optimize the cost spared to environmental protection. Besides the technology, the share of the cost will be one possible solution. The government can undertake the waste treatment cost for the formal recyclers directly or share the cost through subsidy or tax reduction. It is believed that the reduction of additional cost can be achieved through above solutions, which will result in a better formalization extent of informal collection activities.

Corollary 3. Improving the reluctance of sending WEEE to formal recyclers will enlarge the payoff cost of both the formal recycler and informal recycler. And the change of payoff cost from the formal recycler is as twice as the one from the informal recycler.

Proof of Corollary 3.

It can be derived that $\frac{\partial P_f}{\partial v} = \frac{\beta P - c_1}{(3-2v)^2} > 0$, $\frac{\partial P_i}{\partial v} = \frac{\beta P - c_1}{2(3-2v)^2} > 0$, and $\frac{\partial P_f}{\partial v} = 2\frac{\partial P_i}{\partial v}$.

Meanwhile, it is unclear whether $\frac{\partial q_f}{\partial v}$ or $\frac{\partial q_i}{\partial v}$ is bigger than zero or not $\left(\frac{\partial q_f}{\partial v} = \frac{(P-c_1-c_2)-(\beta P-c_1)}{4(1-v)^2}, \frac{\partial q_i}{\partial v} = \frac{4(14v^2-35v+22)(\beta P-c_1)-(3-2v)(P-c_1-c_2)}{4(3-2v)^2(1-v)^2}\right)$.

The formal and informal recyclers compete with each other for the WEEE. To gather more WEEE from the consumer directly, more money should be paid to consumers. The one who offers a higher payoff will succeed in gathering more WEEE. Such competition will be a vicious circle, which is bad for both the formal and informal recyclers from their economy point of view. This leads to an increased cost for both the formal and informal recyclers, and their profit will be squeezed at last given the constant revenue from the market. Therefore, such vicious circle will damage the sustainable development of reverse logistics. It is noteworthy that the change of the overall recycled amount of WEEE by either the formal or informal recyclers is not clear.

13.4 Conclusion

Informal collection and recycling of WEEE are prevailing in many developing countries. The informal collection and recycling activities are particularly serious in China.

Informal collection activities can help waste collection and provide jobs to the low-income population. However, if not recycled properly, the WEEE will undoubtedly generate serious environmental pollution and damage the health of human beings. From these points of view, one constructive proposal is to incorporate these informal collectors into the formal collection and treatment channel while expelling informal recyclers. To analyze the possibility of incorporating informal collectors into formal channel, reveal the influence on the formalization process, at the meantime, provide efficient countermeasures for a better formalization extent, a dual-channel collection and recycling activities model is established in this study. Considering the formal recycler as the Stackelberg leader, the optimal solutions are solved and the concerned problems are studied.

Under a certain condition, it is indicated that the bargaining power of the informal collector has a negative influence on formalization extent of the informal collector. Besides, it is concluded that the additional cost hinders the formalization extent. Moreover, the results reveal that improving the reluctance of sending WEEE to formal recyclers will enlarge the payoff cost of both the formal recycler and informal recycler. However, it is unclear whether the overall amount of WEEE recycled by the formal recycler increases or not.

Accordingly, this study proposed several countermeasures to promote the formalization of informal collectors. On one hand, developing and adopting advanced technology related to formal recycling in order to decrease the additional cost of the formal recycler comparing to informal recyclers. On the other hand, perfecting management system mainly from two perspectives, one is regulating the sell source of recycled materials from the market side so as to decrease the bargaining power of informal recyclers, while the other is providing subsidy or establishing a cost-sharing system to reduce the recycling cost of formal recyclers.

References

Babbitt CW, Williams E, Kahhat R (2011) Institutional disposition and management of end-of-life electronics. Environ Sci Technol 45:5366–5372

Besiou M, Georgiadis P, Wassenhove LNV (2012) Official recycling and scavengers: symbiotic or conflicting? Eur J Oper Res 218:563–576

Bhuie AK, Ogunseitan O, Saphores J, Shapiro A (2004) Environmental and economic trade-offs in consumer electronic products recycling: a case study of cell phones and computers. In: Proceedings of the 2004 IEEE international symposium on electronics and the environment. IEEE, pp 74–79

Campos HKT (2014) Recycling in Brazil: challenges and prospects. Resour Conserv Recycl 85:130–138

Chi X, Wang M, Reuter MA (2014) E-waste collection channels and household recycling behaviors in Taizhou of China. J Clean Prod 80:87–95

Duygan M, Meylan G (2015) Strategic management of WEEE in Switzerland—combining material flow analysis with structural analysis. Resour Conserv Recycl 103:98–109

Linzner R, Lange U (2013) Role and size of informal sector in waste management—a review. Waste Resour Manag ICE Proc 166:69–83

Li RC, Tee TJC (2012) A reverse logistics model for recovery options of e-waste considering the integration of the formal and informal waste sectors. Procedia Soc Behav Sci 40:788–816
Medina M (2000) Scavenger cooperatives in Asia and Latin America. Resour Conserv Recycl 31:51–89
Oguchi M, Murakami S, Sakanakura H, Kida A, Kameya T (2011) A preliminary categorization of end-of-life electrical and electronic equipment as secondary metal resources. Waste Manag 31:2150–2160
Ramzy K, Junbeum K, Xu M, Allenby B, Williams E, Zhang P (2008) Exploring e-waste management systems in the United States. Resour Conserv Recycl 52:955–964
Robinson BH (2009) E-waste: an assessment of global production and environmental impacts. Sci Total Environ 408:183–191
Tang AJ, Wang L (2014) Discussion on the development and trend of dismantling industry of waste electric and electronic products. Renew Resour Circ Econ 35:21–24
Tsydenova O, Bengtsson M (2011) Chemical hazards associated with treatment of waste electrical and electronic equipment. Waste Manag 31:45–58
Widmer R, Krapf HO, Khetriwal DS, Schnellmann M, Boni H (2005) Global perspectives on e-waste. Environ Impact Assess Rev 25:436–458
Wilson D, Velis C, Cheeseman C (2006) Role of informal sector recycling in waste management in developing countries. Habitat Int 30:797–808
Wilson D, Rodic L, Scheinberg A, Velis C, Alabaster G (2012) Comparative analysis of solid waste management in 20 cities. Waste Manage Res 30:237–254
Xue M, Yang Y, Ruan J, Xu Z (2012) Assessment of noise and heavy metals (Cr, Cu, Cd, Pb) in the ambience of the production line for recycling waste printed circuit boards. Environ Sci Technol 46:494–499

Chapter 14
Ecodesign and the Circular Economy: Conflicting Policies in Europe

Carl Dalhammar, Leonidas Milios, and Jessika Luth Richter

Abstract The number of policies that address the various life cycle environmental impacts of products are increasing, especially in Europe. With the rise of the Circular Economy (CE) concept, the traditional product policies have been supported by new ones that pursue new policy objectives such as increasing product lifetimes and encourage more repairs. In this contribution we discuss principles for improving synergies and reduce conflicts among different product policies. We then outline some of the recent conflicts between policies that have emerged in the context of the CE. We conclude that often the conflicts can be mitigated, or that it is possible to reach a compromise. However, we need to accept that in many cases there is no solution that will satisfy all involved stakeholders.

Keywords Circular economy · Policy mix · Product policy · Ecodesign

14.1 Introduction

The number of public policies that address the life cycle impacts of products are increasing (Faure and Dalhammar 2018; Milios 2018). Thus far, these policies have addressed a number of environmental aspects including: (1) chemicals in products (2) collection and recycling of used products and (3) energy efficiency of products. New policies are emerging to address issues such as product life span/durability (Maitre-Ekern and Dalhammar 2016), or to incentivize product repairs (Svensson et al. 2018). The Circular Economy (CE) concept has increased the interest for new types of policies.

Policies can be both compulsory and voluntary, and be adopted both at the EU level and Member State level. Table 14.1 outlines examples of the various policies at the EU and Member State levels.

New policies are also being proposed or under discussion, such as right-to-repair (R2R) laws and labelling schemes that provide consumers with information

C. Dalhammar (✉) · L. Milios · J. L. Richter
IIIEE, Lund University, Lund, Sweden
e-mail: carl.dalhammar@iiiee.lu.se

Y. Kishita et al. (eds.), *EcoDesign and Sustainability I*, Sustainable Production, Life Cycle Engineering and Management, https://doi.org/10.1007/978-981-15-6779-7_14

Table 14.1 Public policies that address the life cycle impacts of products in the EU and Member States

Environmental aspect	Examples of European Union laws and policies	Examples of Member State policies
Chemical and material content	Horizontal legislation (e.g. REACH) Rules related to conflict minerals Sector oriented laws on chemical restrictions (e.g. packaging, electronics, toys)	Green public procurement (GPP) criteria for e.g., chemicals and conflict minerals, or procuring bio based products Eco-labels Taxes on chemicals
Collection and recycling of waste products	General rules and guidelines (e.g., Waste Framework Directive) Sector oriented EPR laws (e.g. WEEE Directive; Waste and Packaging Waste Directive)	Waste related taxes Infrastructure for re-use & recycling Re-use parks/shops for re-used products/repair activities Mandatory re-use obligations for white goods (Spain)
Energy efficiency	Mandatory energy performance standards (MEPS) (set under the Ecodesign Directive) Mandatory energy labelling (set under the Energy Labelling Directive) Energy performance requirements for buildings Voluntary labelling (Energy Star)	Eco-labels Green public procurement criteria The use of life cycle costing (LCC) GPP Promoting energy efficient products through public procurement LEED and other certification schemes for buildings
Durability, lifetime and reparability	***Direct incentives***: Mandatory lifetime requirements set under the Ecodesign Directive for vacuum cleaners and lighting products Ecodesign Directive standards supporting disassembly and repairs (several product groups) Proposal: providing information about expected lifetime to consumers through mandatory information scheme Voluntary eco-design agreement, imaging equipment (e.g. duplex printing standard) ***Indirect incentives***: Minimum rules on consumer guarantees	***Direct incentives***: Banning planned obsolescence (France) ***Indirect incentives***: Incentivizing the provision of spare parts (France) National rules on longer consumer guarantees and/or changed rules for burden of proof is transferred from seller to consumer (several EU Member States) Lower VAT on repair services (Sweden) Public procurement of remanufactured furniture and computers (Sweden)

Source Amended version of the table found in Faure and Dalhammar (2018)

on product life span and durability. Further, the CE has triggered new criteria in established policy instruments; for instance, we can expect more criteria related to resource efficiency and product durability in eco-labelling and green public procurement (GPP) in the future. While this development can be considered positive from a sustainability perspective, the sheer complexity of the policy mix means that we need to carefully consider the interaction of policies. There is an increasing recognition that some policies may be in conflict with each other, and in literature there are an increasing number of examples of such conflicts (Technopolis 2016; Tojo and Thidell 2018). Thus, issues related to 'policy coordination', 'policy integration' and 'policy harmonization' (these concepts are discussed later in this paper) are of increasing importance. The literature related to these concepts—and related practices—are quite scattered among disciplines, and includes research on 'environmental policy integration' in political science, and 'legal harmonization' in legal research. The aim of this paper is to analyze what policy coordination and harmonization means in the context of the environmental product policy mix, and discuss examples of conflicts emerging related to new CE policies that aims to support resource efficiency.

This topic is of great importance for ecodesign as these policies influence design practices both directly and indirectly (cf. Table 14.1). A direct influence occurs when designers must adhere to a law or policy and thus integrates it into design considerations (e.g. rules on chemical content). An indirect influence is when a law does not directly affect the product design per se, but may nevertheless have an influence over design choices. An example of this would be the French criminalization of planned obsolescence.

The main methodology applied for this contribution has been literature reviews related to product policy, the policy mix, and the CE. Specific keywords where used when searching for literature, such as 'ecodesign' Circular Economy', and 'product policy'. The main focus have been on European policies.

In the next section we will discuss the theory and framework and how we use various concepts such as 'policy integration'. In Sect. 14.3 we outline a number of examples of policy conflicts described in literature, and—when applicable—various solutions proposed to these conflicts. Section 14.4 contains a concluding discussion.

14.2 Theory and Framework

14.2.1 The Circular Economy (CE) and Resource Efficiency Policies

The Circular Economy (CE) is meant to achieve a gradual decoupling of economic growth from the consumption of finite resources, and lead to designing waste out of the systems of production and consumption (Murray et al. 2017). Planing (2015) states that 4 building blocks are required for the necessary transition to a CE: transition to a CE: (a) materials and product design; (b) new business models; (c) *global reverse networks; and (d) enabling conditions (i.e. policies and infrastructure)* (Planing 2015). *Moving towards the CE without policy interventions will not be possible. It requires both changes in existing policies (e.g. existing rules on recycling, eco-labels* etc.*) as well as new policies* (Milios 2018; European Commission 2015). *Further, these policy interventions should take place at several levels: local, regional and national authorities are encouraged to support the transition, while the EU has the main responsibility for developing the right regulatory framework for the development of Circular Economy in the EU single market.*

14.2.2 Policies and Laws

In this paper we are primarily interested in 'policy instruments', i.e. interventions made by governments/public authorities at various levels (including the EU level as well as national/regional/local levels) which are intended to achieve outcomes which conform to the objectives of public policy. Different authors use different classifications of policy instruments, but all classifications tend to at least include the categories (a) administrative (regulatory or mandatory); (b) economic (market-based), and (c) informative instruments.

Laws are typically used to implement these policy instruments. In some cases, laws influence how policy instruments can be applied. An example concerns EU rules on public procurement: these do not regulate which sustainability criteria EU member states may use in GPP, but sets the framework for all procurement practice, which can also influence GPP practices and criteria. For instance, member states should not apply GPP criteria that favors domestic manufacturers over manufacturers in other EU member states.

14.2.3 Policy Coordination and Harmonization, and Related Terminology

With a growing number of policies, it is important to coordinate between policies. A first question concerns how we use terms like 'policy coordination' and 'policy harmonization', as these concepts tend to be used in different ways. Table 14.2 outlines how we define these concepts, as well as some additional ones, for the purposes of this paper.

An important aim is to design a policy mix where policies are integrated and/or harmonized, and when this is not possible, at least coordinated. Faure and Dalhammar have discussed some key principles for design of a policy mix (Faure and Dalhammar 2018), which include the identification and management of trade-offs between different policy goals, and using realistic assumptions when choosing the appropriate policy instruments.

Table 14.2 Definition of concepts

Legal/policy harmonization	The process of creating common standards among countries or regions Harmonization aims to (1) create consistency of laws, regulations, standards and practices, so that the same rules will apply to businesses; and (2) reduce compliance and regulatory burdens for businesses operating nationally or transnationally
Positive harmonization (in the EU)	National laws and policies are replaced by common EU rules and policies
Negative harmonization, e.g. in the EU and related to the WTO agreements	National laws and policies are not allowed as they are in conflict with EU rules or international agreements
Coordination of laws and policies	While harmonization aims to create common standards, coordination rather accepts that there are different rules or policies, and instead aim to make these interact as smoothly as possible
Policy integration	This concept has many different understandings. We define it as the process of identifying and addressing synergies and trade-offs between various public policies
Regulatory overlap	Different laws regulate the same issues (e.g. the same type of product or the same life cycle phase in the product life cycle); it can also be that different legal fields (e.g. environmental law vs. corporate law) regulate the same issue. The laws can be in conflict or contradict each other, or work in synergy and complement each other
Conflicting objectives	Different laws and policies have conflicting aims
Conflicting rules and procedures	The aims of different laws and policies may not conflict, but the rules will conflict in practice. For instance, while one law may promote recycling, other rules on levels of chemicals in products may mean that producers do not want to make use of recycled materials in their products as they fear being in breach on these rules
Sequential issues	The sequencing may matter and influence whether rules are complementary or conflicting. For instance, rules that restrict chemicals in products will improve future recycling, as it improves both recycling processes and the health and safety of the recycling environment. However, in the short term stringent rules on chemicals may mean that producers do not want to make use of recycled materials in their products as they fear being in breach on these rules; this undermines the economic case for recycling

The table is based on UNDESA (2015), Dalhammar (2007), Faure (2000), Aagaard (2011)

14.2.4 Improving Synergies Between Different Product Policies

Coordination should strive to find synergies between policies. However, there are numerous dimensions to what 'coordination' and 'synergy' really mean. Therefore, in Table 14.3 we outline examples of conceptual approaches.

Thus, synergies can be related to e.g. policies covering different like cycle phases, or addressing different actors, or providing different types of incentives (Faure and Dalhammar 2018). Policies can be in synergy in relation to quite different objectives, e.g. covering the full product life cycle, or be combined to together provide a comprehensive policy package to achieve specific policy outcomes (e.g. more repairs). An increasingly important issue in the EU is to ensure that EU policies are coordinated with EU member states' national policies.

In reality, we can never expect perfect coordination, as policies are often adopted ad hoc, when there is strong political momentum for addressing a specific issue. Further, policymakers will usually not be able to foresee all potential interactions with other policies when a specific policy is adopted.

We will now discuss some emerging policy conflicts.

14.3 Examples of Conflicts and Potential Solutions in Relation to Circular Economy Policies

14.3.1 Introduction

While there are many relevant conflicting or potentially conflicting policies, here we will focus on some conflicts that seem to be especially important emerging issues in the context of the Circular Economy. We will first discuss the clash between policies that promote recycling and policies that address chemicals in products and materials. Then we will review the potential conflicts between different sustainability criteria applied in GPP, which are relevant in the context of the CE. As a third example, we exemplify how prolonging the product lifetime can lead to conflicts between resource efficiency aspects and energy-related environmental impacts.

14.3.2 Recycling and Rules Related to Chemicals

Toxic-free resource flows is one of the prerequisites for a truly Circular Economy, and thus rules on chemicals in products and materials are beneficial for desired resource efficiency strategies such as remanufacturing and recycling in the longer run. However, in the short run, we see some conflicts emerging. As a first example,

Table 14.3 Conceptual approaches to policy coordination

Policy coordination to improve synergy	Explanation and examples
Coordination for addressing different environmental aspects, different life cycle phases, different actors	Typically, different policies influence different environmental aspects (e.g. product chemical content or product energy efficiency). But policies can also address various actors in different life cycle stages (e.g. producers and consumers)
Combining mandatory and voluntary policies	Often we need to combine mandatory and voluntary policies, for several reasons, including: – Some environmental and social aspects are difficult to address through legal requirements set in the EU, as EU laws only have jurisdiction in Europe. Then we need to make use of voluntary approaches such as GPP – The 'laggards' on the market tends to be mainly affected by mandatory requirements (e.g. the Ecodesign Directive), whereas the 'leaders' who are going beyond legal compliance are more influenced by other policies (e.g. eco-labels, consumer subsidies, or GPP)
Horizontal versus vertical policies and laws	Horizontal rules (which cover several product groups) are supplemented by vertical laws covering specific industrial sectors or product groups. For instance, it may be possible to regulate chemical content for a large number of product groups (e.g. the RoHS Directive), whereas rules on product energy efficiency and energy labeling are best addressed on a product group basis
European Union and Member State policies	Certain standards, e.g. rules on eco-design, are set in Brussels and applies in the whole of the EU. Other policies can be adopted at the member state level or by regions and municipalities, including consumer subsidies, labeling schemes and GPP
Combining policies in a well-designed policy mix	Policies of different kinds, adopted at different levels, can aim at the realization of a specific objectives. For instance, if we want to promote product repairs we can make use of: – Ecodesign rules that force manufacturers to make products easy to disassemble, repair and reassemble – Force manufacturers to provide spare parts and repair tools at reasonable cost – Reduce VAT for the repair sector – Provide information for consumers about the environmental impact of electronics and environmental gains from repairs – Support to do-it-yourself activities like repair cafés

recycling practices and targets may be compromised due to rules on chemicals. For example, many companies are hesitant to use recycled materials in new products, out of fear that these products will not comply with rules on chemicals in products, e.g. the rules found in the RoHS Directive, the Toy Safety Directive, and the REACH Regulation (Tojo and Thidell 2018). This means that there is not sufficient 'pull' on the market of recyclates (for use in new production) that would result in increasing the recycling output—and thus the recycling input as well, with collected waste often redirected to other treatment options (Milios et al. 2018).

More recently, proposed rules on allowed levels of persistent organic pollutants (POPs)—and especially decaBDE—in recycled plastics may imply that it will not be economically viable to recycle such plastics from electronics and vehicles, as the cost for compliance is too high. One proposed level of allowed decaBDE—10 ppm—would probably not be possible to guarantee (EuRIC 2018). This means that the proposed rules on POPs are in direct conflict with EU's vision that by 2030, more than half of plastics waste generated in Europe will be recycled (European Commission 2018).

Can such conflict be resolved? Clearly, there is no perfect solution. Stringent rules on chemicals in products and recycled materials will compromise recycling targets; whereas striving for high recycling targets means that higher levels of chemicals than desired from a health/safety perspective must be allowed. Yet, some compromises are possible (Sevelius 2019). For instance, one solution would be to apply strict levels of POPs in recycled materials to be used in toys, kitchen utensils, and other applications where the users may be vulnerable (e.g. children) or where the risk of contamination is high. In contrast, it may be less risky to allow higher levels of POPs in applications where this risk is low (e.g. heavy machinery).

14.3.3 Conflicting Criteria in GPP

GPP has been practiced for a couple of decades in environmental policy front-runner countries (Jänicke 2005; Dalhammar and Mundaca 2012). During that time period, the number of sustainability criteria applied have increased to cover an increasing number of sustainability aspects, including chemical content, energy efficiency requirements, recyclability, and raw materials. The CE has led to an increasing interest for addressing various raw material and resource-related aspects, and to apply GPP criteria that relate to inter alia product lifetime, rare earth elements, and bio based materials. Examples of concrete practices in Sweden include the procurement of bio based (as opposed to previously fossil based) healthcare products (Leire and Dalhammar 2018), and procurement of remanufactured ICT products (Crafoord et al. 2018) and remanufactured furniture (Öhgren et al. 2019).

Here, we will exemplify some trade-offs using the case of remanufactured furniture, building on a study by Öhgren (2017).

Design furniture typically is durable, and it can often be upgraded, repaired, refurbished, washed, and reconditioned to a 'like new' condition. Buying reconditioned furniture can be a good deal for the public sector as the price is much lower than new furniture, and procuring reconditioned furniture can also save a lot of resources compared to procuring new ones. However, procuring reconditioned furniture has proven to be far from straightforward in practice. A first problem is that reconditioned furniture may be good for the environment, but purchasing them may mean that other GPP criteria may have to be compromised. And why is that? It is because the public sector has a lot of criteria that relate to e.g. chemical content and sustainable sourcing of raw materials (e.g. FSC certified wood). These requirements can be put on new furniture as the relevant information to prove compliance with such criteria is usually available. This is typically not the case for older, reconditioned furniture and thus purchasing such furniture requires that other criteria are compromised. An alternative solution is to design a procurement process that sets two contracts: one for new furniture and one for remanufactured furniture. This is however administratively complex, and thus shows the importance of flexibility and innovation in procurement processes to allow innovative solutions. The rather rigid rules on public procurement may be a barrier for such innovative thinking, but the front-runner agencies have shown that it is possible.

14.3.4 Promoting Product Durability: Conflicting Environmental Aspects

Increasing product durability is one way to save resources: if a product has a longer lifetime, fewer new products need to be produced. However, this resource saving may be compromised by other environmental impacts that may increase instead (Technopolis 2016).

One issue that has come under increasing scrutiny concerns the fact that if a product is kept in use longer, and we can assume that a newer product is more energy-efficient, then making it durable implies that the resource savings come at the expense of energy savings (Bakker et al. 2014). However, whether this is the case in reality relies on several factors. Here we will build on a recent study of LED lamps (Richter et al. 2019) to exemplify some of the issues.

Between 2012 and 2017 there was a rapid development of LED lamps to serve as replacements for incandescent bulbs which were phased out in several markets, including the EU (Bennich et al. 2015). LED lamps offered several advantages: improved energy efficiency, mercury-free, and long lifetimes (up to 50,000 h). A life cycle assessment study (Richter et al. 2019) found that the long lifetimes of LED lamps could result in trade-offs between climate and energy-related impacts and material and toxicity-related impacts when considering that shorter life products allow replacement products with improved energy efficiency and lighter material designs. The study however noted that not all products on the market were in fact more energy-efficient, making the role of minimum performance standards also salient. Moreover, the trade-offs were minimized when considering decarbonized electricity mixes, like those in Norway and (to a lesser extent) Sweden. In these contexts, durability is not a trade-off with energy efficiency to the same extent as in countries with a higher share of fossil fuels in the energy mix. A further important factor concerns the rate of technology change and the maturity of the technology: for technology that is quickly becoming more energy-efficient it makes more sense to replace a product for a new one. Once a technology becomes mature, which means that newer products will not be significantly more energy-efficient, it makes sense to go for long-lasting products as the resource savings do not come at the expense of energy savings. The implication for policy is that once technologies become mature, it may be a good idea to enforce durability through policy instruments such as eco-design requirements.

14.4 Concluding Discussion

The examples explored in this paper highlight the need for more sophisticated policy mixes that take into account emerging conflicts and trade-offs. In the case of POPs in recycled plastic, there is a need to separate the horizontally applied regulations between low-risk products that can make use of the recycled plastic and high-risk products where use of secondary materials should be avoided.

The case of GPP for reconditioned furniture illustrated the trade-offs between use of different sustainability criteria and that weighing the criteria against each other may be necessary, in order to decide what to prioritizes. Further, there is reason to introduce more flexibility in the GPP process and criteria-setting, allowing procuring parties to consider the overall lifecycle impacts of new products versus remanufactured in a synergetic way.

The case of the LED lamps illustrated that there can be trade-offs between environmental impacts which have implications for policies promoting durability. However, the case also highlights the need to consider the influence of both energy policies and local contexts, which may require some member states to supplement EU laws with their own requirements or policies that push for more ambition when that is desirable in the national context.

The various policies can affect eco-design practices directly and indirectly. The example of LED lamps illustrates that we need to consider a number of factors—i.e. LCA studies, expected technological developments, and the costs of new technology—when we consider adopting a mandatory durability requirement at the EU level. However, the experienced trade-offs are often temporary and will most likely diminish over time. For LED lamps, we can expect a situation in the near future when the technology cannot become more energy efficient (when we have 1 W or 2 W lamps), and the costs of these lamps will also go down. Then, adopting mandatory durability requirements will not lead to trade-offs.

References

Aagaard TS (2011) Regulatory overlap, overlapping legal fields, and statutory discontinuities. Virginia Environ Law J 29(3):237–303

Bakker C, Wang F, Huisman J, den Hollander M (2014) Products that go round: exploring product life extension through design. J Clean Prod 69:10–16

Bennich P, Soenen B, Scholand M, Borg N (2015) Updated test report—clear, non-directional LED lamps. https://www.energimyndigheten.se/Global/F%C3%B6retag/Ekodesign/Produktgrupper/Belysning/Report%20on%20Testing%20of%20Clear%20LED%20lamps%20v5%205.pdf. Accessed 14 June 2019

Crafoord K, Dalhammar C, Milios L (2018) The use of public procurement to incentivize longer lifetime and remanufacturing of computers. Proc CIRP 73:137–141

Dalhammar C (2007) An emerging product approach in environmental law: incorporating the life cycle perspective. Dissertation, Lund University, Lund

Dalhammar C, Mundaca L (2012) Environmental leadership through public procurement? The Swedish experience. In: Rigling Gallagher D (ed) Environmental leadership: a reference handbook. Sage Publications, pp 737–745

EuRIC (2018) Position of the European Recycling Industries' Confederation (EuRIC) on the Recast of the Persistent Organic Pollutants (POPs) Regulation (EC 850/2004). Retrieved https://www.euric-aisbl.eu/images/PDF/EuRIC_Recast_POP-Regulation_Position_Fin_Jul.2018.pdf. Accessed 11 June 2019

European Commission (2015) Communication of 2 December 2015 on closing the loop—an EU action plan for the circular economy, COM (2015) 614 final

European Commission (2018) Communication of 16 January 2018 on a European strategy for plastics in a circular economy, COM (2018) 28 final

Faure M (2000) The harmonization, codification and integration of environmental law: a search for definitions. Eur Environ Law Rev 9(6):174–182

Faure M, Dalhammar C (2018) Principles for the design of a policy framework to address product life-cycle impacts. In: Maitre-Ekern E, Dalhammar C, Bugge HC (eds) Preventing environmental damage from products—analyses of the policy and regulatory framework in Europe. Cambridge University Press, Cambridge, pp 57–86

Jänicke M (2005) Trend-setters in environmental policy: the character and role of pioneer countries. Eur Environ 15:129–142

Leire C, Dalhammar C (2018) Long term market effects of green public procurement. In: Maitre-Ekern E, Dalhammar C, Bugge HC (eds) Preventing environmental damage from products—analyses of the policy and regulatory framework in Europe. Cambridge University Press, Cambridge, pp 303–334

Maitre-Ekern E, Dalhammar C (2016) Regulating planned obsolescence: a review of legal approaches to increase product durability and reparability in Europe. Rev Eur Comp Int Environ Law (RECIEL) 25(3):378–394

Milios L (2018) Advancing to a circular economy: three essential ingredients for a comprehensive policy mix. Sustain Sci 13(3):861–878

Milios L, Christensen LH, McKinnon D, Christensen C, Rasch MK, Eriksen MH (2018) Plastic recycling in the Nordics: a value chain market analysis. Waste Manage 76:180–189

Murray A, Skene K, Haynes K (2017) The circular economy: an interdisciplinary exploration of the concept and application in a global context. J Bus Ethics 140(3):369–380

Öhgren M (2017) Upphandling av rekonditionerade kontorsmöbler - En strategi för att stärka utvecklingen mot en cirkulär ekonomi. Thesis, Lund University, Lund

Öhgren M, Milios L, Dalhammar C, Lindahl M (2019) Public procurement of reconditioned furniture and the potential transition to product service systems solutions. Proc CIRP 83:151–156

Planing P (2015) Business model innovation in a circular economy reasons for non-acceptance of circular business models. Open J Bus Model Innov

Richter JL, Tähkämö L, Dalhammar C (2019) Trade-offs with longer life-times? The case of LED lamps considering product development and energy contexts. J Clean Prod 226:195–209

Saublens C (n.d.) Policy instruments. Retrieved http://www.know-hub.eu/knowledge-base/videos/policy-instruments.html. Accessed 13 June 2019

Sevelius E (2019) Ökad plaståtervinning vs giftfria materialflöden En analys av ändringen av POPs-förordningen. Thesis, Lund University, Lund

Svensson S, Richter JL, Maitre-Ekern E, Pihlajarinne T, Maigret A, Dalhammar C (2018) The emerging 'Right to repair' legislation in the EU and the US. In: Proceedings of Going Green Care Innovation, Vienna, 26–28 Nov 2018

Technopolis et al (2016) Regulatory barriers for the circular economy. Retrieved http://www.technopolis-group.com/wp-content/uploads/2017/03/2288-160713-Regulary-barriers-for-the-circular-economy_accepted_HIres.pdf. Accessed 10 June 2019

Tojo N, Thidell Å (2018) Material recycling without hazardous substances: interplay of two policy streams and impacts on industry. In: Maitre-Ekern E, Dalhammar C, Bugge HC (eds) Preventing environmental damage from products—analyses of the policy and regulatory framework in Europe. Cambridge University Press, Cambridge, pp 253–275

UNDESA (2015) Policy integration in government in pursuit of the sustainable development goals. Report of the expert group meeting held on 28 and 29 January 2015 at United Nations Headquarters, New York. http://workspace.unpan.org/sites/Internet/Documents/UNPAN94443.pdf. Accessed 14 June 2019

Chapter 15
Sustainable Supply Chain Management of Clothing Industry—Current Policy Landscape and Roles and Limitation of Multi-stakeholder Initiatives

Dominika Machek, Caroline Heinz, and Naoko Tojo

Abstract High environmental and social impacts arisen from long and complicated global supply chain of clothing industry has been long recognized. However, a thorough review of the sustainable supply chain management (SSCM) policy development pertaining to clothing industry at the EU level indicates that legislative measures specifically addressing SSCM of clothing industry is currently lacking. Multi-stakeholder initiatives (MSIs) emerged as a non-legislative governance measure engaging various stakeholders, and its role to fill in the governance gap has been highlighted in, among others, the global supply chains of clothing industry. Interviews with nine Swedish brand representatives participating in three selected MSIs—Sustainable Apparel Coalition, Sweden Textile Water Initiative and Textile Exchange—on one hand elucidate tangible contributions of the three case MSIs to the brands' SSCM measures. We meanwhile also observe the trend of various MSIs to consolidate their efforts and harmonise their activities, and that the discourse within the MSIs tend to be dominated by a few large brands. The outcome of such discourse, such as standards which in light of current lack of legislative measures could serve as default global standards, may become a suboptimal compromise from sustainability standpoint. The paper closes with potential future way forward to enhance sustainability of clothing supply chain.

Keywords Clothing industry · Textiles · Sustainable supply chain management (SSCM) · Government interventions · Multi-stakeholder initiative (MSI) · Privatisation of standards · EU · Sweden · Brand's perception

D. Machek
CSR Västsverige, Göteborg, Sweden

C. Heinz
Sofies, London, United Kingdom

N. Tojo (✉)
International Institute for Industrial Environmental Economics at Lund University, Lund, Sweden
e-mail: naoko.tojo@iiiee.lu.se

Y. Kishita et al. (eds.), *EcoDesign and Sustainability I*, Sustainable Production, Life Cycle Engineering and Management, https://doi.org/10.1007/978-981-15-6779-7_15

15.1 Introduction

Textiles have been an essential component of human society, both for the maintenance of humans' basic physical needs, as well as, as a conveyer of various societal signals such as personal identities, wealth, power and social status. As the name "the Silk Road" suggests, textiles have been traded globally over centuries: as of 2017, textiles and clothing constituted 6% of the value of manufactured goods exported globally (World Trade Organization 2018).

15.1.1 Sustainability Problems of Textile Supply Chain

Despite its importance, clothing industry has been also known for its various environmental and social impacts. The EU's Joint Research Centre reported that, clothing contributes to 2–10% of environmental impacts among products consumed in Europe (Beton et al. 2014). Some of the most notable environmental impacts such as those related to carbon footprint, water and chemical usage, as well as social impacts related to poor working conditions, occur during the production phase of textile products (Beton et al. 2014; Roos et al. 2016; Fujita and Nakamura 2018).

The negative impacts arising from the production phase of clothing industry is exacerbated by rapidly increasing consumption and over-production. A global survey in 2013 reported the increase in apparel fibre consumption by almost 80% in less than 20 years, from 38,889 million tons in 1992 to 69,728 million tons in 2010 (Food and Agriculture organization of the United Nations International Cotton Advisory Committee 2013). Recent news highlighted that some of the new garments do not even reach the first user and are incinerated (BBC 2018; Nakamura and Fujita 2018).

The necessity of addressing environmental and social impacts related to upstream of clothing industry has been advocated for the last few decades, as manifested in, among others, Clean Clothes Campaign launched in 1989, Better Cotton Initiative started in 2005 and the like (Clean Clothes Campaign 2018; Better Cotton Initiatives n.d.). Together with the (then) emerging notion of corporate social responsibility, these pushed industry to extend its conventional supply chain management practices and incorporate environmental and social consideration. The concept, often referred to as *sustainable supply chain management (SSCM)*, thereby integrates the triple bottom line approach when brand-owning companies work with the upstream suppliers, in line with system and life-cycle thinking (Chkanikova 2012; Carter and Rogers 2008; Seuring and Müller 2008). In addition, concerns regarding clothing supply chain have induced various discussions and activities at different level of governments (e.g. national, Nordic, EU—see Sect. 15.2).

15.1.2 Lack of Government Interventions and Emergence of Multi-stakeholder Initiatives

Production of garments, however, typically go through a long and complicated web of supply chain comprising actors of various sizes and crossing many national borders (Kogg 2009), posing challenges for individual companies to take effective measures that are economically feasible and arguably requiring multiple actors to collaborate (Van Tulder 2012). Negative impacts found in producing countries indicate lack of effective domestic measures, or effective enforcement thereof, to address such impacts. Meanwhile, fundamental principles governing public international law (e.g. state sovereignty) as well as free trade regime put limits to the type of policy measures that could be introduced at the country where garments are consumed. Without having an international law that both producing and consuming countries adhere to, as of today, we lack government interventions effectively addressing various environmental and social impacts arising from clothing supply chain.

Among the measures that emerged in this context is the formulation of so-called *multi-stakeholder initiatives (MSIs)*. Despite their variations in terms of, for instance, participating stakeholders, scope (e.g. targeted sectors, issues), aims and functions, governance, administrative and enforcement mechanisms (Jastram and Schneider 2015; Baumann-Pauly et al. 2017; Mena and Palazzo 2012), MSIs are generally understood as arrangements consisting of various stakeholders such as companies, NGOs, academia and at times government, who come together on a voluntary basis to "define, implement, and enforce rules that direct corporations' behavior with regard to social and environmental issues" (Rasche 2012). While some hail MSIs for their potential to fill governance gaps especially in complex issues such as SSCM (Searcy 2017), others are more critical due to its voluntary nature (Van Tulder 2012), frequent missing voices of upstream actors and conflicting interests and demands of brands and the like (Lund-Thomsen and Lindgren 2014). Others are also critical towards the potential expansion of global brands' influence over the market and global policy setting, and that MSIs may serve as a vehicle to protect those large brands' interests, or prevent damage to their reputation, rather than actually enhancing sustainability of supply chain (O'Rourke 2006; Locke 2013; Martens 2007). There is limited knowledge related to what extent and in which way companies employ MSIs and what are their perception (Lund-Thomsen and Lindgren 2014).

This paper seeks to explore possible means of governance to enhance sustainable supply chain management (SSCM) of clothing industry, focusing on the case of EU and especially Sweden. The paper consists of two main building blocks:

1. Overview and analysis of recent policy development in the EU and Sweden pertaining to SSCM of clothing industry
2. Exploration of the role of multi-stakeholder initiatives (MSIs) in enhancing SSCM of clothing industry, by analyzing the perception of Swedish brands participating in three selected MSIs.

Among environmental and social impacts occurring from various parts of garments' life, this paper puts its focus on production phase, thus measures concerning sustainable supply chain management (SSCM) of clothing industry. Other measures taken in relation to, for example, closing or slowing the material loop, will be mentioned when relevant. EU and Sweden has been selected as the case due to their significance in textile consumption, various recent policy development (see Sect. 15.2) and authors' familiarity with the context. As of 2017, EU was the world biggest importer of both textiles and clothing (World Trade Organization 2018), and the spending on purchasing clothes per person in EU in real term increased by 34% between 1996 and 2012 (European Environmental Agency 2014). The amount of clothing and home textiles put on the Swedish market per capita increased from less than 9 kg in 2000 to 12.8 kg in 2015 (Carlsson et al. 2011; Elander et al. 2017; Statistics Sweden 2018).

The paper is written in connection with Mistra Future Fashion, a cross-disciplinary research programme funded by the Mistra, the Foundation for Strategic Environmental Research. It is based primarily on two M.Sc. theses written by the first and second authors of this paper (Machek 2018; Heinz 2018), supplemented by some additional research. The study on policy development was mainly a desktop research consisting of literature review and discourse analysis, supplemented by interviews with eight experts and attendance in relevant conferences. For the study on multi-stakeholder initiatives (MSIs), following literature review, information related to nine case Swedish brands as well as three case MSIs were collected through interviews and document review (detailed description regarding case selection is found in Sect. 15.3), followed by its analysis based on the framework developed through the literature review.

Following this introductory section, Sect. 15.2 provides an overview of the recent development of relevant government policy at the EU and Sweden, concluded by a short reflection. Section 15.3 describes and discusses findings from the study on the perception of Swedish brands regarding the three selected MSIs. The paper concludes with reflection and discussion on the findings presented in Sects. 15.2 and 15.3 and a few suggestions for potential future pathways to advance more sustainable clothing supply chain.

15.2 Recent Policy Development in the EU and Sweden

There exist a growing number of policy initiatives relevant to SSCM of clothing industry directly or indirectly in recent years, both at the EU level and at the national level.

15.2.1 EU

SSCM has been taken up in several different streams of European policy arena, including development and environment. In development arena, European Commission's communications related to *corporate social responsibility* from 2001 and 2011 brought up SSCM as part of responsible business practices, which also help companies to minimise risks in the long run (European Commission (EC) 2001b, 2011). In environmental policy discourse, supply chain issues appear in relation to management and verification of information on environmental impacts arising from products' life cycle. For instance, European Commission's green paper on integrated product policy from 2001 provided examples of initiatives where information collection methods related to materials in automotive and electronic sectors are standardized across the supply chain (EC 2001a). In the early 2010s, the necessity of collecting information on the life-cycle environmental performance was brought up to facilitate comparison of environmental footprint of products and organisations (EC 2013).

While the aforementioned examples of policy documents related to SSCM do not address clothing industry specifically, the Commission staff's working document on *sustainable garment value chain* from 2017 specifically addresses the responsible management of garment supply chain (EC 2017b). The document listed a number of global guidelines and principles related to responsible business practices developed by entities such as the UN, OECD and ILO, the implementation of which the EU promotes. It also highlights the EU's existing commitment of financing various bilateral, regional and global programmes and projects related to the reduction of environmental and social impacts arising from garment supply chain. The document sets out three priority areas (women's economic empowerment, decent work and living wages and transparency and traceability in the supply chain), and also three areas of intervention (provision of financial support, promotion of social and environmental best practices, and consumers' awareness raising).

Other voluntary policies addressing the upstream environmental impacts of clothing industry include *EU eco-labelling scheme* and *green public procurement*. In addition to a number of criteria related to environmental impacts from production process, requirements specific to fibre types, restricted chemicals and the like, the latest eco-labelling criteria document for textile products from 2014 included criteria related to social issues such as protection of various workers' rights (EU 2014). The development of "model" criteria for textile products (EC 2017a) at the EU level is also an EU's attempt to enhance the integration of green criteria in the public purchasing of textile products in Member States.

In addition to these voluntary policy measures, several EU laws do provide specific mandates in relation to, among others, clothing sector. The so-called *REACH Regulation* (Regulation (EC) No. 1907/2006), since 2006, mandates registration, evaluation, authorization and restriction of chemicals put on the EU market, depending on the risks the chemical in question may pose to human health and the environment. The Regulation in its Annex XVII lists chemicals whose use and/or sales (including products containing such chemicals) are restricted, unless meeting the conditions specified under the same Annex (Article 67). According to the European Chemical Agency, a number of restricted substances concern textiles (European Chemical Agency n.d.). Many of the restricted substances are used during the production process of textile products.

The 2008 *framework Directive on waste (WFD)* (Directive 2008/98/EC), in its revision in 2018 based on the introduction of the EU Circular Economy Package in 2015 (EC 2015), also incorporated several requirements specific to textiles, along with a few other product streams such as electronics and furniture (Directive (EU) 2018/851). The revision requires EU Member States to set up systems promoting repair and recycling activities for textiles (Article 1 (10)) and to separate textile waste from other municipal solid waste from 2025 onwards (Article 1 (12) (b)). Article 1 (12) (e) of the revision further puts mandates on the European Commission to consider preparation for reuse and recycling targets for textile waste by 2024. Though those new requirements do not address upstream of clothing industry directly, they all aim at closing textile material loop, which in turn would reduce environmental impacts related to production process.

Finally, triggered by the 2011 Communication on Corporate Social Responsibility, the EU introduced in 2014 a new mandate on the disclosure of non-financial information (Directive 2014/95/EU). Large companies with more than 500 employees have to publish their policy they implement regarding environment and social matters, as well as diversify company board members.

An in-depth analysis of actors directly engaged in the EU governance structure reveals that the European Parliament has been a strong advocate for having concrete legislative measures (Heinz 2018). They pushed for, for instance, the development of 2014 non-financial information disclosure Directive, as well as the mandatory separation of textile waste under the WFD (Directive (EU) 2018/851; European Parliament 2019).

15.2.2 Sweden

Policy development related to environmental impacts of textiles in Sweden during the last decade was triggered by the mandate set at Article 29 of (then new) 2008 WFD (Directive 2008/98/EC) that required all the EU Member States to establish a waste prevention programme. In search for the content of such a programme, Nordic countries identified textile waste as one of the priority product groups, and started to explore potential policy measures through their policy collaboration mechanism of the Nordic Council of Ministers (Tojo et al. 2012). The main issue the countries sought to address was not directly related to waste, but production phase of environmental impacts via closing and slowing the material loops. The Nordic Council of Ministers have been also active in exploring the potential of setting standards for textile products within the framework of the so-called EU Eco-design Directive, whose use is currently limited to energy related products (Directive 2009/125/EC; Bauer et al. 2018).

In addition to waste prevention work led primarily by the Swedish Environmental Protection Agency (EPA), the Swedish Chemical Agency also set textiles as one of their priority products using hazardous chemicals (Kemikalieinspektionen 2011), and have been active in exploring possibilities of closing material loops while avoiding hazardous substances.

In 2016, the Swedish EPA put together a policy proposal for more sustainable handling of textile and textile waste in response to government commission. In addition to setting an overarching waste reduction and reuse/recycling targets, they also provided two policy options for three life cycle phases of textiles—production, consumption and waste management—respectively. The four policy options they suggested for sustainable production and consumption were (1) facilitation of dialogue with clothing industry sector, (2) financial support for SMEs to develop alternative business models, (3) Green public procurement to enhance reuse and recycling, and (4) information to consumer. Concerning option 1, the Swedish EPA and Chemical Agency have been organising "textile dialogue day" twice a year, which enjoy the participation of a large number of various stakeholders.

15.2.3 Reflection

As seen, a number of policy measures and documents recently developed both at the EU level and in Sweden highlight textiles as a priority product group. Many aims to address upstream environmental and/or social impacts of clothing industry in one way or the other. Meanwhile, except for the mandates in waste related laws, the 2006 REACH Regulation and 2014 non-financial information disclosure Directive, they are essentially voluntary policy measures such as provision of financial support and information dissemination. The fact that the vast majority of clothing companies in Europe are SMEs with less than 50 employees reduce the relevance of the non-financial information disclosure Directive.

15.3 Roles of Multi Stakeholder Initiatives for SSCM of Swedish Clothing Brands[1]

Given this policy context, we now seek to explore the experiences of Swedish clothing brands in participating in three MSIs selected for the study—Sustainable Apparel Coalition (SAC), Sweden Textile Water Initiative (STWI) and Textile Exchange (TE).

MSIs were selected based on the following criteria: (1) if they address environmental challenges in apparel supply chains, (2) participation of Swedish fashion brands, (3) solid establishment of the MSIs judged based on the duration of establishment and participation of leading brands. Concerning the first criteria, to gain

[1] The content of this section is primarily based on (Machek 2018), unless otherwise mentioned.

Table 15.1 Focal areas and number of participating brands in selected MSIs

MSI	Focus within SSCM	# of participating Swedish brand	# of brands participated in the study	Total# of participating actors[a]
Sustainable Apparel Coalition (SAC)	Various env'l issues across the supply chain, development of Higg-Index	3[b]	3	234[b]
Sweden Water Textile Initiative (STWI)	Water-related challenges	18[c]	8	29[c]
Textile Exchange (TE)	Identification of more sustainable materials	6[d]	4	289[d]

[a]Including apparel brands, home textile brands, manufacturers, NGOs and other additional actors
[b]apparelcoalition.org/members/ (figures as of July 2018)
[c]stwi.se/members/ (figures as of July 2018)
[d]textileexchange.org/members/ (figures as of July 2018)

several perspectives across different initiatives it was envisaged that the three MSIs covered different types of challenges.

All the 21 Swedish brands that participate in at least one of the three MSIs were contacted and asked to participate in the study. In total nine of them agreed to take part.

Table 15.1 presents the three MSIs, their main focus, number of Swedish brands as well as total number of actors participating in the respective MSIs, and number of brands participated in this study.

15.3.1 Benefits and Reasons of Participating in MSIs

Interviews with the nine brands revealed that participation in the case MSIs is overall beneficial for the enhancement of their sustainability work. First, MSIs facilitates *access to relevant and reliable information, knowledge and tools* related to the SSCM work, such as environmental laws in the production country and environmental property of materials they use. Instead of having to obtain and update such knowledge themselves, MSIs serve as an information hub. The three MSIs also facilitate the channeling of contacts with experts—for instance, participants of SAC includes academia, and STWI is connected with the Stockholm International Water Institute—and provide a meeting place with other brands who went through similar problems related to SSCM in the past. MSIs thus facilitate access to expertise as well as peer learning. All these in turn reduces the *transaction cost* related to such information gathering exercise.

The collaboration with other brands within the MSIs also *creates leverage* and helps them exert more *influence* on suppliers, especially those with whom they do not have direct contract. Similarly to the overall European context, the Swedish clothing industry consists predominantly of SMEs—95% of the total of 17,060 fashion companies employ less than 10 people (Tillväxtverket 2015). Participation in MSIs helps brands overcome the struggle pertaining to the *size* of these individual actors. It also allows them to cover the *cost* for concrete measures (e.g. improvement of water and chemical usage in the production process) which would have been too *expensive* to tackle by a single brand.

A handful of interviewees also mentioned that participation in MSIs added *credibility* to their SSCM work in the eyes of stakeholders.

Concerning why or why not the brands joined an MSI, all the interviewees indicated that the content of the MSI needs to *match the brand's focus/strategy/ambition*, and that MSI must fulfill a certain need. Moreover, they assess the *cost* of joining the MSI in light of potential *benefits*. One interviewee commented on the *availability of activities* which give the brand an *access to specific expertise and competence* as an important criteria for selection. In regard to content, one interviewee expressed preference on MSIs that present practical measurable changes over those that serves as a platform of dialogue.

These findings reduce the concern by some critics presented in Sect. 15.1, suggesting that brands participate in MSIs to prevent damage to their reputation (Locke 2013).

The perceived benefits found in this study were overall in line with motives and benefits for brands to participate in MSIs as identified in previous studies (Segerson 2013; Börjeson 2017; Oelze 2017; Kogg and Mont 2012). Moreover, the study found that MSIs provides solutions for some of the previously identified barriers that pose challenges for individual brands to practice SSCM, such as *size* of the company, *cost, time, and lack of knowledge/complexity* (Seuring and Müller 2008; Locke 2013; Börjeson 2017; Oelze 2017).

In terms of perceived value of collaboration in general, interviewees considered that sustainability issues is a common problem and is an area of collaboration, not competition. This resembles the reaction of industry when developing an extended producer responsibility system for closing the material loop of their products.

15.3.2 Challenges Related to MSIs

Despite the benefits of cutting time and costs for participating brands, one of the perceived challenges for participation was *resource required* to take part (including time, financial support and human power). This makes it difficult for SMEs to take part in MSIs even when they find concrete benefits of participation.

Another issue that came up was the *financial stability* of an MSI. It was raised particularly in relation to STWI. After its initiation, their activities have been supported by the Swedish International Development Corporation Agency (SIDA). It is perceived difficult to continue receiving additional funding for activities, however, thus STWI needs more support from members. Due to, among others, limited resources, other priorities set by the top management, and limited success achieved by STWI so far, however, securing financial contribution from participating brands has not been an easy task.

A further issue also raised in relation to the limited resource o SMEs was the *power asymmetry* among brands. This came up especially in relation to SAC. While SAC's fee structure for corporate members is differentiated based on their annual revenue, the fact still remains that there is limitation for SMEs to find resources to take part in important decision making process, such as Higg Index. Higg Index is meant to serve as a benchmark tool for sustainability performance of brands (Sustainable Apparel Coalition n.d.), manufacturing facilities and products, which also enables sustainability improvements. Although non-brand actors, such as academia and NGOs, are also engaged in the development of Higg Index, the final decision is made by the so-called Core Team, which consists of only certain brands. Some interviewees revealed that there has been occasions where the level of standards have been compromised at a suboptimal level from sustainability perspective so as not to jeopardise the members of Core Team. The situation is in line with some of the criticisms raised against MSIs in previous studies (Lund-Thomsen and Lindgren 2014; Locke 2013; Martens 2007), as introduced in Sect. 15.1. *Lack of clarity in the decision making process* was another concern raised in relation to SAC.

15.3.3 Additional Findings

The study further found that there is general trend where textile MSIs started to harmonise the standards and tools they are providing. For instance, both TE and STWI are now members of SAC. This makes the role of a larger and long-established MSI like SAC even more significant.

15.4 Conclusions and Suggestions for the Future

15.4.1 Conclusions

The review of recent policy development pertaining to SSCM of clothing sector at the EU level and in Sweden reveals that in both cases, textile has attracted significant attention in the policy discourse. However, despite their intention of addressing upstream environmental and/or social impacts of clothing industry, participation of the brands is voluntary in most cases—exception are the mandates in waste related laws addressing the closure of material loops as well as the REACH Regulation that is to do with the specific property of products. The requirement posed in the

2014 non-financial information disclosure Directive is also limited to a large player, exempting the vast majority of clothing brands in Europe.

This is not surprising given the limitation of introducing legal requirements on activities that take place and affect people and the environment outside of the national border (see Sect. 15.1).

Meanwhile, an in-depth study of the perception of Swedish clothing brands on three case MSIs provides a mixed message regarding the roles of MSIs. On one hand, interviewed brands clearly acknowledge the contribution the case MSIs have been making in their SSCM practices. The interviewees also clarify that they make rather down-to-earth assessment (alignment of the MSI activities to their need, strategy and ambition, benefit and cost of joining the MSI), and negate the likelihood of joining MSIs for the mere purpose of green washing.

On the other hand, the study also highlights concerns regarding the power asymmetry and unclear decision making process observed in some of the MSIs, as well as the overall trend of smaller MSIs joining the larger ones. This, in light of lack of effective public policy, might mean *privatization of standards*, which has been raised as one of the primary criticisms against MSIs (see Sect. 15.1).

15.4.2 Potential Future Pathways

A few pathway could be considered to advance sustainability in the clothing supply chain. Below these ideas are presented, together with areas that might require further consideration for the respective cases.

Prohibition of import of illegally produced clothing, with the use of the producing country's law for the determination of illegality

While the importance of state sovereignty should be respected especially considering the colonial past and associated injustice, consuming country could still introduce the law *restricting the importation of products whose production process violates the relevant national environment/labour law of producing country.*

Such legislation does exist today in the area of timber products, introduced in the US, EU and Australia (Lacey Act of 2008; Regulation (EU) No. 995/2010; Illegal Logging Prohibition Act 2012; Illegal Logging Prohibition Regulation 2012). While details differ, what they have in common is the determination of *illegality based on the country of harvest*, as well as the requirement of going through *due diligence procedure.* Another area in which the due diligence system is used for supply chain quality assurance is conflict minerals.

The three timber laws came into force relatively recently, which makes it difficult to determine the level of success, but there are already some experiences that could be of use to consider:

- The relatively *heavier impacts on SMEs* was experienced at least in the case of US and Australia (Shelley 2012; Rynne and Corden 2015). How to reduce the burden of SMEs could be one of the key areas to look into given the large proportion of SMEs in the clothing industry.

- The timber sector also went through a long struggle of developing such laws. In their absence, some certification schemes, such as Forest Stewardship Council and Programme for the Endorsement of Forest Certification, played a major role in enhancing sustainable harvesting practices. These certifications schemes continue to play a significant role even after the introduction of the three laws, as the criteria in the certification schemes tend to provide much more comprehensive sustainability criteria than addressing only illegal harvesting. Actors in charge of establishing a due diligence scheme could utilize the certification scheme as a way of verifying compliance with the law of harvesting country. Some existing certification schemes in the textile area could be utilized in a similar manner in event a law of this type comes into place. Similarly, various financial assistance and informative measures could co-exist with such legislation to enhance transition.
- Similarly to conflict minerals, the OECD developed a due diligence guideline for clothing sector. As has been used in the case of conflict minerals, national/regional governments could utilize the guideline when developing their laws.

Enhance transparency of existing MSIs

The study shows that, despite their short comings, MSIs certainly have been playing many important roles in enhancing the sustainability of clothing supply chain. An essential step to remedy some of the issues raised is to have a more transparent decision making process, especially concerning the development of standards. In this regard, the German Partnership for Sustainable Textiles have been mentioned as promising concerning how the governance system is designed (Jastram and Schneider 2015), which could offer some food for thoughts for other existing MSIs.

References

Bauer B, Watson D, Gylling A, Remmen A, Hauris Lysemose M, Hohenthal C, Jönbrink A-K (2018) Potential ecodesign requirements for textiles and furniture. TemaNord series 2018:535. Nordic Council of Ministers, Copenhagen

Baumann-Pauly D, Nolan J, Heerden A (2017) Industry-specific multi-stakeholder initiatives that govern corporate human rights standards: legitimacy assessments of the fair labor association and the global network initiative. J Bus Ethics 143(4):771–787

BBC (2018) Burberry burns bags, clothes and perfume worth millions. [Online] Available www.bbc.com/news/business-44885983 [8 July 2019]

Beton A, Dias D, Farrant L, Gibon T, Le Guern Y, Desaxce M, Perwueltz A, Boufateh I (2014) Environmental improvement potential of textiles. Publications Office of the European Union, Luxembourg

Better Cotton Initiative (n.d.) BCI History. [Online] Available www.bettercotton.org/about-bci/bci-history/ [8 July 2019]

Börjeson N (2017) Toxic textiles: towards responsibility in complex supply chains. Environmental science, Natural sciences, technology and environmental studies. Södertörns högskola, Huddinge

Carlsson A, Hemström K, Edborg P, Stenmarck Å, Sörme L (2011) Kartläggning av mängder och flöden av textilavfall [Mapping of quantity and flow of textile waste, in Swedish]. SMED Rapport Nr 46 2011. Sveriges Meteorologiska och Hydrologiska Institut (SMED), Norrköping

Carter C, Rogers D (2008) A framework of sustainable supply chain management: moving toward new theory. Int J Phys Distr Log 38(5):360–387

Chkanikova O (2012) Sustainable supply chain management: theoretical literature overview. IIIEE, Lund University, Lund

Clean Clothes Campaign (2018) Who we are. [Online] Available www.cleanclothes.org/about/who-we-are [8 July 2019]

Commission Decision of 5 June 2014 establishing the ecological criteria for the award of the EU Ecolabel for textile products (2014/350/EU). OJ L 174 13.6.2014, pp 45–83

Directive 2008/98/EC of the European Parliament and of the Council of 19 November 2008 on waste and repealing certain Directives. OJ L 312, 22.11.2008, pp 3–30

Directive 2009/125/EC of the European Parliament and of the Council of 21 October 2009 establishing a framework for the setting of ecodesign requirements for energy-related products. OJ L 285, 31.10.2009, pp 10–35

Directive 2014/95/EU of the European Parliament and of the Council of 22 October 2014 amending Directive 2013/34/EU as regards disclosure of non-financial and diversity information by certain large undertakings and groups. OJ L 330, 15.11.2014, pp 1–9

Directive (EU) 2018/851 of the European Parliament and of the Council of 30 May 2018 amending Directive 2008/98/EC on waste. OJ L 150, 14.6.2018, pp 109–140

European Commission (EC) (2001a) Green paper on Integrated Product Policy. COM (2001) 68 final

European Commission (EC) (2001b) Green Paper. Promoting a European framework for Corporate Social Responsibility. COM(2001) 366 final

EC (2011) Communication from the Commission to the European Parliament, the Council, the European Economic and Social Committee and the Committee of the Regions. A renewed EU strategy 2011–14 for Corporate Social Responsibility. COM (2011) 681 final

EC (2013) Communication from the Commission to the European Parliament and the Council. Building the Single Market for Green Products. Facilitating better information on the environmental performance of products and organisations. COM (2013) 196 final

EC (2015) Communication from the Commission to the European Parliament, the Council, the European Economic and Social Committee and the Committee of the Regions. Closing the loop—an EU action plan for the Circular Economy. COM/2015/0614 final

EC (2017a) Commission Staff Working Document. EU green public procurement criteria for textiles products and services. SWD (2017) 231 final

EC (2017b) Commission Staff Working Document. Sustainable garment value chains through EU development action. SWD (2017) 147 final

Elander M, Tojo N, Tekie H, Hennlock M (2017) Impact assessment of policies promoting fiber-to-fiber recycling of textiles. Mitra Future Fashion report number 2017:3. Mistra Future Fashion, Stockholm

European Chemical Agency (n.d.) Q & As. [Online] Available echa.europa.eu/support/qas-support/browse/-/qa/70Qx/view/scope/reach/restrictions [11 July 2019]

European Environmental Agency (2014) Environmental indicator report 2014. Environmental Impacts of Production-Consumption Systems in Europe. European Environmental Agency, Copenhagen

European Parliament (2019) Environmental impact of the textile and clothing industry. What consumers need to know. European Union, Brussels

Food and Agriculture organization of the United Nations International Cotton Advisory Committee (2013) World apparel fibre consumption survey. July 2013. [Online] Available www.sewitagain.com/wp-content/uploads/2013/12/FAO-ICAC-Survey-2013-Update-and-2011-Text.pdf [10 July 2019]

Fujita S, Nakamura K (2018) The shadow of Putit Pla clothes—hourly wage 400 yen, pushed with delivery [Pura fuku ga kaaeru yami Jikyuu 400 yen, Nouhin Sekasaretuzuke, in Japanese]. Asahi Shimbun Company, Tokyo

Heinz C (2018) Governing sustainability in the garment sector. The European Union's action agenda. IIIEE Thesis 2018:2. IIIEE, Lund University, Lund

Illegal Logging Prohibition Act 2012. No. 166 (Australia)

Illegal Logging Prohibition Regulation 2012. No. 217, 2012 (Australia)

Jastram S, Schneider A (2015) Sustainable fashion governance at the example of the partnership for sustainable textiles. Umwelt-Wirtschafts Forum 23(4):205–212

Kemikalieinspektionen (2011) Kemikalier i varor. Strategier och styrmedel för att minska riskerna med farliga ämnen i vardagen [Chemicals in products. Strategies and policy instruments for reducing the risks with hazardous substances in products, in Swedish]. Rapport Nr 3/11 Kemikalieinspektionen, Sundbyberg

Kogg B (2009) Responsibility in the supply chain: Interorganisational management of environmental and social aspects in the supply chain—case studies from the textile sector. IIIEE Dissertation, 2. IIIEE, Lund University, Lund

Kogg B, Mont O (2012) Environmental and social responsibility in supply chains: the practise of choice and inter-organisational management. Ecol Econ 83:154–163

Lacey Act of 2008, 16 USCS § 3371-3378 (USA)

Locke RM (2013) Can global brands create just supply chains? Boston Rev 38(3):12–29

Lund-Thomsen P, Lindgren L (2014) Corporate social responsibility in global value chains: where are we now and where are we going? J Bus Ethics 123(1):11–22

Machek D (2018) The role of multi-stakeholder initiatives in Swedish apparel brands' sustainable supply chain management. Exploring the cases of: Sustainable Apparel Coalition (SAC), Sweden Textile Water Initiative (STWI) and Textile Exchange (TE). IIIEE Thesis 2018:23. IIIEE, Lund University, Lund

Martens J (2007) Multistakeholder partnerships—future models of multilateralism?. Friedrich-Ebert-Stiftung, Berlin

Mena S, Palazzo G (2012) Input and output legitimacy of multi-stakeholder initiatives. Bus Ethics Q 22(3):527–556

Nakamura K, Fujita S (2018) 100 million pieces of new clothes per year, including popular brands, cannot be sold and discarded [Shinpin no fuku, urezu ni haiki "nen 10-oku ten", ninki brand mo, in Japnaese]. Asahi Shimbun Company, Tokyo

O'Rourke D (2006) Multi-stakeholder regulation: privatizing or socializing global labor standards? World Dev 34(5):899–918

Oelze N (2017) Sustainable supply chain management implementation-enablers and barriers in the textile industry. Sustainability 9(8):1435

Rasche A (2012) Global policies and local practice: loose and tight couplings in multi-stakeholder initiatives. Bus Ethics Q 22(4):679–708

Regulation (EU) No. 995/2010 of the European Parliament and of the Council of 20 October 2010 laying down the obligations of operators who place timber and timber products on the market. OJ L 295, 12.11.2010, pp 23–34

Regulation (EC) No. 1907/2006 of the European Parliament and of the Council of 18 December 2006 concerning the Registration, Evaluation, Authorisation and Restriction of Chemicals (REACH), establishing a European Chemicals Agency, amending Directive 1999/45/EC and repealing Council Regulation (EEC) No. 793/93 and Commission Regulation (EC) No. 1488/94 as well as Council Directive 76/769/EEC and Commission Directives 91/155/EEC, 93/67/EEC, 93/105/EC and 2000/21/EC. OJ L 136, 29.5.2007, pp 3–280

Roos S, Zamani B, Sandin G, Peters GM, Svanström M (2016) A life cycle assessment (LCA)-based approach to guiding an industry sector towards sustainability: the case of the Swedish apparel sector. J Clean Prod 133:691–700

Rynne B, Corden S (2015) Independent review of the impact of the illegal logging regulations on small business. [Online] Available www.agriculture.gov.au/SiteCollectionDocuments/for

estry/australias-forest-policies/illegal-logging/independent-review-impact-illegal-logging-regulations.pdf [12 July 2019]
Searcy C (2017) Multi-stakeholder initiatives in sustainable supply chains: putting sustainability performance in context. Elem-Sci Anthropocene 5
Segerson K (2013) Voluntary approaches to environmental protection and resource management. Annu Rev Resour Econ 5:161–180
Seuring S, Müller M (2008) From a literature review to a conceptual framework for sustainable supply chain management. J Clean Prod 16:1699–1710
Shelley WR (2012) Setting the tone: the Lacey Act's attempt to combat the international trade of illegally obtained plant and wildlife and its effect on musical instrument manufacturing. Envtl L 42(2):549–576
Statistics Sweden (2018) Population and population changes 1749–2017. Statistics Sweden, Stockholm
Sustainable Apparel Coalition (n.d.) The Higg Index. [Online] Available: apparelcoalition.org/the-higg-index/ [12 July 2019]
Tillväxtverket (2015) Modebranschen i Sverige - Statistik och analys 2015 [Fashion industry in Sweden—statistics and analysis 2015, in Swedish]. Rapport 0176. Tillväxtverket, Stockholm
Tojo N, Kogg B, Kiørboe N, Kjær B, Aalto K (2012) Prevention of textile waste. Material flows of textiles in three Nordic countries and suggestions on policy instruments. TemaNord series 2012:545. Nordic Council of Ministers, Copenhagen
Van Tulder R (2012) Foreword—the necessity of multi-stakeholder initiatives. In: van Huijstee M (ed) Multistakeholder initiatives: a strategic guide for civil society organizations. SOMO, Amsterdam
World Trade Organization (2018) World trade statistical review 2018. World Trade Organization, Geneva

Part III
Circular Production and Life Cycle Management

Chapter 16
Production Planning of Remanufactured Products with Inventory by Life-Cycle Simulation

Susumu Okumura, Nobuyoshi Hashimoto, and Taichi Fujita

Abstract Manufacturers can reduce environmental impact and costs by introducing remanufacturing. In remanufacturing, inventory control of remanufactured and newly manufactured products should be considered for production planning. It is indispensable for predicting the demand quantity of products and the collecting quantity of used products to minimize the total cost incurred by remanufacturing and newly manufacturing. The prediction varies depending on the product characteristics and the time elapsed since shipment. In this paper, we study a remanufacturing system minimizing the total cost by a life-cycle simulation method. In the flow of products, there are two types of inventory: one is a remanufactured and newly manufactured product inventory before shipment to the market, and the other is a collected product inventory for used products. Depending on the time elapsed since launch, products pass through an introductory stage, a growth stage, a maturity stage, and a decline stage in a usual manner. We analyze the impact on the total cost of the prediction accuracy of the product demand and the collected used products to each stage. When the ratio of an ordering cost for newly manufactured products to an inventory holding cost for the finished products has a significant value, or as the life-cycle stage is approaching the decline stage, the predicting accuracy of collection of the used products and the product demand has a high impact on the total cost.

Keywords Remanufacturing · Production planning · Inventory · Life-cycle simulation

---

S. Okumura · N. Hashimoto (✉)
Department of Mechanical Systems Engineering, University of Shiga Prefecture, Shiga, Japan
e-mail: hashimoto.n@mech.usp.ac.jp

T. Fujita
Graduate School of University of Shiga Prefecture, Shiga, Japan

Y. Kishita et al. (eds.), *EcoDesign and Sustainability I*, Sustainable Production, Life Cycle Engineering and Management, https://doi.org/10.1007/978-981-15-6779-7_16

16.1 Introduction

In recent years, manufacturers encounter two primary problems. The first problem is that the life-cycle of various products is shortened owing to the rapid improvement of technological innovation. The demand for products typically increases immediately after launch, and the demand for products declines as new products with new functions emerge. Such situations cause various problems, e.g., excessive inventory, increasing disposal, over-ordering. The second problem is the environmental impact caused by consuming a large number of resources and producing waste. Among the International Organization for Standardization (ISO) stipulations for environmental management systems, environmental audits, environmental labels, and life-cycle assessments are in progress to preserve the global environment. Many manufacturers that adopt ISO 14001 are expected to reduce the environmental impacts caused by them. In there, reusable units extracted from used products are incorporated into new products to reduce both environmental impact and costs. Remanufacturing requires inventory control for remanufactured and newly manufactured products.

Some reviews were published regarding inventory control in remanufacturing (Mahadevan et al. 2003; Konstantaras and Papachristos 2007; Lage and Filho 2012; Bazan et al. 2016). Kiesmüller (2003) studied a recovery system for a single product, in which inventories for collected/reusable items and serviceable items were included, and determined the optimal manufacturing, remanufacturing, and disposal rates. Dobos (2003) found the optimal inventory policies in a reverse logistics system under the assumption that demand was a known continuous function in a given planning horizon and the return rate of used items was a given function. van der Laan and Teunter (2006) studied the push-and-pull remanufacturing policies and showed closed-form formulae approximating the optimal policy parameters. Reiner et al. (2009) developed a new stochastic demand model and investigated optimal service levels in terms of customer satisfaction and life cycle profit. El Saadany and Jaber (2010) developed a production/remanufacturing inventory model, in which the flow of returned items was variable and was controlled by the purchasing price for returned items and their acceptance quality levels, and showed mathematical models for multiple remanufacturing and production cycles. Ahiska and King (2010) investigated the optimal inventory policies over the life cycle of a remanufacturable product and determined the optimal or near-optimal policy characterizations by Markov decision analysis. Chung and Wee (2011) investigated green product designs and remanufacturing efforts based on an integrated production inventory model with short life-cycles, and showed new technology evolution, remanufacturing ratios and system's holding costs were critical factors for decision making in a green supply chain inventory control system. Zhou and Yu (2011) studied the integration of supply and customer demand with inventory management for a production/remanufacturing firm, and optimized total discounted profit over a finite planning horizon by implementing production, remanufacturing, product acquisition, and pricing strategies. Hsueh (2011) studied inventory management policy of remanufacturing in four stages depending on the elapsed time since shipment to the market, i.e., an introductory stage, a growth stage, a maturity stage, and a decline stage. Then, it was shown that the optimal production plan with the minimal cost was different in each stage under the time-series forecast of the demand quantity and the collecting quantity in each period. Hsueh assumed that remanufacturing was performed in continuous production, and ignored the ordering cost of remanufactured products and the inventory cost of collected used products. Therefore, it would be a more realistic solution may be obtained by selecting an appropriate type of production system from continuous production and lot production depending on the characteristics of the product. Furthermore, the effects of product-specific ordering costs and inventory costs on an optimal production planning, and the effects of uncertainty on the forecasted demand quantity and collecting quantity depending on product characteristics were not considered in (Hsueh 2011).

This paper deals with a production plan of products with a short life-cycle and evaluates the environmental impact by a life-cycle simulation (LCS) technique. The simulation model includes an inventory management model with system parameters: ordering cost of newly manufactured products and remanufactured products, and inventory cost of newly manufactured products and collected products. The demand quantity and collecting quantity are predicted by the product characteristics to each stage divided by the elapsed time since shipment containing introductory/growth/maturity/decline stages. Assuming that the predicted quantity is different from the actual value, the influence of the system parameter on the optimal production plan of the newly manufactured products, the influence of the prediction accuracy on the system cost, and the influence of system parameters on the optimal production method of the remanufactured products are investigated to each stage.

16.2 Development of LCS Software

16.2.1 Assumptions About the Life Cycle of Recycling Products

An example of the time-series prediction of demand quantity and collecting quantity used in this study is shown in Fig. 16.1, in which the characteristic four stages that exist in the general product life cycle are included.

- Introductory stage: Demand keeps constant at a small value, $t \in [0, 30]$. It is supposed that there is no collection of the used products at this stage.
- Growth stage: Demand continues to increase significantly after the introductory stage,$t \in [30, 60]$.
- Maturity stage: Demand remains constant at a largest value,$t \in [60, 135]$.
- Decline stage: Demand is decreasing due to the appearance of new products,$t \in [135, 250]$.

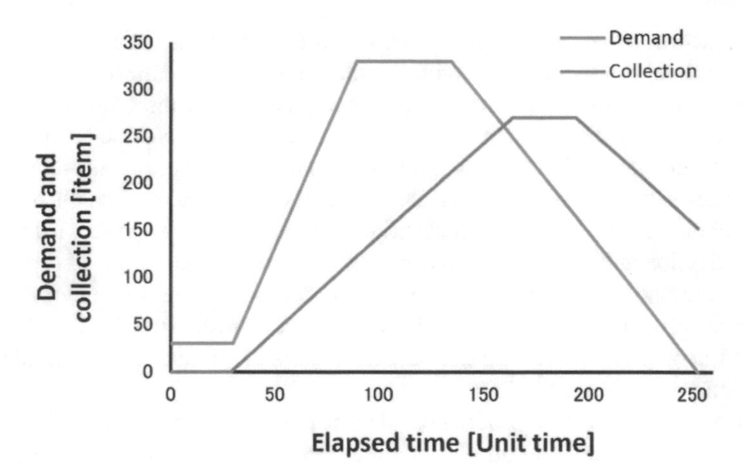

Fig. 16.1 An example of time-series data of demand and collection

The product flow in the LCS is shown in Fig. 16.2. It is assumed that remanufacturing is carried out using the reusable unit extracted from the collected used product. The units of a product are divided into reusable units (RU) intended for reuse and non-reusable units (NRU) intended for disposal for convenient (Okumura et al. 2017). Two types of units are separated in the disassembly block.

Two types of inventory exist in the flow: finished product inventory, in which newly manufactured and remanufactured products are included and collected product inventory. The former inventory keeps newly manufactured products with a new RU

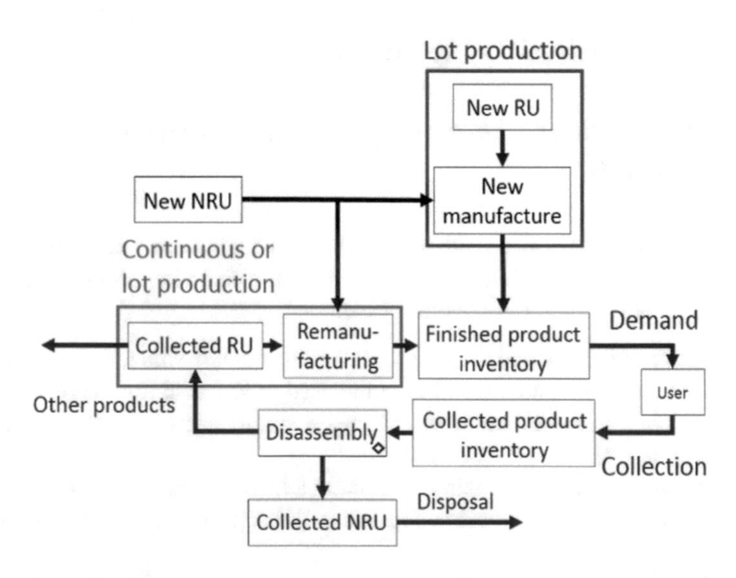

Fig. 16.2 Flow of products

and a new NRU, and remanufactured products with a recovered RU and a new NRU. Hence, the demand quantity generated every time interval is always satisfied. The collected product inventory stores the collected products to procure RUs by disassembling. The collected Rus are employed for remanufacturing, and the collected NRUs are disposed of. Also, when the recovery of RU is more than product demand, they are used for other products. The management costs of the two inventories are simulation parameters.

16.2.2 Assumptions Regarding LCS and Production Methods

The following assumptions for constructing LCS software have been made.

(1) The cost for remanufacturing that includes the cost for collection and disassembly of used products is smaller than that for newly manufacturing. Therefore, remanufactured products are preferentially produced than newly manufactured products. New manufacture is compensated only for the insufficient amount or products that cannot be met by only remanufactured products to satisfy demand.
(2) New manufacture is carried out by lot production. On the other hand, remanufacturing is carried out by either continuous production or lot production. If the finished product inventory is less than the demand at time t, new manufacture is ordered to satisfy demand.
(3) When the number of collected used products is higher than demand, all the products shipped to the market are remanufactured products. Collected RUs that are not shipped to the market are reused or recycled for other products for convenient.
(4) Required time for newly manufacturing, remanufacturing, collection of used products, and disassembly are less than the time interval in LCS.

As for item (2), the lot size for new manufacture is determined by

$$Q_{\mathrm{M}}(t; i_{\mathrm{M}}) = \int_{t}^{t+i_{\mathrm{M}}} (\lambda(t) - \gamma(t))\mathrm{d}t, \tag{16.1}$$

where i_{M} is ordering interval decided in advance, $\lambda(t)$ and $\gamma(t)$ are the demand quantity at time t and the collecting quantity of used products at time t, respectively. It is noted that when $Q_{\mathrm{M}}(t)$ has a large value, the number of ordering of new manufacture can be reduced; however, excessive inventory of newly manufactured products is likely to occur.

As for continuous production in remanufacturing, remanufactured products arise in every time interval of LCS, which yields the production quantity of remanufactured products at time t is the same as the collecting quantity $\gamma(t)$. As for lot production in remanufacturing, an order is placed at the time interval i_{R}. Therefore, the lot size for remanufacturing is determined by

$$Q_{\mathrm{R}}(t; i_{\mathrm{R}}) = \int_{t-i_{\mathrm{R}}}^{t} \gamma(t)\mathrm{d}t. \tag{16.2}$$

Based on the above assumptions, the LCS software has been constructed using the simulation modeling studio WITNESS Horizon.

16.2.3 Objective Function

Inventories for the finished products and the collected used products are managed according to the production plan based on the time-series prediction data of demand and collecting quantity. Also, new manufacture and remanufacturing are directed; both of them incur the cost of ordering. Therefore, we consider the following total cost as the objective function given by the sum of the ordering cost for new manufacture and remanufacturing, and inventory for the collected products:

$$C_{\mathrm{O}} = C_1 + C_2, \tag{16.3}$$

where C_1 is the ordering cost, and C_2 is the inventory cost for collected products.

The ordering cost C_1 is the sum of ordering costs for new manufacture and remanufacturing given by

$$C_1(t; i_{\mathrm{M}}, i_{\mathrm{R}}) = K_{\mathrm{M}} y_{\mathrm{M}}(Q_{\mathrm{M}}(t; i_{\mathrm{M}})) + K_{\mathrm{R}} y_{\mathrm{R}}(Q_{\mathrm{R}}(t; i_{\mathrm{R}})), \tag{16.4}$$

where K_{M} and K_{R} are an ordering cost for newly manufactured products, and an ordering cost for remanufactured products, respectively. The number of new product orders $y_{\mathrm{M}}(\cdot)$ and the number of remanufactured product orders $y_{\mathrm{R}}(\cdot)$ are influenced by lot size $Q_{\mathrm{M}}(t; i_{\mathrm{M}})$ and $Q_{\mathrm{R}}(t; i_{\mathrm{R}})$, respectively. It is noted that $\partial y_{\mathrm{M}}/\partial Q_{\mathrm{M}} < 0$ and $\partial y_{\mathrm{R}}/\partial Q_{\mathrm{R}} < 0$.

Finally, the inventory cost incurred in $[t_1, t_2]$ (t_1: start time of a stage, t_2: end time of a stage) is given by the sum of the costs for newly manufactured products inventory and collected products inventory:

$$\begin{aligned} C_2(t; i_{\mathrm{M}}, i_{\mathrm{R}}) =& h_{\mathrm{F}} \int_{t_1}^{t_2} I_{\mathrm{F}}(t; y_{\mathrm{M}}(Q_{\mathrm{M}}(t; i_{\mathrm{M}})), y_{\mathrm{R}}(Q_{\mathrm{R}}(t; i_{\mathrm{R}})))\mathrm{d}t \\ &+ h_{\mathrm{C}} \int_{t_1}^{t_2} I_{\mathrm{C}}(t; y_{\mathrm{M}}(Q_{\mathrm{M}}(t; i_{\mathrm{M}})), y_{\mathrm{R}}(Q_{\mathrm{R}}(t; i_{\mathrm{R}})))\mathrm{d}t, \end{aligned} \tag{16.5}$$

where h_{F} is an inventory holding cost for the finished products which means the cost when one remanufactured/manufactured product as a finished product is stocked for a unit time in the finished product inventory, $I_{\mathrm{F}}(\cdot)$ is the number of remanufactured/manufactured products in the finished product inventory, h_{C} is an inventory holding cost for the collected products which means the cost when one collected product is stocked for a unit time in the collected product inventory, and $I_C(\cdot)$ is the number of products in the collected product inventory.

16.3 Influence of System Parameters on Optimal Production Plan for New Manufacture

When new manufacture is applied as the lot production, the optimal number of ordering for new manufacture changes depending on demand $\lambda(t)$, collecting quantity $\gamma(t)$, an ordering cost for new manufacture K_M, and an inventory holding cost for the finished products h_F. Therefore, the influence of these parameters on the optimal number of orders is investigated in this section.

16.3.1 LCS Conditions

Concentrating on the optimal production plan of the newly manufactured products, we set parameters for calculating the total cost C_O as shown in Table 16.1. Parameters for the reference conditions as well as cases 1–4 are assigned for comparison. Here, $\lambda_P(t)$ is the predicted quantity of demand, and $\gamma_P(t)$ is the predicted quantity of the collection of used products.

The demand quantity of products and the collecting quantity of used products vary based on Fig. 16.1. Then, it is assumed that remanufacturing is carried out as continuous production to avoid the effect of the ordering cost, K_R, of remanufactured

Table 16.1 Parameters setting for obtaining the effect of optimal production planning for new manufacture

Parameter	Reference conditions	Case 1	Case 2	Case 3	Case 4
$\lambda_A(t)$: the actual amount of demand	$\lambda_P(t)$	$\lambda_P(t)$	$\lambda_P(t)$	$\lambda_P(t)$	$\lambda_P(t)$
$\gamma_A(t)$: the actual amount of collection	$\gamma_P(t)$	$\gamma_P(t)$	$\gamma_P(t)$	$\gamma_P(t)$	$\gamma_P(t)$
K_M: an ordering cost for newly manufactured products	500	250	1000	500	500
h_F: an inventory holding cost for the finished products	0.1	0.1	0.1	0.05	0.2
K_R: an ordering cost for remanufactured products	0	0	0	0	0
h_C: an inventory holding cost for the collected products	0	0	0	0	0

products to focus on the production planning of newly manufactured products. Also, an inventory holding cost for the collected products h_C is omitted to neglect the effect of the collected product inventory.

16.3.2 Results and Discussion

Figure 16.3 shows the influence of the number of ordering for new manufacture on the total cost by varying an ordering cost for newly manufactured products K_M, in which the reference conditions, cases 1 and 2 are applied. Figure 16.4 shows the influence of the number of ordering for new manufacture on the total cost by varying an inventory holding cost for the finished products h_F, in which the reference conditions, cases 3 and 4 are employed. The white markers in both figures represent the optimal number of orders, which has the minimal cost. The gray markers represent the nearly optimal number of orders of new manufacturing that has less than 10% increase over the minimal cost.

In Fig. 16.3a–c, we can see that under the reference conditions, $K_M = 500$, the total cost decreases as time goes from the growth stage to the decline stage. As for

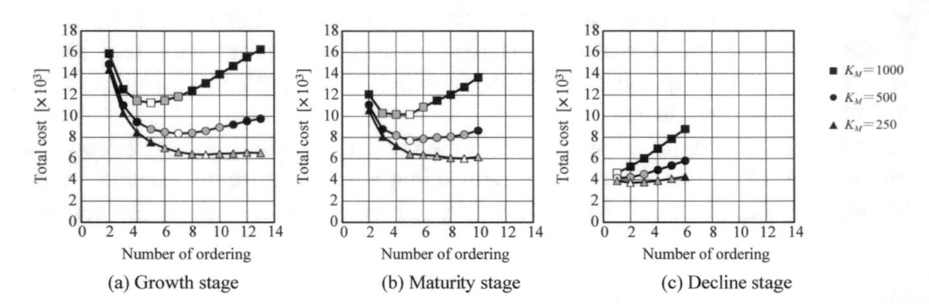

Fig. 16.3 Relationship between the number of ordering for new manufacture and the total cost when K_M changes

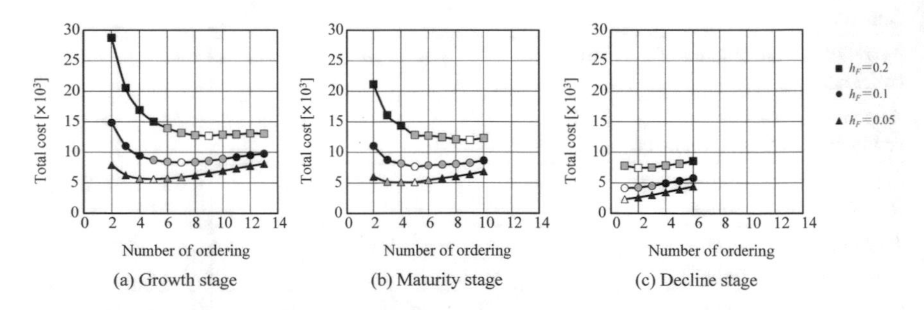

Fig. 16.4 Relationship between the number of ordering for new manufacture and the total cost when h_F changes

the growth stage (a), the number of newly manufactured products required during the stage is remarkable to meet the increasing demand. Therefore, a large number of products are manufactured at a single ordering of new manufacture when K_M has a small value, which leads to an increase in the total cost due to excessive inventory. Also, when the number of ordering has a small value the total cost has a small variation to all K_M; however, the variation of the total cost becomes significant as the number of ordering becomes large. In the maturity stage (b), the increase of demand terminates, and the collection of used products increases. Therefore, the number of newly manufactured products required decreases. Accordingly, the total cost has relatively small values, even if the number of ordering is reduced. During the decline stage (c), the collecting quantity increases, and the demand quantity decreases. Therefore, the newly manufactured product quantity required is further reduced. Moreover, even if all the necessary products are newly manufactured in single ordering (the number of ordering is one), excessive inventory does not occur.

Figures 16.3 and 16.4 show that as K_M and h_F change, different characteristics appear in each stage. The effect of K_M in Fig. 16.3 increases when the number of ordering has a large value; however, the effect of h_F increases when the number of ordering is small. Consequently, it is crucial to focus on the individual values of K_M and h_F as well as the ratio of the two values for the production planning of remanufactured products.

16.4 Influence of Prediction Accuracy of Demand and Collection on the Total Cost

In this section, the influence on the total cost is examined under the condition that the actual values of the demand quantity and the collecting quantity differ from the predicted quantities. As shown in Table 16.2, LCS is carried out by varying the ordering cost K_M, inventory cost for the collected products h_F, and the actual demand quantity $\lambda_A(t)$, and the actual collecting quantity $\gamma_A(t)$ together with the reference conditions in Table 16.1.

Figures 16.5 and 16.6 show the influence of the differences between the predicted demand quantity and the actual one on the total cost, in which Fig. 16.5 is the case varying K_M, and Fig. 16.6 is the case varying h_F. Also, Figs. 16.7 and 16.8 show the

Table 16.2 Parameter setting for obtaining the effect of prediction accuracy of demand and collection

Parameter	Condition assigned
$\lambda_A(t)$	$0.9\lambda_P(t)$, $0.95\lambda_P(t)$, $\lambda_P(t)$ (R), $1.05\lambda_P(t)$, $1.1\lambda_P(t)$
$\gamma_A(t)$	$0.9\gamma_P(t)$,$0.95\gamma_P(t)$,$\gamma_P(t)$(R),$1.05\gamma_P(t)$,$1.1\gamma_P(t)$
K_M	125, 250, 500 (R), 1000, 2000
h_F	0.025, 0.05, 0.1 (R), 0.2, 0.4

R reference condition

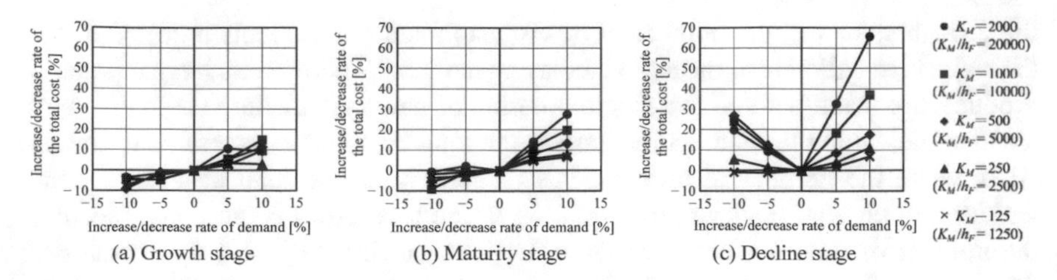

Fig. 16.5 Relationship between the increase/decrease rate of the demand and the increase/decrease rate of the total cost when K_M changes

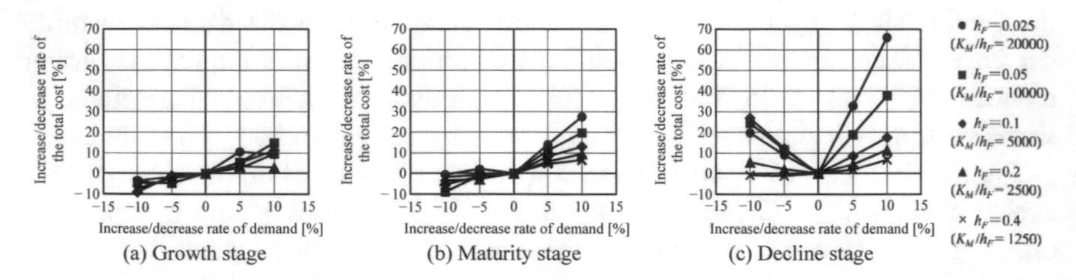

Fig. 16.6 Relationship between the increase/decrease rate of the demand and the increase/decrease rate of the total cost when h_F changes

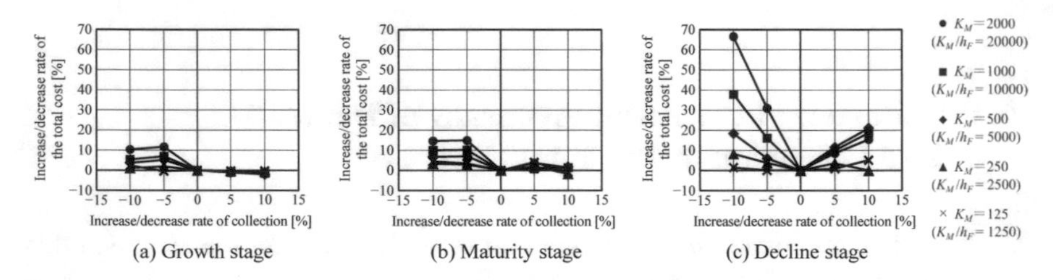

Fig. 16.7 Relationship between the increase/decrease rate of the collection and the increase/decrease rate of the total cost when K_M changes

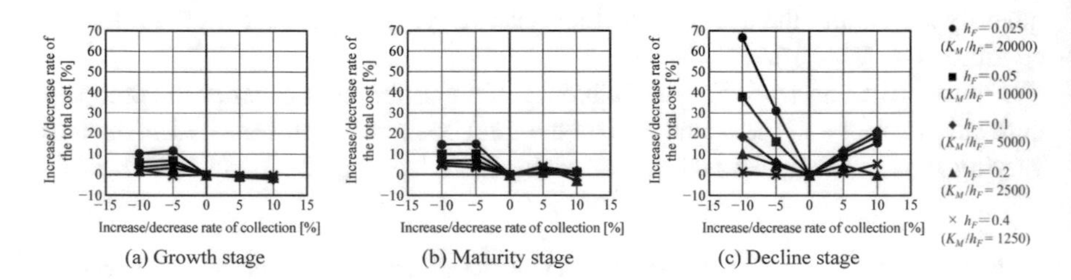

Fig. 16.8 Relationship between the increase/decrease rate of the collection and the increase/decrease rate of the total cost when h_F changes

influence of the differences between the predicted collecting quantity and the actual one on the total cost, in which Fig. 16.7 is the case varying K_{M}, and Fig. 16.8 is the case varying h_{F}. The ratio of an ordering cost for newly manufactured products to an inventory holding cost for the finished products, $K_{\mathrm{M}}/h_{\mathrm{F}}$ ranging [1250, 20000], is also shown in Figs. 16.5, 16.6, 16.7, 16.8. The horizontal axis in Figs. 16.5, 16.6, 16.7, 16.8 represents how much the actual quantity is increased or decreased from the predicted quantity, and also the vertical axis of them represents how much the total cost is increased or decreased compared to the case where the predicted quantity is the same as the actual quantity.

Comparing Fig. 16.5 with 16.6 and 16.7 with 16.8, we can see that the shape of the curves in the figures is the same when the ratio $K_{\mathrm{M}}/h_{\mathrm{F}}$ is the same. Therefore, the influence of the prediction accuracy can be judged based on $K_{\mathrm{M}}/h_{\mathrm{F}}$. In all the life-cycle stages, the larger the K_M/h_F, the higher is the impact on the total cost by the different predictions. Also, the effect on the increase/decrease rate of the total cost increases as time goes from the growth stage to the decline stage. Therefore, when $K_{\mathrm{M}}/h_{\mathrm{F}}$ has a significant value, or as the life-cycle stage is approaching the decline stage, it is crucial to improve the predicting accuracy of the collection of the used products and the product demand.

16.5 Optimal Production System for Remanufactured Products: Lot Production or Continuous Production

16.5.1 LCS Conditions

We investigate the optimal remanufacturing system in this section: lot production or continuous production. In the previous sections, remanufacturing was considered focusing on the newly manufacturing cost under the conditions: an ordering cost for remanufactured products $K_{\mathrm{R}} = 0$ and an inventory holding cost for the collected products $h_{\mathrm{C}} = 0$. In this section, remanufacturing is evaluated under non-zero values of K_{R} and h_{C}.

The ordering interval of new manufacture i_{M} is fixed to each life-cycle stage such that the growth stage, the maturity stage, and the decline stage are 9, 9, and 2 for convenience, respectively, in which all the ordering intervals are the same as the optimal number of ordering in the reference conditions in Table 16.1. The decline stage, $t \in [135, 250]$ in Fig. 16.1, is divided into decline stages 1 and 2. The former is the stage during which the demand quantity is larger than the collecting quantity, $t \in [135, 160]$, the latter is the stage during which the collecting quantity is greater than the demand quantity, $t \in [160, 250]$. The same set of parameters in the reference conditions in Table 16.1 are applied. While parameters K_{R} and h_{C} are assigned as shown in Table 16.3.

Table 16.3 Parameter setting for obtaining the effect of prediction accuracy of demand and collection

Parameter	Assigned value
K_R	0 (R), 10, 50, 100
h_C	0.05, 0.1 (R), 0.2

R reference condition

16.5.2 Results and Discussion

The varying rate of the total cost concerning the order interval for remanufacturing i_R is shown in Fig. 16.9. It is noted that remanufactured products are continuously produced if the total cost for ordering interval of remanufactured products $i_R = 1$; on the other hand, remanufactured products are produced in lots if $i_R \geq 2$. The vertical axis in Fig. 16.9 represents how much the total cost increases or decreases relative to the total cost when the ordering interval i_R is 1, which corresponds to continuous production. The white markers in Fig. 16.9 indicate that the ordering interval that makes the total cost minimize, which means that if white markers appear in the horizontal axis of $i_R \geq 2$ implying that lot production is the optimal production system; while they appear in $i_R = 1$ designating that continuous production is optimum. As for all the stages, the increase/decrease rate of the total cost tends to increase as i_R grows when K_R has a smaller value.

The increase/decrease rate of the total cost is negative when $i_R \geq 2$ in the growth stage (a–c), which indicates that lot production is more profitable than continuous production. As for the maturity stage (d–f), lot production is optimal depending on h_C; however, there is little difference between continuous production and lot production because the increase/decrease rate of the total cost is close to zero. In the decline stage 1 and 2 (g–l), the increase/decrease rate of the total cost has the most significant value compared to those of the growth and maturity stages as the ordering interval for remanufacturing i_R becomes large. As for the decline stage 1, continuous production is optimal when $K_R = 0, 10, 50$; however, lot production is optimal when $K_R = 100$. With regards to the decline stage 2, continuous production is optimal when $K_R = 0, 10$; however, lot production is optimal when $K_R = 50, 100$.

16.6 Conclusions

This paper studies a remanufacturing system with two types of inventories: remanufactured and newly manufactured products inventory before shipment to the market, and collected product inventory for the collected used products. We have investigated an optimal production plan minimizing the total cost by a life-cycle simulation

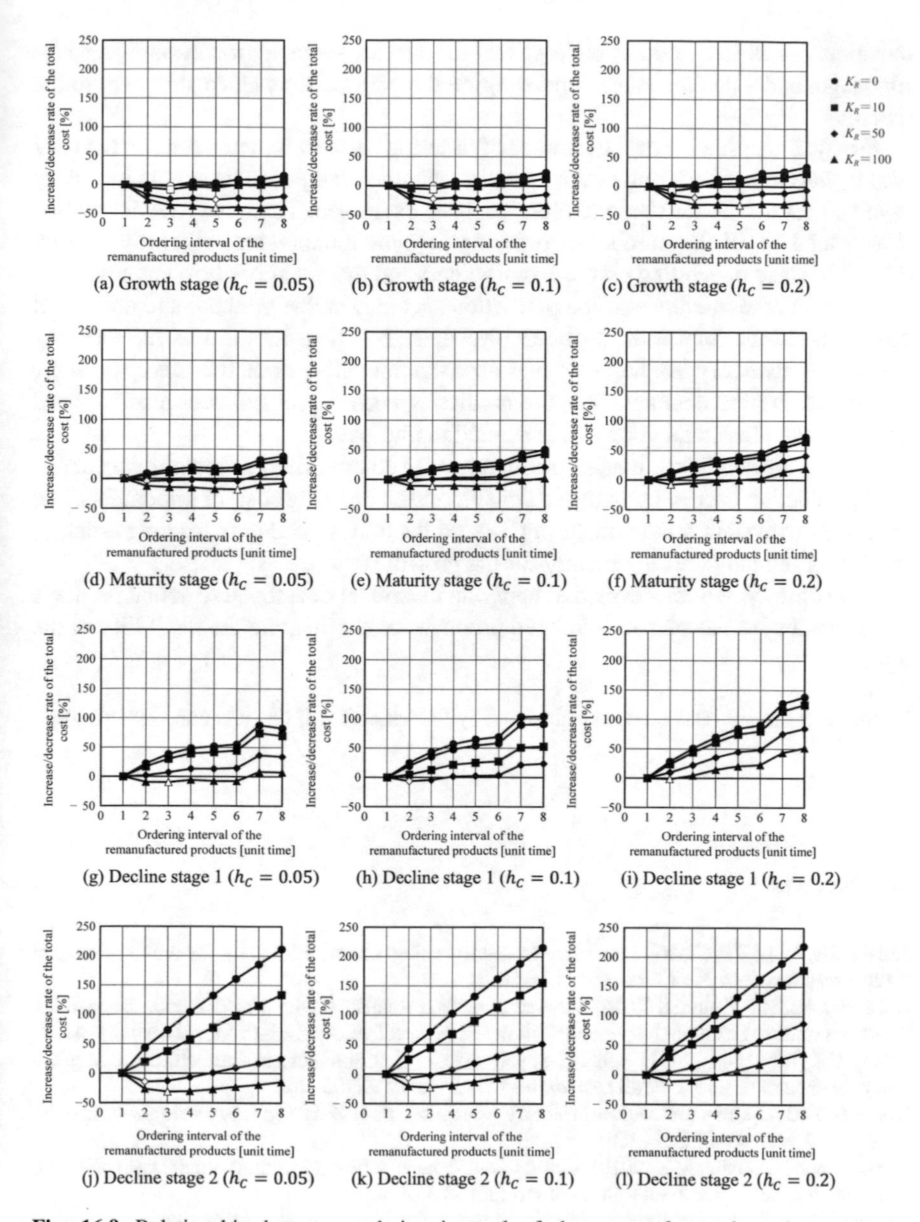

Fig. 16.9 Relationship between ordering interval of the remanufactured products and the increase/decrease rate of the total cost

technique, in which products undergo an introductory stage, a growth stage, a maturity stage, and a decline stage depending on the elapsed time since shipment to the market.

Firstly, the influence of the number of ordering for new manufacture on the total cost by varying an ordering cost for newly manufactured products and an inventory holding cost for the finished products has been examined. Then, we have found that it is crucial to set the number of ordering for new manufacture carefully to each life-cycle stage depending on the ordering cost and the inventory holding cost.

Secondly, the influence of the prediction accuracy of the demand and collection quantities on the total cost has been investigated. When the ratio of the ordering cost to the inventory holding cost has a substantial value, or as the life-cycle stage is approaching the decline stage, the predicting accuracy of collection of the used products and the product demand is needed to improve.

Thirdly, the conditions that lot production has a competitive advantage over continuous production have been analyzed considering an ordering cost for remanufactured products involved in remanufacturing. Lot production is valid only for long ordering interval of remanufacturing mainly for the growth stage.

Future efforts will be devoted to applying the model developed to actual products and extending to the successor/related products to demonstrate the flexibility of the model.

Acknowledgements This work was supported by JSPS KAKENHI Grant Numbers JP26340103 and JP17K00671.

References

Ahiska SS, King RE (2010) Life cycle inventory policy characterizations for a single-product recoverable system. Int J Prod Econ 124:51–61

Bazan E, Jaber MY, Zanoni S (2016) A review of mathematical inventory models for reverse logistics and the future of its modeling: an environmental perspective. Appl Math Model 40:4151–4178

Chung C-J, Wee H-M (2011) Short life-cycle deteriorating product remanufacturing in a green supply chain inventory control system. Int J Prod Econ 129:195–203

Dobos I (2003) Optimal production-inventory strategies for a HMMS-type reverse logistics system. Int J Prod Econ 81–82:351–360

El Saadany AMA, Jaber MY (2010) A production/remanufacturing inventory model with price and quality dependant return rate. Comput Ind Eng 58:352–362

Hsueh C-F (2011) An inventory control model with consideration of remanufacturing and product life cycle. Int J Prod Econ 133:645–652

Kiesmüller GP (2003) Optimal control of a one product recovery system with leadtimes. Int J Prod Econ 81–82:333–340

Konstantaras I, Papachristos S (2007) Optimal policy and holding cost stability regions in a periodic review inventory system with manufacturing and remanufacturing options. Eur J Oper Res 178:433–448

Lage M Jr, Filho MG (2012) Production planning and control for remanufacturing: literature review and analysis. Prod Plan Control 23:419–435

Mahadevan B, Pyke DF, Fleischmann M (2003) Periodic review, push inventory policies for remanufacturing. Eur J Oper Res 151:536–551
Okumura S, Sakaguchi Y, Hatanaka Y, Ogohara K (2017) Effect of a reusable unit's physical life distribution on reuse efficiency in environmentally conscious products. Procedia CIRP 61:161–165
Reiner G, Natter M, Drechsler W (2009) Life cycle profit-Reducing supply risks by integrated demand management. Technol Anal Strateg 21:653–664
van der Laan EA, Teunter RH (2006) Simple heuristics for push and pull remanufacturing policies. Eur J Oper Res 175:1084–1102
Zhou SX, Yu Y (2011) Optimal product acquisition, pricing, and inventory management for systems with remanufacturing. Oper Res 59:514–521

Chapter 17
Stakeholders' Influence Towards Sustainability Transition in Textile Industries

Arpita Chari, Mélanie Despeisse, Ilaria Barletta, Björn Johansson, and Ernst Siewers

Abstract With the rise of global challenges associated with linear models of production, transitioning to more sustainable models has become increasingly important and urgent. However, this transition is not done systematically due to a general lack of organizational knowledge and motivation to apply existing models, metrics and frameworks for sustainability. The current sustainable value proposition in organizations also shows that management rarely has a clear implementation strategy and underestimates what is required for a successful sustainability transition to take place. In addition, few empirical studies exist to corroborate these observations. This research focuses on analyzing the organizational barriers to the long-term sustainable transformation process, by considering the interests of all stakeholders, including the planet. The objective of the paper is to provide guidelines in the form of a decision support framework to textile industries to adopt and implement green technologies in their sustainability transition process.

Keywords Sustainability transition · Stakeholder · Multi-level perspective · Barriers · Textile industry

17.1 Introduction

Among industries that have a large global climate impact, the textile industry has had a significant contribution to climate change, causing severe depletion of the planet's resources. In 2016, the textile industry contributed to 6.5% of global greenhouse gas emissions (GHG), equivalent to about 3.3 billion tons of CO_2 equivalent (Quantis 2008). Additionally, the overall life-cycle of textiles is plagued by several unsustainable issues some of which are: high amounts of water utilization (almost 100 kg to

A. Chari (✉) · M. Despeisse · I. Barletta · B. Johansson
Department of Industrial and Materials Science, Chalmers University of Technology, Göteborg, Sweden
e-mail: arpitac@chalmers.se

E. Siewers
DyeCoo Textile Systems B.V, Weesp, The Netherlands

Y. Kishita et al. (eds.), *EcoDesign and Sustainability I*, Sustainable Production, Life Cycle Engineering and Management, https://doi.org/10.1007/978-981-15-6779-7_17

dye a kg of fabric), dyestuff and other effluents contained in the waste water (Cid et al. 2005; Luo et al. 2018), large transportation related fuel emissions in the supply chain and high energy consumption (Choudhury 2014). These negative effects of the textile industry along with their predominantly linear ("take-make-dispose") model, threaten the limited resources available in our natural ecosystem with a tremendous impact on environmental and societal levels (Ellen MacArthur Foundation 2016). Hence, there is an urgent need for these industries to decouple economic growth from resource utilization and find a balance between the social, economic and environmental dimensions of sustainability through radical innovations.

Sustainable development is a normative and contested concept (Hedenus et al. 2015; Stubbs and Cocklin 2008), one that takes place with varying vested interests and values of the encompassing actors within a defined system. Sustainability transition typically involves this broad network of actors who are dynamically interacting between the different sub-systems.

Research in this field has gained increased traction over the past decade with a number of studies analysing socio-technical transformations into a sustainable economy (Van Den Bergh et al. 2011) from a systems perspective. Various frameworks have also been conceptualized in order to understand these sustainability transitions (Turnheim 2019), namely: the multi-level perspective (Geels 2017; Rip and René 1997; Geels 2002), transition management approach (Loorbach 2010), innovation systems approach (Hekkert et al. 2007; Franco 2002), dialectic issue lifecycle model (Penna and Geels 2012), strategic niche management (Rip and René 1997) among many others. However, many of these models have been criticized for not paying enough attention to the underlying interests of the various stakeholders involved in the transformation process. Farla et al. (2012) in their special issue paper addressed dynamic actor interactions from a systems perspective. They focused on capabilities and strategies that organizations and individuals need to inherently possess in order for successful sustainability transitions to take place. Several studies have analysed that the complexities arising from stakeholder involvement and management commitment have been barriers to the sustainability agenda in domains such as green building and construction (Mok et al. 2018; Hongyang et al. 2019; Pham et al. 2019; Williams and Dair 2007), urban freight transport (Van Duin et al. 2017), facilities management (Elmualim et al. 2010), manufacturing (Moldavska 2016; Orji 2019), circular economy (Houston et al. 2018) and environmental management (Geels 2017; Reed 2008) to name a few.

Epstein and Buhovac (2010) explain that although some organizations address sustainability as part of their business agenda in addition to gains in financial performance, the long-term advantages and opportunities of creating sustainable value for the organization have still been heavily underestimated. Along with developing the Corporate Sustainability Model to measure the drivers of sustainability, they identified the following key challenges of implementing sustainability in organizations:

(a) Setting clear and measurable goals;
(b) Financial incentive pressures;

(c) Stakeholder involvement and reaction to sustainability measures.

This research focuses on the third aspect of sustainability implementation: stakeholder interactions. The aim is to recognize the barriers to sustainability transition at the multi-stakeholder level in the textile industry domain. Several underlying sustainability issues in textile industries are triggered not only by technological advances, linear consumption and production, but also by the surrounding and internal social structures (Farla et al. 2012), namely, the actors. The inclusion of multi-level interactions among different stakeholders in the value chain is critical in realizing the common goal of sustainable development, instead of interests vested with a single industrial organization. With this background and the urgent need for sustainable transition in textile industries, the research aims to address the following questions:

RQ1 How do actors play a role in sustainability transition in textile industries and what barriers hinder this process?

RQ2 What conditions and strategies need to be created in order to realize sustainability transition?

The RQs are addressed through a case study that has successfully adopted and implemented a clean dyeing technology. Results are presented in the form of a decision framework to provide guidance for radical innovation implementation.

In the following section, a literature review on sustainability transitions, innovation implementation and the important role actors play in the organizational change for sustainability have been outlined. Section 3 describes the research methodology adopted in the paper. In Sect. 4, the results of the study along with implementation tactics using a decision framework that can support sustainability transitions have been summarized. The paper ends with discussions in Sect. 5 and conclusions along with future work highlighted in Sect. 6.

17.2 Literature Review

17.2.1 Multi-level Perspective for Sustainability Transitions

Of the several frameworks that have been formulated to understand the multi-dimensional complexities of sustainability transitions, the multi-level perspective (MLP) approach (Geels 2017; Rip and René 1997; Geels 2002) has been explored in this research in order to understand the role of the different players in the textile industry and the barriers hence derived towards sustainability transition. MLP argues that transitions are non-linear and take place due to dynamic interactions among three levels, namely: (a) Niches, which are spaces in which radical innovations or changes in activities take place; (b) Regimes, where rules and practices are already well defined or established and where incremental reformism occurs; (c) Landscape, where if developments occur that put enough pressure on the regime, then it creates opportunities for the niche innovations to emerge. These are represented by levels of

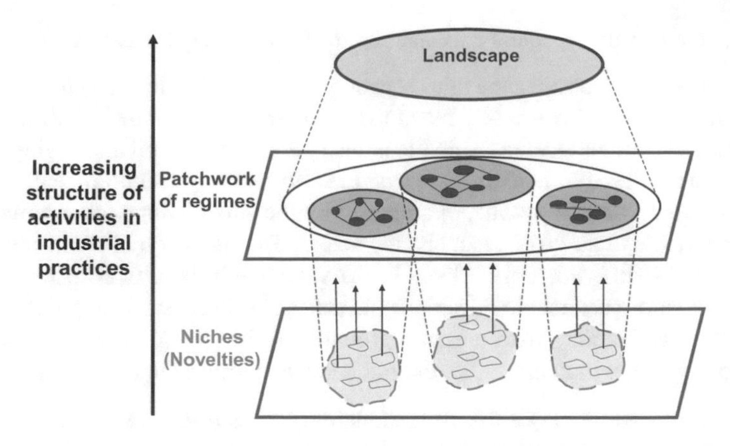

Fig. 17.1 Nested levels of multi-level socio-technical systems (adapted from Geels 2002)

increasing hierarchy where regimes are nested within an exogenous landscape and niches are in turn part of regimes (Geels 2012) (Fig. 17.1)

MLP addresses dynamic stability and radical change, a framework whose development was influenced by neo-institutional theory (where actors are restricted by regulations and shared values), sociology aspect of technology (where innovations are social constructs, developed through the complex interactions of various actors) and evolutionary economics (regimes, niches) (Geels 2002). The interactions between the different levels are enacted by various actors as they progress towards the sustainability transition process. In order for systemic changes (technological advances for example) that occur within niche spaces to be accepted by the highly structured and 'locked-in' regimes, it is important to recognize the role that the actors play within these levels. Although it is important to focus on the impact that these different stakeholders have on the transition process, it is particularly important to adopt a holistic view of the system and the interrelated sub-systems; i.e., if actors want to change one part of the system they may have to comply with rules governing other parts of the system (Farla et al. 2012). The reason of using the MLP framework in this study is to understand how actors at the different levels are affected by sustainability issues and how successful actors can lead the way to sustainability transitions in established domains such as the textile industry.

17.2.2 The Importance of Stakeholder Value

Sustainability transition is a multi-actor process, that includes a range of actors with varying capabilities (Turnheim 2019). Stakeholders or actors are people or groups that have the power to directly affect an organization's future (Ackermann and Eden 2011). Hart and Milstein (2003) in their paper on creating sustainable value in organizations, elaborated that the value embedded in the interconnectedness of stakeholders positively drives sustainable development.

Stakeholder involvement in sustainability transitions is an important stage, one that analyses in which way the stakeholder either influences or is affected when sustainable practices are adopted on a wide, market scale (Welp et al. 2006). Knowing this information is pivotal to ultimately give recommendations to key decision makers in the textile industry: investors on novel production technologies, investors within the apparel and furniture industry, top managements leading those companies as well as influencing conscious demand by consumers. By involving external and internal stakeholders, industries have the opportunity to enhance transparency and trust in their activities, thus increasing their reputation and overall sustainability performance. In order to drive sustainability performance, incorporating and enforcing a sustainability strategy within organizations is important. Epstein et al. (Epstein and Buhovac 2010) elaborated that companies need to integrate both formal systems (e.g. performance measurement metrics and tools) that support the sustainability agenda within their Business-As-Usual activities, as well as informal systems (e.g. leadership, cultural mindset of employees and stakeholder involvement) in the organizational structure as critical drivers of performance, which could lead to more harm than good with a lack of consideration thereof. In the context of this research, individual and collective actors whose roles were stable (Wittmayer et al. 2017) i.e., with fixed responsibilities, were incorporated in the sustainability transition process.

17.2.3 Innovation Implementation

Implementation has been defined by Voss (1988) as "*the user process that leads to the successful adoption of an innovation of new technology*". Sustainable development requires new innovations (Ritzén and Sandström 2017) to develop in the niche area and then adopted and implemented into the fixed business models of organizations within regimes. A concept that is widely addressed in innovation management literature (Van De Ven 1986; Nagji and Tuff 2012), is how novel innovations or solutions could help mitigate the challenges arising from unsustainable activities of industries. This can further be understood from Fig. 17.2 which has been adapted from the innovation ambition matrix published by Harvard Business Review (Nagji and Tuff 2012) based on Igor Ansoff's growth matrix (Ansoff and McDonnell 1986).

The model explains that the extent to which organizations are willing to integrate change and initiatives within their business models along with a good understanding of market potential, will ultimately determine success. In particular, breakthrough innovations that cater to unestablished markets and new customer requirements will create value for the organisation. The case study with DyeCoo illustrates the breakthrough of the innovative dyeing technology in the textile industry domain (their success level has been depicted in Fig. 17.2). Other applications for the technology exist in the fields of extraction (Lang and Wai 2001), impregnation (Üzer et al. 2006)

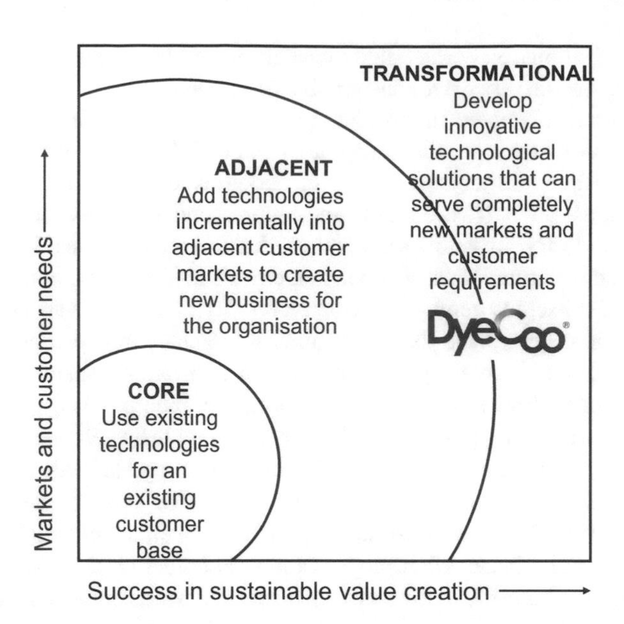

Fig. 17.2 Types of innovations under different market conditions (Nagji and Tuff 2012)

and particle generation (Fages 2018) among others, giving this technology immense potential to make transformational changes in various markets.

A wide field of literature exists that has examined the efforts required to incorporate change at all levels in an organization within the area of organizational development and change (Armenakis 1999; Judson 1991; Klein and Sorra 1996; Kotter 1995; Kotter and Schlesinger 2009) in the past few decades. For instance, Epstein and Buhovac (2010) suggested that the effectiveness of an organization's innovation implementation strategy is based on the organization's strength to adopt the innovation as well as stakeholders' commitment and values to utilize the innovation. Damschroder et al. (2009) in their consolidated framework for implementation model also relate implementation to individual characteristics and responsibilities, priorities, culture and leadership in an organization as well as the needs of stakeholders who are affected by the implementation process. Stål and Corvellec (2018) in their research with Swedish apparel companies explored the extent to which adoption and implementation of sustainable strategies in business models are successful, through organizational institutionalism. To sum up, stakeholders' and organizations' perceptions (Epstein and Buhovac 2010) at multiple levels in the sustainability transition process are therefore extremely important to consider while implementing innovative technologies and go beyond the usual constructs of involving just external experts or scientists.

17.3 Research Design

A case study method as described by Yin (2009) was utilized in this research to identify the barriers to sustainability transition in a real-world environment and explore solutions to deal with those issues. To understand the complexity of multi-actor interaction at the different levels of sustainability transition, several organisations within the textile industry domain were identified as participants for the study and fitted within the MLP framework. Leading researchers in sustainable systems and textiles along with stakeholders such as managers, CTOs and other decision makers in the chosen organisations were selected for the data collection process as they have considerable mandate in strategizing and making decisions regarding sustainability issues in the organizations (Lahtinen and Yrjölä 2019). Among stakeholders within the value chain of an organization, managers in particular perceive sustainable development to be a cost and liability of doing operations (Hart and Milstein 2003), a condition that has existed for the last two decades and one that continues to grow (Oxborrow et al. 2017; Revell and Blackburn 2007). It was for this reason that strategic decision makers were chosen in this study. To ensure that top level management did not face managerial isolation issues (Teece 2007), the organisations confirmed that transparency in communication was maintained within all levels of operations.

17.3.1 Case Description

The aforementioned negative impacts of the textile industry on the environment and society, in particular fresh water consumption and pollution, as well as greenhouse gas (GHG) emissions, have paved the way for new technological development opportunities to reduce the industry's footprint.

Between the fall of 2017 and 2018, we collaborated on a project with DyeCoo, a Dutch company that has successfully commercialized a novel water-free dyeing technique, to estimate the climate implications of the textile dyeing process. The patented and commercially available technology uses reclaimed CO_2 (carbon dioxide) as a solvent instead of water for dyeing polyester fibres and textiles. At a temperature and pressure above the critical point, CO_2 becomes supercritical ($scCO_2$), a state with liquid like density and gas like viscosity. $scCO_2$ is a green solvent with high solvability and permeability which allows dyes to dissolve easily in it. The dyes are then easily absorbed by the fibres. 95% of the CO_2 is recycled in a closed loop system. The process uses no water, no chemicals and produces no waste. Short batch cycles and efficient dye use, without a requirement for water evaporation or waste water treatment all contribute to significantly reduced environmental impact in comparison to traditional water-based dyeing technologies (DyeCoo 2012).

It is known that incumbent actors within regimes have their own innovation agenda (Farla et al. 2012) and strategies for improving performance. The purpose of using the empirical study for this research was to evaluate whether a business case existed for the technology and in turn assess how successful firms such as these (who operate within niches) can influence other industries to break the lock-ins of established technologies and gain momentum towards adopting newer, cleaner technologies. Such radical technologies also have the potential to make industries in the textile industry domain to gradually phase out from being material- and energy-dependent (Stubbs and Cocklin 2008; Hart and Milstein 2003).

17.3.2 Data Collection

To deliver the results from the case study of the project, the paper adopted a qualitative research approach as shown in Fig. 17.3.

Primary data was derived by conducting in-depth semi-structured interviews, a workshop, using questionnaires as well as an onsite visit to DyeCoo's operational facilities. Secondary data was derived from peer-reviewed studies and reports on sustainability in the textile industry issued by established institutions and agencies (Quantis 2008; Houston et al. 2018; EMF 2017). Specifically, data collection was carried out in two phases of the project. During phase 1, a workshop was conducted at DyeCoo's company headquarters, where key experts in the textile industry domain took part in a focus group: DyeCoo, a professor from the University of Borås, Sweden (expert on resource efficient processes for textile dyeing and functionalisation), and a professor from Chalmers University of Technology, Sweden (expert on sustainable energy systems). During the concept mapping, they were asked questions like-'what is your influence and responsibility on/from the adoption of sustainable practices/new clean technologies?'. Additional actors in the textile industry who have

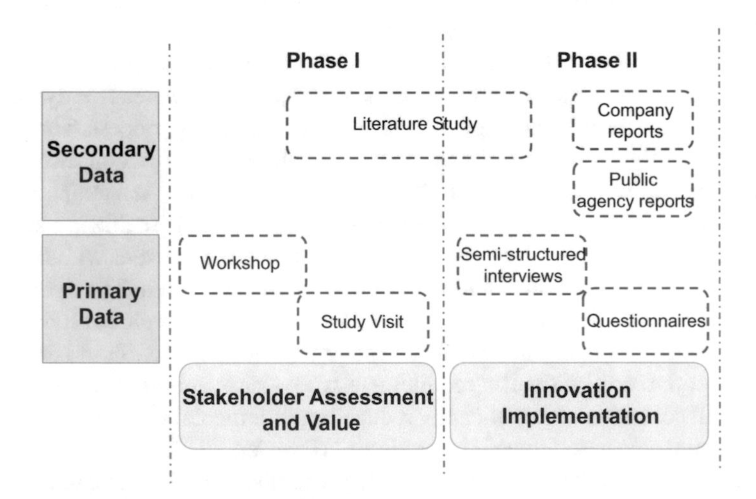

Fig. 17.3 Research methodology followed in the study

implemented/that are interested in the technology implementation were identified during Phase 2 of the project. Semi-structured interviews of about 30–45 min each were conducted with these actors. The questions were developed keeping in mind the primary objective of the project, i.e. to understand the barriers that textile industries face with regards to green technology adoption. When some players did not have sufficient time for an in-depth interview, questionnaires with a similar format were used to gather information. The questionnaire also contained survey-type questions which interviewees were asked to rate on a scale of 1–5, for instance, 'To what extent does your organization/do your customers value: quality, durability ('technology' in the case of DyeCoo), functionality, environmentally friendly (production or recyclability)' on which analysis was performed to understand current priorities and identify the challenges to implement cleaner technologies and methods of production (reflected in Fig. 17.4).

17.4 Key Findings

17.4.1 Barriers to Implementation

Several challenges to technology implementation at different levels of the transition process were identified from the empirical data and previous literature. These barriers have been categorized under different themes and represented in a model based on a PEST (Political, Economic, Social and Technological) analysis framework (Sammut-Bonnici and Galea 2015). PEST analysis tools are generally used to monitor risks involved or factors that can have an impact on an organization. Using such a framework greatly enhanced the understanding of the interesting patterns that emerged whilst studying the challenges in sustainability transitions and categorise these challenges in order to formulate corrective tactics. The technology implementation challenges identified within the PEST framework have further been depicted within a nested view model of the Triple Bottom Line approach (Elkington 1998) in Fig. 17.4.

Here, the 'environmental' factors have been considered to be part of the 'external' influences on an organization. This goes beyond the usual constructs of defining the environment in terms of raw material consumption or environmental impacts. Instead, it has been depicted by 'nature', an important stakeholder in the analysis (Stubbs and Cocklin 2008; Bocken et al. 2013) whose needs have to be met ['eco-centric' low substitutability view of strong sustainable development (Hedenus et al. 2015)].

On a landscape level, these 'environmental' factors along with public policies and regulations under the 'political' factors, put pressure on the already existing internal technical, economic and organizational pressures within regimes, causing changes in their production processes. The industry is still dominant with personal profit margin interests and protection of individual business models. Large players continue to have considerable influence on the end consumer, and this determines the market at the

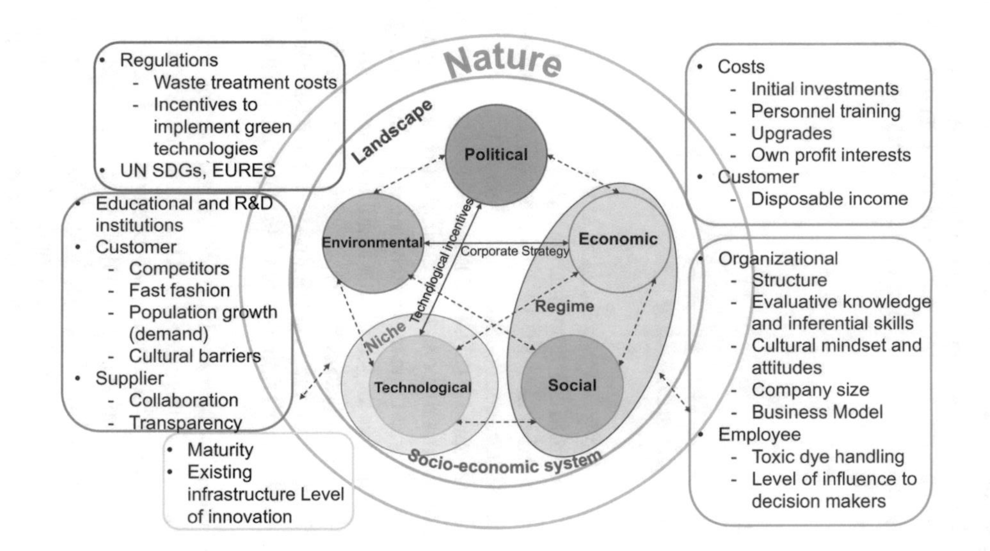

Fig. 17.4 Critical challenges to technological implementation (adapted and modified from (Stubbs and Cocklin 2008; Sammut-Bonnici and Galea 2015; Frambach and Schillewaert 1999; Deradjat and Minshall 2017)

end. As long as these companies are not willing to change their production processes and business models and influence their suppliers and customers in a positive manner, their resistance will continue to be one of the biggest barriers to green technology adoption and sustainability transition.

17.4.2 Architecture of the Decision Framework

Based on the challenges identified from the empirical study, a step-wise decision framework was formulated (Fig. 17.5). It incorporates elements in response to challenges identified within the categorizations of the PEST model. The framework recognises the impact of clean technology adoption in sustainability transitions, so that organisations can 'sense' challenges, 'adopt' better practices and 'transform' their processes, thus creating long-term sustainable value. It acts as a guide for industries to follow the path of sustainability transition. The step-wise processes described in the framework should by no means considered to be linear, but one that should continuously evolve in order to bring about long-term benefits.

Niche actors such as DyeCoo, by means of technology co-evolution and collaboration with supporting institutions, have exemplified the process of successfully shifting from just being a 'story' or in the discussion phase to being in the implementation and operational phases. Similarly, organisations would need to use their resources accordingly to carefully 'scan' market segments and conditions (Teece 2007), understand customer needs and in turn influence their demand towards fast fashion and foresee the competitive advantage from adopting cleaner technologies.

The ability to discover opportunities varies among individuals in organisations and will affect the overall decision for niche technology innovation adoption. This search for innovation implementation should extend beyond local organisational boundaries by engaging in a dialogue among other stakeholders in the textile value chain, bringing about a transparent collaboration that would benefit all parties involved. This would also involve aligning sustainable strategies to the business model and acknowledging the continuously changing decision-making capabilities within organisations.

Contrary to deeply ingrained beliefs of resistance within organisations to adopt new technologies, the scCO_2 technology from DyeCoo is more reproducible and scientific than other conventional dyeing mechanisms which are more 'art' or 'skill' based. The required 'upgrading' of skills and management of knowledge is thus easily transferable.

It was also seen that actors working at operational levels recognise the direct impact of clean technology on their working environment and act as visionaries who can influence higher management to adopt the innovative technologies. This endorsement of technology adoption in the new regimes may even influence landscape developments (Geels 2004) that could in turn support the long-term strategy of the incumbent firms. Organisations can transform their operations into one that is

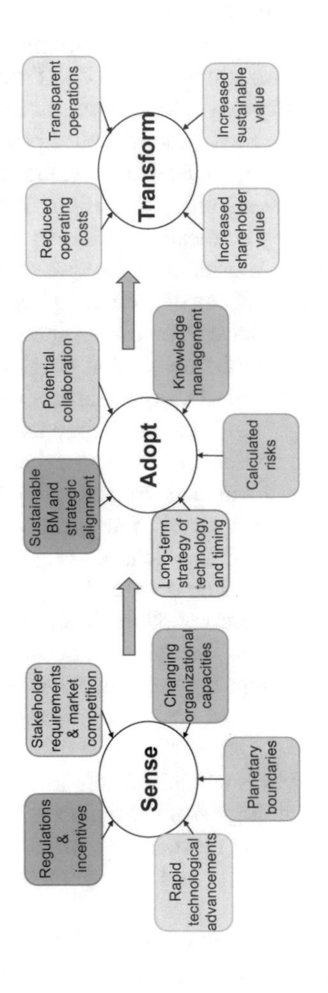

Fig. 17.5 Elements of the decision framework for sustainability transition in textile industries

driven by sustainable value creation if they are able to incorporate long-term changes in their strategic operations and take calculated risks along the transition process.

17.5 Discussion

The MLP has proven to be a useful framework in this empirical research to understand how radical innovations can create windows of opportunities and break through niches to influence the incumbent firms within regimes to adopt them. Although it is a risky proposition to draw on conclusions based on a single case study, different actors who are related to or could benefit from the results were included in the research, giving rise to a rich data set to formulate the results. In the context of the textile industrial domain, the paper has made two contributions to sustainability transitions. The first contribution was to incorporate the complex and dynamic interactions of the different actors to conceptualise successful sustainability transitions. The barriers identified from these different sources resulted in the second contribution: a tactical framework that can enable textile industries to be more sustainable through green innovation implementation. Realistically, the water-free dyeing technology cannot alter the processes of the entire dyeing industry. This is because some applications require dyeing of the fibres at early stages of production (dope dyeing of yarns for example) or applications where volumes of production are very high. However, there is a large potential for this technology in the future. The costs of this radical technology will reduce through economies of scale if products in other areas of application (as mentioned in Sect. 2.3) are produced in large quantities.

The study focused on the textile industry in the EU, but the results could also be applicable in other countries as it is assumed that similar barriers to sustainability transition exist there as well, with different regulations and policies putting pressure on the incumbents.

It is noteworthy to mention that technology cannot be the only enabler to address the negative impacts of textile industries. Sustainability transitions can only be successful if the technological changes are communicated and accepted by the different encompassing stakeholders of this industrial structure who have different values and priorities in their own agendas, by allowing transparent collaborations to bring about mutual benefits.

17.6 Conclusions

The results from this research enabled the formulation of a framework to address the increasing number of sustainability problems such as regulations, costs, technology maturity levels and highly structured organizational processes associated within the regimes of the textile industry domain.

The DyeCoo case study is exemplary to showcase how industries who are willing to implement clean technology innovations, can follow a proactive approach for decision making for sustainability. The prescriptive framework suggests that industries should 'sense' customer and planetary requirements among others and go beyond defensive and reactive sustainability performance measures for regulatory compliance. They should 'adopt' proactive long-term sustainability improvements to align with their sustainability strategies in light of rapid technological advancements. The measures taken should not only impact the organization internally, but also those who are outside direct impact, i.e., the entire value chain, thus bringing about transparent collaborations and operations. It will then be possible for incumbent firms in the textile industrial domain to move away from path dependencies and 'transform' into sustainable value-creating organisations by reducing costs along with building a strong multi-level shareholder network. The framework acts as a foundation in improving an organization's priorities and will be further developed in future work to include data quality and availability for multi-criteria decision analysis in sustainability performance measurement.

Acknowledgements The authors would like to thank DyeCoo, The Swedish School of Textiles at Borås, JOIN and TEKO (Sweden's Textile and Fashion Company) for their valuable input in this study. The research was part of the project ClimaDye, financially supported by Climate-KIC and the ongoing project ReWind, funded by the programme Produktion2030 within the Swedish Agency for Innovation Systems (Vinnova).

References

Ackermann F, Eden C (2011) Strategic management of stakeholders: theory and practice. Long Range Plann 44(3):179–196

Ansoff H, McDonnell E (1986) The new corporate strategy. Sidgwick & Jackson

Armenakis A (1999) Organizational change: a review of theory and research in the 1990s. J Manage 25(3):293–315

Bocken N, Short S, Rana P, Evans S (2013) A value mapping tool for sustainable business modelling. Corp Gov 13(5):482–497

Choudhury AKR (2014) Environmental impacts of the textile industry and its assessment through life cycle assessment

Cid MVF, Van Spronsen J, Van Der Kraan M, Veugelers WJT, Woerlee GF (2005) Excellent dye fixation on cotton dyed in supercritical carbon dioxide using fluorotriazine reactive dyes, pp 609–616

Damschroder LJ, Aron DC, Keith RE, Kirsh SR, Alexander JA, Lowery JC (2009) Fostering implementation of health services research findings into practice: a consolidated framework for advancing implementation science. Implement Sci 4(1):1–15

Deradjat D, Minshall T (2017) Implementation of rapid manufacturing for mass customisation. J Manuf Technol Manag 28(1):95–121

DyeCoo (2012) DyeCoo CO_2 dyeing technology

Elkington J (1998) Partnerships from Cannibals with forks: the triple iottom line of 21st Century Business, pp 37–51

Ellen MacArthur Foundation (2016) Intelligent assets: Unlocking the circular economy potential. Ellen MacArthur Found. pp 1–25

Elmualim A, Shockley D, Valle R, Ludlow G, Shah S (2010) Barriers and commitment of facilities management profession to the sustainability agenda. Build Environ 45(1):58–64
EMF (2017) A new textiles economy: redesigning fashion's future. Ellen MacArthur Foundation
Epstein MJ, Buhovac AR (2010) Solving the sustainability implementation challenge. Organ Dyn 39(4):306–315
Fages J et al (2018) Particle generation for pharmaceutical applications using supercritical fluid technology, no. 01668423
Farla J, Markard J, Raven R, Coenen L (2012) Sustainability transitions in the making: a closer look at actors, strategies and resources. Technol Forecast Soc Change 79(6):991–998
Frambach RT, Schillewaert N (1999) Organizational innovation adoption : a multi-level framework of determinants and opportunities for future research 3004(814):163–176
Franco M (2002) Sectoral systems of innovation and production. Res Policy 31(2):247–264
Geels F (2002) Technological transitions as evolutionary reconfiguration processes: a multi-level perspective and a case-study. Res Policy 31:1257–1274
Geels FW (2004) From sectoral systems of innovation to socio-technical systems: Insights about dynamics and change from sociology and institutional theory. Res Policy 33(6–7):897–920
Geels FW (2012) A socio-technical analysis of low-carbon transitions: introducing the multi-level perspective into transport studies. J Transp Geogr 24:471–482
Geels FW, B, Sovacool BK, Schwanen T, Sorrell S (2017) The socio-technical dynamics of low-carbon transitions. Joule 1(3):463–479
Hart SL, Milstein MB (2003) Creating sustainable value. Acad Manag Exec 17(2):56–67
Hedenus F, Persson M, Sprei F (2015), Sustainable development, nuances and perspectives
Hekkert MP, Suurs RAA, Negro SO, Kuhlmann S, Smits REHM (2007) Functions of innovation systems: a new approach for analysing technological change. Technol Forecast Soc Change 74(4):413–432
Hongyang L, Fang Y, Yan H (2019) A study on the stakeholder influences during green building operation in China. ICCREM 2018:113–117
Houston J, Casazza E, Briguglio M, Spiteri J (2018) Enablers and barriers to a circular economy, no 730378, pp 1–30
Judson AS (1991) Invest in a high-yield strategic plan. J Bus Strategy 12(4):34–39
Klein KJ, Sorra JS (1996) The challenge of innovation implementation. Acad Manag Rev 21(4):1055–1080
Kotter JP (1995) Leading change, Why transformation efforts fail. Harvard Bus Rev (73):59–67
Kotter J, Schlesinger LA (2009) Choosing Strategies for Change—HBR.org, pp 1–8
Lahtinen S, Yrjölä M (2019) Managing sustainability transformations: a managerial framing approach. J Clean Prod 223:815–825
Lang Q, Wai CM (2001) Supercritical fluid extraction in herbal and natural product studies-a practical review. Talanta 53(4):771–782
Loorbach D (2010) Transition management for sustainable development: a prescriptive. Complex-Based Governance Framework 23(1):161–183
Luo X et al (2018) Novel sustainable synthesis of dyes for clean dyeing of wool and cotton fibres in supercritical carbon dioxide. J Clean Prod 199:1–10
Mok KY, Shen GQ, Yang R (2018) Stakeholder complexity in large scale green building projects: a holistic analysis towards a better understanding. Eng Constr Archit Manag 25(11):1454–1474
Moldavska A (2016) Model-based sustainability assessment—an enabler for transition to sustainable manufacturing. Proc CIRP 48(2):413–418
Nagji B, Tuff G (2012) Managing your innovation portfolio, vol 90
Orji IJ (2019) Examining barriers to organizational change for sustainability and drivers of sustainable performance in the metal manufacturing industry. Resour Conserv Recycl 140:102–114
Oxborrow L, Goworek H, Claxton S, Cooper T, Hill H, McLaren A (2017) Managing sustainability in the fashion business exploring challenges in product development for clothing longevity, vol 2017

Penna CCR, Geels FW (2012) Multi-dimensional struggles in the greening of industry: a dialectic issue lifecycle model and case study. Technol Forecast Soc Change 79(6):999–1020
Pham H, Kim SY, Van Luu T (2019) Managerial perceptions on barriers to sustainable construction in developing countries: Vietnam case. Environ Dev Sustain (0123456789)
Quantis (2008) Measuring Fashion 2018: Environmental Impact of the Global Apparel and Footwear Industries Study Full report and methodological considerations
Reed MS (2008) Stakeholder participation for environmental management: a literature review. Biol Conserv 141(10):2417–2431
Revell A, Blackburn R (2007) The business case for sustainability in UK's construction and restaurant sectors. Bus Strateg Environ 16(October):404–420
Rip A, René K (1997) Technological change. Trade Unions Technol Chang 327–399
Ritzén S, Sandström GÖ (2017) Barriers to the circular economy—integration of perspectives and domains. Procedia CIRP 64:7–12
Sammut-Bonnici T, Galea D (2015) PEST analysis
Stål HI, Corvellec H (2018) A decoupling perspective on circular business model implementation: illustrations from Swedish apparel. J Clean Prod 171:630–643
Stubbs C, Cocklin W (2008) Conceptualizing a sustinable business model. Organ Environ 21(2):103–127
Teece DJ (2007) Explicating dynamic capabilities: the nature and microfoundations of (sustainable) enterprise performance. Strateg Manag J 28(13):1319–1350
Turnheim B et al (2019) An agenda for sustainability transitions research: state of the art and future directions. Environ Innov Soc Transit 1–66
Üzer S, Akman U, Hortaçsu Ö (2006) Polymer swelling and impregnation using supercritical CO_2: a model-component study towards producing controlled-release drugs. J Supercrit Fluids 38(1):119–128
Van De Ven AH (1986) Central problems in the management of innovation. Manage Sci 32(5):590–607
Van Den Bergh JCJM, Truffer B, Kallis G (2011) Environmental innovation and societal transitions: introduction and overview. Environ Innov Soc Transit 1(1):1–23
Van Duin R, Slabbekoorn M, Tavasszy L, Quak H (2017) Identifying dominant stakeholder perspectives on urban freight policies: a Q-analysis on urban consolidation centres in the Netherlands. Transport 33(4):867–880
Voss CA (1988) Implementation: A key issue in manufacturing technology: the need for a field of study. Res Policy 17(2):55–63
Welp M, Vega-Leinert A, Stoll-Kleemann S, Jaeger CC (2006) Science-based stakeholder dialogues: theories and tools. Glob Environ Chang 16(2):170–181
Williams K, Dair C (2007) What is stopping sustainable building in England? Barriers experienced by stakeholders in delivering sustainable developments. Sust Dev 147:135–147
Wittmayer JM, Avelino F, Van Steenbergen F, Loorbach D (2017) Actor roles in transition: insights from sociological perspectives. Environ Innov Soc Transit 24:45–56
Yin RK (2009) Case study research : design and methods, 4th edn. SAGE

Chapter 18
The Environmental Implications of Digitalization in Manufacturing: A Case Study

Xiaoxia Chen, Mélanie Despeisse, Patrik Dahlman, Paul Dietl, and Björn Johansson

Abstract The potential of digital technologies is widely discussed and recognized, mainly focusing on process automation, efficiency improvements and quality control. The implications on environmental sustainability is mostly considered from a theoretical perspective rather than operational. This study aims to reveal the potential of digital technologies to reduce the environmental impact in industrial practice. A case study was carried out at two manufacturing sites of an international manufacturing company producing mechanical components for the global markets. Data and insights were collected through interviews and observations. This paper presents the findings in terms of real life implementations and potentials to reduce the environmental impact using digital technologies along the manufacturing value chain, which can serve as input and guideline for the manufacturing industry.

Keywords Digital technology · Environmental sustainability · Case study · Manufacturing

18.1 Introduction

Manufacturing companies increasingly prioritize reducing their environmental impact due to growing international concern about global warming (Xu et al. 2013) and stricter environmental regulations (Mathiyazhagan et al. 2013). Meanwhile, in the manufacturing environment, vertical networking, end-to-end engineering and horizontal integration across the entire value network of increasingly smart products and systems is set to bring in the fourth stage of industrialisation—Industrie 4.0 (Kagermann et al. 2013).

X. Chen (✉) · M. Despeisse · B. Johansson
Industrial and Materials Science, Chalmers University of Technology, Gothenburg, Sweden
e-mail: xiaoxia.chen@chalmers.se

P. Dahlman
AB SKF, Gothenburg, Sweden

P. Dietl
SKF Österreich AG, Steyr, Austria

Y. Kishita et al. (eds.), *EcoDesign and Sustainability I*, Sustainable Production, Life Cycle Engineering and Management, https://doi.org/10.1007/978-981-15-6779-7_18

The environmental impact of implementing digital technologies in the context of Industrie 4.0 initiated controversial debates between optimistic and pessimistic assessments. It is positive because information is generally considered to be distinct from material and energy, and to act as a substitute for the use of material resources. On the other hand, it generates negative environmental impacts due to resource consumption from computers and other hardware, especially the fast-increasing waste of electrical and electronic equipment (Berkhout and Hertin 2004).

Since the first industrialization began at the end of the eighteenth century, the manufacturing industry played a crucial role in the technological revolution and betterment of the human condition. Applying digital technologies in reducing environmental impact directly enforces the possibility to reach a win-win situation.

The implication from digital technology to environmental sustainability is mostly from research perspective. This paper seeks to identify opportunities for realizing sustainable manufacturing by implementing digital technologies from real company cases. The purpose is to reveal the possibilities of integrating digital technologies to reduce the environmental impact in the manufacturing industry. Practices of using digital technology through the entire value chain, including both vertical and horizontal integration, are identified through the lens of environmental sustainability.

18.2 Literature Study

18.2.1 Digital Technologies in Manufacturing

The implementation of three features of Industrie 4.0 was targeted (Kagermann et al. 2013):

- Horizontal integration through value networks
- End-to-end digital integration of engineering across the entire value chain
- Vertical integration and networked manufacturing systems.

Digital technologies in manufacturing enables algorithms and technologies at each system layer to collaborate across a unified structure and realize the desired functionalities of the overall system for enhanced equipment efficiency, reliability and product quality (Lee et al. 2015).

In practice, the horizontal integration through value networks can be the Manufacturing Execution System (MES) (Kletti 2007), which links departments such as logistics and procurement, demand chain, manufacturing, maintenance and product application. The information flows fast and digitally through different functions.

End-to-end digital integration of engineering across the entire value chain can be represented as the monitoring of the complete manufacturing flow, which provides a holistic picture of the operation and engineering as the basis for improvement. Robots and automated guided vehicles realized the efficiency increase in manufacturing and transportation.

Vertical integration and connected manufacturing and logistics systems interlink suppliers, manufacturing unit and customers, mainly through Enterprise Resource Planning (ERP) (Mabert et al. 2003) and warehouse management systems.

18.2.2 Digital Technology and Sustainability

Along with the development of technology, its environmental impact is widely discussed. A study by Kiel et al. explored the industrial internet of things related to economic, ecological, and social sustainability. It concluded that higher resource efficiency was the major benefit from technology towards ecological sustainability (Kiel et al. 2017). Bonilla et al. investigated the sustainability implications of Industry 4.0, and linked to the Sustainable Development Goals with mostly positive impacts (Bonilla et al. 2018).

The vision of future factories was showed with the purpose of reaching eco-effectiveness in (Herrmann et al. 2014). Stock and Seliger identified the opportunities of sustainable manufacturing in Industry 4.0 from Macro and Micro perspectives (Stock and Seliger 2016).

Positive environmental implication of digital technology implementation can be categorized as following (Berkhout and Hertin 2004):

Improved efficiency: Improved production efficiency means less processing time required per product, as well as lower rate of scrap, loss and rework. These are KPIs for measuring production performance. Consumption of energy, cooling fluid and material will decrease accordingly, and contributes to environmental sustainability KPIs.

Dematerialization: Digitalization including 3D CAD systems, CAD-CAM connection, digital documentation, electronic instruction or information for the workers, and bar-code traceability enables an increasingly paperless manufacturing environment. Dematerialization means less material consumption and less waste, which indicates better performance on the KPIs of material and resource efficiency.

Virtualization detection and monitoring of environmental change: Digital or virtual meetings or trainings increase the efficiency of communication and reduce the frequencies of traveling. Sensors and microprocessors support detecting and monitoring of environmental changes, which include measuring temperature, colour, volume of fluids and power consumption. It shows potentials on identifying bottle-necks of environmental performance by tracking the peak processes and moments of KPIs.

Transport and distribution: Digital technology enables efficient communication and coordination between different functions, and reduces transports by optimizing vehicle utilization rate and travel routes. It contributes directly to the KPIs of energy consumption and CO_2 emissions. This will become even more important in the future when using more electric vehicles, due to their limited distance range. The closer relationship between customers and suppliers reduces inventory level and storage requirement, which are the KPIs of throughput time as well as energy consumption.

The literature shows that digital technologies can reduce the environmental impact along the manufacturing value chain. The case study presented in this paper focuses on how digital technologies support sustainability performance in practice.

18.3 Methodology

Any case study finding or conclusion is likely to be more convincing and accurate if it is based on several different sources of information, following a corroboratory mode (Yin 2009). In this case study, qualitative research method was applied. Multiple sources of evidence were used during the case study, including interviews, observation (direct and participant) and documents, as Fig. 18.1 shows.

18.3.1 Interviews

Focused interviews were taken with people from two manufacturing sites from production, process development, environment, health and safety, and human resource, including both white collars and operators. Focused interviews remain open ended and assume a conversational manner, with a certain set of questions to follow, which are derived from the case study protocol (Yin 2009).

Ten interviews were carried out with the same set of questions and each of them lasted for at least one hour. Two interviews were through virtual web-meetings due to long distance, and the rest was done face-to-face. All of the interviews were performed with sharing of the question list in advance, and audio recorded with the permission from the interviewees.

The data was reviewed and analysed through audio transcription by NVivo, a software for interview data analysis. The results were reviewed again afterwards with the interviewees to avoid bias and misunderstanding.

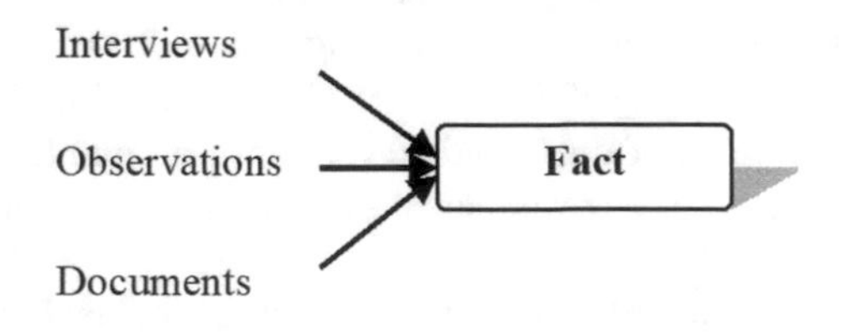

Fig. 18.1 Multiple sources of evidence

18.3.2 Observation (Direct and Participant)

Direct observation

Onsite visits were taken in the case company several times for different purposes. To understand how the MES works among different departments, the process engineer guided a tour by walking through the manufacturing processes along the material flow. To see how the control centre of facility management runs, the interview was carried out in the control room with demonstrations shown by the operating engineer from maintenance department. To check how digital tools facilitated work in production line, a tour was done together with the leading operator from production.

Participant observation

Quick and in-depth understanding of the case company's strategy, culture and activities cannot live without the author's previous working experience in this company. It also provided access to talk with the right person and gain valuable information. Meanwhile, the new role of being an external observer avoided the potential biases as a participant in the organization.

18.3.3 Documents

Two major sources were collected in the study: one was from the official internal website from this company, another one was directly given by the interviewees.

The official internal website provides up-to-date policy, strategy and announcement and can be accessed to all employees, which verified and corroborated information from other sources. The documents from interviewees supports on presenting the results in a visualized manner.

In addition, literature study provided a solid theoretical backbone to this study.

18.4 Results

The case company is an international company with 103 manufacturing sites all over the world. Technology development accounts for an important role along the 110 years' history. Although it is a single case study, it generalizes patterns and findings, which could be applicable for other countries and areas outside of Sweden.

Along manufacturing processes, digital technologies indicate improvement potentials on environmental impact reduction, as it shows in Table 18.1. Four categories of environmental impacts reflect contributions from individual digital technology (Berkhout and Hertin 2004):

- Improved efficiency (represented with sign of)

Table 18.1 Practices of digital technology support to reduce environmental impact in this case study

Operation processes		Improved efficiency	Dematerialization	Virtualization, detection, monitoring of environmental change	Transport
Logistics		Digital signal	ERP		MES AGV
Manufacturing	Production planning	Sensor ERP	MES Sensor	ERP Sensor	ERP
	Process and material efficiency	ERP MES Robot GCM	Sensor Robot GCM	Sensor	
Design and application	Design support		VR	VR	VR
	Customer support	COMO		COMO	COMO
	Anti-counterfeit		Digital signal		
Facility management	Energy and flow	Sensor		Sensor	
	Energy recovery			Sensor	

AGV Automated Guided Vehicle. *COMO* Condition Monitoring System. *ERP* Enterprise Resource Planning
GCM Grinding Cycle Monitoring. *MES* Manufacturing Execution System. *VR* Virtual Reality

- De-materialization (represented with sign of)
- Virtualization, detection, and monitoring of environmental change (represented with sign of)
- Transport and distribution (represented with sign of).

The following sessions explain the practices of implementing digital technology on reducing environmental impacts along the manufacturing processes from the case company: logistics, manufacturing, design and application, facility management.

18.4.1 Logistics

The positive impact on environment brought by digital technology in logistics and supply chain are mainly due to optimized transport and distribution.

Loading of vehicles, e.g. electric Automated Guided Vehicle (AGV) can be optimized with the support of information transparency enabled by Manufacturing Execution System between supply chain and production. The material feeding and Work-In-Process transportation can be planned in the manner of combining time requirement and routes optimization. Less frequent transportation requires less energy consumption and CO_2 emission, which are the main KPIs from environmental sustainability.

Digital signal bridges manufacturing processes and supports a pull system, which leads to minimal inventory kept in hand. Less space is required and unnecessary energy consumption can be avoided accordingly, contributing to the KPIs of energy and CO_2 emissions.

In addition to replacing traditional energy resource with green energy in car and truck fleet, smart planning to travel routes reduces energy consumption and emissions.

18.4.2 Manufacturing

Manufacturing processes includes scheduling, process and material efficiency. In the contribution to environment, improved efficiency, virtualization, detection and monitoring of environmental change, de-materialization and transport and distribution can be realized through the support of digital technology.

Production Planning

The heat treatment process is the most energy intensive process according to LCA studies in the case company. Without smart planning, the furnaces' temperature varies along with the random type of products. Enterprise Resource Planning (ERP) system allows intelligent scheduling by concentrating the lot types and sizes with similar requirement for heat treatment process. With the reduction on frequencies of heating up furnaces, the energy consumption decreases correspondingly.

By monitoring the temperature of furnaces and amount of electricity through sensors, which also provides information to better control of product quality, therefore to reduce the rate of Scrap, Loss and Rework.

The system also supports on scheduling the delivery to customers with an optimized transport load and a pre-defined route, to reach a lower energy consumption and CO_2 emission.

Process and Material Efficiency

MES, robots and a so-called Grinding Cycle Monitoring software support on improving process and material efficiency.

MES: The Manufacturing Execution System is managing production/manufacturing efficiency, material and non-conformance. It is placed in a layer between the Enterprise environment and shop floor. MES receives pushed information from the Enterprise needed on the shop floor, by the machines and operators. Meanwhile, MES also pulls data from the shop floor, which visualized and stored as process data, for traceability reasons and customer demands.

MES monitors and visualizes the energy consumption on each single process, which provides a good basis for observing and trouble-shooting abnormal situations, as well as for improvement. Status of running, lack of material, buffer overflow and others are the four machine conditions that monitored and categorized. The system sends instant reports to corresponding operator, channel manager and maintenance operator, in case of disturbances. It ensures that the situation will be notified and taken care of within a short notice. With the reduction of idle running time and Mean Time to Repair, energy and efficiency loss can also be reduced.

In addition to efficient communication, MES supports on achieving a paperless environment. Before, it was paper attached on each pallet enabled the communication between each process, which indicated the produced pieces, the notification to next process, and the quality issues. Nowadays, operators no longer need to register on papers along each pallet. Along with the process, operators register and record information directly into the system. Barcode attached on each pallet tells the details along the whole process of this batch of products by a single scan.

Robots: Fast resetting, stable quality and higher efficiency motivated the use of robots. Robots can replace heavy and repetitive work tasks with higher and stable efficiency and quality, which leads to lower risk of scrap loss and reworks and higher energy efficiency. It results in a lower environmental load per product by avoiding unnecessary use of energy and waste of material.

Grinding Cycle Monitoring: It is a system to identify improvement potentials by monitoring grinding cycles, which enables reduction of processing time of each cycle without compromising quality. There are four stages of each grinding cycle, where the first stage has the biggest potential to be shortened, as it is the period when the grinding tool approaches the work-piece. Grinding Cycle Monitoring helps to identify the positions of grinding tool and work-piece, as the distance between the two can be different due to the variety of parameters from incoming work-pieces.

Therefore, the Grinding Cycle Monitoring system can optimize the translational speed (feed rate) of the grinding tool according to the detected distance.

Besides higher efficiency, it can also reduce the rate of scrap loss and rework, as crashes between grinding tool and work-pieces can be avoided. The monitoring of the whole grinding cycle provides opportunity to improve quality, as the appearance of work-pieces indicate improvement potentials reflecting to the corresponding grinding phase. Consequently, lower rate of scrap loss and rework contributes to lower consumption of energy and material.

18.4.3 Design and Application

Digital technologies enable effective and efficient communication between products develop engineers and customers, including virtual reality, condition monitoring system and barcode system.

Design support

Virtual reality (VR) is a popular information technology area that provides an indirect experience by creating a virtual space that interacts with the human sensory systems and overcomes spatial and physical constraints of the real world (Electronics and Telecommunications Research Institute (ETRI) 2001). Virtual reality technology offers products develop engineers a better understanding of how the products are going to be applied in customer side. Working conditions, required rotating speed, appropriate load weight, applied facility and installation limitations are illustrated in virtualized environment, which gives a live picture to design engineers. Design engineers will propose product parameters according to this comprehensive picture, and can test it again within this virtual environment. It is no longer needed to produce product prototypes from version to version. In this case, VR saves communication and manufacturing regarding prototype products between production and designer, which leads to a recognizable save on lead-time, resource, energy and cost, which are important KPIs on productivity and environmental sustainability.

Customer support

Condition Monitoring System (COMO) supports on reflecting application condition from customer side to application engineers. With the proper sensors to supply the critical operating information, the machine operates in a safer condition for both machine as well as the personnel operating the machine.

The monitoring data helps application engineers to predict failures accurately and supports on suggesting grease filling and components replacing in time. The accumulated statistic data also assists design engineers to improve in future product design. The Condition Monitoring System enables failure detection, feedback and support to customers in a shorter lead-time, secures trouble-free application for customers. Meanwhile, it contributes to reducing downtime on customer side, reducing frequencies of traveling to application site, and bringing impact to environment accordingly.

Digital measures for anti-counterfeit

Digital technologies allows anti-counterfeit by marking on the product with digital information. Before, the counterfeit was difficult to distinguish from either packaging or appearance. Marking has made it easier to identify counterfeit products. In some cases, counterfeit products could only be found out when it was installing on the machine, which could generated enormous failure and loss both for manufacturers and for customers. Nowadays, the code on product guarantees the authenticity during the preparation of installation on customer side. It helps to build up the credibility of products and avoid the rampant growth on counterfeits. It also contributes on reducing waste of resource and energy under circumstances of using counterfeits products.

18.4.4 Facility Management

As shown in Fig. 18.2, one major task of technical building services is to ensure the needed production conditions in terms of temperature, moisture and purity through cooling/heating and conditioning of the air (Despeisse et al. 2012). With the support of digital technology, the energy and fluid flow management, as well as energy recovery can be visualized and realized.

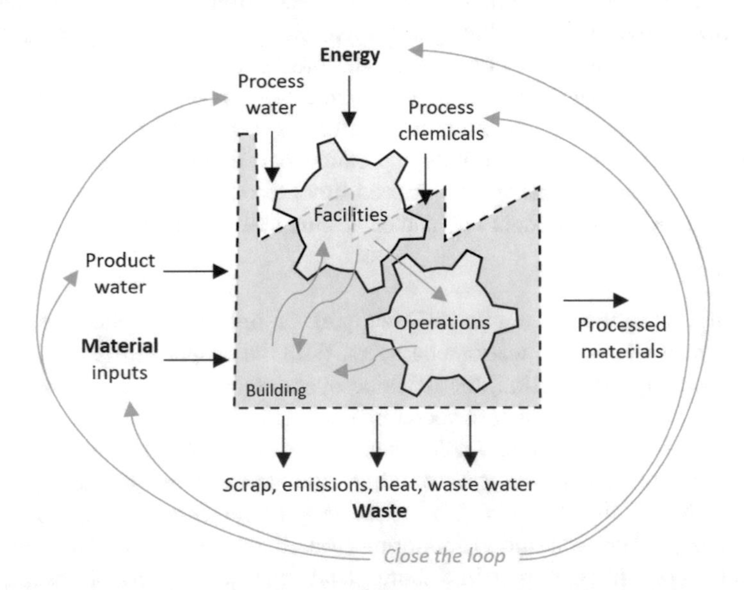

Fig. 18.2 Production facility management illustration (Despeisse et al. 2012)

Energy and fluid flow management

The energy consumption includes electricity or other power sources for heating up ovens, furnaces and building, supplying compressed air and running machines. Fluid flow mainly consists of cooling liquid.

Sensors collect real time data on energy and fluid consumption and indicate abnormal situations instantly. The central control room collects and visualizes energy consumption data from critical processes, e.g. induction, soft annealing and heat-treatment. These processes are the most energy intensive processes based on the LCA studies. With the awareness of huge consumption of energy in furnaces and ovens, optimized planning with fully loaded batches is motivated in order to increase utilization.

Energy recovery

Energy losses also provide opportunities to recover energy for those major consumers. Heat is collected mainly from the following sources:

1. Cooling fluid from the hot rolling mill through water heat exchanger.
2. Heat recovery from air compressors through lubrication oil/water heat exchanger.
3. Energy in hot oil from hardening furnaces.
4. Other sources: hydraulic oil, honing oil, grinding water.

Other solutions for energy recovery are through the ventilation systems with heat recovery, mainly installed in shower-rooms, bathrooms, toilets, etc.

Besides, heat collected for heating up offices and shop floor in wintertime, and heat exchanged with river water for cooling down those areas in summer time. Compared to the situation of 10 years ago, the consumption of electrical energy is around 50% less today in one manufacturing site, despite growing production volume.

To sum up, the potentials of environmental impact reduction from digital technologies shows in Fig. 18.3 along the value chain from the case company. The figure reshaped the "Micro perspective of Industry 4.0", proposed by Stock and Seliger (Stock and Seliger 2016), and visualized possibilities of reducing environmental impact by digitalizing operation processes. Facility management contributes tremendously on increasing energy efficiency, and shows as Smart Factory in Fig. 18.3.

In addition to the support from digital technology implementation, company policy on energy and CO_2 reduction from sourcing department on machine purchase also contributes to environmental sustainability.

The machine purchase should include the environmental requirement in the specification. As indicated in technical specifications in Environmental Standardization Group, the amount of energy consumption and CO_2 emission should not exceed the limit regulated in company policy, including refurbished machines. It is an important indicator when purchasing a new equipment, and cannot be compromised in any circumstances, including lower price.

The distance of material supplier is an important KPI when developing a new supplier, as the distance can influence the lead-time of raw material and components,

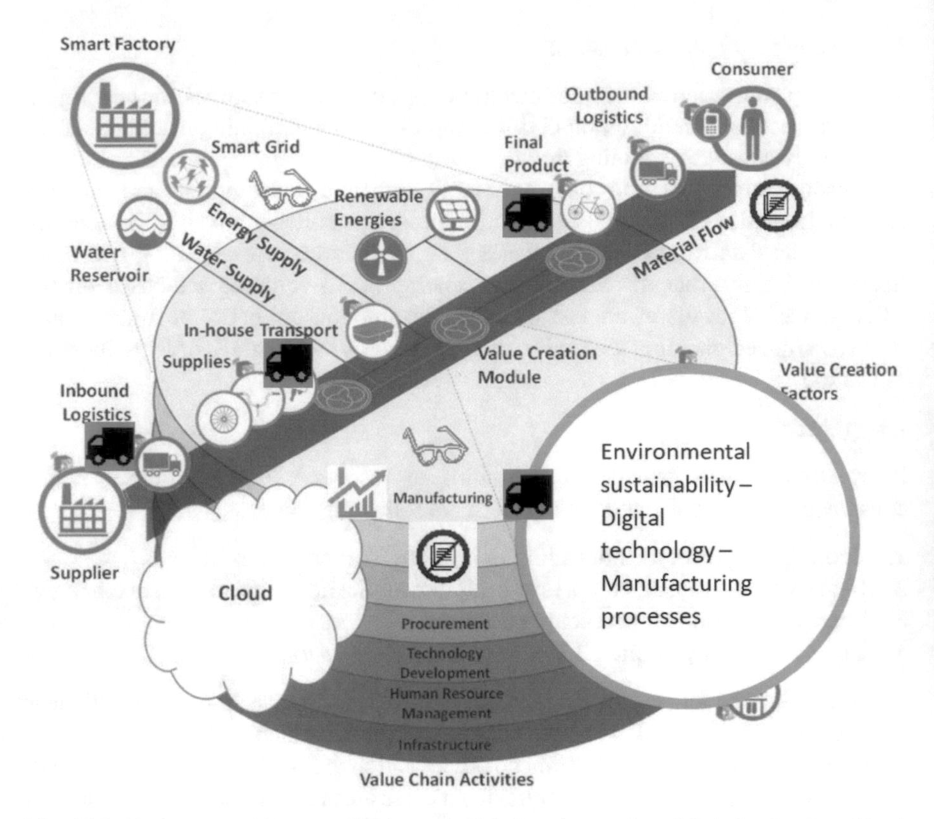

Fig. 18.3 Environmental impact reduction potentials from integration of digital technology (Stock and Seliger 2016)

as well as the transportation distance required and CO_2 emission. It is calculated into the environmental impact per product in the internal LCA studies.

18.5 Discussion

18.5.1 Implications

Value sustainability up to a strategic level helps to implement environmental impact reduction with digitalization. Sustainability goal is set from Group strategic level, and put into practice in each factory and each channel by breaking down the big goal in this case company. Being sustainable at a Group strategic level helps increase the awareness of integrating digital technology with environmental impact reduction. As one of the interviewees responded with "digital technology and environmental care is very much integrated into the whole system", when asked the question *How do you use digital technology to reduce the environment impact*. Energy and CO_2 are the key performance indicators, which is a well-known fact to director, department managers, and operators. Each factory has a specific reduction goal on energy and CO_2 emission, and needs to track and report the progress to the Group every year. Therefore, the investment and implementation of digital technology can consider environmental impact from the very beginning.

Digital technology can support environmental impact reduction along the complete manufacturing value chain. There are certain positive potentials from each application of digital technology to environmental sustainability. This can assist on decision making when planning new investments on digital technologies, alongside efficiency improvement and quality stability. For different environmental impact reduction along manufacturing value chain, such as improved efficiency, dematerialization, virtualization, detection and monitoring of environmental change or transport and distribution, different digital technologies can be adopted accordingly. On the other side, when implementing a new digital technology, it also brings opportunities on exploring potentials on reducing environmental impact.

However, digitalization is not always coming along with positive impacts to environment. For instance, robots' implementation may require more power consumption comparing with manual operations. In addition, as one interviewee argued, the investment to building insulation itself might bring even more direct and bigger reduction to energy consumption, comparing with investing on digitalization. Finding the balance between efficiency increase and environmental impact reduction is the challenge while investing on digital technologies. It also indicates needs on further exploring potentials to environmental sustainability from digitalization.

The acceptance of digitalization from people side requires attention as well. The attitude towards technology acceptance is not always positive, as the fact of threatening certain jobs' replacement is happening. It is challenging to reach consensus now on how to harness technology, which requires deep understanding on the impact to productivity performance, human perspective and environmental sustainability.

18.5.2 Limitations

Proper presentation of the acquired knowledge to expert users supports the correct decision to be taken (Lee et al. 2015). There is no universal model found so far can indicate the potentials of reducing environmental impacts by digital technology along manufacturing value chain. The model adopted to illustrate the practices from digital technology to support environmental impact reduction along manufacturing processes was based on Stock and Seliger's figure of Micro Perspective of Industry 4.0 from (Stock and Seliger 2016). The point of choosing this one is value the combination of digital technology implementation along manufacturing value chain. The impact to environment therefore can be simply added as a third layer to this model.

As a single case study from qualitative analysis, the generalized findings what we believe worked in Sweden, could be applied and valid for other countries and areas as well, which requires further verification in a broader area. The proposed pattern of environmental impact reductions from digitalization along manufacturing value chain needs validations on various cases by quantitative analysis.

18.6 Conclusion

This study presents an initial step towards a framework of using digital technology to reduce environmental impact along manufacturing value chain by showcasing a case study in an international manufacturing company. It provides a viable and practical guideline for manufacturing industry of integrating digital technology on environmental impact reduction. It also supports on decision making to explore opportunities of being more sustainable when investing on digital technology in manufacturing industry.

Acknowledgements The author would like to thank all the interviewees from the case company, for their precious time and valuable input. Great gratitude to the colleagues used to work together, for their contribution on reviewing the technology explanation.

References

Berkhout F, Hertin J (2004) De-materialising and re-materialising: digital technologies and the environment. Futures 36(8):903–920

Bonilla H, Silva O, Terra M, Franco R, Sacomano B (2018) Industry 4.0 and sustainability implications: a scenario-based analysis of the impacts and challenges. Sustainability 10, 3740

Despeisse M, Ball PD, Evans S, Levers A (2012) Industrial ecology at factory level—a prototype methodology. Proc IMechE, Part B: J Eng Manuf 226(10):1648–1664

Electronics and Telecommunications Research Institute (ETRI) (2001) Virtual reality technology/market report. Daejon, 30 December pp 12–29

Herrmann C, Schmidt C, Kurle D, (2014) Sustainability in manufacturing and factories of the future. Int J Precis Eng Manuf-Green Tech 1:283

Kagermann H, Wahlster W, Helbig J (2013) Securing the future of German manufacturing industry: recommendations for implementing the strategic initiative INDUSTRIE 4.0. acatech, Final report of the Industrie 4.0 working group

Kiel D, Müller J, Arnold C, Voigt K-I (2017) Sustainable industrial value creation: benefits and challenges of industry 4.0. Int J Innov Manag V21, 1740015

Kletti J (2007) Manufacturing execution system (MES). In: Manufacturing execution system—MES, pp 13–28

Lee J, Bagheri B, Kao H-A (2015) A cyber-physical systems architecture for industry 4.0-based manufacturing systems. Manuf Let 3:18–23

Mabert V, Soni A, Venkataramanan MA (2003) Enterprise resource planning: managing the implementation process. Eur J Oper Res 146(2003):302–314

Mathiyazhagan K, Govindan K, Noorulhaq A, Geng Y (2013) An ISM approach for the barrier analysis in implementing green supply chain management. J Clean Prod V47:283–297
Stock T, Seliger G (2016) Opportunities of sustainable manufacturing in Industry 4.0. Proc CIRP 40:536–541
Xu L, Mathiyazhagan K, Govindan K, Noorul HA, Ramachandran NV, Ashokkumar A (2013) Multiple comparative studies of green supply chain management: pressures analysis. Resour Conserv Recycl V78:26–35
Yin RK (2009) Case study research: design and methods, 4th edn. Sage, Thousand Oaks, CA

Chapter 19
Life Cycle Simulation System as a Tool for Improving Flow Management in Circular Manufacturing

Keito Asai, Dai Nishida, and Shozo Takata

Abstract Circular manufacturing is a system that manufactures items with multiple circulations, that is, reuse or recycling to use these items efficiently in their life cycles. To manage such complex systems and design circular manufacturing scenarios, an accurate circulation flow management method and life cycle scenarios are necessary. In this study, we developed a life cycle simulation system that is versatile enough to improve the flow management method. Further, we applied it to lithium-ion batteries in electric vehicles, evaluated its life cycle scenario, and improved the circulation flow management method.

Keywords Life cycle simulation · Circular manufacturing · Multitiered circulation · Material flow management · Value provision

19.1 Introduction

In the past, arterial and venous industries were developed separately with different goals. The former was developed to manufacture maximum number of good quality products within competitive costs, and the latter was developed for disposing waste safely and efficiently (Kimura 1999). However, to achieve sustainable development in manufacturing industries along with economic growth, we need to create values with less resources and environmental impact. Thus, we should integrate the arterial and venous industries in the form of circular manufacturing for using resources efficiently. Circular manufacturing is a manufacturing system in which products, modules, parts, and materials (collectively called items) are reused and recycled as much as possible so that we can decrease the amount of resources and environmental load for providing functionalities to the customers and increase profits. To realize circular manufacturing, the life cycles of the items should be designed properly. Accordingly, various types of life cycle options, such as closed-loop reuse, cascade reuse, and recycling, should be introduced. However, there are numerous factors to

K. Asai (✉) · D. Nishida · S. Takata
School of Creative Science and Engineering, Waseda University, Tokyo, Japan
e-mail: keimino030526@akane.waseda.jp

Y. Kishita et al. (eds.), *EcoDesign and Sustainability I*, Sustainable Production, Life Cycle Engineering and Management, https://doi.org/10.1007/978-981-15-6779-7_19

be considered for introducing them effectively. For example, we deliver not only new products but also reused products to users; thus, we must consider items' conditions as well as users' requirements. Furthermore, we must consider the transitions among multiple circulation loops.

In designing complex systems like circular manufacturing, it is essential to have an evaluation method. Accordingly, life cycle simulation (LCS) has been developed (Kimura 1999; Umeda et al. 2000; Takata and Kimura 2003). LCS is a method to simulate items' circulations and evaluate their economic and environmental impacts during their life cycles. So far, many LCS systems with various characteristics have been developed. For example, Kawakami et al. proposed an LCS model for simulating a system with multiple product life cycles considering their interactions (Kawakami et al. 2017). Nassehi and Colledani proposed a multimethod simulation model that combined a multi-agent model with a system dynamics model (Nassehi and Colledani 2018). Komoto et al. applied LCS to product-service system design (Komoto et al. 2005).

Most LCSs have been developed for the purpose of life cycle planning. They simulate the items' flow through various processes. However, there are few systems which can be used for the management of the circular manufacturing system.

In order to use LCS for improving the circulation flow management, the LCS system should be able to check when and where each item is processed and used by which user and in which usage mode. In this study, we developed an LCS system that satisfies such requirement. The developed LCS system can manage the characteristics and behavior of an individual item and user, control the individual item's flow through the processes, and track the history of all items.

We applied the developed LCS system to lithium-ion batteries (LIBs) in electric vehicles (EVs) and demonstrated its effectiveness in improving the circulation flow management methods.

The remainder of this paper is organized as follows. In Sect. 19.2, we explain the problems in management of circular manufacturing. In Sect. 19.3, we evaluate multi-tiered circulation. In Sect. 19.4, we describe the required functions and features of the developed LCS system. In Sect. 19.5, we present an application of the LCS to LIBs in EVs by describing the procedure of improving the circulation flow management scenarios. We conclude the study in Sect. 19.6.

19.2 Problems in Design and Management of Circular Manufacturing

19.2.1 Uncontrollable Supply in Terms of Quantity and Quality

In linear manufacturing, the supply is controlled to synchronize it with the users' demand in the market.

However, in circular manufacturing, it is difficult to control the supply because the supply is determined by the timing of termination of usage, which is on user's discretion. To identify the problems more explicitly in circular manufacturing, we explain them by taking an example of reuse, which is an important life cycle option for realizing circular manufacturing.

Reuse is one of the methods to circulate resources and improve utilization rate of life span. By matching the requirements of users and the realized functions of items, the items are used repeatedly with several users or methods of use. There are two main problems in performing reuse efficiently in circular manufacturing.

The first problem is controlling the quantitative balance of demand and supply (Fig. 19.1a). After users finish using items, those items are collected and certain number of collected items are reused. However, the number of collected items and the rate of items that can be reused fluctuate with time. The number of users requesting items fluctuates with time as well. For performing reuse efficiently, controlling the balance between demand and supply is necessary.

The second problem is appropriate matching of a required function of users and a realized function of items (Fig. 19.1b). Items of different generations realize different level of functions, while for items manufactured in the same generation, the deterioration rate of each items realize different level of functions. Furthermore, each user has different requirement in functions. Therefore, proper matching of each user and item is essential in circular manufacturing.

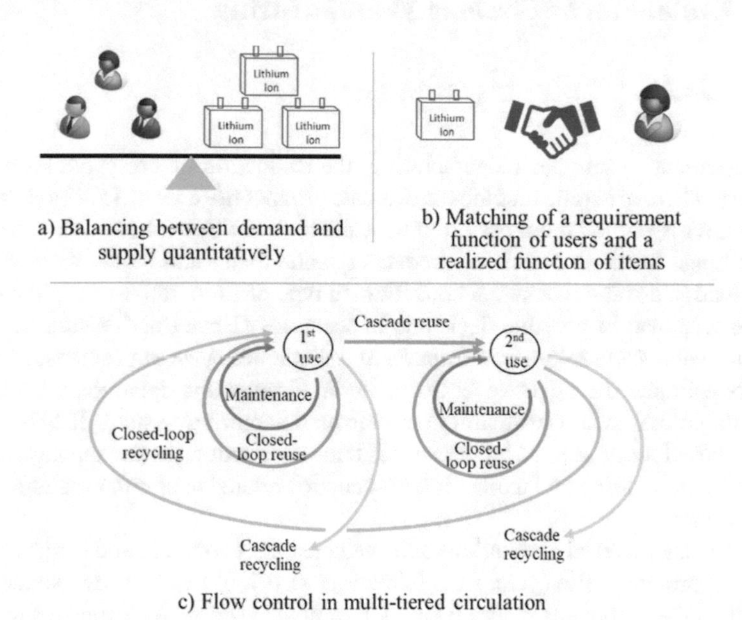

Fig. 19.1 Problems of design and management in circular manufacturing

19.2.2 Multitiered Circulation

Problems of management in circular manufacturing are that circulation is not a single-loop circulation, but a multi-loop one. Circulation flow management for multi-loop needs to overcome very complex systems (Fig. 19.1c). There are many life cycle options, such as maintenance, closed-loop reuse/recycling, and cascade reuse/recycling; thus, we must consider all their characteristics and design to combine the life cycle options appropriately.

Furthermore, to manage various life cycle options properly, we must consider the right timing of transferring the items from the inner loop to the outer loop. Generally, items are circulated in the same loop unless an item can't realize a function that is demanded by the user. However, some items are better to be transferred from the inner loop to the outer loop because supplying a reasonable number of items to the outer loop based on users' requirements may provide more overall values to users than oversupplying items to the inner loop. Therefore, controlling the timing of transfer is important. It is necessary to overcome these problems in multitiered circulation along with that mentioned in Sect. 2.1. Thus, it is inevitable to build a mechanism for properly evaluating the life cycle scenario and overcoming these problems.

19.3 Evaluation of Circular Manufacturing

19.3.1 LCS

In management of circular manufacturing, the evaluation of life cycle scenario is necessary. There are some methods to evaluate product life cycles. Life cycle assessment (LCA) is one of the most prominent methods. LCA evaluates an entire or a specific range in the life cycle of products quantitatively and yields their environmental load in a static manner. A conventional type of LCA can estimate the differences in response to possible decisions in scenarios (Ekvall and Weidema 2004). LCA can evaluate environmental impacts of a single life cycle of products; however, it cannot estimate the effectiveness of material circulations dynamically, which is an inevitable feature for evaluating circular manufacturing system. If behavior of material circulations cannot be perceived, then conducting proper management of circular manufacturing and realization of scenarios related to circular manufacturing is impossible.

To calculate material circulations and evaluate their economic and environmental impacts of products' life cycle, LCS (Umeda et al. 2000) has been developed. LCS is a method that simulates material circulation flow based on a discrete event simulation method. From late 1990s the concept of LCS was developed in the academic field (Kimura 1999; Umeda et al. 2000; Takata and Kimura 2003) and various LCS systems have been developed thus far (Kawakami et al. 2017; Nassehi and Colledani 2018; Komoto et al. 2005; Ekvall and Weidema 2004; Takata et al. 2019). Murata et al. discussed the flow management method for global reuse. They introduced a management algorithm that efficiently realized a circulation and controlled the directions and quantities of the material flow of multitiered circulation especially focusing on global reuse (Murata et al. 2018). In the next subsection, we define the evaluation items that are indispensable for conducting an evaluation of life cycle scenarios in circular manufacturing.

19.3.2 Evaluation Items Required to Design the Effective Circular Manufacturing Scenarios

There are various aspects that needs to be considered in designing the effective circular manufacturing scenarios. We have organized them in the following three perspectives.

1. Evaluation of indices related to the whole life cycles of all items processed during the simulation period
   First, we need to evaluate the overall values from both environmental and economic aspects. Overall means to evaluate all items and processes which are addressed in the simulation. The total cost, resource consumption, and environmental load are the key indicators for overall evaluation. From the economic aspect, the cost of the provider and the value provision to users must be considered. To evaluate different usage mode simultaneously, we use average value provision of each item as a tool to evaluate overall perspective. To evaluate average value provision, we use willingness to pay (WTP) for each usage mode. WTP is the maximum price at or below which a user will pay for an item or one unit of usage (Status Survey on Energy Consumption 2015). We can estimate the value provision of items provided to users in life cycle by calculating the sum of the WTP for each usage per item.
2. Evaluation of indices related to item and customer
   To consider improvement in the overall values, several indicators can be taken into consideration for evaluating measure indicators. For example, life utilization rate, customer satisfaction, and provided value per item.
3. Process and item flow
   To evaluate efficiency of circulation of items, we can use the following. For example, the number of items processed in each process, the number of items in the buffer.

To conduct evaluation of above items, The LCS systems developed so far is unable to evaluate all items explained above; therefore, a new LCS system is required. In the next section, we describe the required functions and features of developed LCS system.

19.4 LCS System Equipped with Functions to Support and Improve Circulation Flow Management

19.4.1 Required Functions to Support and Improve Circulation Flow Management

As mentioned above, circulation flow management is important in circular manufacturing. Management needs to properly deal with, "controlling the quantitative balance of demand and supply" and "appropriate matching of a required function of user and realized function of item". To handle these issues, each item and user must be managed individually. Further, if requirement of users and realized function of items are examined and managed, then matching of items and users should be conducted according to realized function of items and requirement of users. Moreover, to enable the provision of items to users on proper timing, we need to control the items' flows in the system.

To improve a life cycle scenario, the LCS system should have function to recognize the behavior of the items and its users. For example, we check data of actual condition of items and characteristics of provided users in matching process. If it concludes that the conditions of some items are beyond the requirement of user, we can change recovery criteria for items and circulation flow to deliver more items that are in lower conditions. The final evaluation data is insufficient to find the problems and improve the circulation flow management methods easily. To summarize the requirements, the LCS for supporting the design and improvement of life cycle scenarios should have the following functions:

1. Managing the life cycle of each item and users including their characteristics individually
2. Matching mechanism between users and items
3. Flow control mechanism based on items' condition and demands of users
4. Visualization tools to ensure the requirement of the users, the way to use items, and realized functions of items related to each process.

19.4.2 Features of Developed LCS System

We developed the LCS system that satisfied the requirements for improving the circulation flow management scenarios. The LCS system has been developed on the basis of GD.findi,[1] which is a discrete event type production system simulator. GD.findi can model the production line by processing the items. However, modeling of the reuse and recycling processes, which are included in multitiered circulation,

[1] GD.findi, Production system simulation, http://www.lexer.co.jp/en/product/gd_findi. (in Japanese).

are not undertaken. Therefore, we extended GD.findi and added the required features described in Sect. 4.1.

Followings are the features of the developed LCS system that fulfill the requirements described in Sect. 4.1. The numbers of the following list correspond to the numbers of the requirement list in Sect. 4.1.

1. We developed a user model module that can generate and manage the designated number of users according to user distribution for defined market. All users have different characters and features of the functional requirements and the way of use, which are generated randomly according to the distribution. Moreover, different usage modes, such as LIBs used for EVs and stationary battery (SB) used in general household, can be individually designated in this module.
2. We developed a usage model that can calculate deterioration of items according to the method of use defined in the user module.
3. We developed flow control functions using the agent module equipped in GD.findi. The simulation system determines the branch to which the items should be transferred according to the functional state of items and defined criteria.
4. The system maintains the following data related to the users, items, and processes for enabling the analysis of the simulation results.

    (a) The behavior of the items and users were recorded during the simulation so that the state quantities of the items, requirement functions, and method of use of the users can be identified for each individual.
    (b) Processing in each process is recorded in the execution log because of which it is possible to examine the type of process, user, kind of usage, and how the item's state quantity has changed as a result of processing.
    (c) For all processes, the numbers of items existed for each state quantity as stocks can be checked based on the data of execution log.

19.5 Application for Supporting Circulation Flow Management

19.5.1 Life Cycle Scenario of LIBs for EVs

In Sect. 19.5, we conducted LCS taking an example of LIBs for electric cars to verify the effectiveness of the developed LCS system in improving circulation flow management.

We incorporated several reuse methods into the life cycle. The life cycle model of the scenario is shown in Fig. 19.2. It is assumed that the batteries are reused when the functions realized by the item satisfies the functions required by the users.

We considered the closed-loop and cascade reuse methods. In closed-loop reuse (primary use), vehicles for reuse were separated into LIBs and body before EVs were provided to the users to reuse the items efficiently. Some users want to drive long

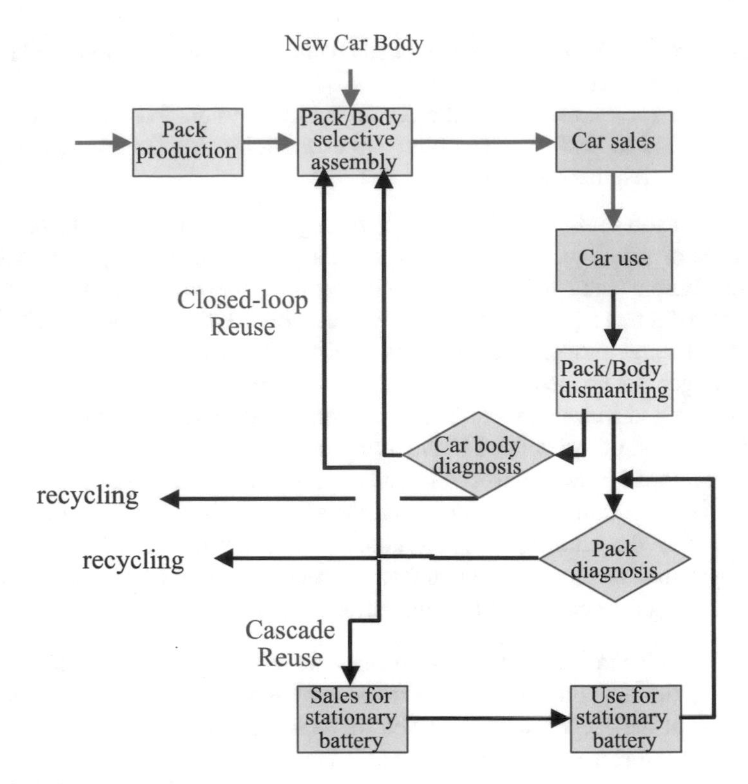

Fig. 19.2 Life cycle model

distance without charging the batteries; thus, they require LIBs with high capacity. Moreover, some users prefer LIBs over good appearance or high-performance functions for body. By checking the adaptabilities of LIBs and bodies to the users' requirements separately, we can selectively combine and provide them to the users with high probability.

Regarding cascade reuse, LIBs are reused as SBs for households (secondary use). LIBs are removed from used and dismantled EVs, and are provided to households. For example, these LIBs charge the electricity generated by solar panels installed in Japanese households.

The state quantity of a LIB is defined by its state-of-health (SOH), which shows the current maximum capacity to the rated one. Those of a vehicle body are defined by mileage and accumulated usage time. We set the following characteristics for the primary users: (1) required distance to empty, (2) annual mileage, and (3) period until the user ceases to use the vehicle. Those for the secondary users are as follows: (1) electricity consumption and (2) period until the user ceases to use the battery. We identified the distribution of each user characteristic based on the questionnaire survey with users in Japan (Market Trend Survey of Passenger Automobile 2017; Tsuchiya et al. 2014; Fukuyo 2011; Status Survey on Energy Consumption 2015).

Regarding LIBs in an EV, we calculated the user requirement of SOH that would enable the user to drive the distance to empty till the user ceases to use the vehicle. We could not obtain the reliable data about the deterioration of LIBs in EVs; thus, we adopted the data of hybrid cars, which shows the deterioration of SOH depending on the intensity of the usage in three levels. If the required distance to empty exceeds the distance that an EV with a new battery can drive, we assumed that the required SOH is 100% and provided the user a newly manufactured battery. For the secondary use of LIBs, the required SOH is calculated based on the break-even point between the total amount of money obtained by power saving and the installation cost of the SBs. The former is evaluated from nighttime power consumption that can be suppressed by using the power generated by the solar panels and that stored in the batteries during the daytime.

Regarding the WTP for primary use, the value is calculated as the product of the distance traveled by car and the price per kilometer estimated from the car share rate. Similarly, the WTP for the secondary use is calculated as the product of the cost of the electricity per kWh and nighttime power consumption, which can be suppressed by using the power stored in the battery. Further, we assume that the value of the battery in the EV contributes 50% to the value provided by the EV to the users.

The number of users we assumed for primary use was 1000 people, and that for secondary use was 300 people. The ratio between the number of EV users and that of solar panel users in Japan was estimated by surveys (Tsuchiya et al. 2014; Fukuyo 2011). We set 35 years as the simulation period.

19.5.2 Simulation Results and Issues Raised by Them

We performed the simulation by setting the conditions for deciding the destination process after a user ceases to use the battery, as presented in Table 19.1. The simulation was initiated assuming that the assembly process of batteries and bodies has an initial inventory of batteries. Their conditions were determined assuming that they have been used once by the users randomly generated using the user characteristic distribution.

Figure 19.3 shows the average value provided by each battery during the simulation period for each condition. Each horizontal bar with different color represents a value provision by each user who used the battery. Figure 19.3a shows most of the

Table 19.1 LIB reusability

Flow control node	Destination	Condition
LIB reusability [SOH (%)]	EV	≧70
	Stationary B	≧65
	Recycling	<65

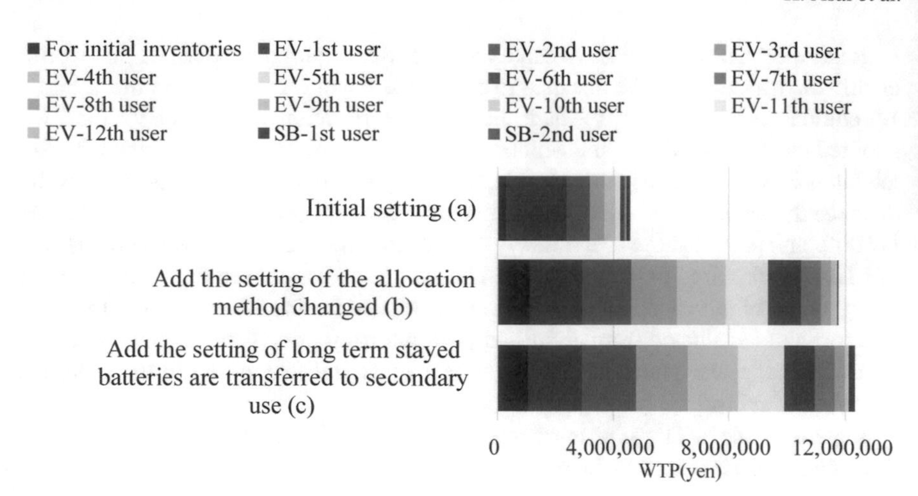

Fig. 19.3 Average value provision (yen)

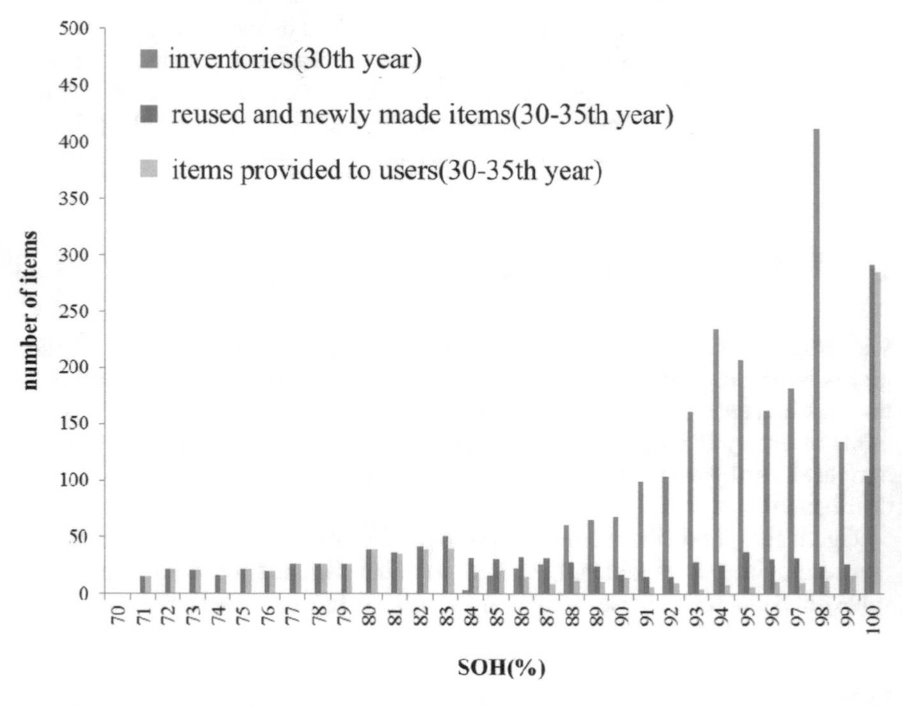

Fig. 19.4 Items' flow and inventories in the assembly process for "initial setting (a)" (30–35th year)

value generated by primary use. We explain the other conditions in the next subsection. In Figs. 19.3, 19.4, 19.5, 19.6, 19.7, 19.8a, b, and c show the same condition settings, respectively.

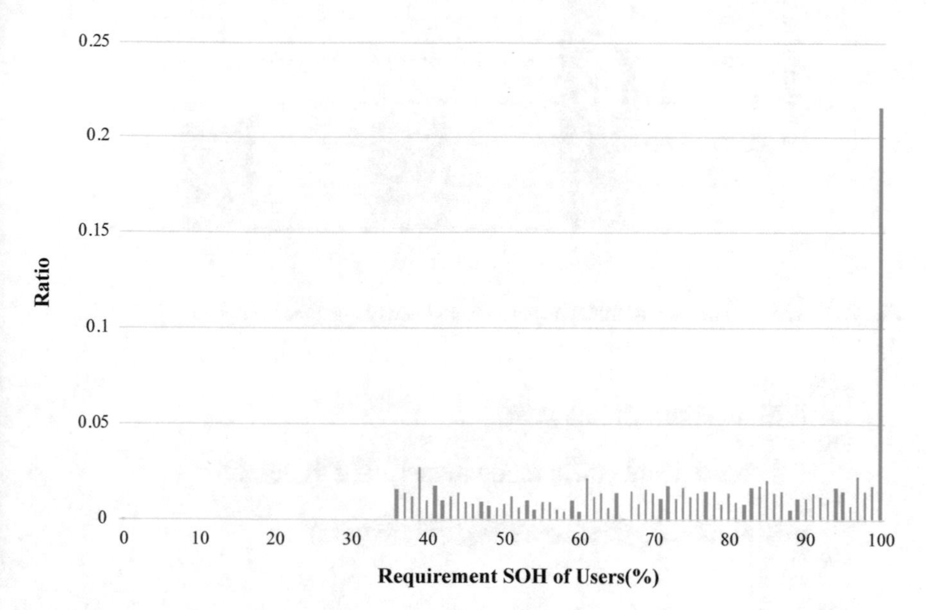

Fig. 19.5 Distribution of the required SOH for EV users

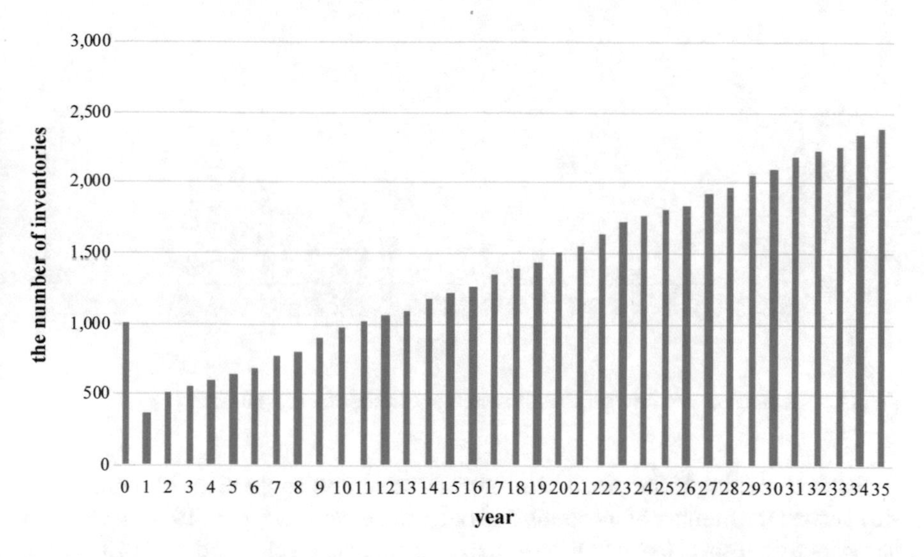

Fig. 19.6 Number of inventories in the assembly process with respect to time for "initial setting (a)"

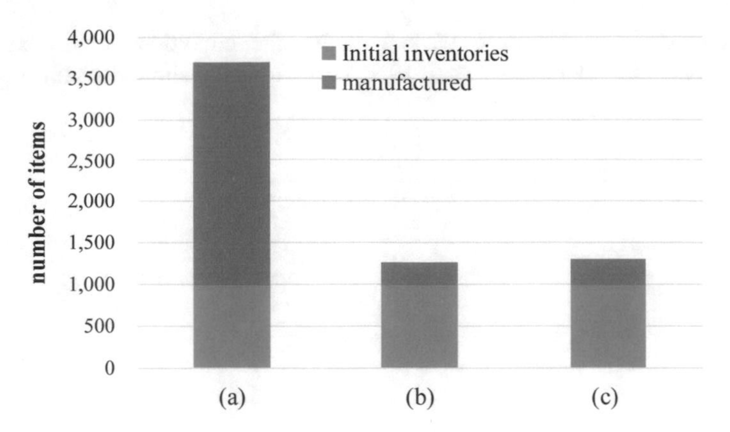

Fig. 19.7 The total number of initial inventories and newly manufactured items

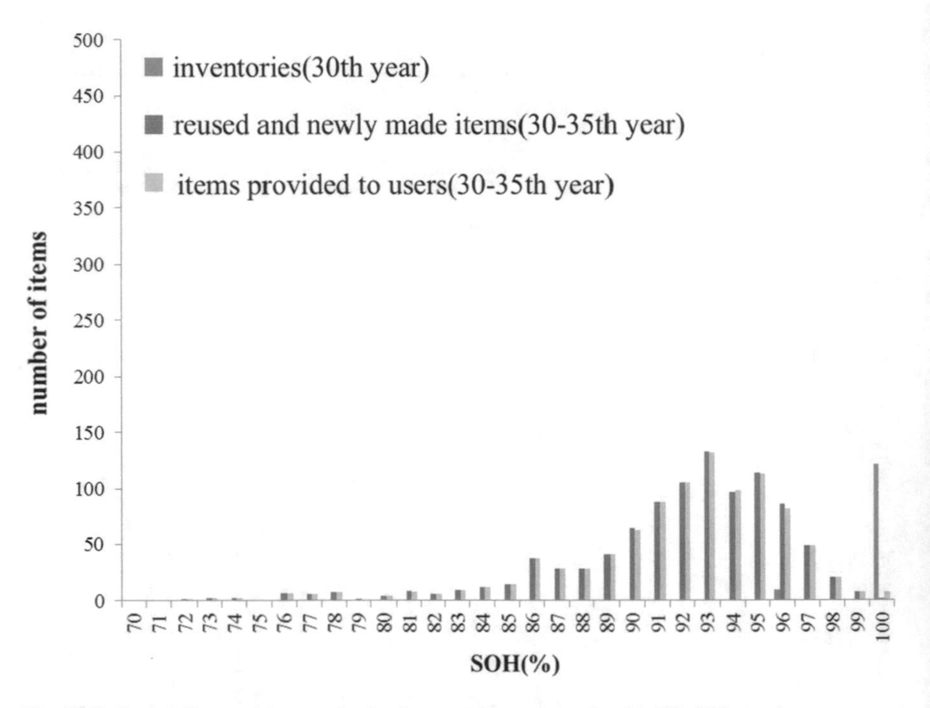

Fig. 19.8 Items' flow and inventories in the assembly process for (b) (30–35th year)

For determining if the proper flow management was adopted in this simulation, we checked the number of incoming and outgoing batteries. Figure 19.4 shows these numbers with respect to the SOH in terms of accumulated values from 30 to 35 years. The figure also shows the number of the batteries in the inventory in the 30th year. As shown in Fig. 19.4, there are numerous LIBs that have SOH close to 100% and many of these LIBs were stocked in the 0th year. This is because there are many users who require LIBs with 100% SOH and only accept newly manufactured LIBs, as shown in Fig. 19.5.

We also checked the change in the number of inventories with respect to time. The result is shown in Fig. 19.6. The figure shows that the number of the inventories increases over time. We consider that some inventories become stagnant and make no value for users.

Based on these results, we adopted two approaches in a step-by-step manner. First, we changed the way of items' allocation to users. Many users require the distance to empty more than the distance that can be driven by the LIBs with 100% SOH. In this case, a user must charge the LIBs at least once during the drive. In this sense, there is not much difference between using a 100% SOH LIB and a LIB with SOH slightly less than 100%. Therefore, we relaxed the restriction on the allocation of LIB to the users and provided the LIB with SOH of 95–100 reused ones to the users that required 100% SOH.

Second, we transferred the inventories that stayed in the buffer for five years to use them for the secondary use. By using these two approaches in a buildup manner, we attempted to increase the value proposition of each LIB and decrease the number of LIBs to satisfy the users for a certain period of time.

19.5.3 Improvement of Item's Flow Management in the Life Cycle Scenario

We calculated the value provisions when adopting these two approaches in a step-by-step manner. Figure 19.3b shows the result when the allocation rule of the LIB to the user changed, and Fig. 19.3c shows the result when the stagnated inventories are transferred to the secondary use in addition to the change in the allocation rule.

As shown in Fig. 19.3, both the change in allocation and the alteration of sending long-term stocked items increased the value provision for users. This was because we minimized the number of items input in simulation, which increased the value provision of each item for the users, thereby increasing the average value provision for the users. In addition, Fig. 19.3c shows a higher value provision in secondary use as compared to that shown in Fig. 19.3b. This is because items that were stagnated in the primary use process were transferred to the secondary one and were used as SB to make value for the secondary users. Figure 19.7 shows the number of initial inventories and newly manufactured items in the simulation period. In fact, we can see that the number of newly manufactured items of (b) and (c) is decreased compared to (a).

Moreover, Fig. 19.8 shows the items' flow and inventories in the assembly process for (b). We can see that the number of stocks is almost nothing except the stocks with SOH of 100%, which are intentionally controlled to keep the stock size to be about 100 during the simulation period.

19.6 Conclusion

In this study, we developed an LCS system that can be used for improving the management method of the circular manufacturing system. The developed LCS system is versatile enough to evaluate various life cycle scenarios which have multitiered circulation. In the system, we incorporated functions such as management of an individual item and user, flow control of each process, and tracking the history of an individual item so that we can check and track when and where each item provides value to the users. We applied this LCS system to evaluate the life cycle scenario of LIBs for EVs that incorporated close-loop reuse and cascade reuse for stationary batteries to illustrate its practical effectiveness.

In the application, we adopted average value provision for users as an evaluation index in this simulation. In the simulation, we checked the requested SOH distribution and items flow/inventories in the assembly process to identify the problems occurring in its circulation flow management and take appropriate measures to improve the result. In fact, we succeeded to eliminate the problems and increase the value provision.

For future work, LCS should be improved to enable the evaluation of various operation management indices for considering the circulation management methods. Regarding items, we need to consider products' line-up and their model changes due to advances in technology.

Acknowledgements This study was supported by the JST-Mirai Program JPMJMI17C1.

References

Ekvall T, Weidema Bo P (2004) System boundaries and input data in consequential life cycle inventory analysis. Int J Life Cycle Assess 9(3):161–171

Fukuyo K (2011) Household attributes and electricity consumption of photovoltaic system owners. J Environ Eng AIJ 76(666):741–750 (in Japanese)

Kawakami K, Fukushige S, Kobayashi H (2017) A functional approach to life cycle simulation for system of systems. Proc CIRP 61:110–115

Kimura F (1999) Life cycle design for inverse manufacturing. In: Proceedings of the 1st international symposium on environmentally conscious design and inverse manufacturing, pp 995–999

Komoto H, Tomiyama T, Nagel M, Silvester S, Brezet H (2005) Life cycle simulation for analyzing product service systems. In Proceedings of fourth international symposium on environmentally conscious design and inverse manufacturing, pp 386–393

Market Trend Survey of Passenger Automobile in 2017, Japan Automobile Manufacturers Association, Inc. (in Japanese)

Murata H, Yokono N, Fukushige S, Kobayashi H (2018) A lifecycle simulation method for global reuse. Int J Automation Technol 12(6):814–821

Nassehi A, Colledani M (2018) A multi-method simulation approach for evaluating the effect of the interaction of customer behaviour and enterprise strategy on economic viability of remanufacturing. CIRP Ann Manuf Technol 67(1):33–36

Status Survey on Energy Consumption in 2015, MRI (in Japanese) http://www.meti.go.jp/meti_lib/report/2016fy/000032.pdf
Takata S, Kimura T (2003) Life cycle simulation system for life cycle process planning. CIRP Ann Manuf Technol 52(1):37–40
Takata S, Suemasu K, Asai K (2019) Life cycle simulation system as an evaluation platform for multitiered circular manufacturing systems. CIRP Ann Manuf Technol 68:21–24
Tsuchiya Y, Ito F, Tagashira N, Baba K, Ikeya T (2014) Analysis of purchase preferences for electric vehicles and its determinant factors. CSIS Discussion Paper, http://www.csis.u-tokyo.ac.jp/blog/research/dp/ (in Japanese)
Umeda Y, Nonomura A, Tomiyama T (2000) Study on life-cycle design for the post mass production paradigm. AI EDAM 14(2):149–161

Chapter 20
Part Agents for Exchanging Modules of Manipulators

Yuki Fukazawa, Yuichi Honda, and Hiroyuki Hiraoka

Abstract To realize the reuse of mechanical parts for the development of a sustainable society, it is necessary to manage individual parts throughout their life cycles. For this purpose, we develop a part agent system using network agents and RFID (radio-frequency identification) technology. To reuse machine parts effectively, we have developed functions of a part agent to predict its deterioration based on the operation history and deterioration information of the part, and to select an appropriate reuse partner based on consumer preference information. To verify the functions of the part agent system, we develop a three-degree-of-freedom (3-DOF) manipulator with modules that can be exchanged, and perform a module exchange experiment. This manipulator is composed of multiple modules that we designed to be easily exchanged. A part agent assigned to a module controls the module, collects its deterioration information, and manages its data. In addition, a part agent that manages a manipulator cooperates with part agents of other manipulators to select a module that is suitable for exchange with its own module. In this paper, we propose a module exchange method for manipulators using part agents. We also discuss the issues involved in exchanging modules and the solutions to these issues.

Keywords Part agent · Reuse · Life cycle · Manipulator · Circulating society

20.1 Introduction

In recent years, excessive environmental load caused by mass production, mass consumption, and mass disposal has become a problem. It is urgent to promote the efficient utilization of resources from production to distribution, consumption, and disposal, and to form a circular type of society with a small load on the environment (Hauschild et al. 2005). The promotion of 3R (Reduce, Reuse and Recycle) is indispensable in realizing this type of society. In particular, reuse should be strongly promoted because it is efficient to reuse used products and parts as they are. However,

---

Y. Fukazawa · Y. Honda (✉) · H. Hiraoka
Department of Precision Mechanics, Chuo University, Tokyo, Japan
e-mail: honda@lcps.mech.chuo-u.ac.jp

Y. Kishita et al. (eds.), *EcoDesign and Sustainability I*, Sustainable Production, Life Cycle Engineering and Management, https://doi.org/10.1007/978-981-15-6779-7_20

in order to reuse parts efficiently, it is necessary to save and utilize part operation histories and deterioration information, and to propose appropriate actions according to the state of the parts.

For this purpose, we develop a part agent (Tanaka and Hiraoka 2009) that manages the information of each part over the entire life cycle using network agent technology. Parts are managed by associating them with part agents using RFID (radio-frequency identification) tags (Borriello 2005). A part agents have a function to predict deterioration based on the operation history and the deterioration information of parts, and to replace them appropriately to reuse machine parts efficiently. In order to clarify the problems and the effects of the function, we developed manipulators using part agents and carried out an experiment to replace their modules for the reuse.

In this paper, we propose a method to exchange modules of manipulators using part agents, and discuss issues involved in the module-exchange experiment.

20.2 Basic Concept of Module Exchange

We call a network agent that corresponds to and manages individual parts a part agent. Figure 20.1 shows the conceptual scheme of the experimental system for the replacement of modules. An RFID tag is attached to a module of a product. This is used to identify the module and associates the module with its corresponding part agent.

The computer sends the module information to the corresponding part agent. The part agent expands the life cycle model of the corresponding module and calculates the expected value in cooperation with other part agents. The part agent proposes to the user an appropriate countermeasure for the corresponding part from this expectation value.

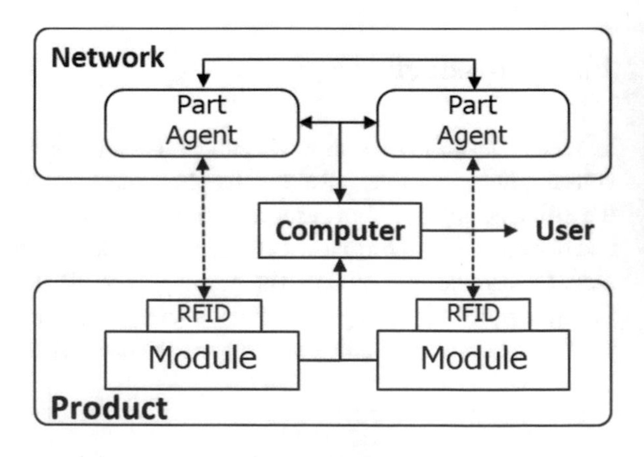

Fig. 20.1 Basic concept of module exchange

20.3 Part Agent System

20.3.1 Part Agent to Support Life Cycle of a Part

In a part agent system, a part agent manages the part information from production to disposal. The purpose of the system is to promote a society that circulates products and parts over the life cycle and to save resources. As described previously, the reuse of parts is a measure that utilizes the functions of parts with lower cost and fewer resources than recycling. A part agent system that manages the deterioration information and operation history of the part and proposes to the user an appropriate judgment on the reuse of the part is developed. This promotes the reuse of parts.

An overview of the part agent is shown in Fig. 20.2.

The part agent autonomously performs various functions through the network such as communication with other agents and the collection of parts information in the database. The agent presents the collected information to the user and stores it in the RFID of the part.

Typically, a part moves among sites in its life cycle as shown Fig. 20.3 and the part agent system handles this life cycle.

In this paper, the life cycle of parts is represented by six stages: Produce, Sell and Buy, Use, Replace, Repair, and Dispose. Changes of life cycle stages corresponding to the sites shown in the figure can be described as follows. In Produce stage, parts are manufactured at the production site and assembled in assembly site. After that, the product moves to the market, is sold to a user. This is Sell and Buy stage. Then,

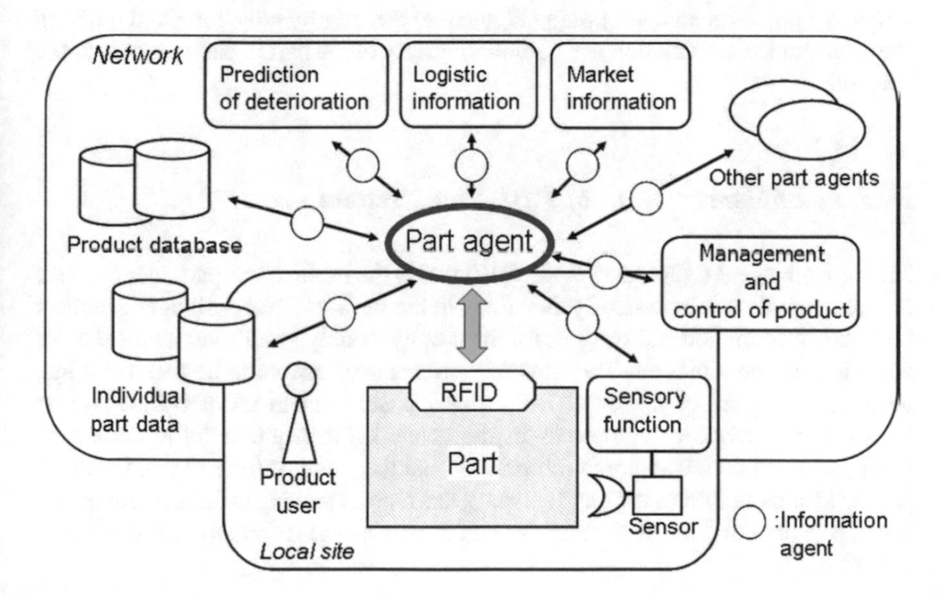

Fig. 20.2 Part agent system

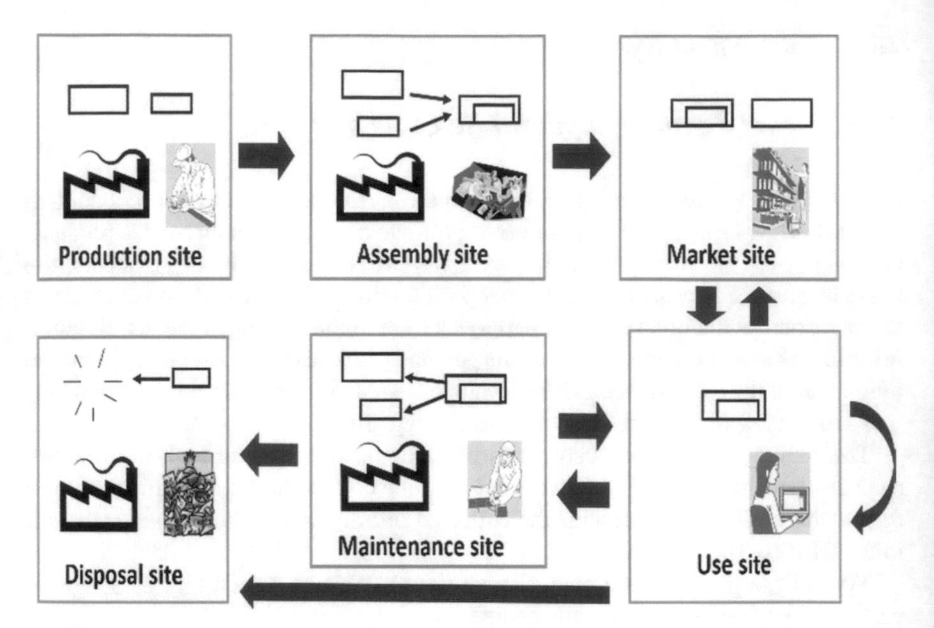

Fig. 20.3 Sites in life cycle of parts

the user uses the product in Use stage. The user repairs the product (Repair stage), disposes it (Dispose stage), sells to the market (Sell and Buy stage), or replace it with the part of other users (Replace stage) depending on the condition of the part. The part agent proposes an appropriate judgment to the user by collecting and utilizing the deterioration information and operation history of the part in each process of the life cycle.

20.3.2 Implementation of Part Agent System

The network agent (Nagasawa et al. 2017) that is the basis of the part agent system is a program that autonomously functions in the network. Network agents collect necessary information and services for the user by flexibly modifying their behavior according to the situation. The flow of movement of agents is as follows. First, an agent that receives a request from a user is sent out to the network. Agents autonomously move to computers in the network for required information and services. Based on the acquired information and the result of the service, the agent searches for more information and services, and repeats moving to another computer. Then, it returns to the user's computer and reports on the information and services collected.

Fig. 20.4 JADE platform

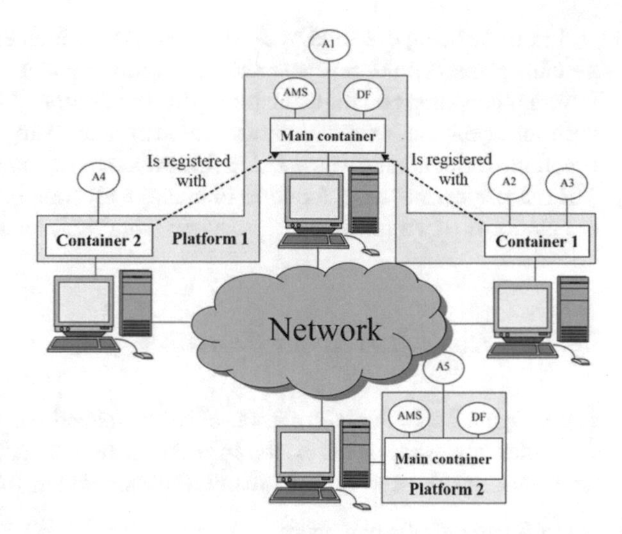

20.3.3 Information Management Using RFID

RFID is a generic name for authentication technology that uses radio waves (Borriello 2005). It is widely and effectively used in our daily life such as in tags for inventory control of goods and passenger cards for public transportation.

In our part agent system, the information of an individual part is managed by associating the part agent with the actual part using RFID. This is acquired through the agent corresponding to the part when the information of a specific part is required.

Since a part agent exists in the network and the corresponding part exists in the real world, a part agent in a network and a part with an IC tag attached in the real world are associated with each other by using RFID and are moved synchronously. In the part agent system, at the stage of manufacturing a part that is a starting point of a product life cycle, an IC tag is attached to the manufactured part, and a specified ID is recorded in the tag. At the same time, an agent having the same ID information is generated. The agent identifies the location of the part by recognizing the ID of the IC tag of the part even if the part moves to other stages, and synchronously moves to the place where the part resides.

20.3.4 JADE

We use the network agent platform JADE (Java Agent Development Framework).[1] JADE is an open-source agent platform written in Java. A schematic diagram of JADE is shown in Fig. 20.4.

[1] Java Agent DEvelopment Framework. http://jade.tilab.com/ (reference: 26 June 2019

Instances being executed by JADE are called containers, and one container stores several agents. A platform consists of multiple containers, and agents can be moved to different containers on the same platform. In addition, it is possible to communicate with an agent that exists in another platform by using the agent communication function. The container where JADE first creates an agent is called the main container. The main container has a function to manage all containers and agents that exist on the platform, and a function to find agents that exist on the platform.

20.3.5 Basic Part Agent Function

A part agent holds the information of the managed parts and the IP address of the computer where the agent exists. In order to realize the plan of the part agent, the necessary functions were defined and implemented as follows:

(1) Movement of a part agent
An agent needs to move in the network with the managed parts. This is the basic function of the part agent. It moves by obtaining the IP address of the computer where the part is transferred.

(2) Trace of the managed part
A part agent on the network traces the target management part. The part agent must exist in the container where the corresponding real part is located, so the part and the agent must be associated. The same part ID as the RFID tag attached to the control part is assigned at the time of generation, and it is compared with the part ID read from the RFID tag when the part is moved.

(3) Check of product status
For a part agent to make an appropriate proposal to the consumer, it is necessary for the agent to know the state of the corresponding part. To realize this function, a part agent has the ability to record the state of products and parts. Based on this information, the agent takes appropriate action for the situation.

(4) Communication between agents
A part agent communicates with other agent the operation history and deterioration information of the managed parts. Using this function, parts can be managed and performance can be evaluated based on the state of the parts.

20.4 Expansion of Life Cycle of Part

In the real world, parts move through life cycle sites such as Production site, Assembly site, Market site, Use site, Maintenance site, and Disposal site. Part agents follow these parts and move over the network. A part agent makes the correspondence between a life cycle stage such as use, repair, replacement, and disposal and one of these life cycle sites. A part agent also provides the user with information appropriate to the life cycle stage and encourages appropriate maintenance action.

In previous research (Yokoki et al. 2015), to realize this function, a system was developed that encourages the user to take appropriate action based on the information of the life cycle model, deterioration of the part, and failure.

20.4.1 Life Cycle Model

Figure 20.5 shows a life cycle model of machine parts. This model is used as an example in this paper. We define the life cycle of a part as consisting of stages connected by paths. The circles represent the stages: produce, sell, use, repair, and dispose, while the arrows represent the paths.

A part agent expands the life cycle of the part over time into an expanded life cycle. This represents possible changes in the life cycle of the part over time. Figure 20.6 shows an example of the expanded life cycle of a part that is expanded starting from the use stage.

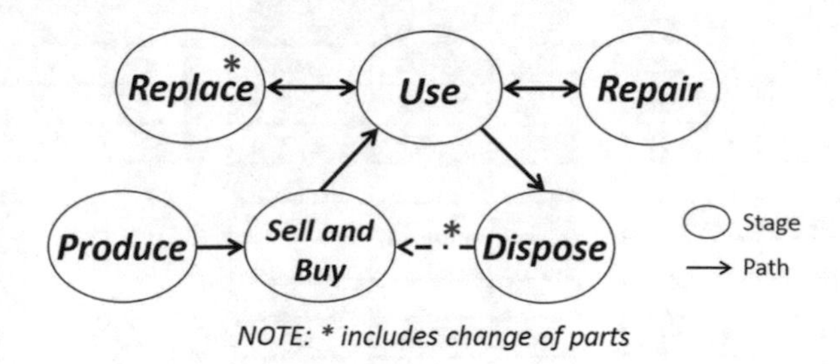

Fig. 20.5 A life cycle model

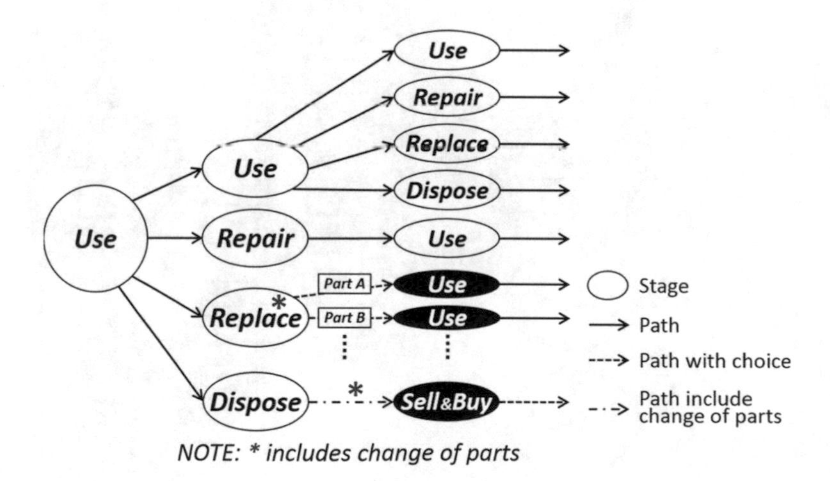

Fig. 20.6 Expanded life cycle model visible to users

Over the life cycle of a part, it is possible for the user to evaluate the life cycle when the user replaces it by connecting the life cycle of another part in the market or a part used by another user to the stage of replacement or disposal.

20.4.2 Prediction and Assessment

A part agent expands the life cycle model by using the state of the part and a deterioration model. Figure 20.7 illustrates how a part agent selects the next stage using the expanded life cycle of the part. The figure shows a situation where the current stage is the use stage, and the next possible stages are Stage1, Stage1′, and Stage1″. A circle denotes an expanded life cycle stage with its property value of profit, cost,

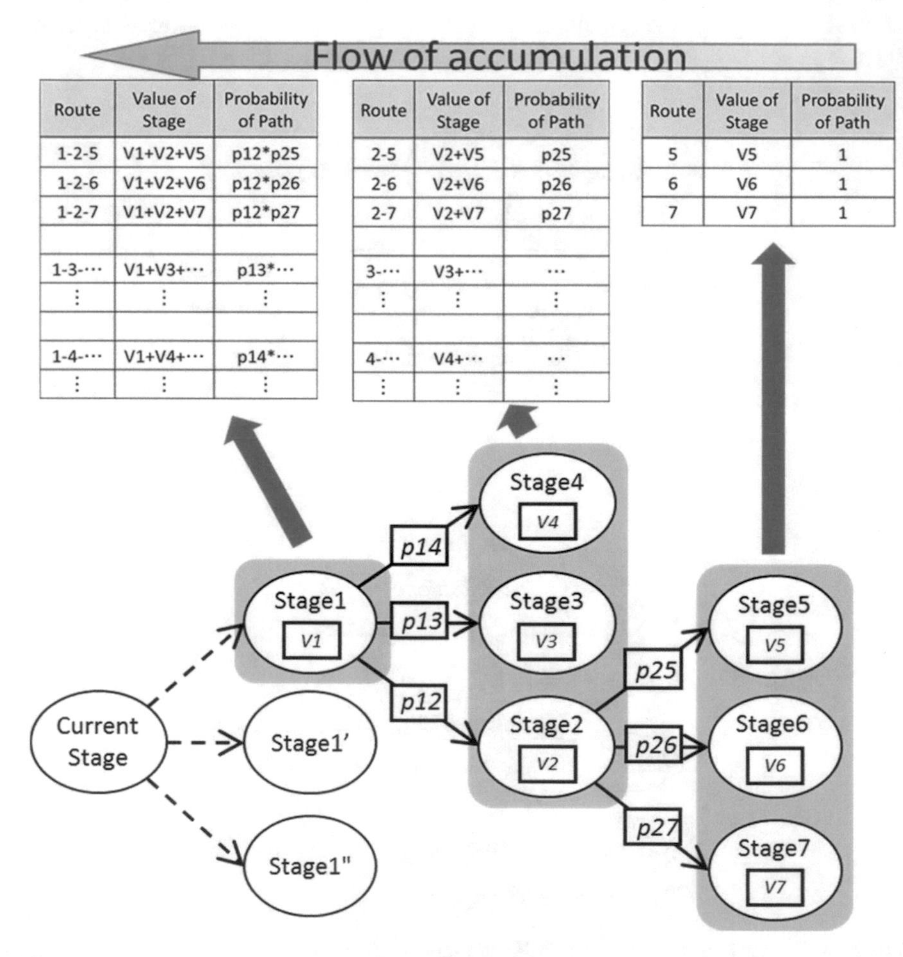

Fig. 20.7 Process of accumulating values of life cycle

or environmental load of the stage (such as V1, V2, and V3). A line denotes an expanded life cycle path with its probability (such as p12, p13, and p14). To evaluate each possible stage, the expected values of properties are calculated by considering the series of paths in the future. We define a series of stages connected to the paths as a route.

The property values for the next stages and their probabilities are collected along all routes that can occur in the future. Then, the expected value (EV) is calculated by multiplying the sum of property values and the product of probabilities, as shown in Eq. (20.1).

$$EV = \sum_{Route} \left(\sum_{Stageinroute} V * \prod_{Pathinroute} P \right) \tag{20.1}$$

where EV is an expected value of a candidate next stage with the following future stages. V is the property value of the stages in the route, and P is the probability of the paths in the route. For example, the expected property value of Stage1 in Fig. 20.7 is calculated by Eq. (20.2).

$$\begin{aligned} expectedvalue(Stage1) =& (V1 + V2 + V5) * (p12 * p25) + (V1 + V2 + V6) * (p12 * p26) \\ &+ (V1 + V2 + V7) * (p12 * p27) + (V1 + V3 + \cdots) * (p13 * \ldots) + \cdots \end{aligned} \tag{20.2}$$

Similarly, EV of Stage 1′ and Stage 1″ can be obtained.

20.4.3 Selecting an Exchange Part

In order for the life cycle of a part in use to be connected to the life cycle of other parts after stages such as replacement and disposal, it is necessary to expand and predict these life cycle models beforehand. All part agents provide the information when there is an inquiry from each part agent.

As described above, the life cycle is expanded and the EVs are calculated in all stages. When EV of the stage for exchange with another part becomes the highest, the part agent asks the part agent of the other part to make an exchange. A part agent which received exchange request checks the EV of the part agent that made the exchange request. Exchange parts when both EV are the highest.

20.5 Module Exchange Experiment

20.5.1 Outline of the Experiment

We plan an experiment to exchange modules of 10 manipulators performing a task to carry loads in order to verify the function of the part agent system. By creating differences in the contents of work to be done by the robot, differences are created in the degrees of progress of deterioration of each robot. A part agent detects the deterioration of the module and expands the life cycle model at fixed time intervals to calculate the EV. Modules are replaced or discarded based on the calculated values. If there is a module on the manipulator that has been marked for discard, the operation of the manipulator is stopped.

This experiment is carried out until all 10 manipulators stop. The part agent system is evaluated by comparing the actual deterioration of the equipment. In this experiment, the stages that a part agent can select are use, replace, and dispose.

20.5.2 Module Exchange Manipulator

A simple robotic manipulator with three degrees of freedom is designed for use in the experiment. Figure 20.8 shows a schematic of the manipulator. Rectangles with dotted lines in the figure denote the modules of the manipulator.

The manipulator consists of four modules: base module, first link module, second link module, and end-effector module. Each module is designed for easy assembly and disassembly by the manipulator.

The base module provides rotational motion around the axis perpendicular to the ground. The first and second link modules provide planar motion in a plane perpendicular to the ground. These link modules are identical and can be interchanged. By employing the same modules in the robot, the number of replaceable modules is increased, and the exchange of modules is promoted.

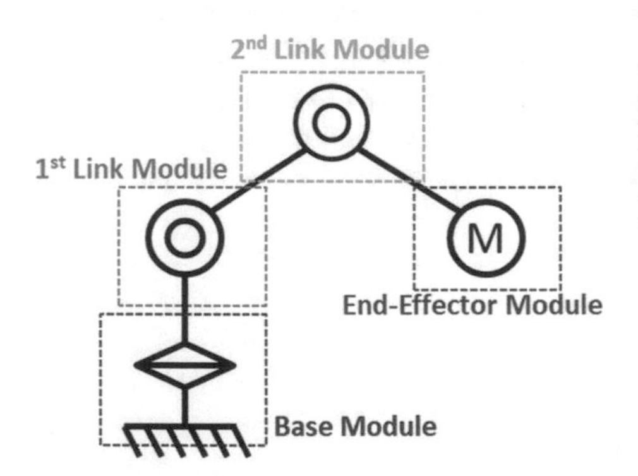

Fig. 20.8 Schematic plan of the manipulator

Figure 20.9 shows a manipulator model that is designed based on the structure shown in Fig. 20.8. The base module shown in Fig. 20.10 and the link module shown in Fig. 20.11 have a built-in DC motor and an angular sensor. The end-effector module shown in Fig. 20.12 has an electromagnet mounted at its tip. These modules are fully

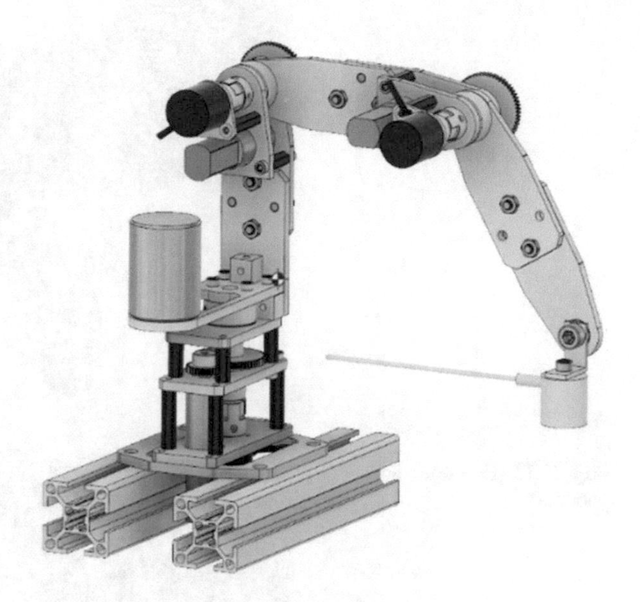

Fig. 20.9 Model of modularized manipulator

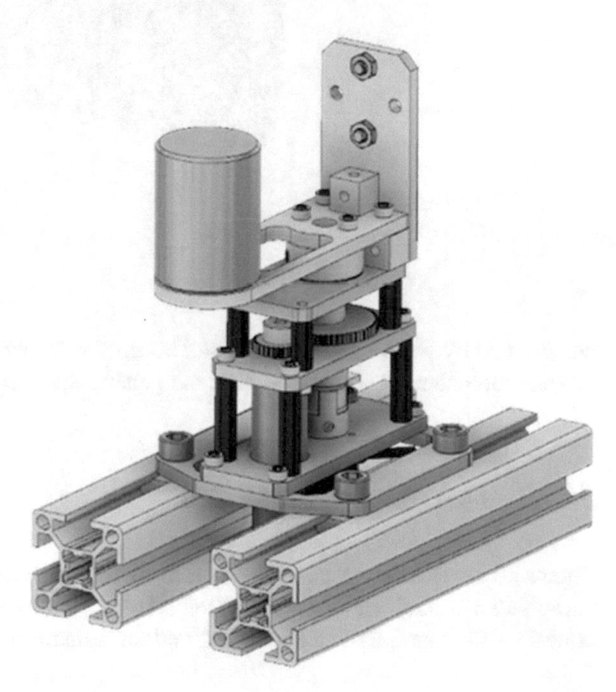

Fig. 20.10 Base module

Fig. 20.11 Link module

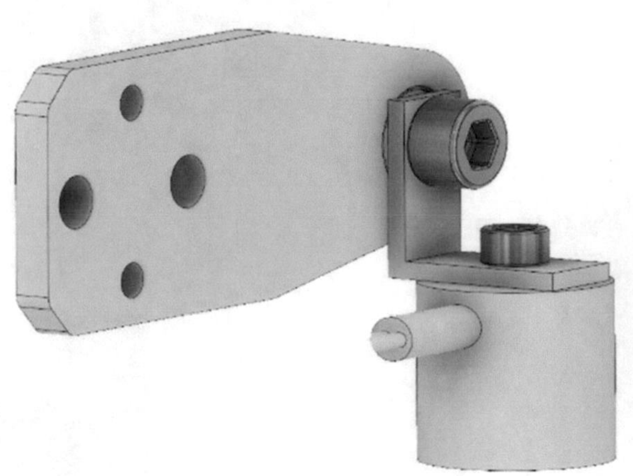

Fig. 20.12 End-effector module

constrained by two-pin positioning and bolting, as shown in Fig. 20.13, to facilitate module replacement. Table 20.1 lists the general specifications.

20.5.3 Deterioration Detection

Figure 20.14 shows the control diagram of this manipulator. A part agent of the end-effector module holds the target angle information of each module necessary for control of the manipulator. The target angle is transmitted to a part agent of the base

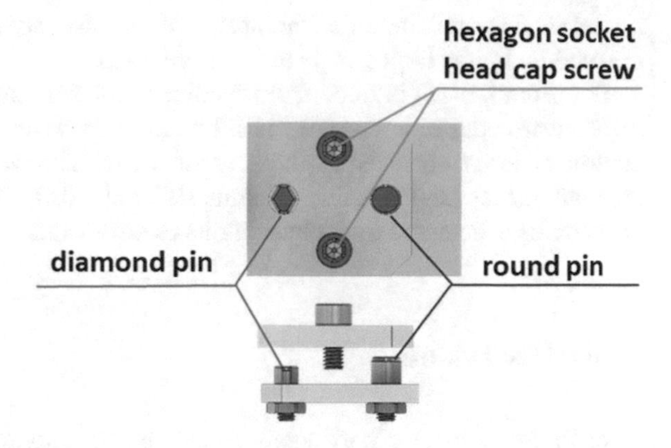

Fig. 20.13 Attaching portion

Table 20.1 General specifications of the manipulator

Link length l[m]	0.1
Weight of the 1st link m_1[kg]	0.25
Center of gravity of 1st link c_1[m]	0.05
Moment of inertia of 1st link $I_{G1}[\text{kg m}^2]$[a]	0.21×10^{-3}
Weight of 2nd link m_2[kg]	0.20
Center of gravity of 2nd link c_2[m]	0.05
Moment of inertia of 2nd link $I_{G1}[\text{kg m}^2]$[a]	0.25×10^{-3}
Target finger speed v[m/s]	0.05
Transportable weight M [kg]	0.1

[a] $I_{Gi} = m_i l^2/12\ (i = 1, 2)$

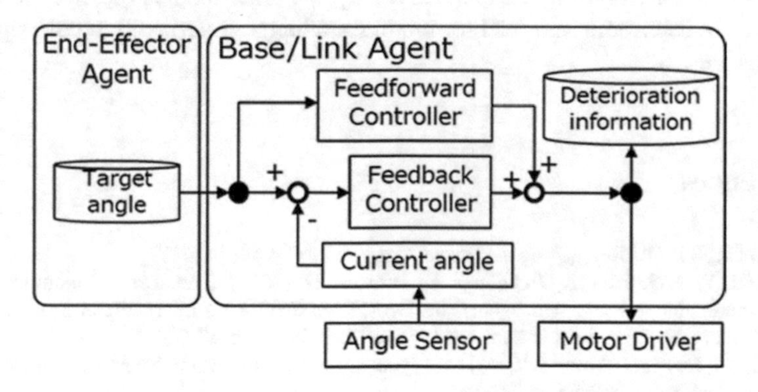

Fig. 20.14 Control diagram

module and the link module. Part agents of the base module and link module control the module by the target angle and current angle.

Deterioration is obtained from feedback control and feedforward control. The feedforward control is given a control quantity in advance. In feedback control, the control quantity changes by the present angle. The control value rises when the manipulator repeats its action (Fukumashi et al. 2017). Therefore, the deterioration of modules is detected by monitoring its control value.

20.6 Discussion

In order to evaluate a part agent system by the module exchange experiment, a deterioration model is required to extend the life cycle model of the part agent. A 3-DOF manipulator is manufactured, and deterioration experiment is carried out to create a deterioration model. Deterioration is obtained from feedback control and feedforward control. The feedforward control is given a control quantity in advance. However, when the control quantity changes due to disturbance, the deterioration value becomes inaccurate. Therefore, the deterioration value can be accurately detected by updating the control quantity of feedforward control from the state of deterioration of parts.

20.7 Conclusion

In this paper, we proposed a method to replace module of manipulator using future prediction of deterioration by expanded life cycle model and manipulator model. We will create a deterioration model for module exchange experiment to develop a part agent system.

References

Borriello G (2005) RFID: tagging the World. Commun ACM 44(9):34–37

Fukumashi Y, Nagasawa A, Fukunaga Y, Hiraoka H (2017) Exchange of modules among robot manipulators using part agents. In: EcoDesign 2017 10th International Symposium on Environmentally Conscious Design and Inverse Manufacturing A7-2

Hauschild M, Jeswiet J, Alting L (2005) Life cycle assessment to sustainable production: status and perspectives. Ann CIRP 54(2):535–555

Nagasawa A, Fukumashi Y, Fukunaga Y, Hiraoka H (2017) Disassembly support for reuse of mechanical products based on a part agent system. In: EcoDesign 2017 10th international symposium on environmentally conscious design and inverse manufacturing, D6-3

Tanaka A, Hiraoka H (2009) Life cycle simulation of parts individuals supported by part agents—Factors that promote parts reuse. JSPE 665–666

Yokoki Y, Yamamori Y, Hiraoka H (2015) User model in the life cycle simulation of a part based on Prospect theory. In: EcoDesign 2015 9th International Symposium on Environmentally Conscious Design and Inverse Manufacturing pp 184–189

Chapter 21
Closed Loop Tolerance Engineering Modelling and Maturity Assessment in a Circular Economy Perspective

Kristian Martinsen, Carla Susana A. Assuad, Tomomi Kito, Mitsutaka Matsumoto, Venkata Reddy, and Sverre Guldbrandsen-Dahl

Abstract Decisions made in the development stage of a new products will affect the whole lifecycle of the product. Manufacturing costs, product performance, maintainability and customer satisfaction in the use phase are parameters the engineers need to consider. In a sustainability perspective can durability and a potential long lifetime with less breakdowns be regarded as positive. Additionally, in a circular economy perspective the potentials for easy disassembly, recyclability and remanufacturing or reuse at the end of life are important. The selection of precision levels and tolerance limits on geometry and material properties in the design phase of mechanical components are decisive for these aspects. While tolerance selections traditionally focused most on meeting customer requirements and interchangeability of parts for assembly, the product development engineers are now facing several "Design for X"—challenges where tolerance selections and distributions are one of the key issues. This paper describes a Closed Loop Tolerance Engineering (CLTE) model describing information flow for tolerance engineering throughout the product lifecycle. The model includes feed forward and feedback of data and information between functional requirements description, tolerance synthesis and analysis, manufacturing process capabilities, measured product performance and end-of-life considerations.

Keywords Tolerancing · Quality assurance · Lifecycle · Closed loop tolerance engineering

K. Martinsen (✉) · C. S. A. Assuad
Norwegian University of Science and Technology (NTNU), Trondheim, Norway
e-mail: kristian.martinsen@ntnu.no

T. Kito
Waseda University, Tokyo, Japan

M. Matsumoto
National Institute of Advanced Industrial Science and Technology (AIST), Tsukuba, Japan

V. Reddy
Indian Institute of Technology Hyderabad (IITH), Hyderabad, India

S. Guldbrandsen-Dahl
SINTEF Manufacturing, Vestre Toten, Norway

Y. Kishita et al. (eds.), *EcoDesign and Sustainability I*, Sustainable Production, Life Cycle Engineering and Management, https://doi.org/10.1007/978-981-15-6779-7_21

21.1 Introduction

Tolerance Engineering (TE) is the process of specifying allowed variations a.k.a. tolerances to components and products. This is usually an activity in the product development phase, with a major goal to ensure interchangeability of parts and to ensure that the product quality and function will meet the customer demands. The selected tolerances will, however also impact manufacturing and inspection processes and thus manufacturing costs. Too tight tolerances "to be on the safe side" regarding assembly and product function which do not take manufacturing capabilities into consideration might lead to selection of an over-qualified manufacturing process leading to more expenses than necessary. In contrast, under-qualified processes could lead to problems to meet the quality requirements increasing scrap production. Literature reports many examples on this; Zhang (Zhang and Wang 2007) states that "*many parts and products are certainly over-toleranced or haphazardly toleranced, with predictable consequences*". Singh (Singh 2002) point at the negative effects of inappropriate tolerances of increased cost and lacking product quality. Ali et al. (Durupt and Adragna 2013) and Krogstie and Martinsen (Krogstie and Martinsen 2013) point at the costs and efforts to change tolerances at a later stage. Watts (Watts 2007) states; "*all industry is suffering, often unknowingly, of the lack of adequate academic attention on tolerances*". Srinivasan et al. states that tolerancing has been "*kept in a high degree of technical focus*" with focus on norms and standards (Srinivasan 2008; Srinivasan 2012) and thus a lack of attention to organizational challenges.

Nevertheless, there are many different product development methodologies and approaches where TE are addressed, such as Robust design (Zhang et al. 2010), Design for Manufacturing (or DfX) (Holt and Barnes 2010; Zhang et al. 1992) as well as digital twins for TE (Söderberg et al. 2017). A comprehensive listing of models and management control of product development shows, however, a lack of focus on TE (Brown and Eisenhardt 1995; Horváth 2004; Richtnér and Åhlström 2010). Moreover, only few of these models take the whole lifecycle of the product and the concept of circular economy (Wang et al. 2018; Nagel and Meyer 1999) into consideration. The CIRP keynote by Shu et al. (Shu et al. 2017) do mention the importance of TE for reduced resource consumption, but on the other hand is TE only briefly mentioned by Tolio et al. (Tolio et al. 2017) on their keynote paper on demanufacturing and remanufacturing systems. Umeda et al. (Umeda et al. 2012) does not mention TE or variation management in their keynote on Life Cycle Engineering. The authors of this paper claims that with a future circular economy (CE) and increased reuse and remanufacturing of products and components, future TE models need to reflect on the circularity paradigm. This paper will address this challenge.

21.2 Closed Loop Tolerance Engineering

Krogstie and Martinsen (Krogstie and Martinsen 2012) developed a conceptual model of Closed Loop Tolerance Engineering (CLTE). CLTE (Fig. 21.1) is a model for "*systematic and continuous re-use and understanding of product-related knowledge, with the aim of designing robust products and processes with the appropriate limits of specifications*".

CLTE sees TE as activities not limited to the traditional activities of tolerance-specification, allocation, modelling/optimization and synthesis, but also an organizational process, with information flow and ability to collect, use and reuse data. Preventing problems from occurring, attention to and understanding of tolerances in the whole value chain, and fact (data) based tolerance engineering are some of the benefits expected. Good tolerance engineering practice includes a collective ability to detect critical situations in the product development phase (Badke-Schaub and Frankenberger 2004) and the decision-making between future desirable vs. negative consequences from variations on the product and processes. CLTE has been applied for analyzing tolerance engineering practices in different companies, including a high-precision aerospace company (Krogstie et al. 2014). The CLTE - model has feed forward and a feedback information flow dimensions. It contains four interconnected activities: (1) Defining functional requirements (FR), (2) TE, (3) Manufacturing Process Capabilities Assessment (PC) and (4) Product Performance Assessment (PP) Furthermore, six pairs of closed loops of information flow (1a/b etc. see Fig. 21.1) passing information forward in the project flow, as well feedback of data to new product development collected from manufacturing and performance of existing products. The ability to prepare and utilise information and data from both feed forward and feed-back dimension is a main key element.

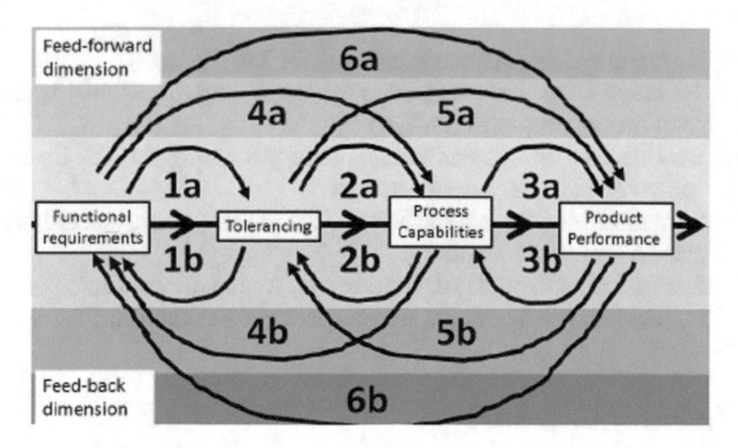

Fig. 21.1 The CLTE-model (Krogstie and Martinsen 2012)

CLTE maturity assessment

Based on the CLTE, Martinsen (Martinsen 2018) developed a maturity assessment tool used to evaluate and plan improvements in TE management. The tool consists of two parts; first part is assessing how well the company is performing in the twelve information flow relation loops visualised in Fig. 21.1. Grades are from (1) not applied, (2) poor, (3) medium, (4) good and (5) excellent to the following questions (see (Martinsen 2018) for more details);

How well are;

1a FR transformed to TE, by whom, and which tools?
1b decision basis from TE feedback to aid FR description in following projects?
2a TE fitted to the PC? – and tolerance stack-up (Bjørke 1989) critical tolerances and reference surfaces?
2b PC data used in TE for new products?
3a PC influence on PP described?/Variation in PP understood?
3b Can it be traced back PC?
4a FR used in manufacturing? critical parameters known?
4b PC feedback (and make an influence) on FR?
5a Relations between TE and the PP variations understood?
5b Variations in PP on existing products feedback to and used on TE on new products?
6a The PP according to the FR? Did we get what we wanted?
6b PP in existing products feedback to aid definition of functional requirements in following projects?

Second part of the maturity assessment is the information and data exchange assessment dividing the maturity into 5 stages (Martinsen 2018) described by the following:

Stage 1: No organised information exchanges for TE.
Stage 2: TE based on expert's subjective opinions. Cross-functional teams using semi-quantitative tools (FMEA).
Stage 3: Real data on ad hoc and manual data processing for TE. Some use of computer-aided TE decision support.
Stage 4: Systemized regular data and information exchange, analysis and computer-aided TE decision support.
Stage 5: Instant and autonomous data exchange. Automatic translation of data and information to adapt to the specific user, project team and TE challenge.

21.3 Industry Case Study

The case study was made at the Kongsberg Automotive (KA) plant at Raufoss, Norway. KA is a Tier 1 supplier to automotive industry designing and delivering connectors for pressurized air brake systems. The company owns several product

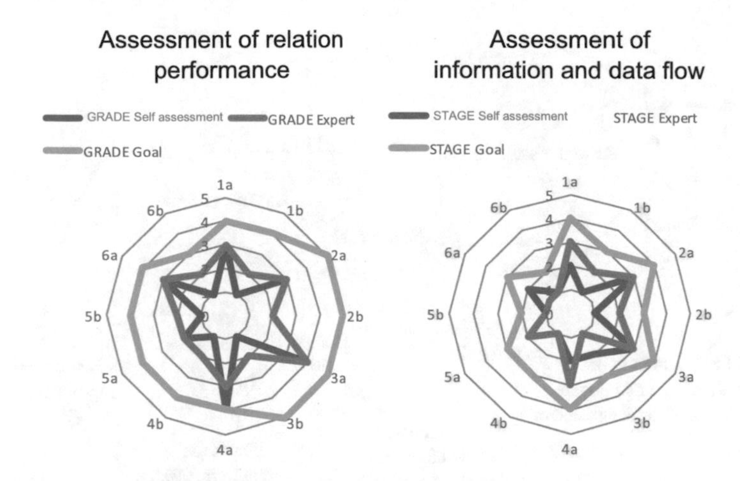

Fig. 21.2 CLTE maturity assessment at KA

patents and develop, manufacture and assemble a complete range of connectors. The charts in Fig. 21.2 show the results from the assessment made by KA and an external expert evaluation team using the maturity assessment tool. The charts show a typical picture where the feed forward (1–6a) are more advanced than the feed-back loops (1–6b). There are some deviations on the expert vs. company self-evaluation. This is quite common, since companies are in some cases more "hard" on themselves on the self-assessment than the expert's assessment. The green line shows the performance goals stated by the company. They might seem somewhat ambitious, but it is long-term goals where the company mean they have to be.

Optimised tolerances casual loop analysis

As the case of KA pointed up, the core challenge of the CLTE model is the implementation of feedback loops (1–6b) in TE. One of the main reasons for such limitations is the lack of storage infrastructure and information analysis capabilities. However, dynamics complexity is probably a more powerful inhibitor to implement and manage all the learning loops visualized in the CLTE model. The time delays between making the design decisions and its effects on process capabilities and product performance slow the learning process. Delays also reduce the learning gained on each loop or cycle, for instance variables can change simultaneously, confusing the interpretation of the system behaviour (in this case the effects tolerances have on product functionality and durability) (Bjørke 1989). Figure 21.3 shows a simplified casual loop diagram (CLD) used to analyse the dynamics involved in tolerance specification. A CLD is a system thinking tool, used to map the mental models behind the understanding of a system. CLD's dynamic hypothesis of a problem to be tested for example with system dynamic or agent-based simulations models (Sterman 2000). In this paper we present a CLD that was obtained with the help of experts in the topic of tolerances. Optimised tolerances lead to less use of resources, which leads

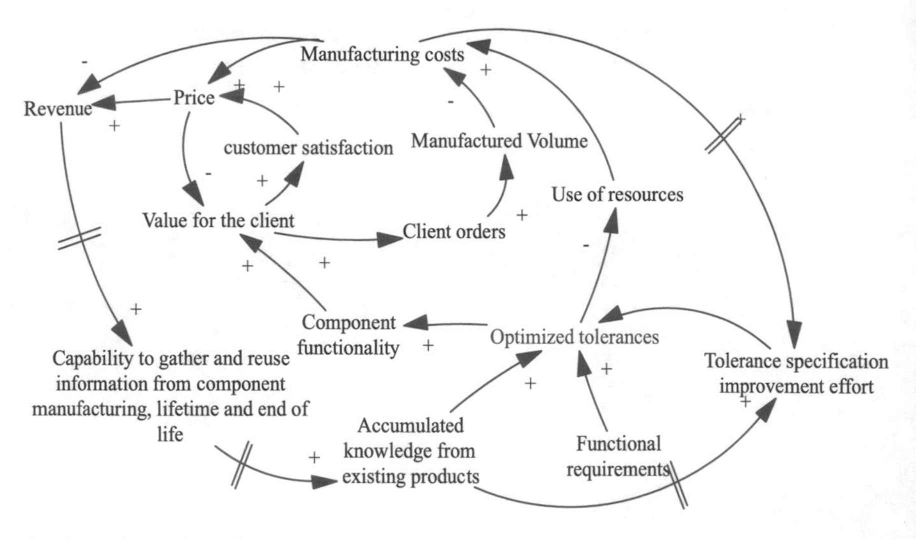

Fig. 21.3 Causal loop diagram at KA

to less manufacturing cost and more revenue. The revenue affects the capability to collect and reuse information which leads to more accumulated knowledge to increase the optimisation of the tolerances. Optimised tolerances also increase revenue by increasing the component/product functionality and adding more value to customer, which allows for a higher price. Accumulated knowledge is represented as an important variable affecting positively optimised tolerances. The delays in the system are visualized as two short lines across the causal link.

In the process of setting tolerances several types of uncertainty are present (Morse et al. 2018). The analysis on this paper deals mainly with epistemic uncertainty which refers to any lack of knowledge. The CLD in Fig. 21.3 shows that in order to implement the learning loops necessary to improve tolerances by minimizing the lack of knowledge, revenues are necessary to invest in infrastructure and capabilities to gather and reuse the accumulated knowledge. However, revenues are only possible with low manufacturing cost and high customer satisfaction which are variables that are highly sensitive to tolerances specifications. Since this is a reinforcing loop a possible leverage point (Meadows 1997) could be to make an effort to create an initial infrastructure and capability to gather and reuse information and implement actions to minimize the delays presents in the learning loop. Additional actions to affect the willingness to use accumulate knowledge by the designer will also be beneficial.

21.4 Extended Clte in a Circular Economy Perspective

The CLTE model is currently mainly focus on TE and processes within one company. Customer demands are expressed through the functional requirements and product performance, but the influence on TE and PC at suppliers are not reflected. Information and data collection from the product use phase and end-of-life and possibly reuse and remanufacturing of products are also not reflected. One of the current trends are manufacturers liability of products after end-of-life (EOL) as well as the extension of the product to a product-service system. Products containing sensors open for new business models, but the data collected could also be used for future CLTE activities, such as functional requirement definitions. These trends will be even more emphasized in a circular economy perspective and increased reuse and remanufacturing of products and components rather than re-melting or disposal at the EOL. Future tolerance engineering and variation management need to take this into consideration.

Figure 21.4 shows a simplified process flow diagram of an extended CLTE (without the information flows arrows).

The information flows are visualized in Fig. 21.5, similar to the original CLTE model where the value chain and the circular perspective is included. Blue arrows are feed forward information flow, red arrows feedback of historical data acquired. Here, the PP is both the functional performance of the product as well as the durability, maintainability and ability to prolong the lifetime of the product as much as possible. The TE will, of course affect all these, as predicted by the Taguchi loss function (Pedersen and Howard 2016). When the product reaches its End Of Life (EOL) (Nagel and Meyer 1999) whatever the reason, a decision must be made on possible disassembly, recycling, reuse or remanufacturing of the product or components (Lieder and Rashid 2016). Here again the original TE will be an important factor both for the durability and state of the components and the potential for reuse and remanufacturing. Ultimately when remanufacturing used components, the TE of the new product with remanufactured components needs to take the total cycle of tolerances into consideration.

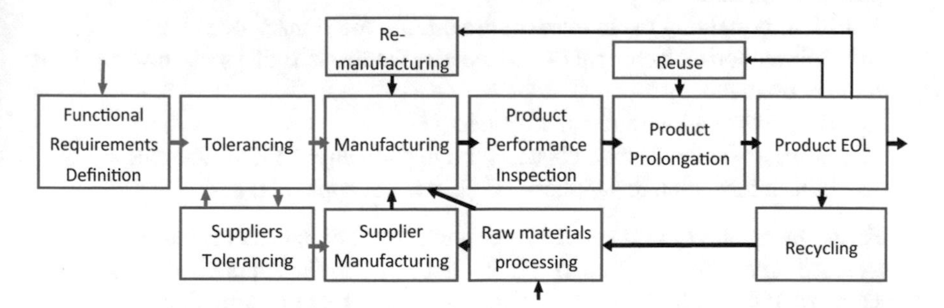

Fig. 21.4 Extended CLTE simplified process flow

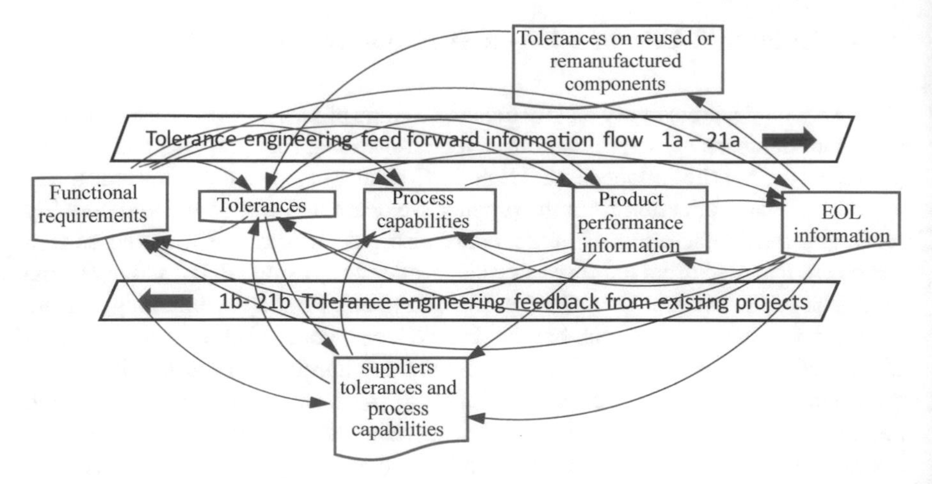

Fig. 21.5 Extended CLTE simplified information flow

The addition of three new information flow nodes as indicated; EOL information, Tolerances on reused or remanufactured components and supplier's tolerances and process capabilities adds potentially 18 new pairs (in total 26 pairs) of closed loop information flows in addition to the 1a/b to 6 a/b in the original model. For example, information flow to and from the EOL process will be distributed through the following closed loop information pairs;

7a FR information flow to make optimal EOL decisions
7b Information on historical EOL decisions feedback to FR for new products development
8a TE information flow to make optimal EOL decisions
8b Information on historical EOL decisions feedback to TE of new products development
9a PC information flow to make optimal EOL decisions
9b Information on historical EOL decisions feedback to PC decisions for new products manufacturing
10a PP information flow to make optimal EOL decisions
11b Information on historical EOL decisions feedback to PP assessment methods for new products
12a EOL information to remanufactured TE
12b Information feedback on historical remanufactured TE to EOL decisions
13a EOL information to suppliers TE (12b is not applicable).

Furthermore, there be potentially six pairs for "Tolerances on remanufactured and reused components" (13a7/b–19a/b) and si pairs for supplier tolerances and process capabilities (20a/b–26a/b), although some of these information flows might be obsolete. Similar to the maturity mapping tool based on original CLTE-model, an extended maturity mapping tool can be developed, although this is probably a bit premature for most companies since CE is still mostly a vision more than reality. A crucial point is however, to be able to optimize the selected tolerances not only for simultaneous maximised product performance and minimised manufacturing costs, but also the maximised lifetime of a product through maximised product durability and maintainability, furthermore how to deal with a constant flow of remanufactured and reusable components. Ultimately the selected tolerances at the Beginning Of Life (BOL) of a product should reflect that a component should be allowed to live in several loops in a Circular Economy manufacturing system.

Figure 21.6 shows a simplified Casual Loop Diagram which reflects this possible future system. Adding the EOL processing feedback loops requires that manufacturing companies look outside their production processes and include additional type of information in their tolerance specification. Such challenge increases the complexity of the system, since the information delays are longer, data contains more noise, and there are more actors in the system to interact with (that dynamic hypothesis is represented in the CLD in the variable "effort to gather information from external actors"). With a CE perspective manufacturing cost will also be affected by environmental effects that are amplified by poor tolerance specification like the increase in scrap and pollution. The analysis of the dynamics of the extended CLTE model brings up the following additional questions for the CLTE maturity assessment tool: (i) Are the delays involve in the information loops accounted for when using accumulated knowledge? (ii) Are the designers involved in cultural change programs, so accumulated knowledge is added in the tolerance specification routine? (iii) Are actors in the supply chain involved in data accumulation? and (iv) Are the noises in the data from EOL accounted for in tolerances specification?

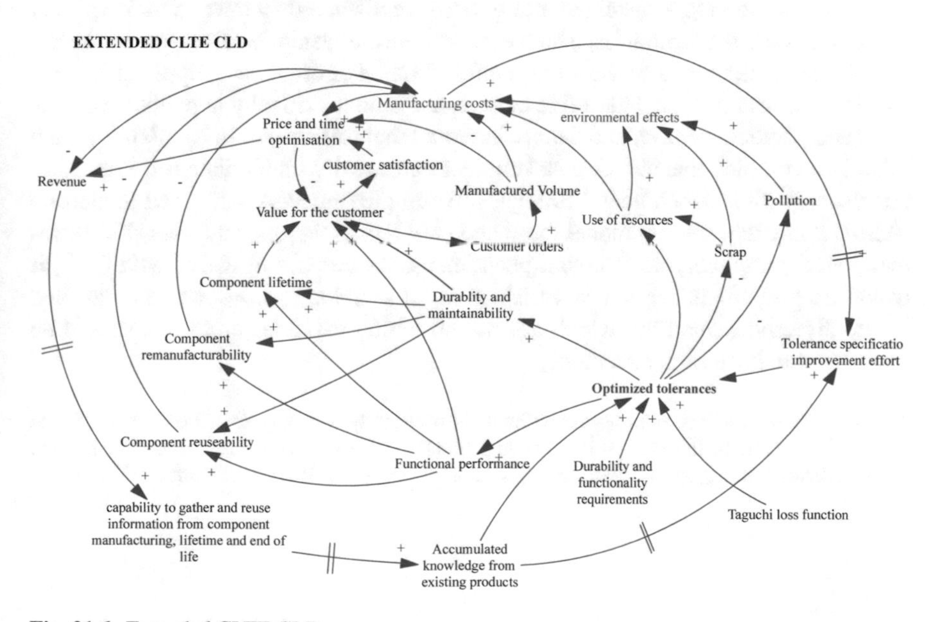

Fig. 21.6 Extended CLTE CLD

21.5 Discussions, Conclusions and Further Work

The CLTE maturity assessment tool is a semi-quantitative tool for mapping a company and to point at possible improvements. The case company is measuring and storing large quantities of data along product development and manufacturing phases. To transfer the data and extract information useful across the lifecycle is currently difficult, though, especially the feedback loops. For example, currently data from statistical process control are stored, but to translate these charts to generic process capabilities and make it easy accessible and usable for the product designers and tolerance definition is still not straight forward.

The extended CLTE with a review on how CLTE or similar approaches might play role to achieve a circular economy with prolonged product lifetime, increased reuse and remanufacturing is still a vision for future manufacturing system. The central aspect of this paper is to raise the awareness of the importance of tolerance engineering processes. In order to start to take a more holistic and systemic view where the potentials for product prolongation and several loops of reuse and remanufacturing of components could play a role in tolerance optimisation. Further work would see longitudinal results from industries using the tool with measured improvements, as well as making a simulation model for the Extended CLTE with a Circular Economy perspective in order to understand its effects and environmental impacts.

Alternative model using agent based modelling (ABM)

CLD can give us a good visual image of what affects what, helping us with understanding how the complex circular manufacturing system works. Then, it will be very useful if this can help a stakeholder (e.g. producer) when they have to make critical decisions. But one essential issue here is that, since the system involves various stakeholders (depending on the case—they could be OEMs, component suppliers, service providers, consumers,…), the value of each decision is critically dependent on how other stakeholders behave, and their behaviours themselves would be affecting each other in a complex manner. One stakeholder can't really manipulate (or, even fully understand) other stakeholders' strategies. By employing an Agent based Modelling (ABM) approach, we can model these stakeholders as agents, and treat the abovementioned complexity as emergent phenomena. By combining ABM with CLD, it would be possible to see which variables of CLD belong to (= can be controlled by/are directly defined by) which agent, and which variables and links should be considered to have high uncertainty.

Acknowledgements The authors wish to thank the case study company KA. The work was based on activities within the INMAN project funded by the Intpart program contract number 275156 and the SFI Manufacturing, funded by the Norwegian Research Council contract number 237900.

References

Ali SA. Durupt, Adragna P (2013) Reverse engineering for manufacturing approach: based on the combination of 3D and knowledge information. In: Abramovici M, StarkSmart R (ed) Product engineering. Springer Berlin Heidelberg, p 137–146

Badke-Schaub P, Frankenberger E (2004) Management kritischer Situationen-Produktentwicklung erfolgreich gestalten Heidelberg, Springer Verlag

Bjørke Ø (1989) Computer-aided tolerancing, New York, ASME Press. xiii, 216 s

Brown SL, Eisenhardt KM (1995) Product development: past research, present findings, and future directions. Acad Manag Rev 20(2):343–378

Holt R, Barnes C (2010) Towards an integrated approach to "Design for X": an agenda for decision-based DFX research. Res Eng Des 21(2):123–136

Horváth I (2004) A treatise on order in engineering design research. Res Eng Des 15(3):155–181

Krogstie L, Martinsen K (2012) Closed loop tolerance engineering—a relational model connecting activities of product development. Procedia CIRP 3:519–524

Krogstie L, Martinsen K (2013) Beyond lean and six sigma; cross-collaborative improvement of tolerances and process variations-a case study. Procedia CIRP 7:610–615

Krogstie Lars, Martinsen Kristian, Andersen Bjørn (2014) Approaching the devil in the details. Surv Improving Tolerance Eng Pract Procedia CIRP 17(1):230–235

Lieder M, Rashid A (2016) Towards circular economy implementation: a comprehensive review in context of manufacturing industry. J Cleaner Prod 115:36–51

Martinsen K (2018) Industry 4.0 closed loop tolerance engineering maturity evaluation. In: Proceedings of the international workshop of advanced manufacturing and automation. https://link.springer.com/book/10.1007/978-981-13-2375-1

Meadows D (1997) Places to intervene in a system. Whole Earth 91(1):78–84

Morse Edward, Dantan Jean-Yves, Anwer Nabil, Söderberg Rikard, Moroni Giovanni, Qureshi Ahmed, Jiang Xiangqian, Mathieu Luc (2018) Tolerancing: managing uncertainty from conceptual design to final product. CIRP Ann 67(2):695–717

Nagel C, Meyer P (1999) Caught between ecology and economy: end-of-life aspects of environmentally conscious manufacturing. Comput Indus Eng 36(4):781–792

Pedersen SN, Howard T (2016) Data Acquisition for Quality Loss Function Modelling. Procedia CIRP 43:112–117

Richtnér A, Åhlström P (2010) Top management control and knowledge creation in new product development. Int J Oper Prod Manag 30(10):1006–1031

Shu LH, Duflou J, Herrmann C, Sakao T, Shimomura Y, De Bock Y, Srivastava J (2017) Design for reduced resource consumption during the use phase of products. Ann CIRP 66(2):635

Singh N (2002) Integrated product and process design: a multi-objective modeling framework. Robot Comput Integr Manuf 18(2):157–168

Söderberg R, Wärmefjord K, Carlson JS, Lindkvist L (2017) Toward a digital twin for real-time geometry assurance in individualized production. CIRP Ann 66(1):137–140

Srinivasan V (2008) Standardizing the specification, verification, and exchange of product geometry: research, status and trends. Comput Aided Des 40(7):738–749

Srinivasan V (2012) Reflections on the role of science in the evolution of dimensioning and tolerancing standards. In: 12th CIRP Conference on Computer Aided Tolerancing, Elsevier Ltd. Huddersfield, UK

Sterman J (2000) Business dynamics: systems thinking and modeling for a complex world. Irwin/McGraw-Hill, p 21–23

Tolio T, Bernard A, Colledani M, Kara S, Seliger G, Duflou J, Battaia O, Takata S (2017) Design, management and control of demanufacturing and remanufacturing systems. CIRP Ann 66(2):585–609

Umeda Umeda, Yasushi Shozo Takata, Kimura Fumihiko, Tomiyama Tetsuo, Sutherland John W, Kara Sami, Herrmann Christoph, Duflou Joost R (2012) Toward integrated product and process life cycle planning—an environmental perspective. Ann CIRP 61(2):681

Wang P, Kara S, Hauschild MZ (2018) Role of manufacturing towards achieving circular economy: the steel case. CIRP Ann 67(1)

Watts D (2007) The "gd&t knowledge gap" in industry. ASME Conf Proc 48051:597–604

Zhang C, Wang HPB (2007) Tolerancing for design and manufacturing, in handbook of design, manufacturing and automation. Wiley, p 155–169

Zhang C, Wang HP, Li JK (1992) Simultaneous optimization of design and manufacturing—tolerances with process (machine) selection. CIRP Ann 41(1):569–572

Zhang JJ, Li SP, Bao NS, Zhang GJ, Xue DY, Gu PH (2010) A robust design approach to determination of tolerances of mechanical products. CIRP Ann Manuf Technol 59(1):195–198

Chapter 22
Analysis of Electric Vehicle Batteries Recoverability Through a Dynamic Fleet Based Approach

Fernando Enzo Kenta Sato and Toshihiko Nakata

Abstract The aim of this study is to propose a dynamic model for forecasting the changes in the number of batteries recovered from end of life electric vehicles considering different power trains. To achieve a sustainable society, the dependency of the energy on fossil fuels must be overcome. One of the first steps to manage this objective is through the reduction of its direct consumption by the wide-scale adoption of EV (HV/PHV/BEV/FCV). Low cost and stable production of lithium ion batteries (LiB) are expected to be a key element for the electrification of the transportation. For this reason, an efficient cascade use of electric vehicle batteries (EVB) to minimize its raw material supply risk, disposal risk, environmental impact and material cost/consumption in its production process become essential. Additionally, by the promotion of a closed loop life cycle, cost reduction in the end of life batteries treatments can be also expected. However, to grab this opportunity and create a sustainable market, balance between the demand and recoverability of LiB must be clarified to propose reliable second life projects. This study proposes a method based on system dynamics modeling for forecasting the vehicle fleet, sales and end of life vehicles by power train considering data of scrapping rates of vehicles by year of use. Moreover, the supply potential of scrapped batteries from a reverse logistic scheme is analyzed. Here, the Japanese vehicle market is considered as a case study and a timeframe of 2018 to 2050 forecasted. Results indicate that the amount of scrapped EVB will increase 45 times from 2020 to 2050. Moreover, a complete closed loop of them can be expected around 2050 only if the exportation of used electric vehicles is hardly diminished.

Keywords Electric vehicle batteries · Closed loop · Forecasting · Dynamic modeling

F. E. K. Sato (✉)
Cyclical Resource Promotion Division, Honda Motor Co., Ltd, Wako, Japan
e-mail: kenta_b_sato@hm.honda.co.jp

F. E. K. Sato · T. Nakata
Department of Management Science and Technology, Graduate School of Engineering, Tohoku University, Sendai, Japan

Y. Kishita et al. (eds.), *EcoDesign and Sustainability I*, Sustainable Production, Life Cycle Engineering and Management, https://doi.org/10.1007/978-981-15-6779-7_22

22.1 Introduction

It is inevitable the advance of the electrification of vehicles considering that the transportation area accounts 25% of the total the energy consumption (U.S. Energy Information Administration 2016) and CO_2 emission (International Energy Agency 2009) of the word, and the several global efforts to combat climate changes.

Compare to the internal combustion engine vehicles (ICEV) which depend totally to fossil fuels, the electric vehicles, which includes hybrid electric vehicles (HEV), plug-in hybrid electric vehicles (PHEV), battery electric vehicles (BEV) and fuel cell vehicle (FCV) depends partially or totally on the electricity. Currently, the EVs account 32.9% of the vehicle sales in Japan (Next Generation Vehicle Promotion Center 2018), and it is expected a rapid increase in its share in the following years. Here, the size and weight of its batteries vary depending on the electrification level, driving range of vehicle and its technology. Many of the current EV use LiB. However, Nickel-metal hybrid batteries (NiMH) are still available, and this study considers as EVB both of them.

Previous studies (Argonne National Laboratory 2012) estimate the weighs of the LiB for electric vehicles as 19 kg for HEV, 89 kg for PHEV and 210 kg for BEV; and the increment of the dependency of the transportation sector on this technology in a middle and long term seems inevitable.

As is well known, sustainable production of the LiBs in the upstream of the supply chain is indispensable; however, an adequate collection, treatment, recycling and reusing of those batteries in the downstream stage is also necessary considering the following aspects:

- Security: Electrical, fire-explosion, and chemical hazard potential of the LiBs (Diekmann Jan et al. 2018). The correct treatment and disposal, including restricting the inappropriate second use of them, will avoid high scale accident.
- Legal: Considering the aspect mentioned above, many national-level governments request to the automakers be responsible for the collection and the adequate treatment of the LiBs (European Union 2013).
- Environmental: Manufacturing phase will dominate environmental impact across the life cycle of the LiB. Here, the carbon intensity of the electricity used in the production of its cells is the most impact-intensive, and the cascaded use system appears significantly beneficial (Ager-Wick Ellingsen Linda et al. 2013) (Leila et al. 2017).
- Economic: Currently, the processing and transportation cost of scrapped LiB is approximately 10 to 15 thousand yens per unit of battery for HV in Japan (Honda Motor Co. 2017); being a critical amount when the total weight of a scrapped EVB is considered. Moreover, valuable critical metals such as Co and Ni can be recovered with an adequate recycling process (Olivetti et al. 2017). Additionally, batteries are applied in different fields of energy storage and the reusing of them could also promote the use of renewable energies.

For those aspects, planning the future production/supply and the respective collection/processing/recycling (recoverability) of the batteries is going to be essential for creating a sustainable electric vehicles market and reach a circular supply chain.

Previous studies analyze the recycling process of LiB (Zheng et al. 2018) (Gaines 2014) (Winslow and Laux 2018) enhancing its importance. However, none of them consider possible inconvenient related to the supply of the scrapped batteries, where the -time- and -quantity- are representatives for its economic viability.

The aim of this study is to propose a dynamic model for forecasting the changes in the number of batteries recovered from the end of life electric vehicles considering different power trains.

22.2 Methodology

Figure 22.1 shows the concept of our model, where the entire vehicle market is considered. This approach divides the vehicle market into three parts: sales, ownership/aging, and scrap, also considering, the different power trains (ICEV, HEV, PHEV, BEV) sold by type of vehicle (mini passenger cars, mini Trucks, standard passenger cars, small passenger cars, standard trucks, small trucks & large buses and small buses).

The total number of vehicles in a region (ownership) can be annually updated considering the number of vehicles sold and scrapped in a year (1). Here, the term vehicle scrapped includes the ELVs that are dismantled but also the ones that spent its second life in foreign countries.

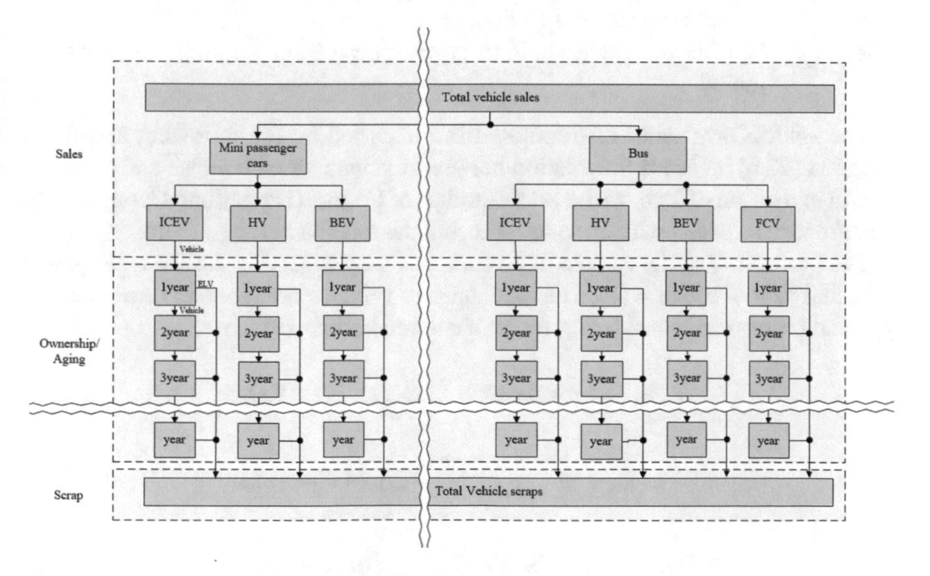

Fig. 22.1 Concept of the dynamic model for forecasting

$$V_{t+1} = V_t + VSa_{t+1} - VSc_{t+1} \tag{22.1}$$

V_t Number of vehicles at the end of the year *t* [units].
V_{t+1} Number of vehicles at the end of the year *t + 1* [units].
VSa_{t+1} Number of vehicles sold during the year *t + 1* [units].
VSc_{t+1} Number of vehicles scrapped during the year *t + 1* [units]

The above flow is separately analyzed considering the type and power train of the vehicles (2)-(5).

$$V_t = \sum_i \sum_p \sum_l V_{t,i,p,l} \tag{22.2}$$

$$V_{t+1} = \sum_i \sum_p \sum_l V_{t+1,i,p,l} \tag{22.3}$$

$$VSc_{t+1} = \sum_i \sum_p \sum_l VSc_{t+1,i,p,l} \tag{22.4}$$

$$VSa_{t+1} = \sum_i \sum_p VSa_{t+1,i,p} \tag{22.5}$$

$V_{t,i,p,l}$ Number of vehicles of type *i*, power train *p* and year of life *l* at the end of the year *t* [units].
$V_{t+1,i,p,l}$ Number of vehicles of type *i*, power train *p* and year of life *l* at the end of the year *t + 1* [units].
$VSc_{t+1,i,p,l}$ Number of vehicles scrapped of type *i*, power train *p* and year of life *l* during the year *t + 1* [units].
$VSa_{t+1,i,p}$ Number of vehicles sold of type *i*, power train *p* during the year *t + 1* [units]

The vehicle ownership of the model is forecasted by the following approach. Equation (22.6) indicates the relation between the number of vehicles and the GDP growth of a country. Here, we based on studies of Dargay (Dargay and Gately 1999), who propose an s-shape function to represent the relation between them.

The probability of the vehicle to be scrapped vary depending on the type, power train, and year of life of it (7). The total sales of vehicles per year can be divided by type and powertrain, considering future share predictions (8).

$$VO_{t+1} = \gamma * \theta * e^{\alpha * e^{\beta \, GDP}} + (1 - \theta) * VO_t \tag{22.6}$$

$$VSc_{t+1,i,p,l} = \sum_i \sum_p \sum_l V_{t,i,p,l} * Psc_{t+1,i,p,l} \tag{22.7}$$

$$VSa_{t+1,i,p} = VSa_{t+1} * \sum_p \sum_l Ss_{t+1,i} * Ss_{t+1,p \in i} \tag{22.8}$$

VO_{t+1} Vehicle ownership at the end of the year $t + 1$ [units per 1000 people]
VO_t Vehicle ownership at the end of the year t [units per 1000 people]
γ Saturation level of the number of vehicles [units per 1000 people].
α Parameter alpha related to the shape of the function.
β Parameter beta related to the shape of the function.
θ Speed of effect between the variables ($0 < \theta < 1$).
$Psc_{t+1,i,p,l}$ Probability of a vehicle type i, powertrain p and year of life l to be scrapped during the year $t + 1$.
$Ss_{t+1,i}$ Sale share of vehicle type i during the year $t + 1$.
$Ss_{t+1,p\in i}$ Sale share of vehicles with powertrain p in the market of the vehicle type i during the year $t + 1$.

Here, the amount of returning flow of EVB can be estimated by the following Eq. (22.9).

$$Lib_t = \sum_i \sum_p \sum_l VSc_{t,i,p,l} * Lib_{i,p,l} * Exp_{t,i,p,l} \tag{22.9}$$

Lib_t Amount of LiB scrapped during the year t [kwh].
$Lib_{i,p,l}$ Amount of LiB of a vehicle type i, power train p and year of life l [kwh/unit]
$Exp_{t,i,p,l}$ Percentage of vehicle scrapped and exported of vehicle type i, power train p and year of life l during the year t.

22.3 Japanese Vehicle Market as a Case Study

Our study analyzes the Japanese vehicle market considering the importance of it in term of production and sales, but also, taking in account that Japan is one of the countries that lead the electrification of the vehicles.

Firstly, Fig. 22.2 shows the share of the future vehicles sales in Japan by type and power train; which was elaborated considering data from the Ministry of the Environment (Ministry of the Environment Government of Japan 2010). It is worthy of mentioning that the share from 2009 to 2017 represents the actual values calculated basing on reports from Next Generation Vehicle Promotion Center (2019).

Secondly, the number of vehicles (ownership) has been estimated through Eq. (22.6). Here, forecasts of the Japanese GDP and the population growth of the country, presented by the OECD (OECD 2018) and The world bank (The world bank 2019) have been considered. Figure 22.3 shows the Japanese vehicle fleet estimated through the method mentioned above. Moreover, the current composition of it by type and powertrain was calculated based on reports from the Next Generation Vehicle Promotion Center (2019).

Thirdly, the vehicle scrapping rates were calculated as the percentage of vehicles that are scrapped annually per year of use base on reports of the Automobile Inspection & Registration Information Association (Automobile Inspection & Registration

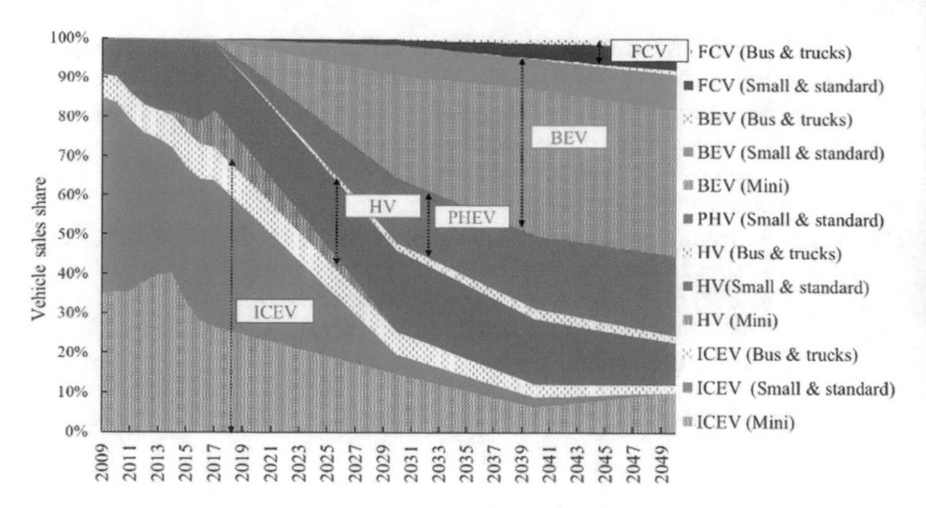

Fig. 22.2 Share prediction of the vehicle sales for the Japanese market

Fig. 22.3 Japanese vehicle fleet forecast

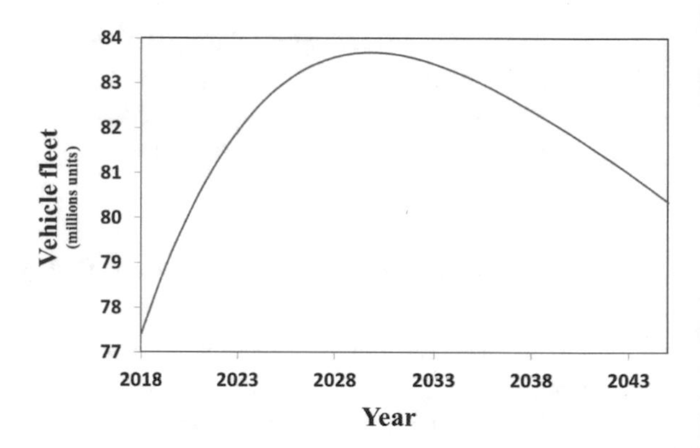

Information Association 2019). Figure 22.4 shows the historical and forecasting of the vehicle scrapping rate considered in this model. It is worthy of clarifying that the scrapping rate of passenger cars varies shapely every two years due to the impact of the automobile inspection requirement of the Japanese government.

Finally, the exportation of used vehicles in Japan represent approximately 28% of the ELV (Ministry of Environment, Government of Japan 2015). The demand for used HV is notorious, wherein 2017, approximately 85% of them were sent overseas to spent its second life (Japan Automobile Recycling Promotion Center 2017). Here, exportation level for PHEV and BEV were considered similar to it.

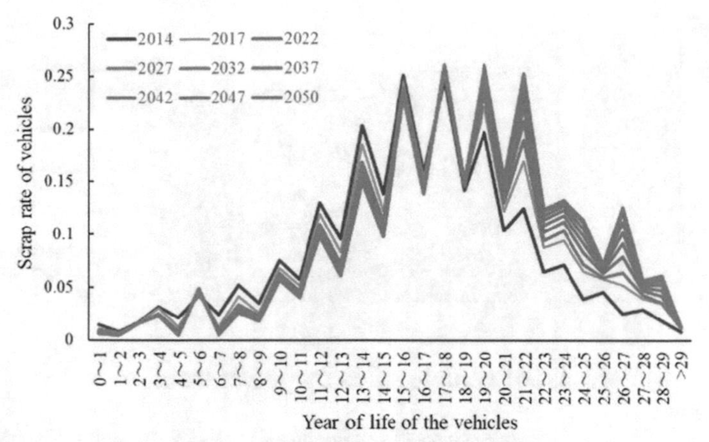

a) Forecast of the passenger cars scrapping rate

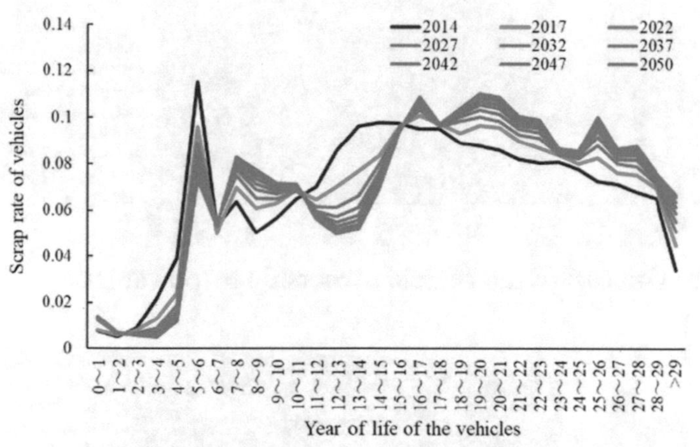

b) Forecast of trucks and buses scrapping rate

Fig. 22.4 Scrapping rate forecast of the Japanese vehicles

22.4 Results and Discussions

The proposed forecasting model was simulated through System dynamics and the software Vensim PLP x32 (Vensim PLP x32 [Computer software] 2019) used for this propose.

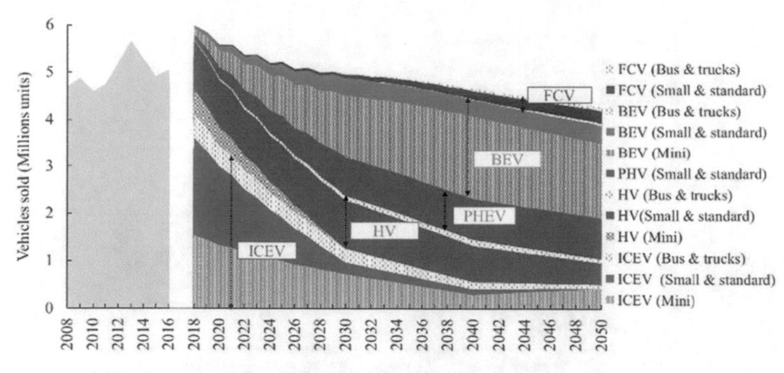

a) Forecast of vehicle sales by power train

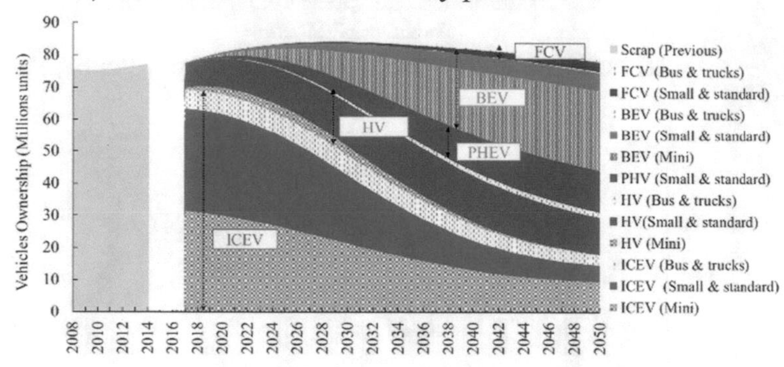

b) Forecast of the vehicle ownership by power train

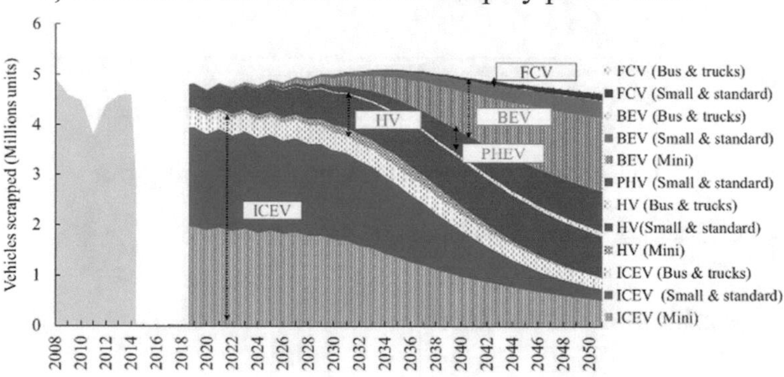

c) Forecast of the vehicle scrap by power train

Fig. 22.5 Forecast of the Japanese vehicle fleet by power train

22.4.1 Forecast of the Japanese Vehicle Fleet

Figure 22.5a shows the forecast of the vehicles sales by power train in the market. Here, it can be observed that the total sales of vehicles decrease moderately in the coming years. This can be explained by the decrease of vehicle ownership in Japan due to the reduction expected in the Japanese population for the following years. The sales of electric vehicles are expected to reach 2.89 million units/year in 2030 and remain almost constant in the following years.

Figure 22.5b shows the vehicle ownership forecast of the vehicles market. Here, it is possible to observe the compatibility of the number of vehicles in the Japanese fleet calculated in Fig. 22.3. Moreover, compared to sales, where it can be observed a drastic change of the share to electric vehicles in the following years, the vehicle fleet itself will be still predominated by ICEV in the next decade.

Figure 22.5c shows the forecast of the vehicle scrap. Here, it can be noted that even substantial quantity of HVs are reaching its end on life, the amount of BEV and PHEV expected to be collected until 2025 seem to be minimal.

Grey parts on the left side of each figure indicate past vehicle sales, ownership, and scrap data, which are compatible with the values forecasted in this study.

22.4.2 Forecast of EVB Supply and Recovery for the Japanese Vehicle Market

Size, weight, and capacity of the EVB vary widely depending on the type, power train, and specifications of the vehicle. In this study, the capacity and weight of the batteries have been considered, 2 kWh and 19 kg for HV, 9kWh and 89 kg for PHEV, and 28 kWh and 210 kg for BEV (Dunn et al. 2012). It is worthly to mention that HVs are powered by gasoline and electricity generated by the car's own braking system, PHEVs can also be recharged plugging into an external source of electrical power, and BEVs are fully electric vehicles. The battery size varies depending on the degree that electricity is used as their energy source.

Figure 22.6a shows the forecast of the EVB supply simulated by this model, here, it can be noted that even though, in term of vehicles sales, HVs represent most of its share, the demand of LiB for BEV is going to dominate the market considering the energy required for it. Moreover, the EVB demand is going to increase rapidly in the following years; however, it is expected to reach maturity near 2030 and meet the peak in 2040.

Figure 22.6b shows the forecast of the EVB scrapped and the number of batteries recovered from the ELVs. The uncolored section of the curve represents the number of batteries that are supposed to be exported as used vehicles. The quantity of EVB for recycling and reusing is expected to be minimal compared to the quantity of battery to be supplied in the market in the following years.

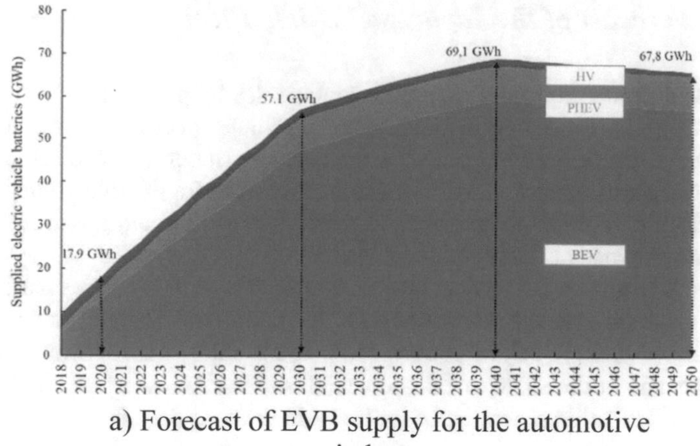

a) Forecast of EVB supply for the automotive industry

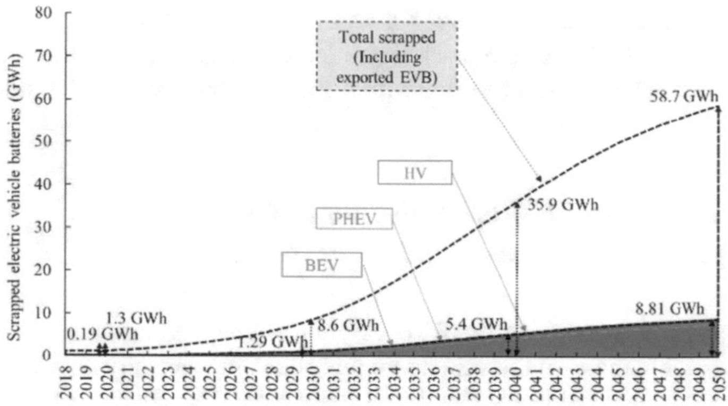

b) Forecast of EVB scrapped and recovered from the ELV

Fig. 22.6 Forecast of EVB supply and recoverability for the Japanese vehicle market

Table 22.1 Supply and recover of EVB from the Japanese vehicle market

Lithium ion battery flow	Year *Incl. exports						
	2020	2025	2030	2035	2040	2045	2050
Supply to the automotive industry (GWh)	17.9	38.5	57.1	64.1	69.1	67.6	65.8
Recovered from the vehicle fleet (GWh)	0.2	0.5	1.3	3.1	5.4	7.5	8.8
Scrapped from the vehicle fleet * (GWh)	1.3	3.4	8.6	20.4	35.9	49.9	58.7
Recovered/supply (%)	1.1%	1.3%	2.3%	4.8%	7.8%	11.1%	13.4%
Scrapped/supply * (%)	7.2%	8.9%	15.1%	31.8%	52.1%	73.9%	89.3%

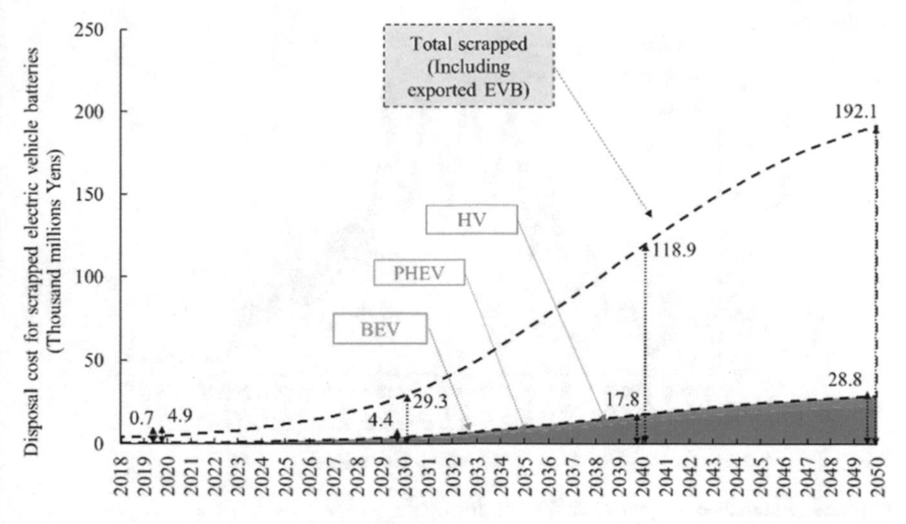

Fig. 22.7 Forecast of the processing cost of the EVB

Table 22.1 summarizes the quantity of EVB expected to be supplied, scrapped, and recovered from the market, but also its balance is calculated in the last two rows. Here, the percentage of batteries for the Japanese vehicle industry that theoretically can be covered by reused parts and recycling materials are calculated. It can be noted that in 2030 the recoverability will be approximately 2.3%; but also rapid growth of those values are expected in the following years. Furthermore, if the percentage of electric vehicles exported as used cars decrease considerably, the recovered LiB can support 8.9% of the EVB production in 2025, and a total closed loop can be achieved near 2050.

Finally, Fig. 22.7 shows a forecast of the fee necessary for the processing of the LiB if the current cost is kept. Here, average processing and transportation cost per kg of LiB has been considered (Honda Motor Co. 2017). Even contemplating a high level of ELV exportation, the total processing cost will be 4.4 thousand million yens in 2030, 17.8 thousand million yens in 2040, and 28.8 thousand million yens in 2050. This results, reinforcing the necessity of improving the current downstream scheme for the EVB.

22.4.3 Sensitive Analysis: Changes in the Year of Use of BEV

Considering the current low quantity of BEV in the Japanese vehicle fleet and being a considerably new technology, there is no certainty on the life span of those vehicles and the duration of its batteries. Even technologies of batteries are being advanced, and longer BEV life span may be expected, many studies are carried to predict the

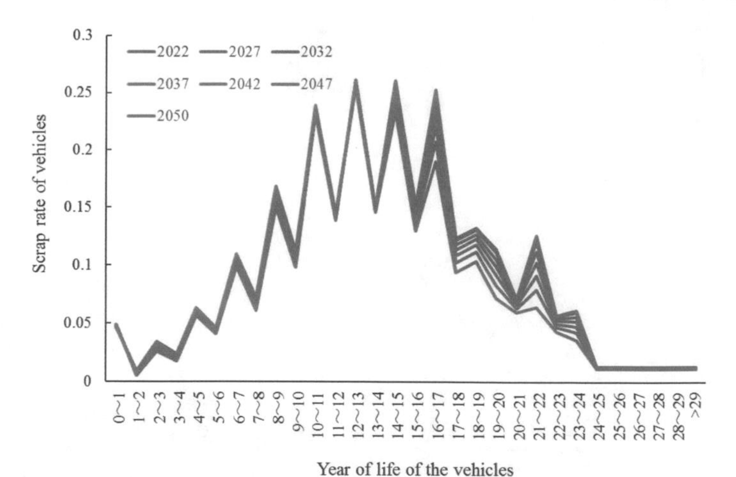

Fig. 22.8 Alternative scrapping rate forecast for BEV

Table 22.2 Differences between the quantity of EVB supplied and recovered

Supply and recovery difference	Year						
	2020	2025	2030	2035	2040	2045	2050
Basic scenario (Gwh)	17.8	38.0	55.8	61.1	63.7	60.1	57.0
Shorter BEV life scenario (GWh)	17.7	38.3	58.0	65.8	70.5	67.7	65.1

aging of its batteries (Bolun et al. 2018; Egoitz et al. 2018). In this sense, this analysis considers the variation of BEV life span as a variable of interest.

In this section, the simulation is re-conducted shortening the life span of BEV (passenger cars). Here, it was considered the scrapping rate of them 5 years lower than the forecasted before (Fig. 22.8).

Comparison of the differences between the supplied and recovered quantity of EVB by the life span of BEV are indicated in Table 22.2. Here, it can be observed that if the life span of the BEVs decrease, more quantity of LiB will be collected, but also production and supply of vehicles are going to increase jointly in order to fulfill the necessity gap generated in the market. In consequence, the supply-demand balance indicates that even more batteries are recycled and returned as material to the market, the shrink of the BEV life is going to generate an increment of material necessary for its production.

22.4.4 Discussion

This study proposed a model to forecast the Japanese vehicles market in order to analyze the recoverability of the EVB.

Different approximations in the inventory analysis for the Japanese vehicle market has been carried in order to conduct the simulation. Here, external analysis, such as the forecasting of the Japanese population, GDP, and vehicles sales share has been considered. Changes on those values could guide us to different results; however, the proposed forecasting model and main conclusions are not going to be affected.

Considering that most of the electric vehicles reach its end of life overseas, those destinations, which are mostly underdeveloped countries, should also settle legal requirement for the adequate treatments and disposal of them in order to avoid high scale accidents with the LiB. Future studies must clarify the final destination of the used EV, also considering the possibility of installing recycling facilities in the countries where ELV concentration is expected.

This study has adopted a constant exportation rate of used vehicles; however in the long term, saturation of those markets and drop of this rate can be expected. In this sense, the proposed approach considered recovered and scrapped volume of LiB separately.

Through the diffusion of EV, CO_2 emission related to the production of EVB will increase drastically in the following years. However, considering a late and low return of them as scrap, possible recycling benefits of them are far from balancing the high impact expected in its production phase.

The results presented in this study gives a whole picture of the ELV market. The model can be applied by automakers and related companies to propose or/and verify the economic feasibility of different battery reusing and recycling business models. Moreover, this model could also be applied to other industries like electrical and electronic equipment, been a complement to the approach proposed by Baldé et al. (Forti V. Gray 2017).

As shown before, the total fee related to the processing of EVB will increase drastically in the following years; however, the quantity of returning batteries seems to be low. In this sense, the development of LiB recycling technologies should be center, in the short term, in developing low cost and efficient technologies. Here, processing cost improvement by the increment of the volume of LiB to be recycled and efficient transportation could be essential keys.

Dismantlers and material recycling companies are going to be able to define optimal plans for the adaptation of its facilities, knowing the time and quantity of EV returning from the market.

Finally, it worthy of mentioning that the proposed model can be easily adapted to other countries but also for different products such as electrical household appliance that needs to be recycled after its use.

22.5 Conclusions

This study proposes a dynamic fleet model for assessing the future flow of used EVB. System dynamic concepts and open data used for the simulation of the Japanese vehicle market.

The main conclusions of this approach are listed below.

- The total sales of vehicles in Japan seem to slow down in the following years because of the decreasing Japanese population. However, sales of electric vehicles will increase to 3.72 million units/year in 2030 and remain almost constant until 2050.
- EVB demand is going to increase rapidly in the following years; however, it is expected to reach maturity near 2030 and meet the peak in 2040.
- Exportation of used EV has a substantial impact on the LiB processing/recycling market. EVB for recycling and reusing is expected to be low compared to the quantity of battery to be supplied in the market in the following years.
- The amount of scrapped EVB will increase 45 times from 2020 to 2050. Moreover, a complete closed-loop of them can be expected around 2050 only if the exportation of used electric vehicles is hardly diminished.
- Processing cost for EVB was calculated as 4.4 thousand million yens in 2030, 17.8 thousand million yens in 2040, and 28.8 thousand million yens in 2050.
- The development of LIB recycling technologies should be center, in the short term, in developing low cost and efficient technologies.
- Shorten the use life of BEV will increase the returning flow of material for recycling; however, the increment of material needed for the production of new vehicles will be higher.

Finally, what needs to be emphasized is that the proposed model can be adapted to other countries but also for different products. Results presented in this study can also guide automakers to propose feasible business models for the reusing/recycling of LiB, but also for dismantles and material recycling companies to adjust its installation for the coming changes in the vehicle market.

References

Argonne National Laboratory (2012) Material and energy flows in the materials production, assembly, and end of life stages of the automotive lithium ion battery life cycle. https://greet.es.anl.gov/publication-lib-lca

Automobile Inspection & Registration Information Association (2019). Tendency of the vehicles ownership in our country. https://www.airia.or.jp/publish/statistics/trend.html

Baldé CP, Forti V. Gray V, Kuehr R, Stegmann P (2017) The global e-waste monitor—2017, United Nations University (UNU), International Telecommunication Union (ITU) & International Solid Waste Association (ISWA), Bonn/Geneva/Vienna

Bolun Xu, Alexandre Oudalov, Andreas Ulbig, Goran Andersson, Kirschen Daniel S (2018) Modeling of lithium-ion battery degradation for cell life assessment. IEEE Trans Smart Grid 9(2):1131–1140

Dargay J, Gately D (1999) Income effect on car and vehicle ownership, worldwide 1960–2015. Transp Res Part A 33:101–138

Diekmann J, Grützke M, Loellhoeffel T, Petermann M, Rothermel S, Winter M, Nowak S, Kwade A (2018). Potential dangers during the handling of lithium-ion batteries. In: Kwade A, Diekmann

J (eds) Recycling of lithium-ion batteries. sustainable production, life cycle engineering and management. Springer, Cham
Dunn JB, Gaines L, Barnes M, Sullivan J, Wang M (2012) Material and energy flows in the materials production, assembly, and end-of-Life stages of the automotive lithium-ion battery life cycle. Argonee National Laboratory ANL/ESD/12–3
Egoitz Martinez-Laserna, Elixabet Sarasketa-Zabala, Igor Villareal Sarria, Daniel-Ioan Stroe, Maciej Swierczynski, Alexander Warnecke, Jean-Marc Timmermans, Shovon Goutam, Noshin Omar, Pedro Rodriguez (2018) Technical viability of battery second life: a study from the ageing perspective. IEEE Trans Ind Appl 54(3):2703–2713
Ellingsen LA, Majeau-Bettez G, Singh B, Srivastava AK, Valøen LO, Strømman AH (2013) Life cycle assessment of a lithium-ion battery vehicle pack. J Indus Ecol 18(1):113–124
European Union (2013) Document 2006L0066—EN—30.12.2013–003.001–1
Gaines L (2014) The future of automotive lithium-ion battery recycling: charting a sustainable course. Sustain Mater Technol 1:2–7
Honda Motor Co (2017) Advanced recycling of lithium ion batteries. https://www.env.go.jp/policy/kenkyu/special/houkoku/data_h28/pdf/3K152013.pdf
International Energy Agency (2009) Transport Energy and CO2: Moving towards sustainability
Japan Automobile Recycling Promotion Center (2017). Vehicle recycling data book 2017. https://www.jarc.or.jp/renewal/wp-content/uploads/2017/07/DataBook_2017.pdf
Leila Ahmadi, Young Steven B, Michael Fowler, Fraser Roydon A, Achachlouei Mohammad A (2017) A cascaded life cycle: reuse of electric vehicle lithium-ion battery packs in energy storage systems. Int J Life Cycle Assess 22(1):111–124
Ministry of Environment, Government of Japan (2015) Industrial structure council, industrial technology environmental working group, waste. Recycling subcommittee, automobile recycling working group. central environment council, recycling social committee, automobile recycling technical committee, joint meeting. Report on evaluation and examination of the implementation status of the automobile recycling system, Sept 2015. https://www.env.go.jp/council/03recycle/y033-43/mat03_2.pdf
Ministry of the Environment Government of Japan (2010) Strategy for diffusion of environmental vehicles. https://www.env.go.jp/air/report/h22-02/06_chpt3.pdf
Next Generation Vehicle Promotion Center (2018) Strategy for diffusing the next generation vehicles in Japan. http://www.cev-pc.or.jp/event/pdf/xev_in_japan_eng.pdf
Next Generation Vehicle Promotion Center (2019a) http://www.cev-pc.or.jp/tokei/hanbai3.html
Next Generation Vehicle Promotion Center (2019 b) http://www.cev-pc.or.jp/tokei/hanbai.html
OECD (2018) GDP long-term forecast (indicator)
Olivetti Elsa A, Gerbrand Ceder, Gaustad Gabrielle G, Xinkai Fu (2017) Lithium-ion battery supply chain considerations: Analysis of potential bottlenecks in critical metals. Joule 2017(1):229–243
The world bank (2019) Population estimates and projections. https://datacatalog.worldbank.org/dataset/population-estimates-and-projections
U.S. Energy Information Administration (2016) International energy outlook 2016. DOE/EIA-0484
Vensim PLP x32 [Computer software] (2019)
Winslow KM, Laux SJ, Townsend TG (2018) A review on the growing concern and potential management strategies of waste lithium-ion batteries. Resour Conserv Recycl 129:263–277
Zheng X, Zhu Z, Lin X, Zhang Y, He Y, Cao H et al (2018) A mini-review on metal recycling from spent lithium ion batteries. Engineering 4:361–370

Chapter 23
Application of Causality Model to Propose Maintenance Action of Parts

Takeru Nagahata, Hiroki Saitoh, and Hiroyuki Hiraoka

Abstract This paper discusses the application of a causality model to propose maintenance actions for the purpose of promoting the reuse of mechanical parts. To realize effective reuse of mechanical parts for the development of a sustainable society, it is essential to manage the individual parts over their entire life cycle. For this purpose, we are developing a "part agent" that is programmed to follow its real-life counterpart throughout its life cycle. A part agent provides the user with appropriate advices on the reuse of its part and promotes the circulation of reused parts. In this study, we propose a method to estimate the probability of failure occurrence and determine the corresponding maintenance action. The probability of the failure occurrence is calculated by the causality among events relating to the part and the conditional probability, which is the probability of occurrence of the resultant event against the occurrence of input events. To calculate the conditional probability of events representing the state of a part, the deterioration of the part is simulated based on its operating and environmental conditions. A causality model including this conditional probability is constructed, and the failure probability of the part is estimated. In this paper, we report a mechanism that updates the conditional probabilities according to the operating and environmental conditions that correspond to the different stages in a life cycle.

Keywords Maintenance action · Causality model · Life cycle simulation · Part agent · Conditional probability

23.1 Introduction

In recent years, the energy resources on the earth are being depleted, and the transition from a consumption type society, which involves processes such as mass production, mass consumption, and mass disposal, to a circulating society, which suppresses the consumption of the resources, becomes urgent. To realize the circulating society, it is

T. Nagahata · H. Saitoh (✉) · H. Hiraoka
Department of Precision Mechanics, Chuo University, Tokyo, Japan
e-mail: h-saitou@lcps.mech.chuo-u.ac.jp

Y. Kishita et al. (eds.), *EcoDesign and Sustainability I*, Sustainable Production, Life Cycle Engineering and Management, https://doi.org/10.1007/978-981-15-6779-7_23

necessary to promote the so-called 3Rs: reuse, reduce and recycle. Especially, reuse, which can reduce the energy consumption and generation of waste is emphasized. For an efficient reuse, it is important to control the life cycle of each part. However, it is difficult to manage the life cycle of used parts because the behavior of the users is diverse, and it is difficult to predict the state of parts. To solve this problem, we are developing a part agent system that manages the state of parts and supports the maintenance action of users (Yokoki and Hiraoka 2017). A part agent system is a system for managing the life cycle of parts by making the network agent follow the actual parts.

The deterioration of parts varies by various factors such as use environment and use time, and their maintenance actions are also diversified. Moreover, it is generally difficult to observe the progress of deterioration directly. Therefore, it is necessary to estimate the progress of deterioration by some method based on the behavior of users and observable events, and on this basis decide the maintenance actions. In this study, we propose a method to determine the maintenance action by creating a causality model between the states of parts and estimating the probability of occurrence of a failure using Bayesian estimation based on the model (Nagahata et al. 2018). The causality model between the state of parts is developed by combining the deterioration process and a deterioration simulation, and the failure probability of parts is calculated supported on it. We then propose a method to determine the maintenance action by utilizing the probability for the life cycle simulation (LCS).

23.2 Part Agent System

A part agent is a system in which an agent dedicated to each product or part collects information such as the state of the product or part, manufacturing information, and usage history, and presents it so that a user can make an appropriate judgment on the maintenance of the product or part.

Figure 23.1 shows a conceptual diagram of the part agent. By attaching an RFID tag (Gaetano 2005) to each part, the part agent follows the movement of the part along the production factory, repair factory, consumer route, retail store, etc. in the network. The correspondence is acquired through the RFID tag, and information such as the deterioration level of the parts is acquired through the network. Based on the information, the agent devises the most appropriate treatment for the corresponding part at that time and proposes it to the user.

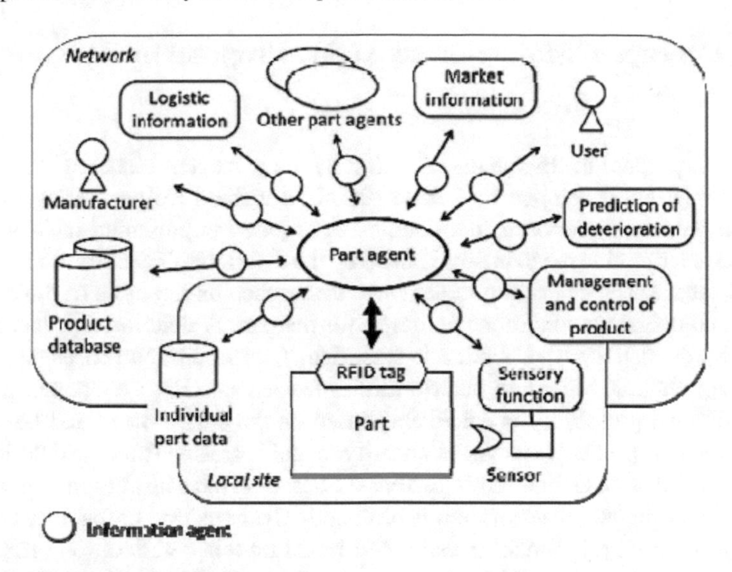

Fig. 23.1 Conceptual scheme of part agent

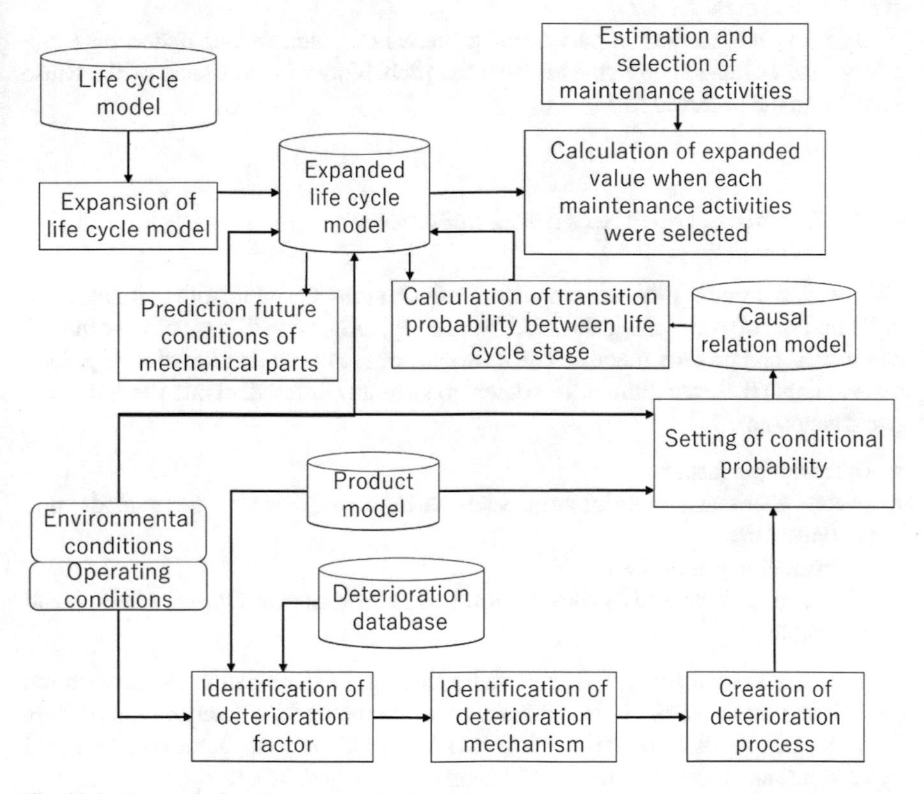

Fig. 23.2 Proposal of maintenance action by part agent

23.3 Concept of Maintenance Action Proposal by Part Agent

The flow to propose the maintenance action by the part agent is shown in Fig. 23.2. Using the system developed by Takata (2006), a method for generating the deterioration process is proposed (the detailed description is presented in Sect. 23.4). First, the deterioration mechanisms caused by the deterioration factors are registered in the database. The deterioration database is searched on the basis of the environmental and operating conditions assumed for the part. A deterioration mechanism caused by the deterioration factors is specified. A deterioration process is created by linking multiple identified deterioration mechanisms. For the created process, the conditional probability is calculated based on the environment and conditions under which the product is used. A causality model is created from the deterioration process. A Bayesian estimation is performed based on the causality model, and the probability of product deterioration is obtained. The transition probability between stages of the life cycle model is estimated based on this deterioration probability, and the expectation value of the next stage is calculated (the detailed LCS methods are described in Sect. 23.7).

By using this method, a part agent proposes the maintenance action by using a Bayesian estimation, which estimates the probability of occurrence of the cause event from the observed facts.

23.4 Creating Deterioration Processes

This section describes the deterioration analysis system. The factors that cause the deterioration of parts are influenced by the operating conditions, environmental conditions, and product models. The existence of factors of deterioration is judged by examining these conditions. The deterioration can be classified into the following two categories:

- Deterioration factors
- Factors that cause deterioration, such as high temperatures, large loads, and prolonged use
- Deterioration mechanism
- Effects on parts caused by the deterioration factors, such as fatigue, fracture, and corrosion.

The deterioration process is created by linking the degradation factor with the degradation mechanism. Figure 23.3 shows an example of the deterioration process in the general corrosion of parts. In this figure, the deterioration factor is represented by a square and the deterioration mechanism is represented by an ellipse. “Water” and “Chloride ion” become deterioration factors, and cause “Wet corrosion,” which is a deterioration mechanism. The “Wet corrosion” generated by this process becomes a

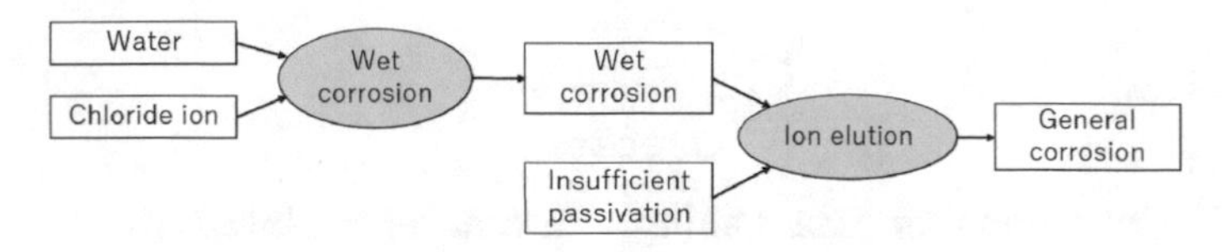

Fig. 23.3 Example of deterioration process in general corrosion

new deterioration factor, which together with "Insufficient passivation film" causes "Ion elution."

In this way, various deterioration processes are represented in a unified way, and the function that proposes the maintenance action by the part agent is developed.

The deterioration factors and the deterioration mechanisms caused by them are identified by searching the deterioration mechanism database by operation condition, environmental condition, and the product model assumed. A deterioration factor should be defined as that occurring when a certain parameter satisfies a certain condition. For example, when the temperature exceeds a certain threshold value, the occurrence of "high temperature" is determined. This threshold is assigned to each input event when the assumed input satisfies the condition. The determined deterioration mechanisms are linked to generate a deterioration process. The deterioration factors determined by the operating and environmental conditions of the user are classified as an event by the input, the observable deterioration factor, and the unobservable deterioration factor.

To estimate the failure probability, a causality model is created by assigning a conditional probability to the created deterioration process (Nagahata et al. 2019). The conditional probability is the probability of occurrence of a resultant event, and it varies according to the occurrence of a causal event that is used to estimate the probability of occurrence of deterioration of a part (the detailed simulation methods are described in Sect. 23.5).

23.5 Calculation of a Conditional Probability

The procedure of deterioration simulation is shown in Fig. 23.4. The probability of occurrence for the parameter of the input event of the deterioration process is given based on the product model, environmental conditions, and operating conditions, and is defined as a prior probability. To that end, data on the occurrence of input events is used, if available. If not, an appropriate value is set. Based on this prior probability, a statistical model using the probability distribution is created.

Figure 23.5 shows how to determine the occurrence of a resultant event. Resultant events occur when the parameters of an input event satisfy a particular relationship. For example, in the case of a causal relationship in which "Stress corrosion cracking" arises from "High ion concentration environment" and "High temperature,"

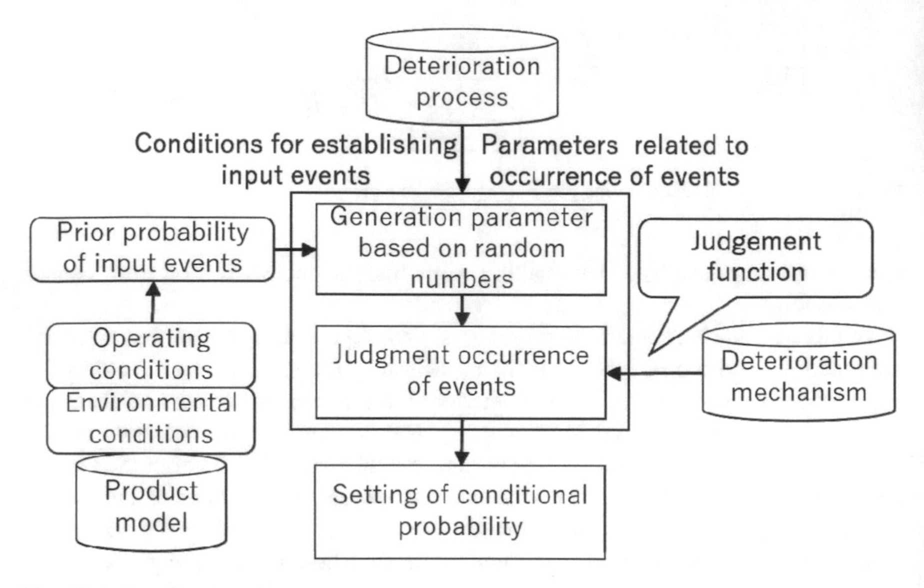

Fig. 23.4 Deterioration simulation procedure

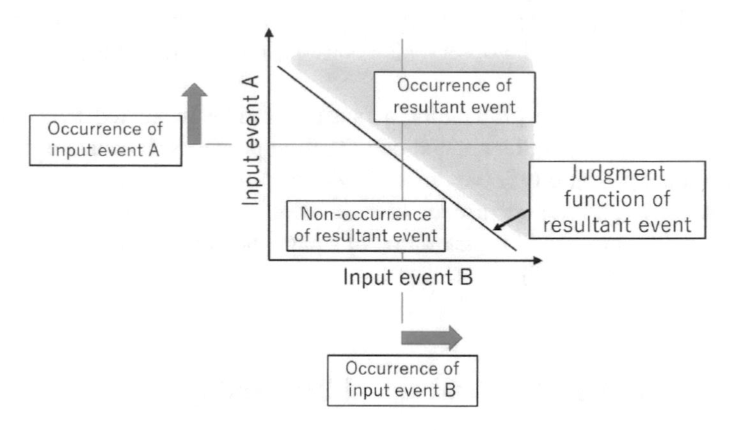

Fig. 23.5 Evaluation of the occurrence of resultant events

the occurrence of "Stress corrosion cracking" is expressed as a function of "Temperature" and "Ion concentration". The judgment function, which determines the occurrence of the resultant event by the parameter of the input event, is registered in the database of the deterioration mechanism in advance. If a judgment function already exists, it is used, and if not, an appropriate function is used. As a method for calculating the occurrence probability of the resultant event, the Monte Carlo method is used (R and Raghavan 1995). The parameter generated randomly according to the probability distribution of the input event is substituted into this judgement function multiple times, and it is judged whether the resultant event occurs. In this way, the

Table 23.1 Conditional probability table

Input event A	Input event B	Conditional probability of the resultant event
0	0	P1
0	1	P2
1	0	P3
1	1	P4

deterioration simulation is carried out by giving the distribution of the parameter value of the input event to this function, and the quantitative function relation is created based on the quantitative causal relation of the degradation process.

Then, the rate at which the resultant event occurs for each combination of input event occurrences (conditional probability) is calculated to prepare a conditional probability table. Table 23.1 shows the conditional probability table, where 1 denotes the occurrence of an input event and 0 denotes the absence of an input event. P1–4 represent the corresponding probabilities.

23.6 Estimation of Failure Probability Using Bayesian Estimation

The failure probability is estimated by a Bayesian estimation based on the conditional probability calculated in Sect. 23.5. Figure 23.6 shows an example of causal network representing a causal relation between the user's operations and failures of a product (Hiraoka et al. 2013). The nodes with a rounded box represent inputs to the network that are, in this case, operations by the user. The nodes with a square box represent events related by causal relations. The events with a shaded box cannot be observed directly, which means that we have to suppose their occurrence based on other observable events. This example model contains three operations for user's choice and two observable states that can be measured via sensory data. It represents the following causal relations: Operation A and B cause Defect 1, which in turn affects Observed state 1. Defect 1 and Operation C cause Defect 2. If Defect 2 occurs, it not only affects Observed state 2 but also requires maintenance.

A conditional probability is assigned to each node in the Bayesian network. It describes to what extent the occurrence of the event is affected by the prior events in its causal relation. For example, in Fig. 23.5, the conditional probability for the node Defect 1 is shown in the table at the top left. It shows the probability of Defect 1 for the combination of Operation A and Operation B, i.e. P(D1|OpA, OpB). It indicates that the probability of Defect 1 is 0.1 if both operations do not occur, 0.4 if only Operation B occurs, and so on.

Here, we assume that we have two reused parts, Part 1 and Part 2. Part 1 has a poor value in Observed state 1 and a good value in Observed state 2. On the contrary, Part 2 has a good value in Observed state 1 and a poor value in Observed state 2. We also

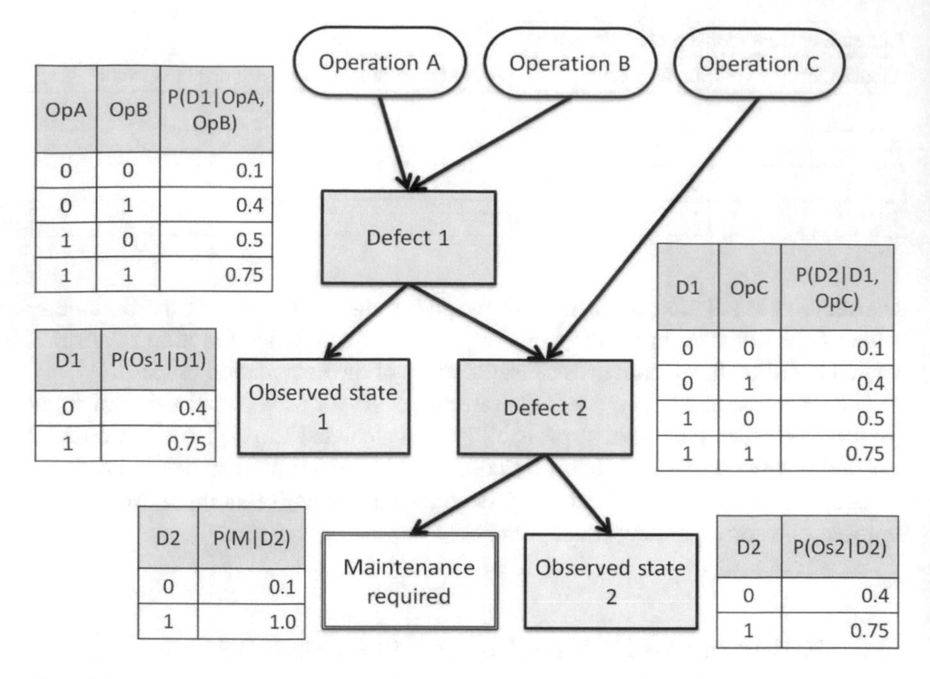

Fig. 23.6 Representation of causal relations with conditinal probability tables

have two users, User 1 and User 2. Their prior probability for Operation A, B, and C, which represents how the two users use the product, is presented in Table 23.2.

Based on this information on parts and users and with the information of the causal network shown in Fig. 23.6, the probability for the requirement of maintenance is calculated as presented in Table 23.3. Note that, for User 1, its probability of maintenance with Part 2 is higher than that with Part 1. For User 2, it is higher with Part 1 than with Part 2. Thus, the part agent recommends Part 1 for User 1 and Part 2 for User 2.

Table 23.2 Prior probability of user operation

	Operation A	Operation B	Operation C
User 1	0.25	0.25	0.25
User 2	0.75	0.75	0.25

Table 23.3 Probability for the requirement of maintenance

	Use part 1 P(M|Os1)	Use part 2 P(M|Os2)
User 1	0.370	0.428
User 2	0.66	0.611

Thus, the combination of the input event pattern and the part state changes the probability of occurrence of a failure event. It is possible to propose a more suitable maintenance action by estimating the failure probability.

23.7 Life Cycle Simulation

The part agent predicts the future state of the part and proposes the maintenance of that part to the user. Therefore, the part agent carries out an LCS on the basis of the acquired information and life cycle model. Figure 23.7 shows a life cycle model (Yokoki and Hiraoka 2017). The life cycle model is constituted by life cycle stages such as "Use," "Repair," "Dispose," "Sell and Buy," and life cycle paths that connect the life cycle stages. The parts are produced in the factory, and the life cycle stage of the part moves from the Produce Stage, which represents production, to the Sell and Buy Stage, which represents market transactions. When a part is purchased by a user, the life cycle of the part shifts to Use, and it is used. Subsequently, it goes through stages such as Repair and Replace, and after proceeding to Dispose, it returns to Use through Sell and Buy.

The part agent predicts the future by developing the life cycle model along the elapsed time, as shown in Fig. 23.8. Redrawing the life cycle model as a one-way model makes it possible to evaluate the life cycle of a part. The developed life cycle model shows the action for the parts per unit time, and it is structured to advance to the next connected stage when the parts are operated for one step. The shaded stage in the figure represents the life cycle stage of a part different from that of the part in use, and the chain line arrows represent the paths, including the selection and exchange of parts. That is, it is connected to the life cycle stage of another part through these paths.

The expected value of profit, cost, and environmental loading is calculated referring to the developed life cycle model, and the life cycle is evaluated. The process of selecting the next stage based on the life cycle in which the part agent is deployed is shown in Fig. 23.9.

Each stage has values for "Benefit," "Cost," and "Environmental Load" of the stage, such as V1 and V2, and the path indicated by the arrow has the probability of selecting the path. This probability is calculated based on the failure probability

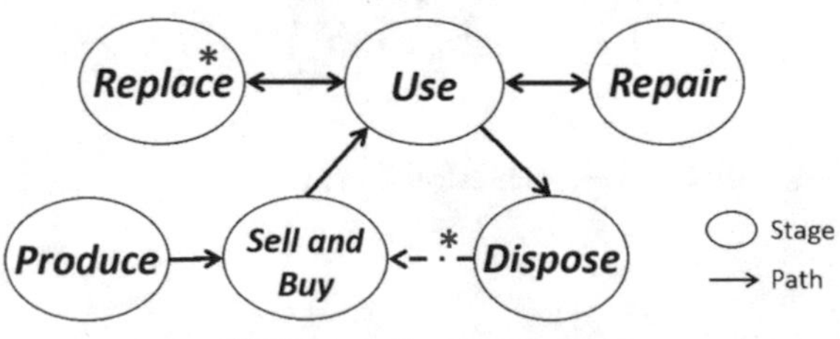

Fig. 23.7 Life cycle model

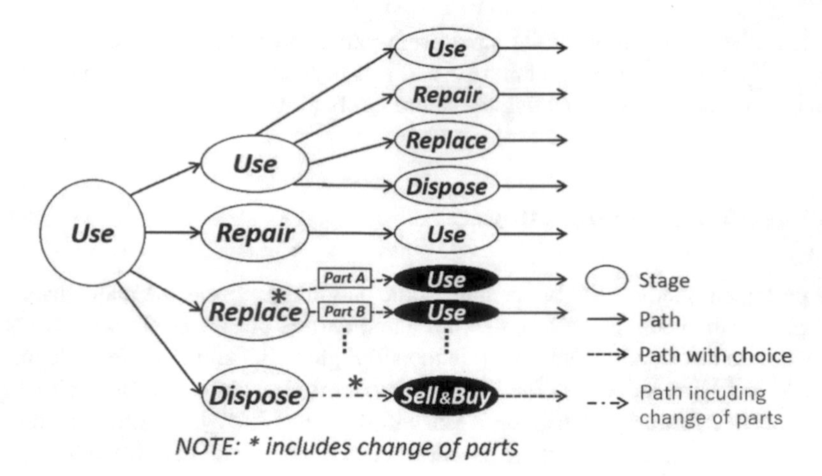

Fig. 23.8 Life cycle model deployed

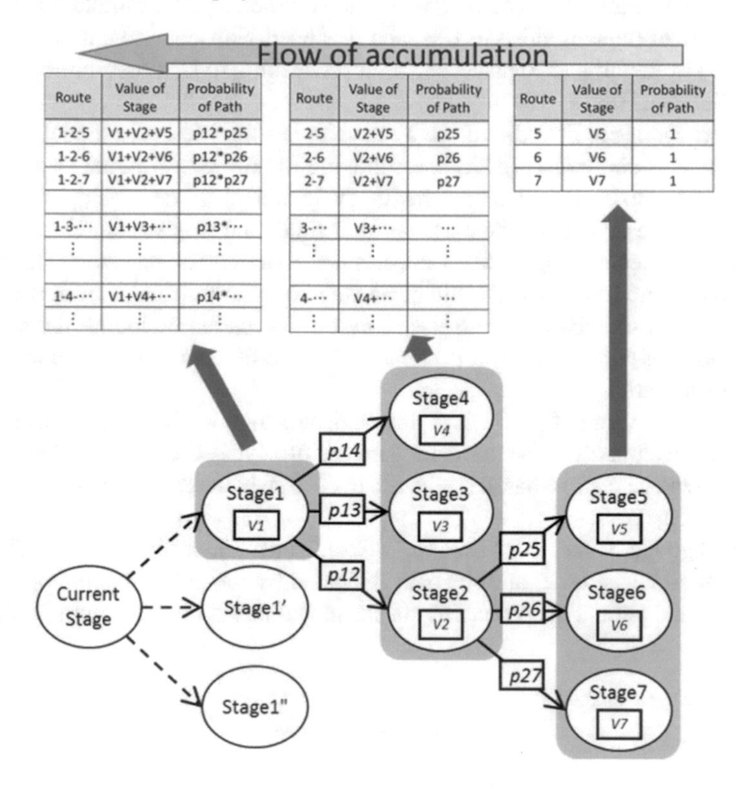

Fig. 23.9 Expected value calculation process

of the parts. A route connected by a path is defined as a "route," and the next stage is evaluated by calculating the expected values of all routes existing ahead of the next stage. From right to left, the table presents the values of each stage from the end of the expanded life cycle and the procedure for aggregating the probabilities of the paths connected to that stage for each route. The expected value for each route is obtained by multiplying the sum of the values of the stages in the route by the probability of the path.

It is necessary to give the probability of the path based on the situation for each of the assumed next stages, and the failure probability affects the value. Therefore, in this study, the stage to be advanced is decided by incorporating the failure probability estimated in Sect. 23.6 as an equation for obtaining the probability of the path.

23.8 Example: A Simple Manipulator

As an example, a method for determining the maintenance action by estimating the failure probability using a single degree of freedom manipulator, as shown in Fig. 23.10, is presented.

The structure of the mechanism comprises the following parts:

1. Motor
2. Angle Sensor
3. Gear
4. Arm (Length: 120 [mm])
5. Weight

A gear is mounted on the shaft of the motor, and another gear is mounted on the shaft of the arm, which has a weight fixed at the tip. When the motor applies torque, the arm rotates around the shaft through the gear. The arm is controlled to move back and forth between 0–45°. When the mechanism deteriorates, the arm falls below 0° owing to the insufficient torque. Thus, we manually set the arm to 0° every 180 s. The

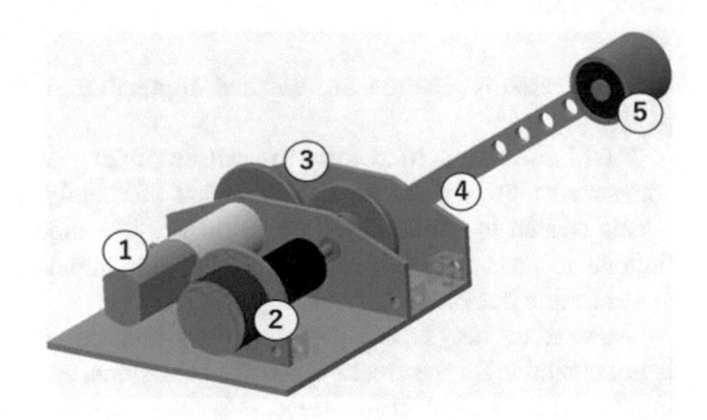

Fig. 23.10 One degree of freedom manipulator

degree of deterioration is determined from the angle measured every 180 s. When the machined arm falls below 0° and reaches a certain angle, it is judged as "failure."

The deterioration factor and mechanism are registered on the database beforehand. The deterioration factors are specified on the basis of the product models, such as individual information of parts and mechanisms, environmental conditions such as "Temperature" and "Humidity," and operating conditions such as "Rotational speed" and "Loads." In this way, deterioration factors such as "Fast contact", "Multiple contacts", and "Heavy load" may be identified. Then, the deterioration mechanism caused by the identified deterioration factor is searched from the database. In this case, "Motion of friction" is identified from "Fast contact", "Multiple contacts," and "Heavy load," whereas "Slide of axis" is identified from "Heavy load" and "Multiple contacts". The deterioration mechanism caused by the deterioration factors, including those generated by the identified deterioration mechanism, is searched again from the database. This process is repeated until no new deterioration factors appear, and the deterioration process is generated by linking the determined multiple deterioration mechanisms. The algorithm for creating the deterioration process is shown in Fig. 23.11.

The factors that cause the deterioration are regarded as input events, and the factors that are caused by the deterioration are regarded as resultant events. Based on this relationship, a deterioration simulation is performed to calculate the conditional probability between events. The judgement on the occurrence of the resultant event is carried out based on the judgment function registered in the deterioration database.

By assigning the conditional probability calculated by the deterioration simulation to the deterioration process, the causal relation model is created. The Bayesian estimation is performed using this causal model to estimate the failure probability.

The estimated failure probability is incorporated into the path probability, as shown in Fig. 23.9. Then, the expectation value in each route is calculated by multiplying this possibility with the value of the stage in the route, and the appropriate maintenance action is decided.

23.9 Discussion

More research is required for practical application of the system proposed in this paper.

First issue is how to identify the causes of deterioration for a product. We need some measures to extract a specific event efficiently from all the existing known events related to deterioration based on product model information. Other issues include how to obtain environmental conditions and usage as well as how to prepare deterioration database.

We need to study how to deal with the situations when enough clear information is not available for the creation of causal relations, such as the prior probabilities of inputs and the judgment functions. It is also a research issue how to determine the cause of deterioration when no related events are observable.

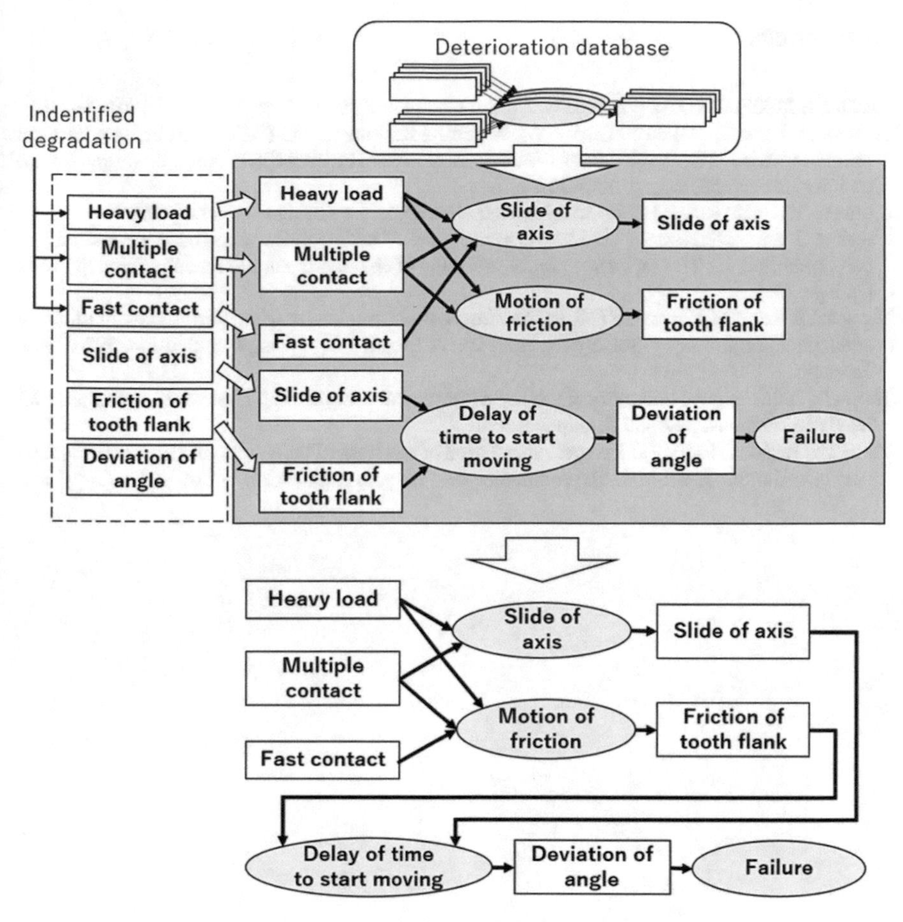

Fig. 23.11 Algorithm for creating deterioration processes (the diagram is redrawn based on (Takata 2006))

23.10 Conclusion

In this work, we propose a method to determine maintenance actions by calculating the failure probability using the conditional probability between events related to the deterioration of parts and reflecting this to the probability in the LCS. Remaining issues to be studied in the future include the implementation of the proposed method, its verification using a real machine, and validation of the simulation.

Acknowledgements We would like to express our deepest gratitude to Ryunosuke Yagi for providing us with the data from the experiment.

References

Gaetano B (2005) RFID: tagging the world. In: Communications of the ACM, vol 44(9):34–37

Hiraoka H, Ueno T, Ookawa H, Arita M, Nanjo K, Kawaharada H (2013) Part agent advice for promoting reuse of the part based on life cycle information. In: 20th CIRP international conference on life cycle engineering, p s337–342

Motwani, R; Raghavan, P (1995) Randomized algorithms, Cambridge University Press

Nagahata T, Saitoh H, Hiraoka H (2018) Determination of maintenance action based on causality of parts condition. In: The Japan society for precision of engineering autumn competition, p 32–33 (in Japanese)

Nagahata T, Saitoh H, Hiraoka H (2019) Creating causality models for part deterioration for life cycle simulation. In: The Japan Society for Precision of Engineering Spring Competition, 597–598(in Japanese)

Takata S (2006) Logical and rational facility management lifecycle maintenance to optimize LCC. In: JIPM-solutions, p 84–89 (in Japanese)

Yokoki Y, Hiraoka H (2017) Life cycle simulation of mechanical parts with part agent considering user behavior. In: The 24th CIRP conference on life cycle engineering, p 116–121

Chapter 24
Life Cycle Simulation of Machine Parts with Part Agents Supporting Their Reuse

Takumi Sugahara, Tsuramichi Tanigawa, and Hiroyuki Hiraoka

Abstract To effectively reuse machine parts for the development of a circulating society, it is essential to manage each part throughout its life cycle. Therefore, we propose a part agent that manages the part information and supports the user in reusing the part. A part agent is a network agent that autonomously acts corresponding to an individual part, manages information regarding a part from manufacturing to disposal, and proposes actions such as disposal and reuse to its user at an appropriate time. This activity promotes part reuse, reduce, and recycle. However, it is difficult to evaluate the effectiveness of a part agent by implementing it in an actual product and operating the product throughout its life cycle. Hence, we developed a system to simulate the behavior of parts and products throughout the entire life cycle with the part agent installed. There are several means to reuse products and the appropriate reuse method for the user differs according to the environment and the user's manner of thinking. In this study, we conducted a life cycle simulation considering different reuse methods using the part agents.

Keywords Life cycle simulation · Part agent · Reuse

24.1 Introduction

Presently, much of society is a consumption type that repeats mass production, consumption, and disposal causing environmental problems such as depletion of energy resources and high environmental load.

To solve these problems, there is an urgent need to shift society to a circulating society that effectively utilizes limited resources and energy by promoting efficient utilization of materials from production to distribution, consumption, and disposal. The promotion of reuse, reduce, and recycle (3R) is indispensable for the transition

T. Sugahara (✉) · T. Tanigawa · H. Hiraoka
Department of Precision Mechanics, Chuo University, Tokyo, Japan
e-mail: sugahara@lcps.mech.chuo-u.ac.jp

T. Sugahara
Hiraoka Laboratory, Chuo University, Tokyo, Japan

Y. Kishita et al. (eds.), *EcoDesign and Sustainability I*, Sustainable Production, Life Cycle Engineering and Management, https://doi.org/10.1007/978-981-15-6779-7_24

to a circulating society. Specifically, reuse should be particularly promoted because it produces less waste and consumes less energy.

Though accurate prediction of collected part quantity and quality is necessary to carry out production considering the reused parts by enterprise, this prediction is difficult for a number of reasons such as the diversity of the utilization form of the consumer. In addition, it is difficult for the consumer who uses the product to conduct the appropriate operation, management, and maintenance of the component unit of the product (Nanjo et al. 2015). These problems hinder the formation of a circulating society.

To effectively reuse machine parts for the development of sustainable society (Hauschild et al. 2005) based on 3R, it is indispensable to manage individual parts throughout their life cycle.

Yokoki et al. (Yokoki et al. 2015) developed a "part agent system" that controls the life cycle of individual parts and supports producer and consumer reuse to realize a circulating society.

A part agent is a network agent that autonomously acts corresponding to an individual part, and manages information regarding a part that constitutes a product from manufacturing to disposal. It also proposes actions such as disposal and reuse to a user at an appropriate time, and promotes 3R.

It is difficult to evaluate the effectiveness of a part agent by implementing it in an actual product and using the product throughout its life cycle. Therefore, Tanigawa et al. (Tanigawa et al. 2018) developed a system that simulates the behavior of parts and products throughout the entire life cycle with the part agent installed. In this study, it was considered that the method of reuse desired by each person was different according to the individual's thinking and the surrounding; thus, we developed a multiple reuse method. By properly conducting the simulation using the setting of the reuse method, a simulation considering various users using the part agent is possible.

24.2 Part Agent System

A network agent associated with a part is termed a part agent. The part agent follows the corresponding part by moving within the network and manages the state throughout the part life cycle. As products and parts move to various site in there life cycle such as production factories, dealers, users, and repair shops in the real world, part agents also move within the network following the real products and parts, providing necessary information according to the life cycle stages when necessary, determining appropriate maintenance actions for parts and users, and proposing these actions to users.

Our proposal assumes the spread of networks and high-precision radio frequency identifier (RFID) technology (Borriello 2005); thus, a part agent is generated during the manufacturing phase when an RFID tag is attached to each part. The part agent identifies the ID of the RFID tag during the part life cycle, tracking the part through the network. We chose an RFID tag for identification as RFIDs have a higher resistance to smudging and discoloration than printed bar codes and will last for the entire life of the part.

The conceptual scheme of the part agent is shown in Fig. 24.1. The part agent communicates with various functions within the network and collects the information required to manage its corresponding part, such as product design, logistical, or market information or predicted part deterioration. It also communicates with local functions on-site, such as sensory functions that detect the part state, storage functions for individual part data, and product management and control functions. This communication is established using information agents that are subordinates generated by the part agents.

A framework for a part agent to advise the user of the necessary maintenance actions based on the life cycle model of its corresponding part is shown in Fig. 24.2. At each time point, the part agent predicts the possible states of the part in the near future and evaluates those options to provide the user appropriate recommendations.

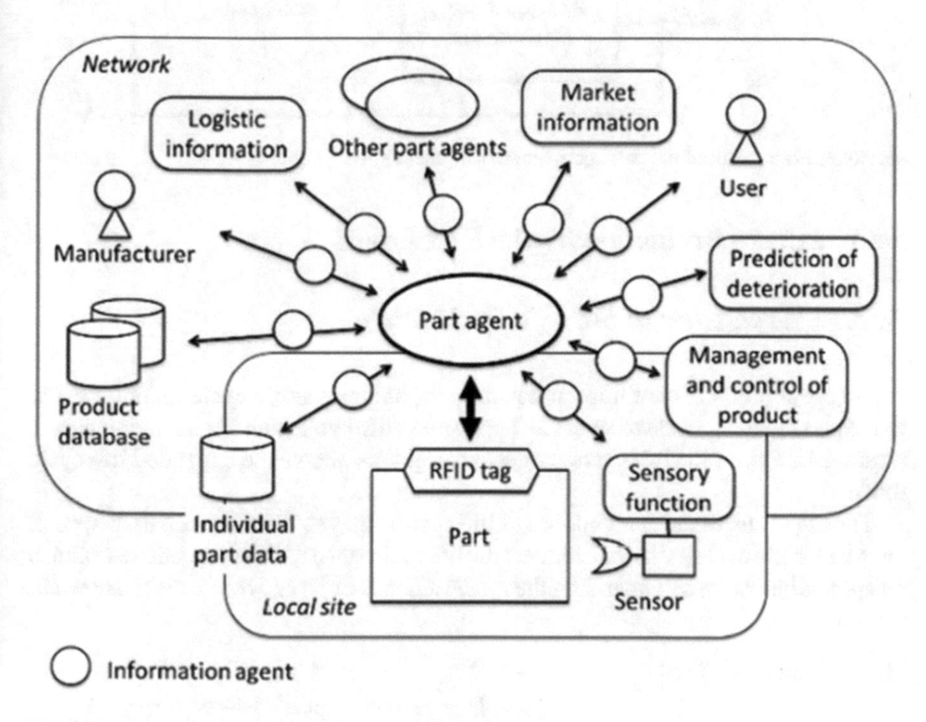

Fig. 24.1 Conceptual scheme of the part agent

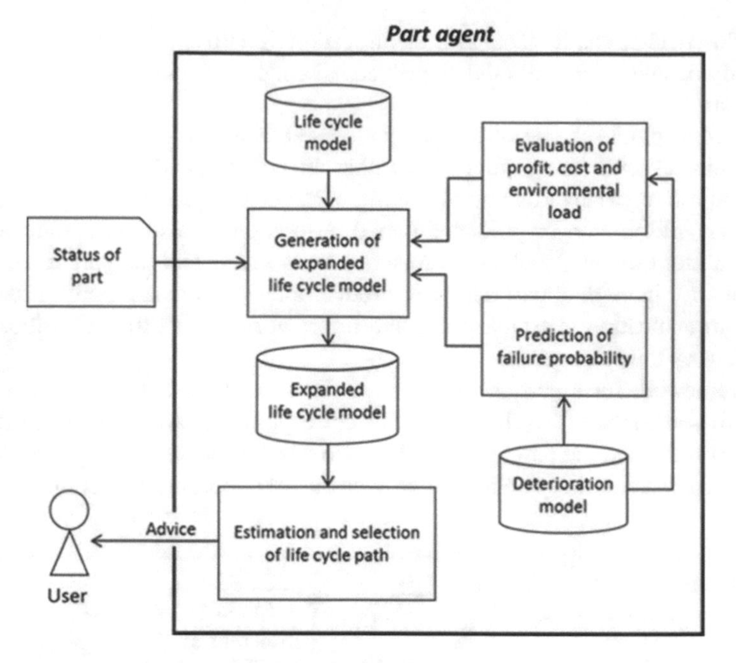

Fig. 24.2 Framework of the part agent for advice generation

24.3 Future Prediction by the Part Agent

24.3.1 Expansion of the Life Cycle Model

A part agent predicts the future of a part by expanding the life cycle model specific to the part along the time to select an appropriate life cycle path for its maintenance. Figure 24.3 shows the life cycle model and Fig. 24.4 shows the expanded life cycle model.

The life cycle model is composed of life cycle stages such as use, repair, disposal, etc. and the life cycle paths that connect the life cycle stages; the part agent determines an appropriate route by expanding the model along every route over time as shown in

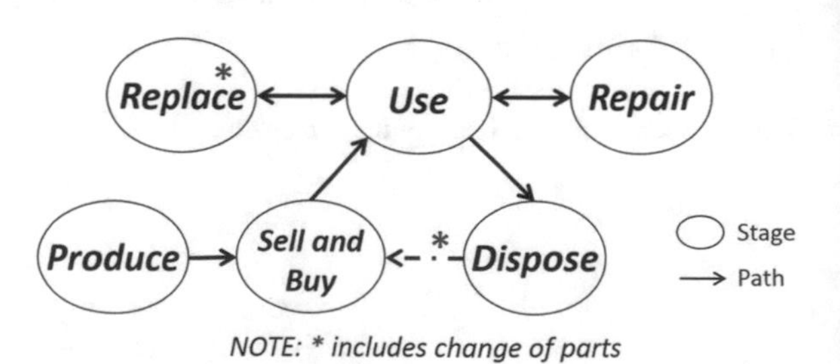

Fig. 24.3 Life cycle model

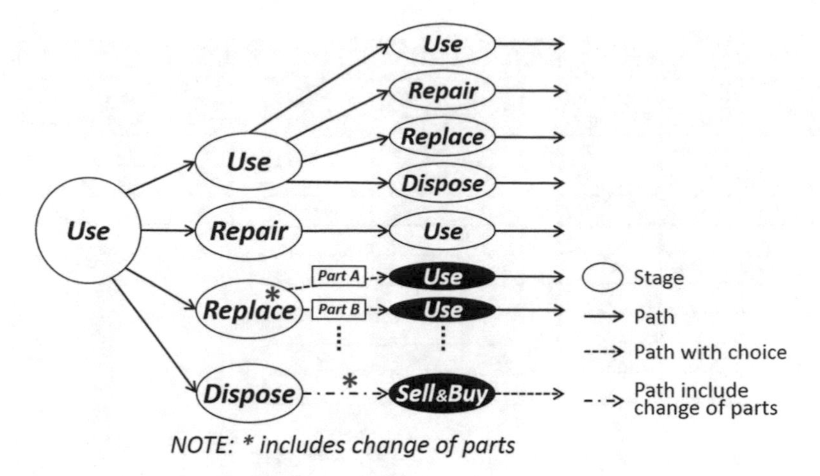

Fig. 24.4 Expanded life cycle model

Fig. 24.4. A stage has a profit, cost, and environmental load according to the action and the path has a probability with which it will be selected.

24.3.2 Prediction and Evaluation of the Life Cycle by the Part Agent

The part agent expands the life cycle model based on the state of the part and the deterioration model described in the following. Figure 24.5 shows the process of selecting the next stage based on the life cycle in which the part agent is deployed. The figure shows a state in which there is a Stage 1, Stage 1', and Stage1" as the next stages that can be selected during the current stage. Each stage has values such as V1 and V2 and the paths indicated by the arrows have probabilities such as p12 and p25. A route is defined by stages connected by a path along a time series and the next stages are evaluated by calculating an expectation value for each route starting from the next stage.

Assuming that the expected value of each route is *EV*, *EV* is calculated as shown in Eq. (24.1), where *V* is the value of each stage and *P* is the probability of the path leading to the stage.

$$EV = \sum_{Route} \left(\sum_{Stageinroute} V * \prod_{Pathinroute} P \right) \tag{24.1}$$

As shown in Fig. 24.5, the expected value of Stage1 is given by Eq. (24.2) as follows:

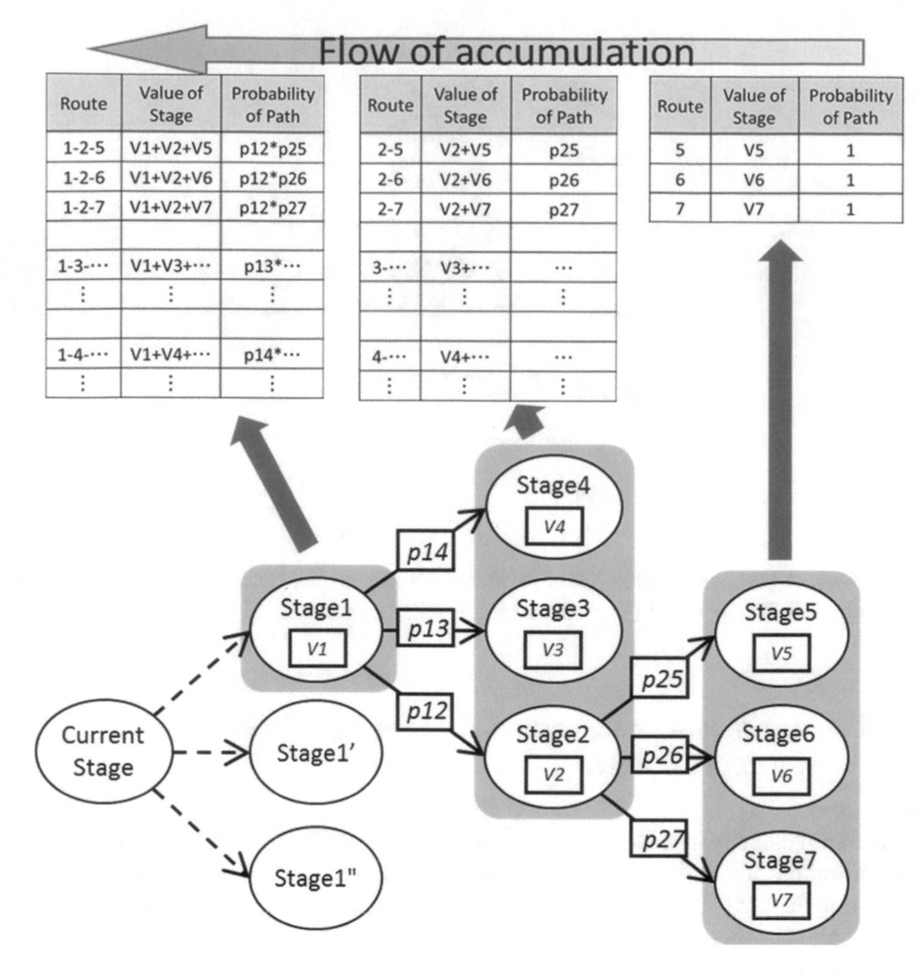

Fig. 24.5 Flow of life cycle expectation value calculation

$$
\begin{aligned}
expected\ value(Stage1) = &(V1 + V2 + V5) * (p12 * p25) \\
&+ (V1 + V2 + V6) * (p12 * p26) \\
&+ (V1 + V2 + V7) * (p12 * p27) \\
&+ (V1 + V3 + \ldots) * (p13 * \ldots) + \ldots
\end{aligned}
\tag{24.2}
$$

Similarly, the expected values of Stage1' and Stage1" can be obtained.

Depending on the user's preferences, the important aspects of the evaluation of the next lifecycle stage differ. For example, there may be users who wish to increase profit, users who wish to reduce cost, and users who wish to reduce environmental load. Hence, which parameter is used to evaluate the next life cycle stage depends on the user. Therefore, the part agent has a function to set the user's preference as to which parameter should be given priority.

Profit, cost, and environmental load are selected as parameters to be generated during each life cycle stage of the part. Part agents evaluate candidate next stages as previously described by combining these three parameters.

Because an appropriate combination of these parameters may differ according to evaluation aspects, the following two equations are provided as evaluation indices of the life cycle stage.

The first is INDEX as follows:

$$\mathrm{INDEX} = profit - cost - environmental\ load \tag{24.3}$$

INDEX is an index that shows how much profit remains by subtracting the cost and environmental load.

The second is the Total Performance Index(TPI) (Kondo et al. 2008) as follows.

$$TPI = \frac{profit}{\sqrt{cost \times environmental\ load}} \tag{24.4}$$

TPI is an index that shows how efficiently profit can be generated given the cost and environmental load.

In these two equations, a, b and c are coefficients for weighting, respectively, for each user, the part agent can evaluate the life cycle stage suitable for various users.

The part agents evaluate the most appropriate stage using these indices.

24.4 Reuse Method

There are several means by which consumers can reuse products, such as buying and selling through the used market and buying and selling between neighboring inhabitants. The reuse method the user desires depends on individual preference and the environment of the region where the user lives.

Three types of reuse methods were provided in this study to carry out the simulation corresponding to the user's manner of thinking and the surrounding environment as follows: (1) reuse through the market, (2) reuse via trade between users, (3) reuse combining these two means. Each is described in detail in the following.

24.4.1 Reuse Through the Market

Some users may have resistance against transactions between users. Therefore, reuse through the market is provided as one of the reuse methods.

The functions of the market for used parts are storage, cleaning, repairing, selling and buying.

When a part agent proposes replacement, the user replaces the part with one of the parts on the market. Using this method, replacements occur by selling the parts owned by the user and buying the parts on the market.

The buying price and selling price are determined by the market according to the part condition, and the transaction is carried out. The parts purchased by the market are stored in the market and are candidates for buying during the next transaction.

If the performance of a part drops below a certain level, and the subsequent use is hindered, repair is performed, and the selling price increases based on the amount of the repair.

24.4.2 Reuse Between Users

When the user cannot find the desired part in the market, or when there is no used market near the user, reuse between neighboring inhabitants may be carried out.

Yamamori et al. (Yamamori et al. 2015) developed a simulation system that supports reuse of consumers using a part agent system to promote reuse of the used parts.

When a part agent proposes reuse between users, the part agent selects an appropriate partner to promote reuse.

The procedure from the suggestion of reuse by the part agent system to the transaction is as follows:

(1) Propose reuse
The part agent that manages the information regarding the part proposes reuse to the user.

(2) Calculate the degree of satisfaction (*DoS*)
Part agents are divided into sellers and buyers. Once they are separated, *DoS* is defined as an indicator for each user to decide who to address. Each agent performs calculation of the *DoS* for the part agents in the opposite position. *DoS* is calculated using the following Eqs. (24.24.5) and (24.24.6):

$$DoS \equiv \frac{1}{a_1 P + a_2 S + a_3 D} + a_4 L \tag{24.5}$$

$$P \equiv \frac{|P_P - P_U|}{P_U} S \equiv \frac{|S_P - S_U|}{S_U}$$

$$D \equiv \frac{D_U}{MD} \tag{24.6}$$

where P is the ratio of the difference between the current price P_P of the evaluated product and the transaction price P_U of the product desired by the user. S is the ratio of the difference between the specification S_P of the evaluation product and the specification S_U of the product desired by the user. the distance limit

(MD) represents the longest distance the user can travel. D is the ratio of the actual travel distance D_U to the MD of each user. L is the remaining life of the evaluated product. a_1 to a_4 are the weighting factors for each item.

(3) Match

After all agents perform calculations on *DoS*, matching is performed using a maximum weight bipartite matching problem and the paired agents perform transactions.

The weighted maximum bipartite matching is a problem to determine a combination in which the sum of the costs of the matched edges is maximized by matching as many vertices as possible, assuming that the cost is set for each connecting edge in the bipartite graph. In this study, a black vertex shown in Fig. 24.6 is a buyer and a gray vertex is a seller. An edge is placed between the vertex of the two agents when the distance from a certain buyer agent to a certain seller agent is within the movable distance of one another.

As a method to solve the weighted maximum binary matching problem, the minimum-cost flow problem (Ahuja et al. 1993) using an algorithm termed the Bermanford method (Bellman 1958) is used.

The minimum-cost flow problem assumes that each edge of a directed network, as shown in Fig. 24.7, is given an integer capacity and a cost resulting from an

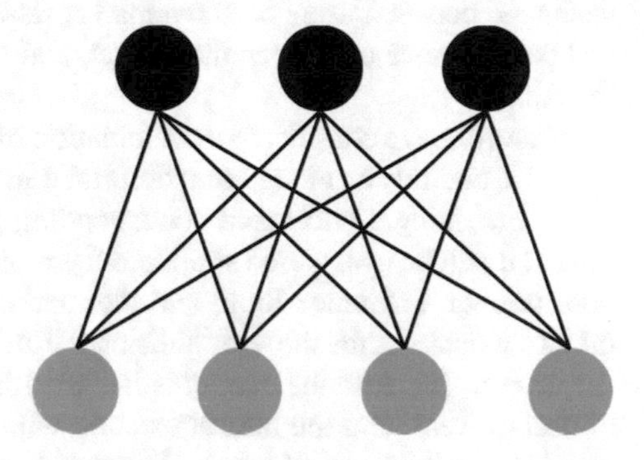

Fig. 24.6 Weighted maximum bipartite matching

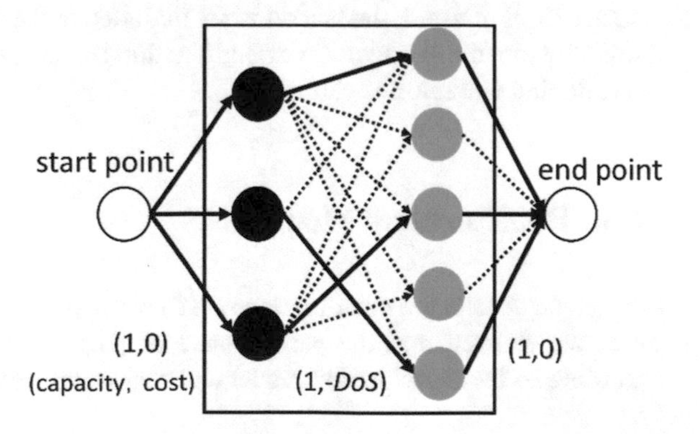

Fig. 24.7 Minimum-cost flow problem

integer flow passing through the edge. It is a problem to determine which edge to pass through to minimize the total cost when a certain flow value is passed from the start point to end point.

Let 0 be the cost and 1 be the capacity of the edge connected to the vertex of the left set and the edge connected to the endpoint of the right set from the viewpoint. Let the capacity of the edge that connects the left side set and the vertex of the right side set in the frame be 1 and let the cost be the value that makes the cost of the edge in the maximum-two part matching problem with the weight negative. With this setting, the combination of the edges whose total cost is the minimum is selected only for the portion within the frame shown in Fig. 24.7. Because the cost is negative, the combination of sides with the largest absolute value is selected and the resulting combination of sides results in the weighted maximum bipartite matching.

In this study, the average value of satisfaction of a seller side agent from the viewpoint of a buyer side agent and satisfaction from the viewpoint of the opposite side is set as the cost.

24.4.3 Reuse Including the Market and Users

Even if there is a used market near the user, the decision regarding whether to exchange between users or in the market depends on the price of parts, the number and performance of parts in the market, and the distance from the user's location to the market.

Therefore, we considered a combination of reuse through the market described in Sect. 4.1 and reuse among users described in Sect. 4.2.

By including a part agent corresponding to a part in the market as a matching object, it can be considered similar to the reuse between users. However, the market does not set a distance limit, and the market participates in matching only if the market exists within the user's distance limit. Figure 24.8 shows reuse including the market. Because the market is included in the distance limit of user1, the parts of user2, user3, and the market existing within the distance limit participate in the matching in the reuse of user1. However, because the market is not included in the distance limit of user6, the market is not included in the matching, and in the case of reuse of user6, user4 and user5 existing within the distance limit, they participate in the matching and reuse is carried out.

24.5 Deterioration Model

Part agents predict the deterioration rate to simulate the part life cycle in the future. Thus, we showed that the performance of parts deteriorated with the use time according to the deterioration model and that it will recover if the part is repaired.

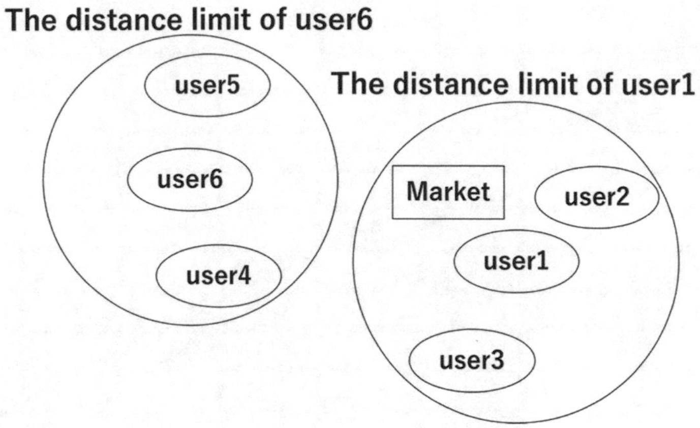

Fig. 24.8 Reuse including market and matching

The deterioration model is an equation that represents the relationship between deterioration factors and performance. For simplicity, here the deterioration factor is only the usage time.

Parts deteriorate with each use and thus the performance decreases. When a part deteriorates, the profit, cost, and environmental load also change. Therefore, a part agent predicts the information regarding the part using not only a life cycle model but also a deterioration model of profit, cost, and environmental load. We set the value of the profit, cost, and environmental load depending on the deterioration model for to each life cycle stage.

In this study, we suppose that the deterioration model differs by part. Thus, we develop the simulation to set the deterioration model for each part.

Figure 24.9 shows an example of the deterioration model. There are three types of deterioration models: a linear deterioration model shown in orange, a deterioration model which is a cubic function and changes from an upwardly convex shape to a downwardly convex shape shown in green, and a deterioration model which is a cubic function and changes from a downwardly convex shape to an upwardly convex shape shown in blue. Performance deteriorates according to the deterioration model corresponding to each part and it recovers when a repair is performed.

In this simulation, such a deterioration model was set for the parts.

24.6 Flow and Evaluation of the Simulation

24.6.1 Flow of the Simulation

The flow of the simulation is as follows.

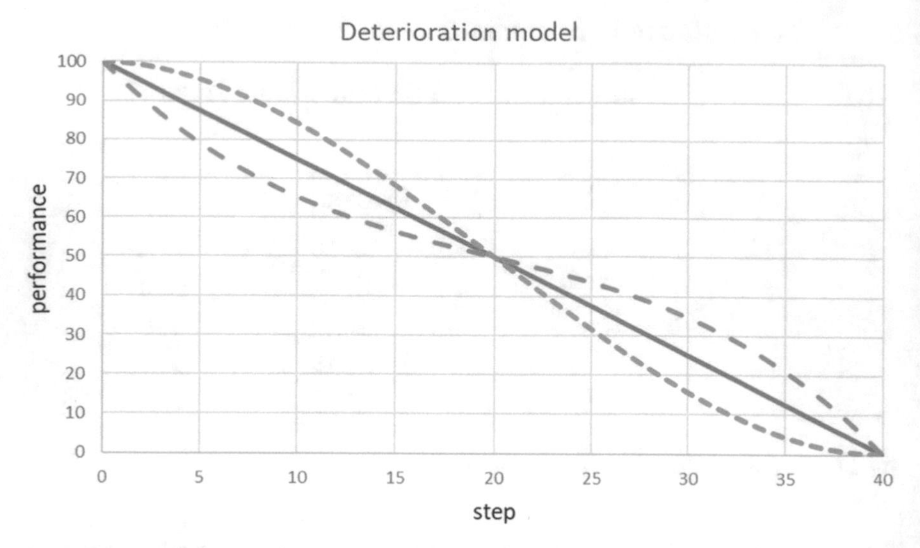

Fig. 24.9 Deterioration model

(1) Set the number of users.
(2) Prepare a part owned by each user.
(3) Set a part agent for each part.
(4) The performance of a part decreases according to the deterioration model when the user uses the part.
(5) The user selects the stage according to the proposal of the part agent and proceeds to the next step.
(6) Finish the simulation after a certain period has passed.

Weighting coefficients a, b, and c are set to the part agent when the user obtains a new part. In the case of used parts obtained via replacement, they are overwritten by the settings of the new owner.

The person who performs the simulation selects the reuse method at the start of simulation.

24.6.2 Evaluation of the Simulation

Profit, cost, and environmental loads generated according to each life cycle stage depend on the part performance. When the simulation finishes, the sum of the INDEX in Eq. (24.3) and the sum of the TPI in Eq. (24.4) based on profit, cost, and environmental load are calculated for each user.

We will simulate with and without part agents using the INDEX and TPI. We will evaluate the part agent based on this result. We will also simulate with and without part agents; compare the sum of the profit, cost, and environmental load in each case; and evaluate in which aspect the part agent can be effectively used.

24.7 Discussion

We have still issues in developing a practical system for life cycle simulation of parts with part agents.

First issue is the evaluation of the life cycle stages. We use three parameters that are profit, cost, and environmental load for that purpose. These are same parameters with elements of TPI except that profit is used instead of value to capture the economic gain at each time step. These parameters are basic ones but other parameters may exist for appropriate evaluation of life cycle stages.

Another issue is the deterioration factors that cause deterioration of products. Many factors may affect the process of deterioration depending the product, its usage, environments, and the situations. Although many researchers have been working on the deterioration mechanisms, the mechanisms are often difficult to specify even qualitatively. Our deterioration model depends on only the duration time for simplicity, we can include other factors to the model when such information is available.

We assume in this paper a simple life cycle model shown in Fig. 24.3. More detailed life cycle is necessary for practical application. It should cover different products, diversified usages in various conditions.

Finally, we should implement the proposed simulation for practical applications and evaluate its effectiveness.

24.8 Conclusion

We designed an integrated simulation in which the part agent predicts the part life cycle and proposes the maintenance action to the user. We proposed three types of reuse methods for the simulation as follows: reuse through the market, reuse between users, and reuse combining these two means. By enable selection of a reuse method from among multiple methods, it is possible to conduct simulations suitable for various users.

Acknowledgements This work was supported by JSPS KAKENHI Grant Number 15K05772. Shogo Hashimoto developed the function described in Sect. 4.2.

References

Ahuja R, Magnanti T, Orlin J (1993) Network flows: theory, algorithms, and applications

Bellman R (1958) On a routing problem. Q Appl Math 16(1):87–90

Borriello G (2005) RFID: Tagging the World. Commun ACM 48(9):34–37

Hauschild M, Jeswiet J, Alting L (2005) From life cycle assessment to sustainable production: status and perspectives. Ann. CIRP, p 535–555

Kondo S, Masui K, Mishima N, Matsumoto M (2008) Total performance analysis of product life cycle considering the uncertainties in product-use stage. In: Advances in life cycle engineering for sustainable manufacturing businesses, p 371–376

Nanjo K, Yamamori Y, Yokoki Y, Sakamoto Y, Hiraoka H (2015) Maintenance decisions of part agent based on failure probability of a part using Bayesian estimation. In: The 22nd CIRP conference on Life cycle engineering, Sydney, 7–9 Apr 2015, S12 108

Tanigawa T, Nagahata T, Masuda N, Hiraoka H (2018) Data structure for life cycle simulation in consideration of part agents promoting the reuse of mechanical parts. In: The 17th ICPE, A-2–3

Yamamori Y, Yokoki Y, Hiraoka H (2015) Seller-buyer matching for promoting product reuse using distanced-based user-grouping. In: EcoDesign 2015 9th international symposium on environmentally conscious design and inverse manufacturing, p 733–738

Yokoki Y, Yamamori Y, Hiraoka H (2015) User model in the life cycle simulation of mechanical parts based on Prospect theory. In: EcoDesign 2015 9th International Symposium on Environmentally Conscious Design and Inverse Manufacturing, p 184–189

Chapter 25
Model-Based Design of Product-Related Information Management System for Accelerating Resource Circulation

Hitoshi Komoto, Mitsutaka Matsumoto, and Shinsuke Kondoh

Abstract In the manufacturing industry, information about products accumulated through the whole life cycle such as design descriptions, usage data, and failure event history is considered crucial to innovate the industry's value propositions. Although such information can be used to increase the efficiency of the operations and decision makings in the end-of-life stage, such information is not effectively shared and used by the potential stakeholders involved in the life cycle of products. In order to analyze the value of information sharing among the stakeholders and design a system to realize desired information sharing, this study proposes a simulation-based approach to specify the external and internal requirements of such a system and evaluate the impact of the system on resource circulation and profitability of the stakeholders.

Keywords Life cycle simulation · Information modeling · Model-based systems engineering · Resource circulation

25.1 Introduction

In the manufacturing industry, information about products accumulated through the whole life cycle is considered crucial to innovate the industry's value propositions. For instance, the usage data of machine tools (incl. alarm and failure event history) are currently used for scheduling of maintenance actions and spare parts delivery. Moreover, they have enabled provision of statistical analysis of machine tool usage and recommendations regarding better tool usage (Plattform Industrie 2018; Robot Revolution Initiative 2018).

Product-related information such as design descriptions and usage data can be also used to increase the efficiency of the operations and decision makings in the end-of-life stage. For instance, design information is crucial in improving the efficiency in the end-of-life processes of electric vehicles (Li et al. 2018). Material composition

H. Komoto (✉) · M. Matsumoto · S. Kondoh
Advanced Manufacturing Research Institute, National Institute of Advanced Industrial Science and Technology, Tsukuba, Japan
e-mail: h.komoto@aist.go.jp

Y. Kishita et al. (eds.), *EcoDesign and Sustainability I*, Sustainable Production, Life Cycle Engineering and Management, https://doi.org/10.1007/978-981-15-6779-7_25

defined in the design stage is a key design variable of end-of-life alternatives that increases recycling rate of products like LCD TV (Umeda et al. 2013).

However, product-related information is not effectively shared and used by the potential stakeholders involved in the life cycle of products. Unless legislative obligation (such as the WEEE directive), the owners of product-related information is not willing to collaborate one another and share product-related information (Aschehoug et al. 2012). Moreover, standards for information models to deal with product-related information across product life cycles are crucial for effective information sharing (Rachuri et al. 2008).

The value of information sharing among stakeholders in a supply chain has been intensively studied with various analytical techniques (e.g. game theory-based (Raweewan and Jr Ferrell 2018) and equation-based (Teunter et al. 2018)). However, these models do not sufficiently express the complex behavior of stakeholders in the various life cycle stages and interactions among them.

The objective of the study is to analyze the value of information sharing among the stakeholders involved in various stages of a product life cycle. To do so, this study proposes a simulation-based approach to specify the external and internal requirements of such a system and evaluate the impact of the system on resource circulation and profitability of the stakeholders.

The content of this paper is divided into two parts.

The first part of the paper (Sect. 25.2) summarizes the various types of interactions among stakeholders involved in a life cycle of electric and electrical equipment in terms of expected product (material) flows, information flows, and monetary flows. These flows have been described by interviews with manufacturers and recyclers of electric and electrical equipment in Japan. Figures 25.1, 25.2 and 25.3 organizes complex flows of product, information, and money on the network of stakeholders considered in the manufacturing and recycling industries in Japan. Although these

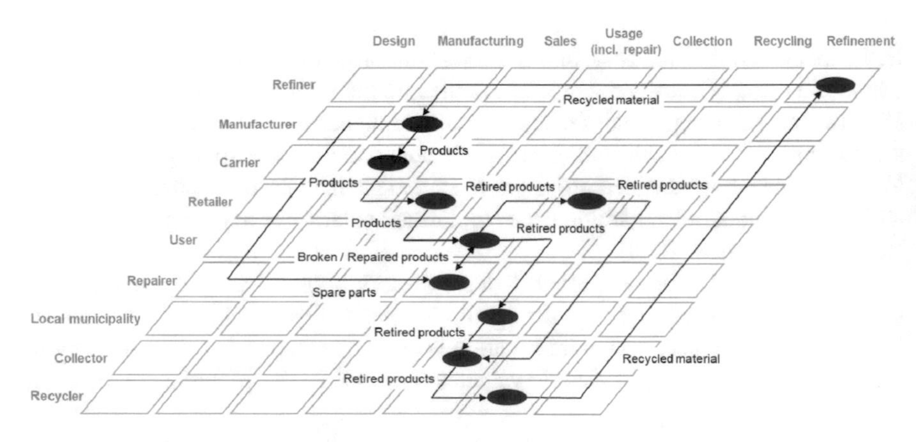

Fig. 25.1 Stakeholder interactions (product flows)

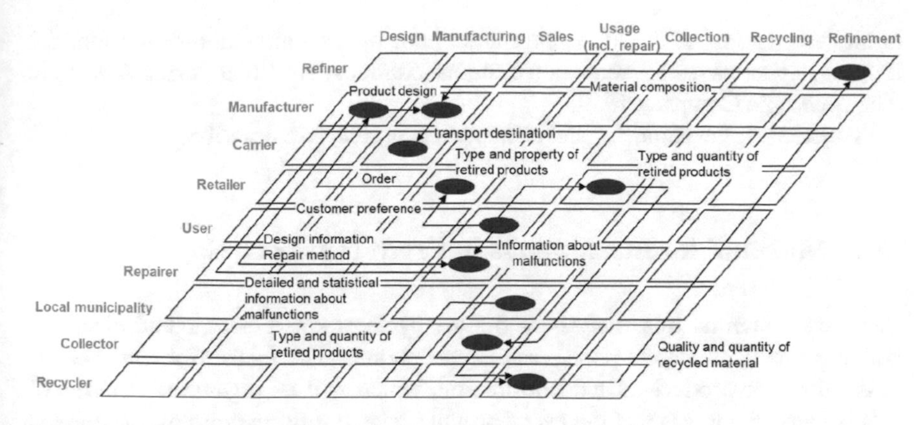

Fig. 25.2 Stakeholder interactions (information flows)

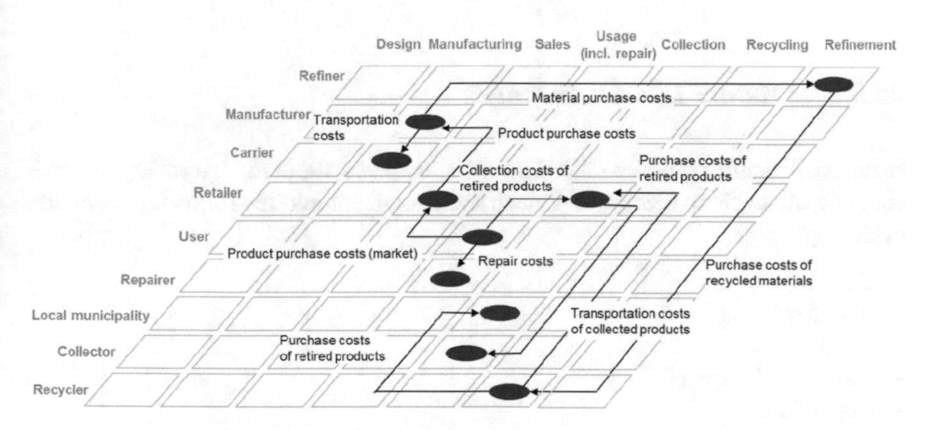

Fig. 25.3 Stakeholder interactions (monetary flows)

flows are not exhaustive, they characterize complexity of the interactions among the stakeholders across product life cycle.

The second part of the paper (Sect. 25.3) introduces the life cycle model that contains the essential parts of interactions among the stakeholders introduced in Sect. 25.2. The life cycle model is briefly explained regarding stakeholders and their interactions. The life cycle model is used to design an information system for collection, management, and analysis of product-related information. The design process consists of two steps. The first step defines the desired inputs to (outputs from) the information system. Then, the next step defines logical procedure in the information system. The model is then evaluated in terms of two performance criteria; resource circulation and dispersion of monetary flows. In this study, the behavior of stakeholders in a life cycle is simulated with life cycle simulation (LCS) (Umeda et al. 2000; Komoto and Tomiyama 2008; Fukushige et al. 2017). This section presents an implementation of the life cycle model built on the simulation framework for designing cyber physical production systems (Komoto et al. 2019), using AnyLogic (The AnyLogic Company 2019).

In Sect. 25.4, the summary and outlook of this study are described.

25.2 Stakeholder Interactions in Product Life Cycle

This section summarizes analysis of the life cycle of small electric and electrical equipment such as mobile phones and smart phones. The focus of the analysis is on the flows of product-related information, which will be organized in terms of stakeholders as sources and targets of product-related information flows generated in a variety of the product life cycle stages.

25.2.1 Product Life Cycle Stages

In order to organize product-related information flows, the study introduces the division of a life cycle of electric and electrical equipment into the following seven life cycle stages.

- Design
- Manufacturing
- Sales
- Usage (incl. repair)
- Collection
- Recycling
- Refinement.

25.2.2 Stakeholders

The study assumes that the stakeholders involved in the life cycle of electric and electrical equipment belong to one of the following stakeholder types with specific roles.

- Manufacturer
- Carrier (or new products)
- Retailer
- User
- Repairer
- Local municipality
- Collector (of retired products)

- Recycler
- Refiner.

25.2.3 Stakeholder Interactions Across Life Cycle Stages

Interactions among the stakeholders across the product life stages were analyzed in terms of product (material) flows, information flows, and monetary flows. Figures 25.1, 25.2 and 25.3 show product flows, monetary flows, and information flows organized regarding the stakeholders and the life cycle stages, respectively. The refiner is placed at the first row for the sake of visibility.

As shown in Figs. 25.1, 25.2 and 25.3, three types of flows do not correspond to one another. For instance, the information flows between the manufacturer in the design stage and the repairer in the use stage have no corresponding monetary flows. Moreover, the manufacturer delivers spare parts, which are produced in the manufacturing stage, to the repairer in the use stage, without receiving money from the manufacturer.

25.3 Model-Based Design of Product-Related Information Management System

25.3.1 External and Internal Requirements

In order to realize information flows for better resource circulation, new stakeholder called product-relation information management system that collects, analyzes, and publishes product-related information, can be introduced to the life cycle of electric and electrical equipment. The information system can be designed following two types of requirements.

- External requirements are concerned with the input and output of the information system that can increase resource circulation, while improving the profitability of the involved stakeholders.
- Internal requirements are concerned with the logic in the information management system that successfully computes the expected output with the given input.

The design process to deal with the external and internal requirements is regarded as the validation and verification process of the information management system, respectively.

25.3.2 Life Cycle Modeling

In the design process to deal with the external requirements, first, a life cycle model, where the system is situated, should be defined. The product life cycle model consists of five types of stakeholders interacting one another. The role of each stakeholder and the inputs and outputs of the information management system is described below.

- E (1): External environment represents sources and targets of products (materials) and information other than the stakeholders listed below.
- F (2): A set of the stakeholders that manufacture and deliver products to users (i.e. Forward supply chain). They include manufacturer, carrier, and retailer introduced in Sect. 25.2. They occasionally send *DesignData* the system I, when new products are manufactured.
- U (3): Users of products. They periodically (and automatically) send *UsageData* to the system I upon agreement of users.
- R (4): A set of the stakeholders that perform maintenance, repair, and end-of-life operations for products (i.e. Reverse supply chain). They include repairer, local municipality, collector, recycler, and refiner introduced in Sect. 25.2. They send *RepairData* and *ReuseData* to the system I. They also send *DesignDataRequest* when reusing and recycling products efficiently.
- I (5): The information management system that collects, analyzes, and publishes product-related data. In this context, the system collects the data from the stakeholders F, U, and R. The system only handles with information flows but not with product flows. It sends *UsageDataSummary* to F when a certain amount of *UsageData* is collected. In responding *DesignDataRequest* from R, the system searches the corresponding *DesignData* and replies to R.

The integer in the bracket following each symbol of the stakeholder types indicate the common index of the row and column of the matrices shown in the formulations 1–2 and 5–6, respectively. Figure 25.4 shows stakeholder interactions under consideration in terms of the product flows and information flows. The monetary flows among the stakeholders are computed based on these flows as described later.

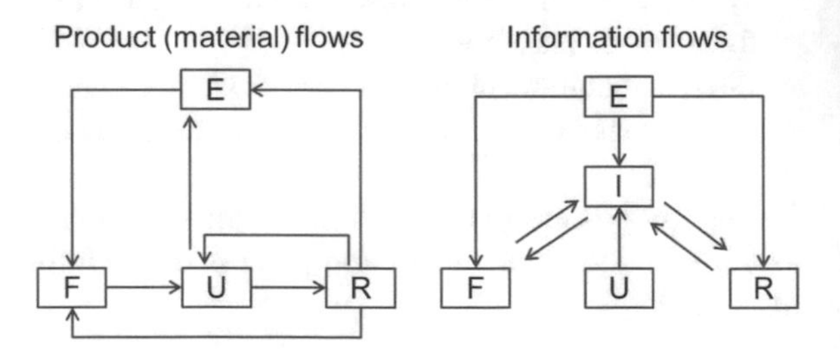

Fig. 25.4 Stakeholder interactions under consideration

25.3.3 *Life Cycle Simulation*

Life cycle simulation (LCS) is based on discrete event simulation with several characteristics suitable to design and analysis of products, processes, and life cycles (Umeda et al. 2000; Komoto and Tomiyama 2008; Fukushige et al. 2017). Products, modules, parts, and components are regarded as inputs and outputs of life cycle processes. The structure of these elements is dynamically reconfigured during their life cycles (e.g. through assembly and disassembly processes).

In LCS, individual stakeholders assigned to specific life cycle processes perform actions. These actions are triggered by receiving of a product or information, or by its own state transition. These actions can cause state transition of the received product, delivery of the received product to a specific stakeholder, and generation of new information to specific stakeholders.

Although an execution of LCS generates the history of events and state transitions, and a lot of statistical figures, this study focuses on the numbers of product flows and information flows on the network of stakeholders in Fig. 25.4. These numbers are organized as the variables of the matrices P and I as follows.

$$P = \begin{pmatrix} - & p_{EF} & - & - & - \\ - & - & p_{FU} & - & - \\ p_{UE} & - & - & p_{UR} & - \\ p_{RE} & p_{RF} & p_{RU} & - & - \\ - & - & - & - & - \end{pmatrix} \quad (25.1)$$

$$I = \begin{pmatrix} - & i_{EF} & - & i_{ER} & i_{EI} \\ - & - & - & - & i_{FI} \\ - & - & - & - & i_{UI} \\ - & - & - & - & i_{RI} \\ - & i_{IF} & - & i_{IR} & - \end{pmatrix} \quad (25.2)$$

where the value of variable at (i, j) represents the (positive) number of product (information) flows from the stakeholder i to the stakeholder j, and—is undefined.

25.3.4 *Performance Evaluation*

The behavior of the product life cycle model has variations regarding the costs and effects of interactions. Considering such variations, the performance of the life cycle model is evaluated in terms of two criteria; resource circulation and profit distribution (to stakeholders).

First, the degree of resource circulation can be estimated by the number of specific product flows. With referring to Fig. 25.4, a life cycle model can gain better resource circulation by increasing p_{RU} (repair) p_{RF} (reuse, remanufacturing, or recycle) and decreasing p_{EF} (material extraction) p_{UE} (disposal by users) and p_{RE} (disposal by recyclers) in P. The value of these variables mainly depends on decision makings performed by U and R. Thus, design of information flows that cause desired product flows with better resource circulation is crucial.

Second, profit distribution can be quantified by analyzing disparity in monetary flows derived from product and information flows shown in Fig. 25.4. The monetary flows are defined in (3).

$$M = P \circ V + I \circ W \tag{25.3}$$

where, $\circ$ is the pointwise product of two matrices with the same size (it is also called Hadamard product). For instance, the amount of monetary flows from the stakeholder i to the stakeholder j is an element of M as defined as follows.

$$m_{ij} = p_{ij}v_{ij} + i_{ij}w_{ij} \tag{25.4}$$

As explained above, the matrices P and I mean the cumulative number of product and information flows during the product life cycle computed by LCS. Variables in the matrices V and W are the unit monetary flow per product flow or information flow from one stakeholder to another as shown below.

$$V = \begin{pmatrix} - & v_{EF} = 0 & - & - & - \\ - & - & v_{FU} > 0 & - & - \\ v_{UE} = 0 & - & - & v_{UR} & - \\ v_{RE} = 0 & v_{RF} > 0 & v_{RU} & - & - \\ - & - & - & - & - \end{pmatrix} \tag{25.5}$$

$$W = \begin{pmatrix} - & w_{EF} = 0 & - & w_{ER} = 0 & w_{EI} = 0 \\ - & - & - & - & w_{FI} > 0 \\ - & - & - & - & w_{UI} > 0 \\ - & - & - & - & w_{RI} > 0 \\ - & w_{IF} > 0 & - & w_{IR} > 0 & - \end{pmatrix} \tag{25.6}$$

The sign of these variables defines the direction of monetary flows. A positive value of a monetary flow means that the corresponding flow element is a *valuable* product or information, and that the direction of the flow is opposed to the direction of the corresponding product (information) flow. The sign of v_{UR} and v_{RU} is undetermined, because flow elements are either valuable (e.g. users can sell products to recycling companies) or valueless (e.g. users must pay recycling fee to recycling companies). Monetary flows from and to E are not considered in order to bind the scope of solution spaces to the design problem under consideration.

The balance of monetary flows from and to a stakeholder is defined by the sum of the monetary flows to the stakeholder subtracted by the sum of the monetary flows from the stakeholder as follows.

$$m_k = \sum_j m_{kj} - \sum_i m_{ik} \tag{25.7}$$

In case of Fig. 25.4, the balance of monetary flows for F, U, R, and I are defined as follows.

$$m_F = p_{FU} v_{FU} - p_{RF} v_{RF} + i_{FI} w_{FI} - i_{IF} w_{IF} \tag{25.8}$$

$$m_U = p_{UR} v_{UR} - p_{FU} v_{FU} + i_{UI} w_{UI} \tag{25.9}$$

$$m_R = p_{RF} v_{RF} + p_{RU} v_{RU} - p_{UR} v_{UR} + i_{RI} w_{RI} - i_{IR} w_{IR} \tag{25.10}$$

$$m_I = i_{IF} w_{IF} + i_{IR} w_{IR} - i_{FI} w_{FI} - i_{UI} w_{UI} - i_{RI} w_{RI} \tag{25.11}$$

25.3.5 Simulation Framework

The simulation framework (see detail in (Komoto et al. 2019)) supports simulation of the life cycle model and animates and evaluates the behavior of the life cycle model illustrated in Fig. 25.4. Figure 25.5 shows the screenshot of the simulation framework that executes the life cycle model.

The life cycle model (system model) is defined on the upper left side. In the model green squares represent five types of stakeholders. Each green square has four ports, which represent input/output of product/information. Branches of flows are defined with diamond nodes. In order to animate product and information flows, presentation of the model is defined on the upper right side, in which the location of each stakeholder is defined by a square with the predefined symbol (e.g. F and E). Simulated information flows and product flows are dynamically animated with envelopes and circles.

The simulation results at the bottom of Fig. 25.5 show the cumulative number of each product and information flow appeared in the amount of monetary flows for each stakeholder (see, formulations 8–11). Moreover, the monetary flows are computed based on the product and information flows and the values of variables in the matrices V and W, which can be adjusted by the sliders at the lower right of Fig. 25.5.

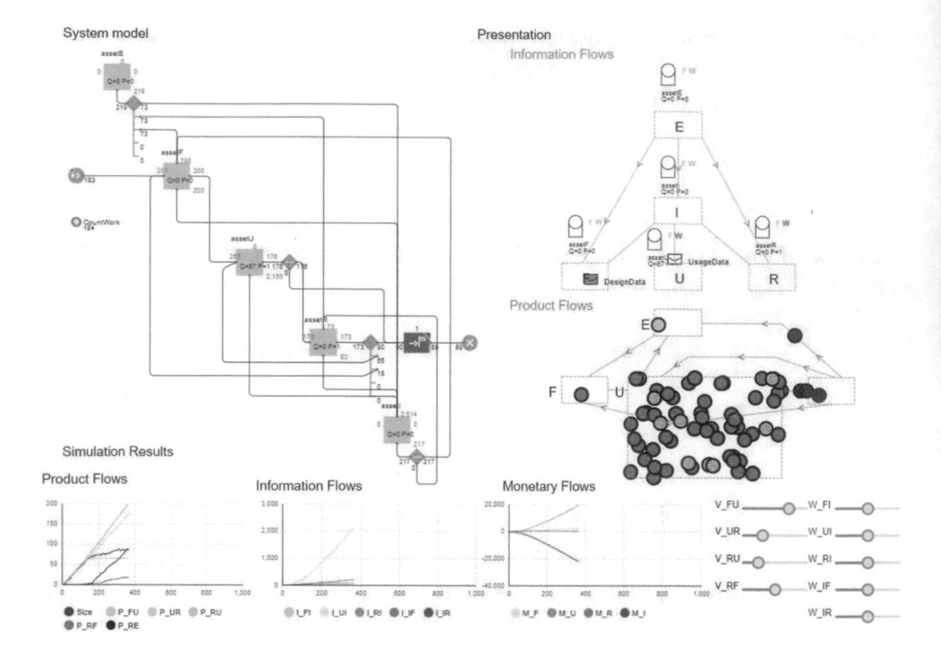

Fig. 25.5 Screenshot of the simulation framework running simulation of the life cycle model

25.4 Summary and Outlook

25.4.1 Summary

This paper has presented interactions among stakeholders in the life cycle of small electric and electrical equipment, which are based on the interview from the manufacturers and recyclers in Japan, which characterizes complexity of the interactions among the stakeholders across product life cycle. Based on the description of the product life cycle, this paper has presented a life cycle model that can analyze a variation on the costs and effects of such interactions. The model has been used to define the internal and external requirements of the information management system to deal with product-related information and evaluate the model regarding the external requirements.

25.4.2 Outlook

Considering limitations of the proposed model, which will be reviewed later, this study can be extended from at least following aspects:

- System boundary and scope: the proposed model describes interactions among the stakeholders in a product life cycle. These interactions can specify requirements of the information management system in terms of input and output flows of items (e.g. products, parts, and materials) and information. But the model regards the system as a black-box. In other words, the model lacks capability to define the internal structure of the system, which partly represents the system's design. Therefore, this capability enables designers to test whether a system with a certain internal structure can realize intended flows of items and information.
- Measures to data unavailability: the availability of detailed data defined and owned by manufacturers to recyclers in the life cycle depends partly on the design of the product model and the information management system. The product model should possess a composition so that the information management system can partly retrieve information consistent with the given authorization and security criteria.
- Model verification procedure: the paper described the development process of the proposed model, but the verification process was out of the scope of this paper. The verification process can be divided in terms of the type of models to be verified; life cycle models, information management system models, and product models. First, a life cycles model can be verified regarding capability of the model to simulate expected (qualitative and quantitative) behavior specified by model users, while assuming the internal structure of the employed information management system and the product model is regarded as black boxes. Second, an information management system model can be verified regarding capability to model the internal structure of the system, and simulate the variations on decision making policies regarding, for instance, access control and security. Finally, a product model can be verified regarding capability of a model to define its attributes intended by model users and store its data and to define a view specified by each stakeholder (e.g. users and stakeholders.).

Furthermore, the model will be used to analyze the value of the variables representing the life cycle model that satisfy the external requirements and identify the appropriate input and output of the information management system. Then, the model will be used to model the detailed logic of the analysis of the collected information so that the quality and quantity of output information satisfy the needs of the stakeholders.

Acknowledgements This paper is based on results obtained from a project commissioned by the New Energy and Industrial Technology Development Organization (NEDO). The authors thank Mr. Asada for supporting the development of the life cycle model on the simulation framework.

References

Aschehoug S, Boks H, Storen C (2012) Environmental information from stakeholders supporting product development. J Cleaner Prod 31:1–13
Fukushige S, Nishioka M, Kobayashi H (2017) Data-assimilated lifecycle simulation for adaptive product lifecycle management. CIRP Ann CIRP 66(1):37–40
Komoto H, Tomiyama T (2008) Integration of a service CAD and a Life cycle simulator. CIRP Ann CIRP 57(1):9–12
Komoto H, Kondoh S, Furukawa Y, Sawada H (2019) A simulation framework to analyze information flows in a smart factory with focus on run-time adaptability of machine tools. Procedia CIRP 81:334–339
Li J, Barwood M, Rahimifard S (2018) Robotic disassembly for increased recovery of strategically important materials from electrical vehicles. Robot Comput Integr Manuf 50:203–212
Plattform Industrie 4.0 (2018) Usage viewpoint of application scenario value-based service. Federal ministry for econimic affairs and energy (BMWi)
Rachuri S, Subrahmanian E, Bouras A, Fenves SJ, Foufou S, Sriram RD (2008) Information sharing and exchange in the context of product lifecycle management: role of standards. Comput Aided Des 40:789–800
Raweewan M, Jr Ferrell WG (2018) Information sharing in supply chain collaboration. Comput Indus Eng 126:269–281
Robot Revolution Initiative (2018) Functional viewpoint of application scenario value-based service
Teunter RH, Babai MZ, Bokhorst JAC, Syntetos AA (2018) Revisiting the value of information sharing in two-stage supply chains. Eur J Oper Res 270:1044–1052
The AnyLogic Company (2019) AnyLogic version 8.4. (https://www.anylogic.com/)
Umeda Y, Nonomura A, Tomiyama T (2000) A study on life-cycle design for the post mass production paradigm. Artif Intell Eng Des Anal Manuf Cambridge Univ Press 14(2):149–161
Umeda Y, Fukushige S, Mizuno T, Matsuyama Y (2013) Generating design alternatives for increasing recyclability of products. CIRP Ann CIRP 62(1):135–138

Chapter 26
Recyclability of Tungsten, Tantalum and Neodymium from Smartphones

Nils F. Nissen, Julia Reinhold, Karsten Schischke, and Klaus-Dieter-Lang

Abstract Despite increasing pressure to reduce the consumption of critical resources, a number of highly relevant elements are still not being recycled from electronics, or only in the few applications, where they are used in bigger units. In high volume products and particular mobile products such as smartphones the content of these materials is low per unit, but still a large quantity when multiplied with millions of units shipped per year. The sustainably SMART projects explores the option to separate components containing tantalum, neodymium and tungsten from disassembled smartphones as these metals are lost in conventional electronics waste recycling processes. In this paper the target components are tantalum capacitors, loudspeakers, and vibration motors. Example quantities per device are explored and mirrored against current and potential recycling processes.

Keywords Recyclability · Critical raw materials · Recycling of mobile devices

26.1 Introduction

The conventional waste electrical and electronics recycling processes do not recover the conflict minerals tungsten (W) and tantalum (Ta) or the rare earth element Neodymium (Nd). Some smelters buy whole units, batteries potentially being removed beforehand, without rewarding a prior separation of individual sub-assemblies and even advising preprocessing companies preferably not to shred units for data destruction. The reason is the potential loss of precious metals when certain material fractions or sub-assemblies are separated, either following shredding (Chancerel et al. 2009) or manually, and shredding potentially leads to losses through dust emissions. Under these conditions the processes of copper and precious metal

N. F. Nissen (✉) · J. Reinhold · K. Schischke · Klaus-Dieter-Lang
Fraunhofer Institute for Reliability and Microintegration IZM, Berlin, Germany
e-mail: nils.nissen@izm.fraunhofer.de

Klaus-Dieter-Lang
Technische Universität Berlin, Berlin, Germany

Y. Kishita et al. (eds.), *EcoDesign and Sustainability I*, Sustainable Production, Life Cycle Engineering and Management, https://doi.org/10.1007/978-981-15-6779-7_26

smelters can be taken as a benchmark when judging the process of separating tungsten, tantalum and neodymium containing components. A simplified process flow of state-of-the-art pyrometallurgic electronics waste recycling is depicted in Fig. 26.1 (own adaption based on (Wölbert 2016)).

A large range of metals can be recovered, but due to the material properties, reactivity and also the typically very low concentration tungsten, tantalum and neodymium cannot be recovered. As these metals do not create volatile oxides and as they melt at very high temperatures (W, Ta) it is assumed that they leave the process as very minor contaminants with the iron silicate sand. Consequently any recovery of tungsten, tantalum, or neodymium requires separation early in the overall recycling process, i.e. at a mechanical separation step before shredding or smelting. Even then it is required to separate these target metals from other recoverable metals.

A separation purely for these target metals—in particular through manual disassembly—is out of the question. The low content cannot offset the costs for any additional disassembly steps, as will be summarized later. Developments for a deeper

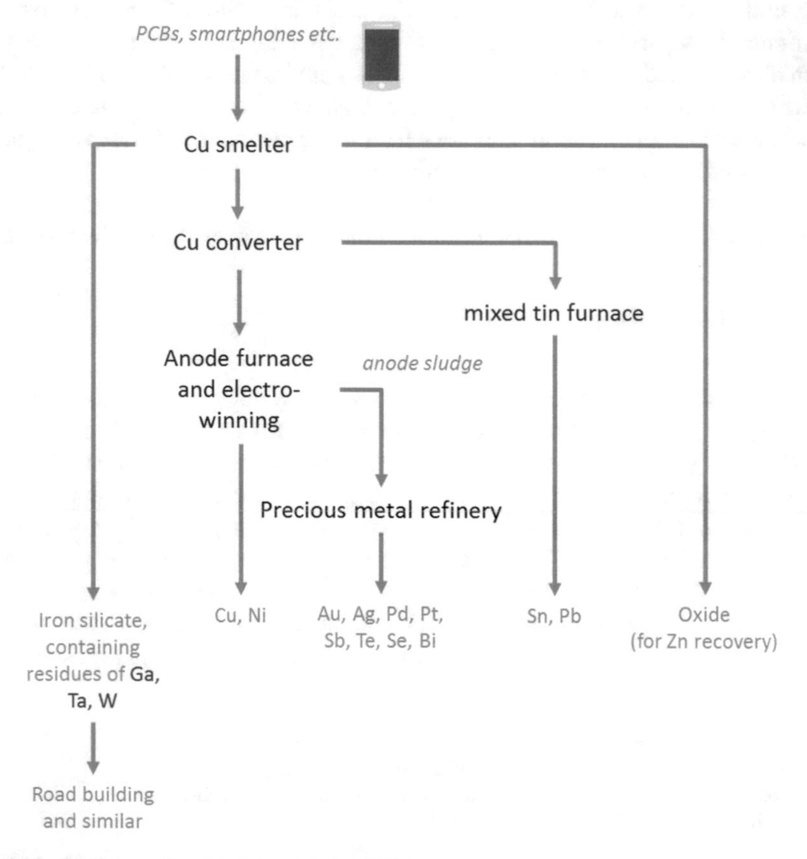

Fig. 26.1 Recovery of metals in a copper smelter

disassembly of even small mobile devices are however on the rise through semi-automated and fully automated disassembly lines, as exemplified by the Daisy robots from Apple (Apple 2019).

Apart from the safe removal of lithium batteries, disassembly robots may also target repair, refurbishment or parts scavenging and therefore use circular economy principles as economic drivers, and improved material separation and recovery as second order benefits.

The goal of this analysis is to provide quantified details from disassembly studies of smartphones regarding the three target metals. Intentionally omitting the steps, costs and problems of the automated separation of parts (covered in other research), the link to currently existing recycling processes is discussed. Future recycling potentials and avenues of further action are addressed.

To complement the fact finding from disassembly case studies on the lab scale and the results from literature research, interviews with recycling companies were carried out. Example findings from these interviews are shown in the results chapter.

26.2 Characteristics and Recycling of Tungsten, Tantalum and Neodymium

26.2.1 Tungsten

On average the earth's crust contains around 1.25 g/ton of tungsten (Willersen 2016). Tungsten is a metal of the refractory group with high hardness, high density (19.25 g/cm^3) and a melting point at 3422 °C. Tungsten is one of the T3G conflict minerals because of the high economic importance, its unique properties and the high supply risk when depending on imports. The major producers are: China, Russia, Bolivia and North America but it is also mined in Austria. In 2014, about 82% of the global tungsten production came from China (Reichl et al. 2016). When combined with carbon to make tungsten carbide, it is almost as hard as diamond. These and other properties make it useful in a wide variety of important commercial, industrial, and military applications (see Fig. 26.2).

The leading use for tungsten is as tungsten carbide in cemented carbides, which are wear-resistant materials used by the construction, metalworking, mining, oil and gas drilling industries. Pure or doped tungsten metal is used for contacts, electrodes, and wires in electrical, electronic, heating, lighting, and welding applications. Tungsten is also used to make alloys and composites to substitute for lead in ammunition, radiation shielding, and counterweights, super alloys for turbine engine parts as well as coatings. Tungsten chemicals are used to make catalysts, corrosion-resistant coatings, dyes and pigments, fire-resistant compounds, lubricants, phosphors and semiconductors. Tungsten has three primary uses in electronic products: vibration motors, integrated circuits and liquid crystal displays (LCD) (USGS 2014).

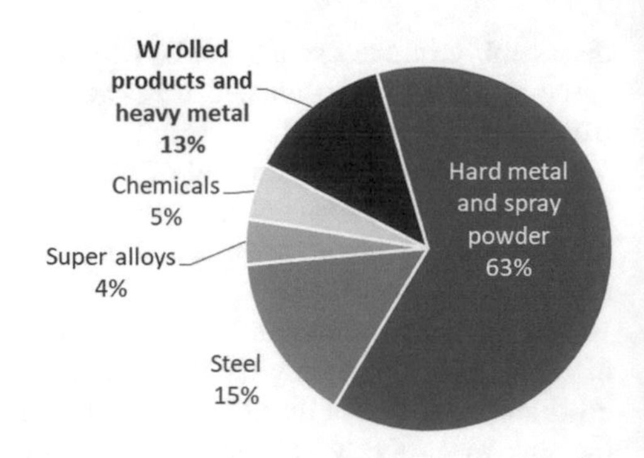

Fig. 26.2 Use of tungsten per application (Gille and Meier 2012)

The use of tungsten as vibration motor in smartphones (and other mobile devices) is only a marginal share of the above stated 13% tungsten heavy metal use.

The tungsten market was in oversupply, because of an economic slowdown in China and weak economic conditions elsewhere, which led to mine production exceeding consumption. Global tungsten prices trended downwards during 2014 and 2015. USGS reports prices for tungsten to have reached 38 USD per kilogram in 2015 whereas in 2016 a price of 17.9 USD per kilogram is reported. In spite of oversupply and low prices, new production entered the market. In 2015, tungsten mines in the United Kingdom and Zimbabwe began production, a large new mine in Vietnam ramped up production and its processing plant started producing ammonium paratungstate and tungsten oxides as well as new ferrotungsten plants in Russia and the Republic of Korea began production (USGS 2016).

Tungsten is economically extractable from the two minerals wolframite and scheelite. The ore content of 0.1 to 2.5% WO_3 is concentrated up to 65–75% for commercial trading as feedstock for further refining (ITIA 2011). The most common intermediate is ammonium paratungstate (APT) for further tungsten products, e.g. tungsten carbide. Besides tungsten ores, about 30% of tungsten-bearing scrap is used as secondary resource, which is directly introduced into the supply chain for the production of high-purity APT (Willersen 2016). In 2015, the estimated tungsten contained in scrap consumed by processors and end users represented 59% of apparent consumption of tungsten in all forms (USGS 2016).

The recycling of tungsten is mostly concentrated on tungsten carbides, because they represent approx. half of the tungsten consumption and as they are comparatively simple to collect. The tungsten used in the vibration motors counts as tungsten heavy metal alloys (Luidold 2017). The requirements on tungsten products are huge wherefore contamination free and homogenous fractions with high purity have to be processed especially when recovering small quantities as they are found in the vibration motor.

26.2.2 *Tantalum*

Tantalum is another element of the group of refractory metals, which are characterized by a high resistance against wear and which typically have a very high melting point—in the case of tantalum 3017 °C—which also determines, how tantalum can be processed metallurgically. Tantalum is one of the T3G conflict minerals, which are mined in civil war affected regions of D.R. Congo. This fact and that for a certain period of time the mining worldwide was limited to only a few regions made tantalum a critical raw material. The related ore coltan is typically mined in open pit mines by artisanal miners. Environmental impacts—besides the social and direct human health impacts—are related to landscape impacts, and potential waste water emissions from mining processes, but at least the character of open pit mining means a rather low energy consumption for mining operations.

Tantalum is used as a material for capacitors in mobile IT devices as it allows to realize a high capacitance in a small volume, thus is crucial for miniaturization: Tantalum forms very thin protective oxide layers, which serve as dielectric layer in between the metallic tantalum and the cathode material. The global production of tantalum yielded 1300 t in 2013, according to DERA thereof 10% for capacitors in the electronics industry (Marscheider-Weidemann et al. 2016). A significantly higher share of 42% for tantalum capacitors is reported in (Gille and Meier 2012) and shown in Fig. 26.3. Projections show a steady increase in tantalum consumption for capacitors, which could even come close to today's total production by 2035.

Recycling of tantalum capacitors is in place at few locations worldwide largely for post-industrial scrap and for post-consumer scrap where tantalum is found in larger amounts, such as turbine blades (Gille and Meier 2012). There are significant internal recycling flows and from raw-material pretreatment, powder/ingot production, and end-product manufacturing a significant amount of the input material is returned as post-industrial scrap to a secondary material input. It is a high theoretical potential to recover tantalum from capacitors, but low concentration, i.e. high dissipation is the main barrier to increase this share of the recycling rate (Gille and Meier 2012). There are some recyclers processing tantalum capacitors from post-industrial scrap,

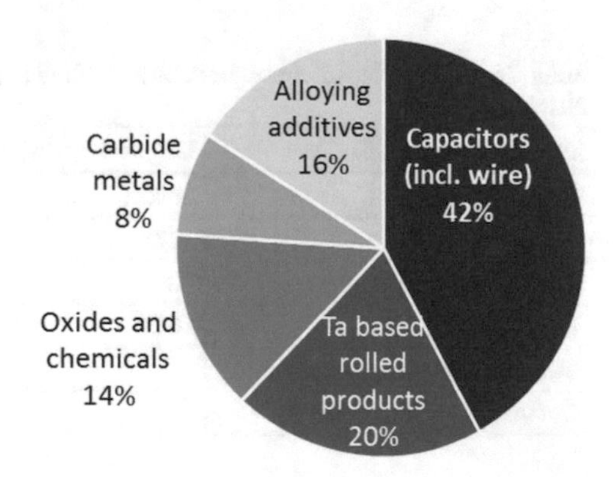

Fig. 26.3 Use of tantalum per application (Gille and Meier 2012)

and there are also some indications that already today also printed circuit boards with a larger number of tantalum capacitors are separated for dedicated tantalum recovery.

26.2.3 Neodymium

Neodymium belongs to the rare earth elements (REE) and possesses high magnetism with a low density (7.003 g/cm^3) and a melting point at 1024 °C. REE occur in the earth's crust in various minerals. Large amounts of energy, water and toxic chemicals are used for the extraction of rare earths and the subsequent elemental separation, which is associated with high environmental pollution (Ökoinstitut 2011). Neodymium is utilized in various technical applications such as electric motors, batteries, catalytic converters, wind turbines and hybrid cars. The largest application is the production of NdFeB magnets, which are used in many information and communication technology devices: Permanent magnets from hard-disk-drives, miniaturized loudspeakers, headphones, smartphones and microphones. Neodymium magnets appear in products where low mass, small volume, or strong magnetic fields are required. With concentrations ranging up to 23–29%, neodymium is an important component of high performance permanent magnets, which may additionally contain the heavy rare earth elements praseodymium and dysprosium (Böni et al. 2015).

The demand for REE is high and continues to rise. Neodymium demand is estimated to exceed global production per year by a factor of 3.8 (Lutz 2010). Table 26.1 indicates the neodymium concentrations of several different products in WEEE recycling and compares these concentrations to the concentration of a primary mine.

Loudspeakers and laptops at the product level already contain concentrations of neodymium similar to a primary mine. If the magnets were separated from the remaining materials by new separation technologies, the neodymium concentration in this material fraction would be higher than the best mineral deposits in the world and up to 200 times higher in comparison to a lower quality primary mine.

Table 26.1 Example Nd-concentrations according to project e-Recmet in Switzerland (Böni et al. 2015)

Products/components	Disposed quantities 2012 (t)	Nd-concentration (ppm)
Desktop PC/server	5096	377
Laptop PC	1275	1024
Mobile phone/smartphone	70	740
Loudspeaker	1075	3325
Magnets in electronic products		28000–250000
Comparison: Primary mine		1200–17600

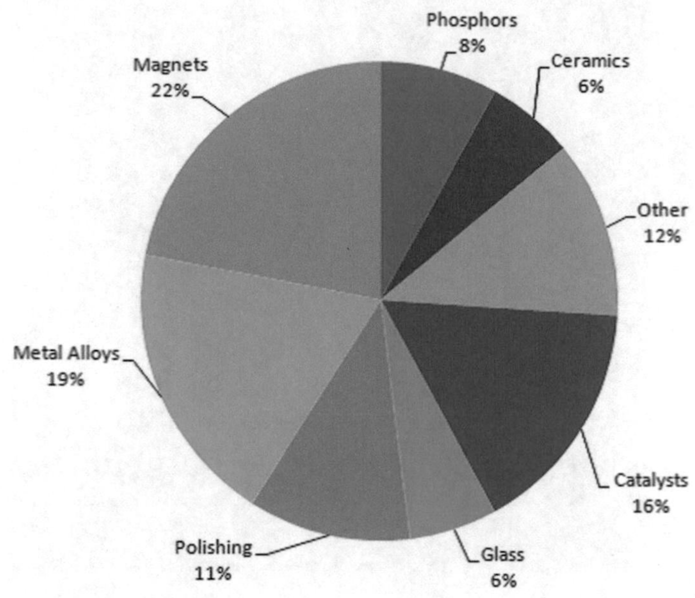

Fig. 26.4 Use of REE per application (Kingsnorth 2012)

There is no complete breakdown of the usage of neodymium. The main applications using REE are depicted in Fig. 26.4. Magnets are a large factor across all REE uses, and presumably have a much higher share of the neodymium usage.

26.3 Investigating Relevant Smartphone Components

The relevant components for tungsten, tantalum and neodymium are not the largest components in smartphones. Figure 26.5 shows example placements and sizes of components in a partially disassembled smartphone.

The variation in size, type and composition of the indicated components across smartphones is of course huge. The use of tantalum capacitors in particular seems to be fluctuating wildly with no clear trend, except in general that they are only used where ceramic capacitors are not sufficient. The other components—loudspeakers and vibration alarms—are often customized for each device generation, although some designs reappear across different models.

A range of smartphones from the years 2012–2016 have been disassembled and analyzed at Fraunhofer IZM to generate a crosscut of variants, which might just start to appear in the waste streams.

Fig. 26.5 Components of interest for tungsten, tantalum and neodymium in a sample disassembled smartphone

26.3.1 Disassembly of Vibration Motors and Loudspeakers

For case study data on vibration motors and micro loudspeakers the smartphone models listed in Table 26.2 were disassembled.

Dismantling is necessary to separate the tungsten-containing component for subsequent material recovery. The tungsten-containing component, the tungsten ring, is mounted on other components inside the metal housing. Most commercially available smartphones contain coin-shaped linear resonant actuators (LRAs). Figure 26.6 shows the disassembled linear actuator and all the components it contains: The metal housing, the tungsten ring, a wave spring, the NdFeB magnet as well as a copper coil and the adhesive foil.

The iPhone 7 contains a different type of vibration actuator called a Taptic Engine (see Fig. 26.7). The Taptic Engine contains copper coils and various vertically oscillating magnets. This technology replaced LRAs used in earlier iPhone models.

Most smartphones of that generation contain cylindrical or almost rectangular micro loudspeakers. Figure 26.8 shows the disassembled components of a near-rectangular loudspeaker from the Samsung S3 series.

The loudspeaker of the iPhone 7 is slightly bigger and more complex internally (see Fig. 26.9).

Table 26.2 Investigated smartphone models

Smartphone model	Year	Total mass (g)	Analysed components
Samsung Galaxy S3 mini	2012	113	Vibration motor, loudspeaker
Samsung Galaxy S3 LTE	2012	133	Vibration motor, loudspeaker
Samsung Galaxy S4 mini	2013	107	Vibration motor, loudspeaker
Samsung Galaxy S5	2014	145	Vibration motor, loudspeaker
Samsung Galaxy S6	2015	138	Loudspeaker
Samsung Galaxy S7	2016	152	Vibration motor, loudspeaker
iPhone 7	2016	138	Vibration motor, loudspeaker
Comparison: Linear actuator	2012	1.7	Vibration motor

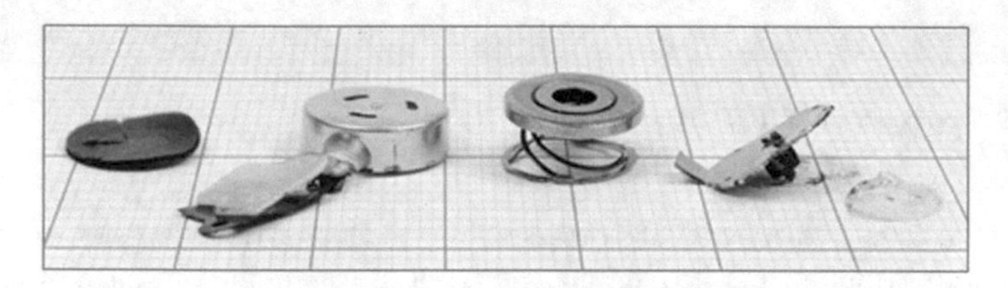

Fig. 26.6 LRA of Samsung Galaxy S3 LTE

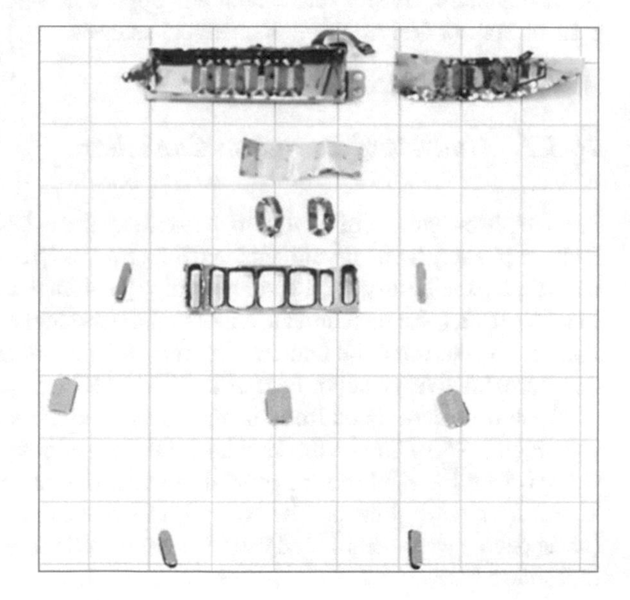

Fig. 26.7 Taptic Engine of the iPhone 7

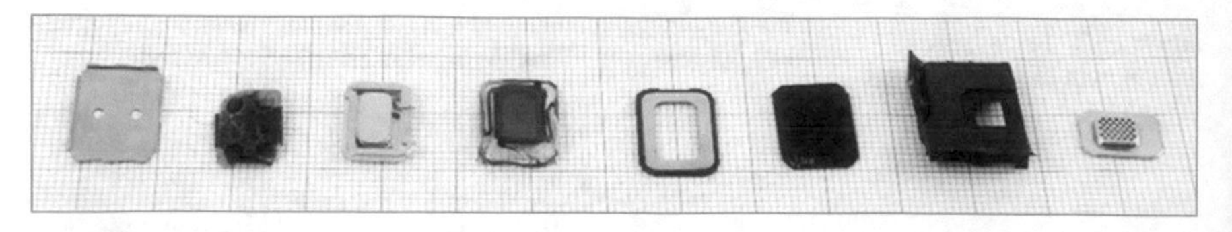

Fig. 26.8 Loudspeaker of Samsung Galaxy S3 LTE

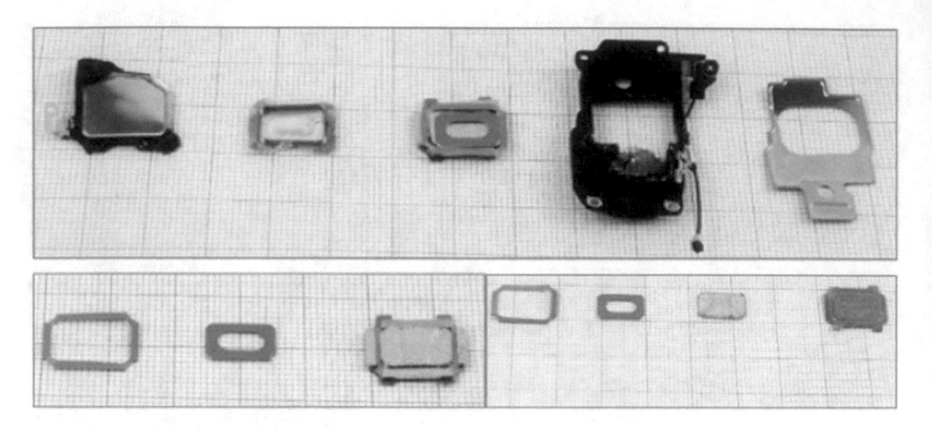

Fig. 26.9 Loudspeaker of iPhone 7

After the maximum mechanical disassembly the magnets and tungsten-containing weights were weighed and further analyzed. For tungsten, the presence of tungsten was verified by density measurement, before sending the parts to an external analysis. For neodymium, density verification was applied as well, but the exact neodymium content was not derived through chemical analysis.

26.3.2 Analysis of Tantalum Capacitors

For the investigation of tantalum capacitors three bestselling examples of the Samsung Galaxy S range were selected (S3, S5 and S7, respectively). The mainboards were first visually inspected for tantalum capacitors using manufacturer markings. In the next step, the tantalum capacitors on the motherboard could be verified using X-rays. To determine the tantalum content and the internal structure more closely, one representative capacitor type of the Galaxy S7 was selected.

These were desoldered from the mainboard and cross sections prepared for analysis. Figure 26.10 shows the standard light microscopy of a cross section. X-ray analysis with REM-EDX was applied to verify the materials, which are principally known from other sources. The volumes of the tantalum rod and the tantalum cube (or "sponge") were determined from the cross sections and additional X-ray images (not shown here).

Fig. 26.10 Microscopy cross section of tantalum capacitor "Js" from Samsung Galaxy S7

26.4 Results

26.4.1 Tungsten Content

The tungsten in the LRA type vibration alarms is usually contained in a ring of tungsten. Here the tungsten content was verified by checking the density of the material, and the weight of the ring is calculated as tungsten. For the slightly different LRA Samsung Galaxy S3 and S4 mini the presumed content of tungsten is embedded in a plastic chip. Here the weight together with the plastic is used. For the taptic engine of the iPhone the materials could be separated sufficiently to get the weight contributions.

Figure 26.11 shows the weight contributions of the disassembled parts of the vibration actuators, plus one LRA type, which is not from one of the disassembled models. The tungsten content based on these models of 2012–2016 ranges between 0.35 and 1.2 g. For the Galaxy S3 LTE tungsten content reaches 56% of the complete actuator assembly.

26.4.2 Neodymium Content

The presence of neodymium could be verified for all magnetic materials through density measurement. Figure 26.12 shows the different material weights in loudspeakers per smartphone model. Neodymium magnet weight is determined in the

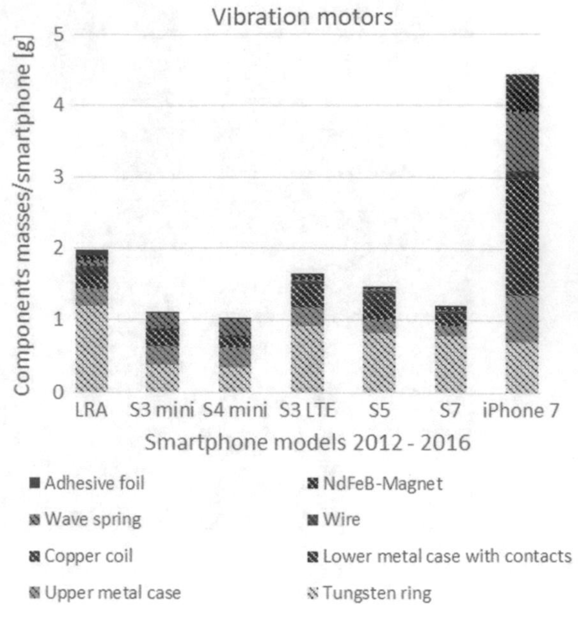

Fig. 26.11 Tungsten content of analyzed vibration motor

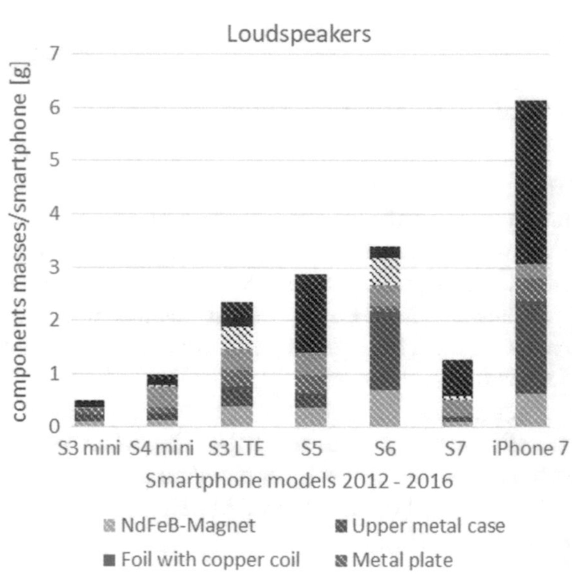

Fig. 26.12 Neodymium content of analyzed loudspeakers

range between 0.11 and 0.7 g. For the Galaxy S6 loudspeaker the neodymium magnet is 20% of the weight of the loudspeaker.

The content of neodymium in some loudspeaker assemblies in absolute and relative terms is so low, that a selection by model type might be necessary to achieve reasonable concentrations for recycling.

26.4.3 Tantalum Content

Figure 26.13 shows the tantalum content for the three smartphone models selected for this part of the analysis. The tantalum weight ranges from 0.012 g to 0.07 g. Tantalum content is therefore very low in absolute and relative terms. The weight of the respective smartphones ranges between 133 and 152 g. For the Galaxy S7 the tantalum concentration in relation to total weight is only 0.05%, and even lower for the other models.

Relatively, the S7 has higher tantalum content due to the higher number of tantalum capacitors. Older devices and other applications have much larger tantalum capacitors, but for these smartphone generations the size of the tantalum capacitors has low variation.

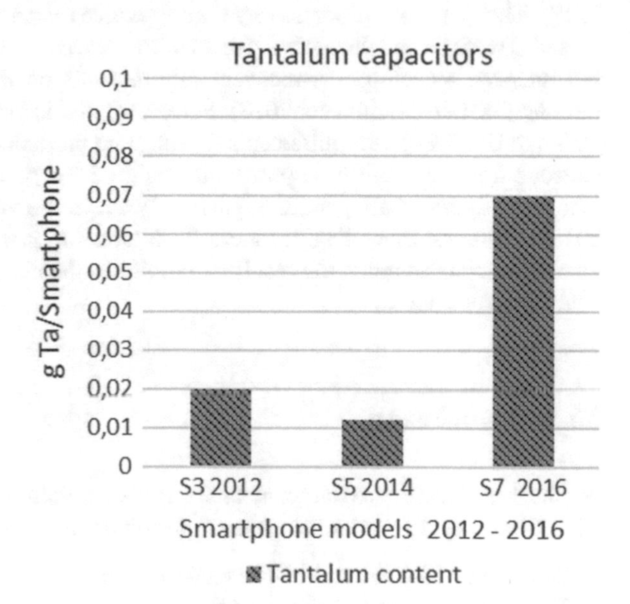

Fig. 26.13 Tantalum content of analyzed capacitors

26.4.4 Interviews with Recycling Companies

Interviews with companies involved in the recycling area should provide an overview of arising and processed quantities. Apart from this, insights into existing recycling processes are obtained. The focus was on the following questions to determine the recyclability of the considered electronic components and to determine recycling strategies:

- the existence of a recycling process
- the quantities arising and processed
- the respective recovery rate
- the resulting quality
- price information about the used metal
- consideration of the economic efficiency of the recycling of small quantities.

All the interviewed companies established recycling processes for scrap containing tungsten, tantalum or neodymium which can be used for micro-loudspeakers and tantalum capacitors, but which have not yet been implemented with vibration motor components. The processing of the vibration motors requires a further process step in the pre-processing of the tungsten rings. According to Mühlbauer, this can be done by mechanical processing into particles or powder (<1 mm^3) (Mühlbauer 2018). The quantities processed by the respective recycling companies are between 10 and 60,000 t/a and the recovery rates of the processes are between >90 and 100%. For tungsten recycling, the recovery rate depends on the degree of leaching and the slagging rate (Mühlbauer 2018). Ferro-tungsten is produced in the quality of W 75% and Fe 25%. The requirements for tungsten products are high, therefore homogeneous fractions with high purity are required for processing, especially for the recovery of very small quantities, such as occur in the vibration motor (Mühlbauer 2018). Tantalum Recycling produces 87% tantalum powder while the other interviewed companies adapt the resulting quality to the desired product by adding the necessary substances.

26.5 Conclusion

For the three investigated target metals tungsten, tantalum and neodymium rather low end-of-life recycling rates have been determined in literature (Graedel et al. 2011):

- Tungsten: >10–25% end-of-life recycling rate
- Tantalum: <1% end-of-life recycling rate
- Neodymium: <1.35% end-of-life recycling rate.

Among the target metals, the recycling rate of 10–25% for tungsten is significantly higher. This refers basically to the recycling of massive tungsten alloy parts (hard metal) with high tungsten content and limited impurities.

To improve the effective recycling rates, on the one hand recycling input streams with higher concentrations of the target metals are necessary, on the other hand novel recycling processes or upscaling and duplication of existing specialty recycling plants might be needed.

The recycling capabilities are ramping up. Current recycling of tungsten needs a minimum of 75% content of tungsten. Parts close to this level going into recovery without shredding should have thicknesses of less than 1 mm in one dimension, according to the interviews with recyclers. This would be true for most tungsten parts from the vibration actuators, but not for the complete actuator assemblies. At these quality levels the tungsten should be added to existing processing operations rather than setting up new recycling plants. Recyclers, who are already processing secondary tungsten, would welcome even smaller quantities, but this is not yet happening in reality.

Similarly, for neodymium the recycling would be feasible, if the magnets are separated at an early stage within the disassembly process. The recovery of neodymium from permanent magnets in large scale is limited by its chemical properties as the contained elements are expensively to recycle. Standardized qualities for the chemical composition of NdFeB magnets and tungsten rings would be beneficial (Burkhardt 2019). Implementing a recycling factor that includes the accessibility of the component as well as the Nd– and O_2– content could lead to improved material recovery (Burkhardt 2019).

For tantalum capacitors there are recycling operators who take even small quantities. Tantalum parts do not have to be separated from the rest of the capacitors. Nevertheless the time or expenditure to separate 1 kg of tantalum capacitors from mobile devices is prohibitive.

Even with improved recycling capacities and processes, the economic value of the target metals from mobile devices is low. Table 26.3 projects global content of the target metals based only on the specific components analyzed in the investigation.

As noted in the introduction of this paper, it is therefore important to gain access to higher concentration waste streams on a component level generated by other

Table 26.3 Potential worldwide W, Nd, Ta content of mobile phones, based on the investigated components

Component	Average content (g)	Price informtion (€/kg)	Value/smart-phone (€)	Potential worldwide (t)
W from vibration motor	0.72	11–36.5	0.01–0.03	1000
Nd from loudspeaker	0.35	58	0.02	500
Ta from capacitors (2–7 per smartphone)	0.01	165.3	0.002–0.01	30–100

Source for globally 1.4 billion smartphones: (Statista 2019)
Sources for price information: (Neumann 2018; USGS 2019; Burkhardt 2019; Gilerman 2019)

economic and legal drivers. Removal of batteries is one such legal driver in many parts of the world, while automated disassembly for circular economy services could be the larger economic driver. When even small devices are opened at end-of-life instead of being shredded, these additional options start to come into play.

For smartphones specifically the accessibility within an already open device could be improved through design, processing and better data provision. The vibration alerts containing the tungsten are often lightly glued to a midframe or backcover cavity in many designs. This already allows approaches to remove the vibration alert assembly with a small vacuum sucker. Automation would seem possible, if the glues are indeed selected not too strong.

Similarly, for neodymium, easy access to the loudspeakers is essential. These components are integrated typically already as a sub-module, which eases separation.

Achieving higher tantalum concentrations would require desoldering of all such capacitors. For an automatic desoldering it would be beneficial to place all tantalum capacitors on the same side, preferably on the same side as ICs potentially desoldered for reuse, and not covered under a soldered shield or shielding frame. Color coding, such as the orange or ochre coloring formerly used for almost all tantalum capacitors might ease the identification of target components, but building up a database with the location of tantalum capacitors would serve a similar purpose. Such information would have to be made available to automated desoldering stations placed in a large number of locations.

From an environmental viewpoint reuse of the assemblies or components would be preferable to the material recycling, but standards for long term reuse are unlikely to emerge.

Separating tungsten, tantalum and neodymium containing components as a separate activity at end-of-life of small devices is not economic; therefore it cannot be driven by straightforward economic reasoning. Keeping critical resources and conflict minerals in use is however already perceived as valuable for some industry players looking beyond the costs of sourcing the material.

Acknowledgements The project sustainably SMART has received funding from the European Union's Horizon 2020 research and innovation programme under grant agreement no. 680640.

References

Apple (2019) Environmental responsibility report https://www.apple.com/environment/pdf/Apple_Environmental_Responsibility_Report_2019.pdf

Böni H, Wäger P, Figi R (2015) Rückgewinnung von kritischen Metallen wie Indium und Neodym aus Elektronikschrott am Beispiel von Indium und Neodym, Schlussbericht, p 10

Burkhardt C (2019) OBE Ohnmacht und Baumgärtner GmbH & Co. Berlin, KG, Personal Interview

Chancerel P, Meskers CEM, Hagelüken C, Rotter VS (2009) Assessment of precious metal flows during preprocessing of waste electrical and electronic equipment. J Ind Ecol 13(5):791–810

Gilerman R (2019) Tantalum recycling. Personal Interview, Berlin

Gille G, Meier A (2012) Recycling von Refraktärmetallen. In: Thomé-Kozmiensky KJ, Goldmann D (eds) Recycling und Rohstoffe-Band 5. TK Verlag, Neuruppin, pp 537–560

Graedel TE, Allwood J, Birat J-P, Buchert M, Hagelüken C, Reck BK, Sibley SF, Sonnemann, G (2011) What do we know about metal recycling rates? J Indus Ecol 15(3)

ITIA (2011) International tungsten industry association information on tungsten: sources, properties and uses, https://www.itia.info/tungsten-concentrates.html

Kingsnorth D (2012) Rare earth supply security: dream or possibility? Curtin University and Industrial Minerals Company of Australia PTY Ltd

Luidold S (2017) Herausforderungen beim Recycling wolframhaltiger Schrotte, In: Recycling und Rohstoffe-Band 10, TK Verlag, Neuruppin, p 3

Lutz F (2010) Die künftige Verfügbarkeit knapper, strategisch wichtiger Metalle, Bonn, p 17

Marscheider-Weidemann F, Langkau S, Hummen T, Erdmann L, Tercero Espinoza L, Angerer G, Marwede M, Benecke S (2016) Rohstoffe für Zukunftstechnologien, DERA-Rohstoffinformationen 28, Berlin

Mühlbauer G (2018) Wolfram Bergbau und Hütten AG, Personal Interview, Berlin

Neumann M (2018) Nickelhütte Aue GmbH, Personal Interview Berlin

Ökoinstitut e.V. (2011) Hintergrundpapier Seltene Erden, Berlin

Reichl C, Schatz M, Zsak G (2016) World-Mining-Data, Austrian Federal Ministry of Science, Research, and Economy, Vienna http://www.wmc.org.pl/sites/default/files/WMD2016.pdf

Statista (2019) https://de.statista.com/themen/581/smartphones/

USGS (2014) Tungsten, minerals yearbook, United States Geological Survey https://s3-us-west-2.amazonaws.com/prd-wret/assets/palladium/production/mineral-pubs/tungsten/myb1-2014-tungs.pdf

USGS (2016) Tungsten, mineral commodity summaries United States Geological Survey https://s3-us-west-2.amazonaws.com/prd-wret/assets/palladium/production/mineral-pubs/tungsten/mcs-2016-tungs.pdf

USGS (2019) Tungsten mineral commodity summaries https://prd-wret.s3-us-west-2.amazonaws.com/assets/palladium/production/atoms/files/mcs-2019-tungs.pdf

Willersen C (2016) Reactive extraction and critical raw materials: industrial recovery of Tungsten http://onlinelibrary.wiley.com/doi/10.1002/cite.201600079/epdf

Wölbert C (2016) Der Weg des Schrotts, Elektroschrott: Was recycelt wird—und was nicht, c't, 14/2016, p 76

Chapter 27
An Economic Evaluation of Recycling System in Next-Generation Vehicles Considering the Risk of Spilled EOL to Overseas

Hiroshi Kuroki, Aya Ishigaki, Ryuta Takashima, and Shinichirou Morimoto

Abstract In recent years, the demand for rare resources used for high-tech products has been increased due to growing populations and industrial developments. Since rare resources are limited in producing countries, countries that cannot produce them at home are expected to become depleted. Therefore, it is important to keep resources by recovering and regenerating rare resources from End-of-life product (EOL). From this, the formation of a recycling-oriented society that considers regeneration to improve resource depletion and secure scarce resources is drawing attention. However, with a recycling business, the expense is required for recovery and reproduction of EOL. Therefore, it is necessary to estimate the cost by determining the location and recovery route of the recycling site. Furthermore, in order to maintain a high recovery rate, it is necessary to prevent and protect the risk of spilled EOL to overseas. Therefore, in order to continue maintaining a recycling business, after evaluating risk and economic efficiency in advance, it is necessary to design a system. Furthermore, the overseas outflow of many used products is caused in the pursuit of high profits in recent years. For that reason, the overseas outflow of EOL also causes the outflow of rare earth indispensable to manufacture of the high-tech machine included in used products. If foreign spills occur, countries that cannot produce rare earth in their own country will run out of resources. This lack of resources can reduce the productivity of high-tech products. In other words, if it not only can pursue the present profits but the risk and economic efficiency of a recycling business can be evaluated by taking future value into consideration, it will become possible to build the maintainable recycling system which foresaw the future. If it is possible to construct a recycling system for next-generation vehicles whose production number is increasing with CO_2 emission regulations and the trend of the times, overseas spills of rare earth such as neodymium are prevented and future recycling Business will be better. This study considers collecting rare earths from a next-generation car, and reproducing, and aims at performing economic efficiency

H. Kuroki · A. Ishigaki (✉) · R. Takashima
Department of Industrial Administration, Tokyo University of Science, Noda, Japan
e-mail: ishigaki@rs.tus.ac.jp

S. Morimoto
National Institute of Advanced Industrial Science and Technology, Tsukuba, Japan

Y. Kishita et al. (eds.), *EcoDesign and Sustainability I*, Sustainable Production, Life Cycle Engineering and Management, https://doi.org/10.1007/978-981-15-6779-7_27

evaluation with the risk of a new recycling business. In this study, some scenarios are set up and the knowledge to a future recycling business is given by evaluating risk and economic efficiency for each scenario.

Keywords Recycling system · Economic evaluation · Facility location · Sustainable supply chain · Risk of spilled overseas

27.1 Introduction

The demand for rare resources used for high-tech products has been increased due to growing populations and industrial developments. Since it becomes more likely to cause resource depletion in the future, a requirement for resource circulation, e.g., recycling, reuse, and reproduction, becomes large, consumption of mineral resources such as rare earth elements (REEs) and other critical materials (Mark Seddon and Argus Media Ltd 2016). Global innovation and invention of new technology have boosted the demand for high-technology products such as batteries, motors, and mobile devices. REEs can significantly improve the properties of alloys and materials by adding only a small amount, so it becomes an indispensable resource in the industry. For this reason, the NdFeB-sintered and -bonded magnets are extensively used as permanent magnet motors in hybrid electric vehicles (HEVs), home appliances, such as air conditioners, washing machines, refrigerators, personal computers, and several mechanical robots. However, there are limited mineral resources, and very few countries produce these resources. Therefore, if there is a supply disruption such as export restrictions from REEs resource countries, it will be difficult for countries that are not REEs resource countries to secure REEs in their own country. One way to solve this difficulty is to recycle from end-of-life (EOL) products. Recovery of REEs from EOL products does not require work on natural resources such as mining, thus reducing the environmental impact. Among them, gold and platinum have low recovery and cost and are relatively recycled. By comparison, REEs such as Nd is difficult to secure stable, so the cost of recycling is high and the recycling business has not progressed.

Therefore, recycling Nd from EVs and HEVs is crucial for achieving a sustainable industry. In order to realize sustainable recycling, the systems considering minimizing a transportation cost of a cyclical form supply chain and forecast of the resources to collect needs to be designed. However, if the recycling systems are not economically inefficient, the firm might not implement the project. Therefore, it is important to evaluate the economic efficiency of supply chain systems. Furthermore, since there is a possibility that regional trends may appear in the generation of waste, it is necessary to introduce recycling equipment at a location close to the collected waste. They focused on the parameter change when determining the product value, but did not perform multi-period change and waste recovery prediction.

This study considers collecting rare earths from a next-generation car (EV and HEVs), and reproducing, and aims at performing economic efficiency evaluation with the risk of a new recycling business. In this study, some scenarios are set up and the knowledge to a future recycling business is given by evaluating risk and economic efficiency for each scenario.

27.2 Literature Review

There are several ongoing studies that evaluate the future demand of REEs (Zhou et al. 2017) and the current status of recycling (Yang et al. 2016; Du 2011). Moreover, Artem investigated the flow of the rare earth supply chain around the world by examining the production and supply of REEs (Golev et al. 2016). However, to make the industry sustainable, it is necessary to evaluate the cost of Nd recycling from EOL products with regard to the rapidly increasing quantities of waste. However, the change in the future demand of REEs will affect the transportation cost that is incurred while collecting the EOL waste. Therefore, in order to consider transportation costs, a recycling system is built using location-routing problems (LRP) (Min et al. 1998).

The LRP is that minimizes the sum of the facility cost which is proportional to the number of sites and the transportation cost which is proportional to the moving distance when input data as collection volume. Therefore, in the LRP dealt with in this research, instead of deciding the location of the facility, it is extended to the plan to add the facility place at each period. In the location routing plan, the collection volume is given at a single period, and the LRP is solved using this volume. On the other hand, in the LRP in multi periods, different collection volumes are given at each period. Therefore, it is necessary to consider the addition of facilities by solving LRP through multi periods. Therefore, this research is planned by solving a simple LRP using the result of solving the LRP for every single year. In order for companies to sustain recycling business, it is important to secure profits. In addition, because the project involving facilities requires a preparation period, it is necessary to evaluate the total cost based on forecasts in the future and plan the establishment of recycling facilities. In the recycling business, one of the means to reduce the total cost is the optimization of the number of recycling facilities, the place of installation and the collection route. This makes it possible to reduce transportation costs (Ballou 2001).

Barreto et al. Proposed a method to solve the problem by generating a traveling route only by the location information of demand places excluding bases as the first stage and inserting bases near the route in the second stage (Barreto et al. 2007). Yu et al. proposed a search method for LRP using a one-dimensional vector consisting of bases and demand points. Then, they proposed a method to easily generate a neighborhood solution by simultaneously changing the setting of the base and the traveling route of the demand place using a one-dimensional vector and finding the best solution (Yu et al. 2010). Rosemary et al. Modelled a location routing problem focusing on distance constraints when designing distribution system wiring (Berger et al. 2007). Moreover, as multi-period LRP, Sung proposed a method to determine the location of the electric station that supplies electricity to electric vehicles etc. in Korea (Chung and Kwon 2015). Xim has designed a multi-period network system that models the delivery route of goods and the location of facilities taking into account the geographical factors of France (Tang et al. 2019). However, these LRP studies are proposing the proposed method and optimizing the location and number of facilities, and only forecasting the amount of recovery required for recycling projects and minimizing the cost of economics was not evaluated. Saman considered the location of the facility with uncertain demand and returns in the supply chain (Amin and Zhang 2013). Luu has designed a recycling network model that minimizes transportation costs to minimize the costs incurred in the recycling process of multiple appliances (Dat et al. 2012). Aras et al. It was modeled that the recycling business to recover e-waste in Turkey takes into consideration the location of the recycling facility and the recycling facility that will take over the processing capacity (Aras et al. 2015). However, in these cases as well, only cost minimization is performed, and economic evaluation is evaluated, and optimal investment timing is not considered. Danijel and Marija are research that evaluated the economics of introducing equipment to facilities (Kovačić and Bogataj 2013). The economic evaluation uses a discount rate, so an analysis was conducted by incorporating the discount rate with reference to Schneider (Schneider et al. 2009).

27.3 Recycling Systems

27.3.1 Conceptual Diagram of Recycling Systems

The recycling system of this research is built for the project to recycle rare earth from the discarded EV motor. Figure 27.1 shows the recycling system handled in this research.

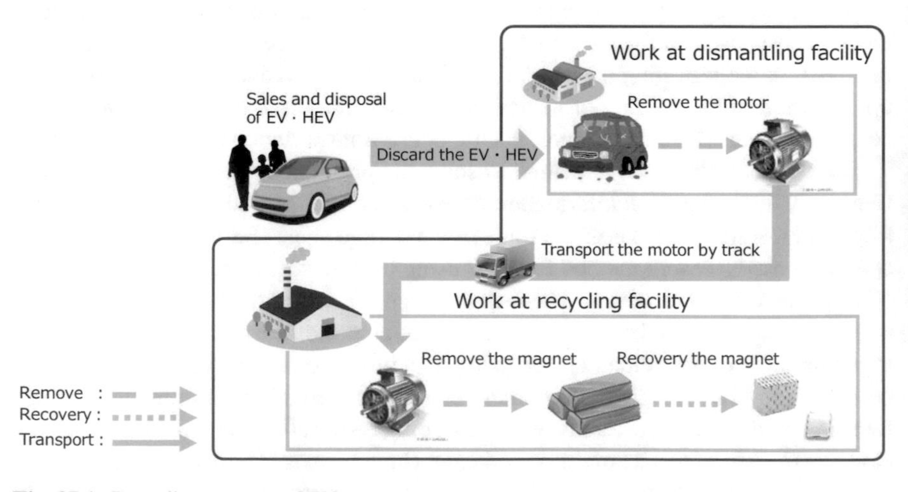

Fig. 27.1 Recycling systems of EV

Firstly, after EV · HEV is used, these are discarded and collected as disposal companies as disposal vehicles. Second, at the dismantling company, take out the motor from the disposal vehicle. The motor is transported by truck to a recycling facility. Finally, at the recycling facilities, the rare earth is regenerated through demagnetization and refining of the motor. In this research, the demagnetization process indicates that the magnet is extracted from the rare earth which became a magnet, and the refining refers to the operation of removing impurities from the rare earth and taking out highly pure rare earth. In addition, since the transport covered by this research covers a collection of motors to be discarded, transportation of magnets regenerated at the recycling facilities is not considered. Therefore, collect only in the area to be recovered and do not consider the consumption of the magnet inside or outside of the area afterward.

27.3.2 Precondition Used in This Research

In this research, the following precondition are used for simplicity.

- It is assumed that existing buildings are used for recycling facilities.
- Only one demagnetization processing device is installed at each recycling facilities, and the service life of the device is set to 10 years.
- The distance between the disassembling base and the recycling base is defined as Euclidean distance $\times$ 1.3, and transportation costs are generated in proportion to this distance.
- Do not consider processing capacity constraints of recycling facilities.
- The motor is transported to the nearest recycling facilities.
- As for the collection route, the same route will always be used after the introduction of the recycling equipment, and it will not be changed.
- Delivery time and limitations on the working hours of the driver are not taken into consideration.

27.3.3 Materials and Methods—Location Routing Problem

In this research setting in Japan, we set only an HEV motor as the target of an EOL product because it is the main product that utilizes an NdFeB-sintered magnet. The future demand for HEV was estimated using the 2001–2015 annual demand for HEVs and by applying it to perform a multiple regression analysis based on the future global population and gross domestic product (GDP) using the SRES B2 scenario (Intergovernmental Panel on Climate Change 2007).

Next, let us consider estimating the number of waste vehicle using the forecasted number of units sold. A similar case to this is the research report on re-use cars, among which the recovery distribution is reported to follow a log-normal distribution (Rai and Singh 2006).

However, in this research, it is assumed that the number of discarded items follows the Weibull distribution of scale parameter 3.6 and shape parameter 19.9 because we consider not to reuse EVs and HEVs but to take out motors from the discarded vehicles. D that shows the annual motor can be estimated using Eq. (27.1).

$$D = P \times N \tag{27.1}$$

where P is the annual EV and HEV production and N is the rate of permanent magnet that is contained in the motor.

The numbers and locations of the recycling facilities were optimized by solving Eqs. (27.2)–(27.7).

$$\max \sum_{t=1}^{T} \frac{1}{(1+r)^t} \left\{ \begin{array}{r} profit \times D_t - u \sum_{i=1}^{I} \sum_{j=1}^{J} c_{ij} x_{ijt} \times rt_{jt} \\ + \sum_{i=1}^{I} F_A y_{it} + LC \times D_t \end{array} \right\} \tag{27.2}$$

$$\text{Subject to} : \sum_{i=1}^{I} \sum_{j=1}^{J} x_{ijt} = 1 (t = 1, \ldots, T) \tag{27.3}$$

$$\sum_{i=1}^{I} y_i = \sum_{i=1}^{I} y_{t-1} (t = 1, \ \ldots, T) \tag{27.4}$$

$$x_{ijt} \leq y_{it} \tag{27.5}$$

$$\frac{\sum_{i=1}^{I} \sum_{j=1}^{J} c_{ij} x_{ijt}}{speed} \leq time (t = 1, \ldots T) \tag{27.6}$$

$$y_{it} \in \{0, 1\}, x_{ijt} \in \{0, 1\} \tag{27.7}$$

where I is the number of candidate locations in a recycling facility (i = 1,…, I); J is the number of car dismantling facilities (j = 1,…, J);T is the period (t = 1,… T); x_{ij} is a binary variable, which is equal to one if the car dismantling facility j is assigned to recycling facility i and is, otherwise, equal to zero; y_i is a binary variable, which is equal to one if the recycling facility is established in candidate location i and is, otherwise, equal to zero; c_{ij} is the distance from the car dismantling facility j to the recycle facility i (km); F_A is equipment cost in faciliteis(JPY/facilities); rt_{jt} is the number of round trips between the car dismantling facility j and the recycling facility; u is the transportation unit cost (JPY/ton km); LC is the personnel and power unit expense per EOL waste (JPY/ton facility); *speed* shows the traveling speed of the truck(km/h); *time* shows a movable time(hour);*profit* is the selling price of REE(JPY/car).

Equation (27.2) maximizes the revenue consisting of sales profit, transportation cost, equipment cost, and labor cost using discount rate. Constraint (3) shows to ensure that each car dismantling facility belongs to exactly one recycling facility in t. Constraint (4) shows that when the equipment is introduced to the recycling facility in the first term, the equipment is introduced continuously from the next year. Constraint (5) is used to restrict the car dismantling facilities that should be assigned to a recycling facility. Constraint (6) limits the time that the driver can work on the track. Constraint (7) specifies the binary variables.

The future annual waste of an NdFeB-sintered magnet was estimated using the parameters of Weibull distribution and was assigned to m areas according to the population and car ownership at a particular prefecture in Japan based on the considered regional characteristics. The transportation cost of an HEV motor was estimated based on the distance from a car dismantling facility to a recycling facility. Each car dismantling facility was installed in each J area, and the recycling facilities were constructed at a particular I. The location of the recycling facilities and the allocation of car dismantling facilities were estimated using the previous model formula. Further, the transportation route were optimized by solving Eqs. (27.8)–(27.12).

$$\max \sum_{t=1}^{T} \frac{1}{(1+r)^t} \left\{ profit \times D_t - u \sum_{i=1}^{I} \sum_{j=1}^{J} c_{ij} x_{ijt} \times rt_{jt} \right\} \tag{27.8}$$

$$Subject\ to: \sum_{i=1}^{I} \sum_{j=1}^{J} x_{depot,jt} = 2m(t = 1, \ldots, T) \tag{27.9}$$

$$\sum_{i,j \in S}^{I} x_{ijt} \leq |S| - N(S) \tag{27.10}$$

$$\sum_{i=1}^{I} \sum_{j=1}^{J} x_{ijt} - \sum_{i=1}^{I} \sum_{j=1}^{J} x_{jit} = 0(\mathrm{t} = 1, .., \mathrm{T}) \tag{27.11}$$

$$x_{ijt} \in \{0, 1\} \tag{27.12}$$

where r is the discount rate; m is the number of tracks that can be used.; $x_{depo,j}$ is shows the location of the depot obtained from Eq. (27.2); S indicates a subset of the dismantler; $N(S)$ shows the number of trucks needed to transport the motor of the dismantler in S. It is necessary to solve the bin packing problem to calculate $N(S)$. Therefore, it use the *cap* for the loading capacity of the truck.

Equation (27.8) maximizes the revenue consisting of sales profit and transportation cost using discount rate. Equation (27.9) shows that m trucks can leave only from the recycling facility determined by Eq. (27.2). Constraint (10) prohibits the truck's capacity and the partial circulation route. Constraint (11) shows that the number of trucks passing through recycling facilities and dismantling facilities. Constraint (12) specifies the binary variables.

27.4 Result and Discussion

27.4.1 Experiment Condition

The analysis period be from 2020 to 2040, and decide the optimum location of the recycling facilities by solving location routing plan. Also, the candidate site for dismantling facilities J and recycling facilities I will be the prefecture of Japan ($I = J = 46$).

However, Okinawa Prefecture is excluded on the ground road connection. The other parameters used in experiments are shown in Table 27.1 shows.

In this research, since the number of vehicles owned by region is considered to be influenced by lifestyle, it is assumed that discarded vehicles are proportional to the number of vehicles owned by prefectures. That is, it is considered that the amount of motor recovery will be generated according to the car ownership rate for each prefecture. P (2020) to $P(2035)$ are the results of the waste vehicle estimated from the future prediction.

Table 27.1 Parameters

Parameter	Value
N: Rate of permanent magnet	0.42
P(2020): Annual motor at 2020	317,382
P(2025): Annual motor at 2025	874,072
P(2030): Annual motor at 2030	1,453,807
P(2035): Annual motor at 2035	1,836,891
cap: Loading capacity	2
u: Transportation unit cost	14.6
F_A: Equipment cost	1,703,492,156
r: Discount rate	0.03
Speed: Traveling speed	60
Time: Movable time	8
LC: Personnel and power unit expense	1,689
Profit: Selling price of REEs	13,710

27.4.2 Result of Scenario Analyzes

In this section explored what policies are effective for implementing a sustainable recycling business. Figure 27.2 shows transport costs, profits, equipment costs, and revenues when the recovery rate is 100% when the recycling business is started in 2020. Equipment costs are assumed to be depreciated over 10 years and are shown as expenses per year. In the future, it is expected that sales of EVs and HEVs will increase, so sales profit will increase year by year. On the other hand, since the number of discarded vehicles also increases, the transportation cost for collecting waste products also increases. Therefore, the profit increased year by year as the profit exceeded the cost of transportation. However, since the revenues will exceed the cost of facilities in 2025, from 2020 to 2025, it will not be realistic because the business will be conducted with the final revenue in a negative state. For the reason, in this study, we constructed and analyzed two scenarios for introducing a sustainable recycling business.

Scenario 1: Introduction of subsidy

In this scenario, we will examine the effects of government and other subsidies on the recycling of scrap vehicles during periods of negative revenue.

As shown in Fig. 27.2, the revenue in 2020 is −2.4 billion yen and converted to −22,000 yen per waste vehicle, so if subsidies of 22,000 yen can be obtained per waste vehicle, the recycling business will start in 2020 It is possible. In addition, if the start time is changed to 2022, the profit in 2022 will be −2.1 billion yen, so it will be −11,000 yen when converted to one discarded automobile. Furthermore, if you change the start time to 2025, no negative earnings will appear. for the reason, in the current stage of Japan, even if the recycling business is immediately started, sufficient profits cannot be obtained, resulting in the outflow of waste products overseas. However,

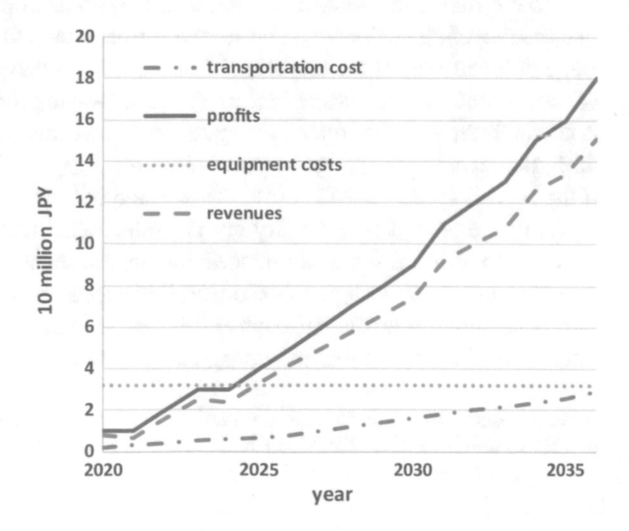

Fig. 27.2 Revenue and each cost in recycling rate 100%

we expect to be profitable in the future recycling business, as it will be positive in about 5 years after the introduction of subsidies, and the increase in profits due to the spread of EVs and HEVs.

Scenario 2: Improvement of recycling rate

In this scenario, we examine the impact of the waste recovery rate on the revenue to determine the recycling rate needed to get a positive return.

We investigated the impact of waste collection rate on profitability when the recycling business was started in 2020. When the recovery rate of waste products was 100%, positive returns occurred after 5 years from 2020, but when the recovery rate was 40%, positive returns were not achieved within 10 years. In Japan, the recovery rate of units containing rare metals is said to be 30–40%, so no recovery can be expected within 10 years with the current recovery rate. On the other hand, positive income is expected within 10 years when the recovery rate reaches 60% or more, so in order to build a sustainable recycling business, it is necessary to take measures to increase the recovery rate of waste products.

27.5 Conclusion

In this study, we examined recovering and regenerating rare earth from next-generation vehicles (EV and HEV), and aimed at performing economic efficiency evaluation with the risk of the new recycling business. We set two scenarios for introducing a sustainable recycling business and evaluated the risk and economic efficiency of each scenario. As a result, in Scenario 1 considering the introduction of subsidies, when starting the REE recycling business in 2020, it was possible to start the recycling business if subsidies of 22,000 yen could be obtained. Furthermore, it was found that if the start date is set to 2025, no negative revenue will appear and that the recycling business can be started immediately. In other words, in the current stage of Japan, even if the recycling business is immediately started, sufficient profits cannot be obtained, resulting in the outflow of waste products overseas. In Scenario 2 in which the effect of the recovery rate was investigated, if the recycling rate was 40%, the revenue would not be positive within 10 years. However, it was found that if the recycling rate exceeds 60%, the revenue will be positive within 10 years.

As future task, it is necessary not to collect REE from next-generation vehicles only, but to estimate home appliances that are the second largest application of REE permanent magnets in Japan. Moreover, it is necessary to develop a model that takes into consideration tariffs, etc., since this research does not take into consideration international recycling that may progress in the future.

Acknowledgements This research was partially supported by the Japan Society for the Promotion of Science (JSPS), KAKENHI, Grant-in-Aid for Scientific Research (C), 16K01262 from 2015.

References

Amin SH, Zhang G (2013) A multi-objective facility location model for closed-loop supply chain network under uncertain demand and return. Appl Math Model 37:4165–4176

Aras N, Korugan A, Büyükozkan G, Serifoglu FS, Erol I, Velioglu MN (2015) Locating recycling facilities for IT-based electronic waste in Turkey. J Clean Prod 105:324–336

Ballou RH (2001) Unresolved issues in supply chain network design. Inf Syst Front 3(4):417–426

Barreto S, Ferreira C, Paixão J, Santos BS (2007) Using clustering analysis in a capacitated location-routing problem. Eur J Oper Res 179:968–977

Berger RT, Coullard CR, Daskin MS (2007) Location-routing problems with distance constraints. Transport. Sci. 41(1):29–43

Chung SH, Kwon C (2015) Multi-period planning for electric car charging station locations: a case of Korean Expressways. Eur J Oper Res 242:677–687

Dat LQ, Linh DTT, Chou SY, Yu VF (2012) Optimizing reverse logistic costs for recycling end-of-life electrical and electronic product. Expert Syst Appl 39:6380–6387

Du X (2011) Global rare earth in-use stocks in NdFeB permanent magnets. Indus Ecol 15(6):836–843

Golev A, Scott M, Erskine PD, Ali SH, Ballantyne GR (2016) Rare earths supply chains: current status, constraints and opportunities. Res Policy 41:52–59

Intergovernmental Panel on Climate Change (2007) The Special Report on Emissions Scenarios (Online). Available at: https://www.ipcc.ch/pdf/special-reports/emissions_scenarios.pdf. Accessed on 20 October 2017

Kovačić D, Bogataj M (2013) Reverse logistics facility location using cyclical model of extended MRP theory. Central European J Oper Res 21(1):41–57

Mark Seddon, Argus Media Ltd (2016) Analysing the changing rare earths supply and demand outlook, 4 August

Min H, Jayaraman V, Srivastava R (1998) Combined location-routing problems: a research directions synthesis and future. Eur J Oper Res 108:1–15

Rai B, Singh N (2006) Customer-rush near warranty expiration limit and nonparametric hazard rate estimation from known mileage accumulation rates. IEEE Trans Reliab 55(3):480–489

Schneider EA, Deinert MR, Cady KB (2009) Cost analysis of the US spent nuclear fuel reprocessing facility. Energy Econ 31(5):627–634

Tang X, Lehuédé F, Péton O, Pan L (2019) Network design of a multi-period collaborative distribution system. Int J Mach Learn Cybernet 10(2):279–290

Yang Y, Walton A, Sheridan R, Güth K, Gauß R, Gutfleisch O, Buchert M, Steenari B-M, Van Gerven T, Jones PT, Binnemans K (2016) REE recovery from end-of-life NdFeB permanent magnet scrap: a critical review. In: Proceedings of journal of sustainable metallurgy 2017 3(1): 122–149

Yu VF, Lin SW, Lee W, Ting CJ (2010) A simulated annealing heuristic for the capacitated location routing problem. Comput Ind Eng 58:288–299

Zhou B, Li Z, Chen C (2017) Global potential of rare earth resources and rare earth demand from clean technologies. Procedia Minerals 7(11):203

Part IV
Green Technologies

Chapter 28
Green Power Generations and Environmental Monitoring Systems Based on the Dielectric Elastomer Transducer

Seiki Chiba, Mikio Waki, Makoto Takeshita, Mitsugu Uejima, Kohei Arakawa, Koji Ono, Yoshihiro Takikawa, Ryu Hatano, and Shoma Tanaka

Abstract Recently, actuators and sensors using dielectric elastomers (DEs) have attracted attention because they are lightweight, low cost, and efficient. Electric power can also be generated without producing carbon dioxide emissions or using rare earth materials. DE generators (DEGs) can generate electricity from various renewable energy sources. DEGs can actively introduce DE power generation systems tailored to areas such as solar power, wind power, water power, and wave power. As a result, it will reduce the large amount of CO_2 generated by conventional power plants and contribute to the improvement of the global environment. In order to build a sustainable social infrastructure in an energy recycling society, methods of using a DE converter system have been studied. In order to succeed commercially, it is necessary to take advantage of the benefits of DE compared with conventional technology.

Keywords Dielectric elastomer · Power-generator · Sensor · High efficiency actuator · Soft transducer · Renewable energy

28.1 Introduction

The energy required in 2040 will be about 40 TWh (about 1.6 times required in 2015) worldwide, and consumption in the Asian region is estimated to comprise about 68% of use (Agency for National Resources and Energy in Japan 2018; REN21

S. Chiba (✉)
Chiba Science Institute, Yagumo, Meguro Ward, Chiba Tokyo,, Japan
e-mail: epam@hyperdrive-web.com

M. Waki
Wits Inc, Oshiage, Sakura, Tochiji, Japan

M. Takeshita · M. Uejima · K. Arakawa
Zeon Corporation, Marunouchi, Chiyoda-Ku, Tokyo, Japan

K. Ono · Y. Takikawa · R. Hatano · S. Tanaka
Aisin AW Co., LTD, Takane, Fujii-Cho, Anjo, Aichi, Japan

Y. Kishita et al. (eds.), *EcoDesign and Sustainability I*, Sustainable Production, Life Cycle Engineering and Management, https://doi.org/10.1007/978-981-15-6779-7_28

2018). Furthermore, due to problems such as the exhaustion of resources and global warming, it is necessary to rapidly expand the proportional use of renewable energy. Currently, its proportion of the world's total electricity generation is only 22%.

In the very near future, the population of the earth will reach 10 billion and the power consumption will be massive. If, for example, a population of 10 billion attempts to maintain the plentiful lifestyle that many enjoy today with frequent consumption of beef, the number of cattle required to sustain that will cause a huge spike in greenhouse gases with their exhalations. We could solve this by lab-grown cultured meat, but that necessitates a great deal of energy. Other advancements such as the growing use of electronic currency will incur greater energy demands and more currency networks link up to monitor one another. In short, humanity is in dire need of highly efficient and renewable power sources.

Renewable energy comes in a myriad of possible forms: photovoltaic power generation, solar thermal utilization, wind power generation, biomass utilization power generation, geothermal power generation, snow and ice thermal power generation, and wave power generation. However, since the power generation cost of common systems remains very high, research and development of more effective materials and those of peripheral equipment continues (Liu et al. 2011; Krajacˇic´ et al. 2011; Rourke et al. 2009; Khan et al. 2009; Segurado et al. 2011).

The dielectric elastomers (DEs) are very efficient as actuators and can be used as accurate sensors and also capable of high efficiency power generation. They have received a lot of attention as a means to help address the growing dangers our world faces. This paper discusses how to construct global smart cities, efficient transportation, environmental monitoring systems, etc., using the DE transducer.

28.2 Back Ground of DE

DEs are a relatively new transducer technology that was first investigated by Pelrine, Chiba et al. in SRI in 1991(Pelrine and Chiba 1992). DEs use rubberlike polymers (elastomers) as actuator materials (Pelrine and Chiba 1992; Pelrine et al. 1999). The basic element of DE is a very simple structure comprised of a thin elastomer sandwiched by soft electrodes, as shown Fig. 28.1. When a voltage difference is applied between the electrodes, they are attracted to each other by Coulomb forces leading to a thickness-wise contraction and plane-wise expansion of the elastomer.

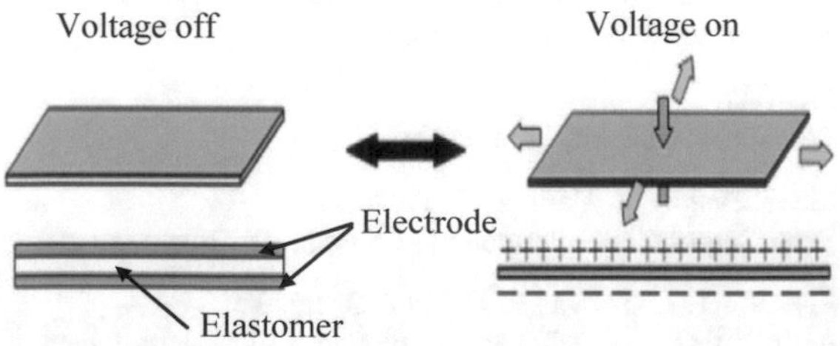

Fig. 28.1 Basic operational principle of DEs

Fig. 28.2 DE actuator having only 0.15 g of DE can lift a weight of 4 kg easily using SWCNTs

At the material level, DE actuator has a fast speed of response (over 100 kHz), with a high strain rate (up to 630%), high pressure (up to 8 MPa), and power density of 1 W/g (Chiba 2016). A DE actuator having only 0.15 g of DE can lift the weight of 4 kg easily using Single wall carbon nano tubes (SWCNT) (Chiba 2019a) (see Fig. 28.2). Consequently, a great number of researchers have been studying it (Pelrine and Chiba 1992; Pelrine et al. 1999; Chiba 2016; Chiba 2019a; Chiba et al. 2012; Pei 2018; Anderson et al. 2012; Zhou et al. 2016; Yuan et al. 2016; Lin et al. 2009).

The current of the DE is very small with respect to the voltage. As shown Fig. 28.3, the electric current was only 0.3 μA at 2400 V (Chiba 2016). A curve tracer was used for the measurement. This proves the actuator may achieve a highly efficient transduction from electric energy into mechanical energy.

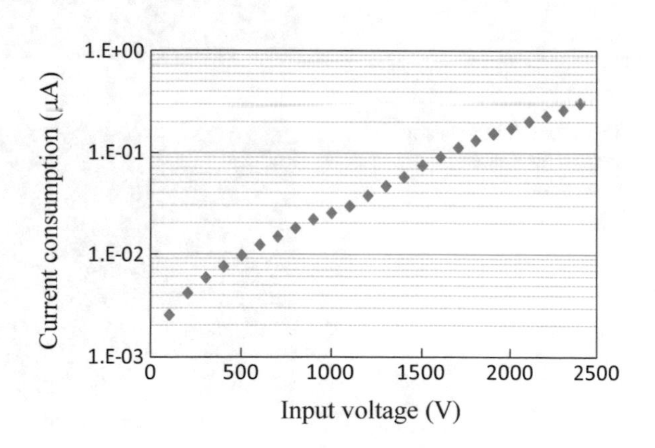

Fig. 28.3 Relationship between electric current and voltage in the DE actuators

It is now possible to make transparent DE actuators using SWCNT electrodes (Chiba et al. 2012; Chiba 2016). The integrity of the actuator is not broken even with a needle (see Fig. 28.4).

Various types of medical and industrial sensors are under development (Chiba 2019c). Examples of medical sensors are shown in Figs. 28.5 and 28.6. Examples of industrial sensors are described in Sect. 28.4.

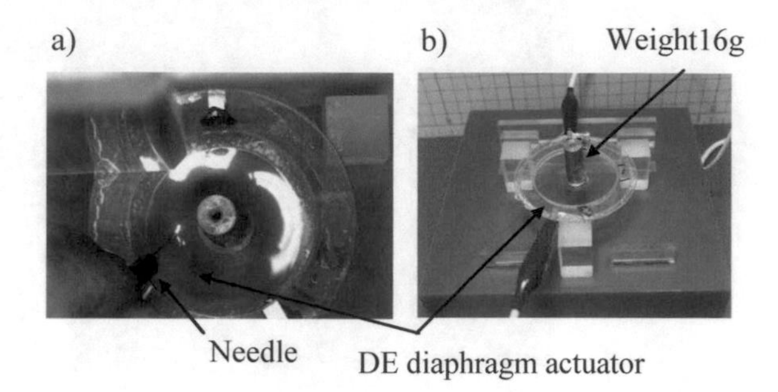

Fig. 28.4 A picture of a transparent diaphragm DE **a** Photo of needle stuck to a diaphragm DE on a transparent electrode, **b** Transparent DE diaphragm having a diameter of 4 cm and weight of 16 g

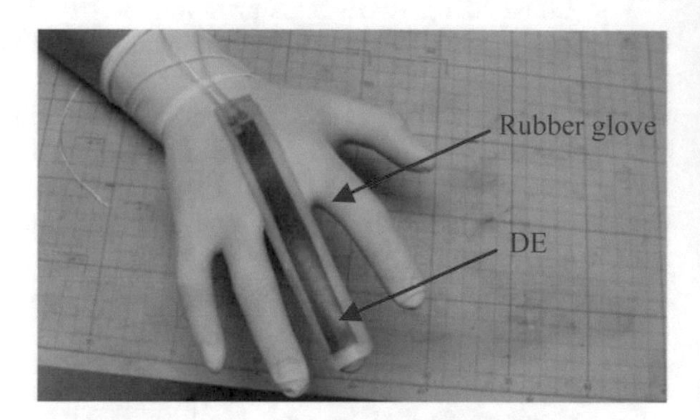

Fig. 28.5 Ribbon form actuators for rehabilitation purposes

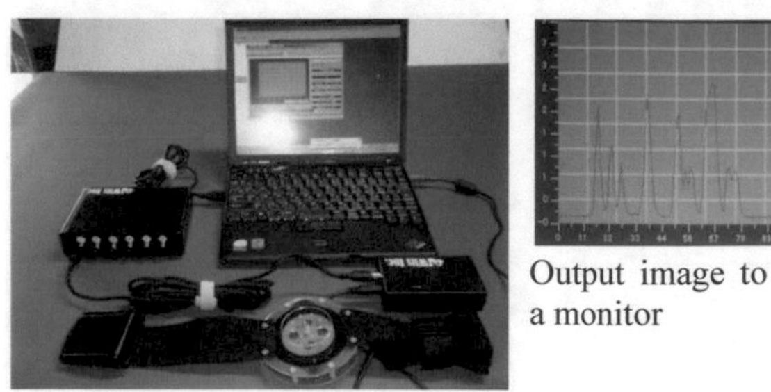

Fig. 28.6 A DE sensor system for dementia prevention

The actuator, shown in Fig. 28.5, can assist human motion. At the same time, it can work as a motion feedback sensor. Such equipment may be used to evaluate a recovery situation precisely. The DE was made from an acrylic elastomer with a carbon black electrode.

Figure 28.6 shows the gauging of the timing and depth of breathing with a PC using the breathing sensor developed in this study. A study into utilizing this sensor system for dementia prevention is underway at some Japanese universities. Additional applications to sports training, mental training and medical treatment are also possible.

More recently, the use of DEs in the reverse mode, in which the deformation of the elastomer by external mechanical work is used to generate electrical energy, has been gaining more attention. Since power generators using DEs do not depend on the frequency, motion and vibration of the low frequency band, which is difficult to utilize for electric power production, they can be easily adapted and their demand is growing considerably (Lin et al. 2009; Chiba 2019; Kovacs 2018; Koh et al. 2009; Huang and Shian 2013; Brouchu et al. 2010; Vertechy et al. 2015; Bortot and Gei 2015; Moretti et al. 2015; Yuan 2009; McKay et al. 2011; Anderson et al. 2012; Van Kessel et al. 2015; Moretti et al. 2019; Arena 2018; Chiba et al. 2013, 2008, 2017a, b, c; Chiba et al. 2007b, 2008; Waki and Chiba 2017).

28.3 Promotion of Local Generation of Dc Electric Power and Its Networking and Coupling with de Generation in the Ocean

In this section, we would like to consider two issues: (1) collaboration between locally produced DE power generation systems and large DE power generation systems in the ocean and (2) efficient consumption of that energy within smart cities with well-arranged smart grids.

28.3.1 Local Generation for Local Consumption Using dE Power Generation

As mentioned above, DC electricity is produced by local DE generators and then is consumed in nearby factories, offices, homes, etc. Surplus electric power is used to produce hydrogen, which is stored and then transported to areas where electric power is needed, and is converted back to DC electricity through fuel cells. An ultra-large DE power plant could also be built in the ocean, and could convert electricity to hydrogen, which could then be carried by tanker and distributed to various places through a hydrogen transport chain on the ground.

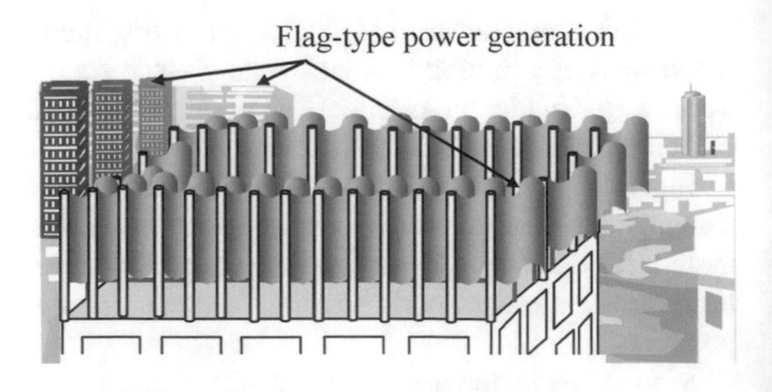

Fig. 28.7 DE wind power system installed on the roof of a building

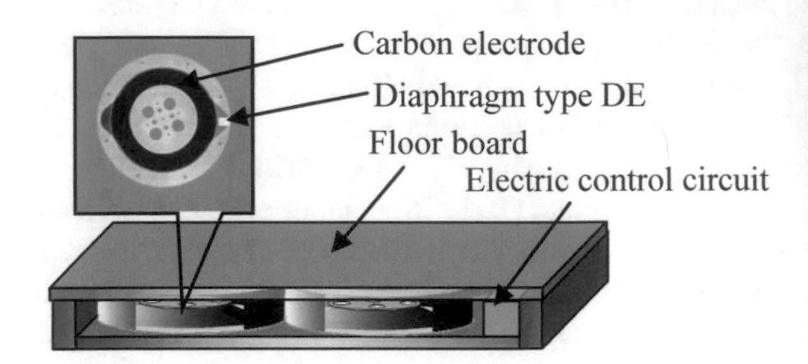

Fig. 28.8 Illustration of DE power generation system and 8 cm DE diaphragm cartridge

For green buildings, more wind power can be generated on the roofs of buildings because of their height, making them optimal for DE wind power generation, as shown in Fig. 28.7. The DE generator is suitable for installation on the roof of a building because it is freely selectable in shape and very light in weight. In this case, by using transparent DEs, it is possible to install a wind power generator without damaging the landscape around it.

We are also researching a combined system of existing wind turbines and DEs. Existing wind generators cannot turn their rotors in the direction of the wind by themselves. Therefore, the direction of the rotor can be changed by the DE actuator. In addition, in order for existing wind power generators to generate electricity efficiently, winds of 8 m/s or more are required. Since the DE can generate the power efficiently even at relatively low winds, this power can be used to drive a DE motor and rotate existing wind power rotors to generate power more efficiently with slower wind.

Moreover, we can get energy recovery from vibration and/or heat accumulation on roads or building floors, waste heat and wastewater from buildings.

As a system to generate electricity from foot traffic, an experiment using a piezoelectric device with a size of 4 m $\times$ 4 m was carried out at the Tokyo station by JR East Japan in 2006. However, it was unfortunately not an efficient system at all (Chiba 2019b). The DE system is considerably more efficient than the piezoelectric system (Yuan et al. 2016; Chiba 2008; Ashida et al. 2000; Zurkinden et al. 2007; Jean-Mistral et al. 2010). In our latest experiments, the power generation was employed using four DE diaphragm cartridges (0.35 g) having diameters of 8 cm. This system

(0.5 m × 0.6 m) can generate about 7 W/day walking at 1 step/1 s (see Fig. 28.8). This value is 150–300 times superior to piezoelectric system.

Figure 28.8 shows an illustration of the DE power generation system and 8 cm DE diaphragm cartridge. Although this principle proof device used a diaphragm type DE cartridge, in the near future, simply enlarging a very thin DE sheet makes it possible to obtain larger amounts of energy easily and efficiently.

The power can be generated from the movements of people, animals, and vehicles using simple devices. By way of another example, attaching the DE power generation unit to the sole and walking, the DE is distorted to generate electricity. There is a difference in power generation capacity due to weight and walking speed, etc. When an adult male walked one step second per second, about 1 W of power was obtained from one shoe (see Fig. 28.9) (Chiba 2007a). In order to increase the generated power, a grid having conical holes is placed on the DE film to increase the amount of deformation of the film (Fig. 28.10).

A more powerful floor generator can be easily realized by arranging the grid type power generation elements described above on the floor (Waki 2017). A principle proof device in which 24 elements are arranged is shown in Fig. 28.6. The energy of 1 W can be obtained by pushing the grid element once per second. When a person walks one step per second, he steps on four elements at a time, producing about 4 W of energy.

Although this proof-of-principle device utilizes a grid type, it will be possible in the near future to efficiently obtain large amounts of energy simply by laying a thin DE sheet like a carpet also.

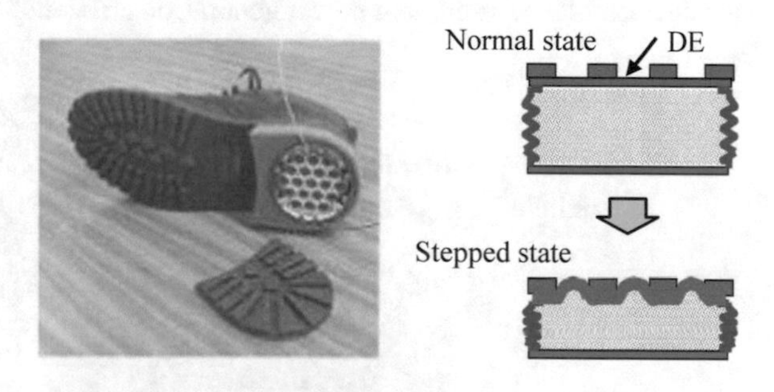

Fig. 28.9 DE shoe power generation

Fig. 28.10 Floor power generation device for proof of principle (Chiba and Waki 2011)

Researchers have advocated for DE floor power generation for more than 20 years, but since DEs also act as a speaker and a sensor, it is usually utilized as a floor speaker. However, if the noise is undesirable, it is possible to reverse the vibrator function and cut off external noise (Chiba 2007a). Users can use the sensor function to take measures against theft while still being able to use the power generation function.

Moreover, it is possible to thin the film of the DE, attach it to a body such as a person or an animal, and generate power by those movements. An artificial film with a width of 5 cm and a length of 25 cm (about 0.5 g), and one rotation (Stretch and contract) per second can take about 55 mJ of energy. As mentioned above, a thin rubbery DE generating sheet can be laid under roads (including passageways of buildings) and can generate power by using the vibration and pressure changes due to vehicle and foot traffic.

It can generate electric power by using solar heat in the form of heat accumulated on the surfaces of buildings or roads in the summer. Also, it is possible to generate electricity similarly with waste heat and drainage. We consider a new type of solar heat generator based on DE which has a simple structure compared with existing solar power generators (Chiba 2016). Figure 28.11 shows a DE solar heat generator. When the air in front of the piston is heated by solar heat, its volume increases. The piston pushes a DE generator (DEG).

2.8 g of acrylic elastomers and carbon black electrodes were used in a DE cartridge. It has a maximum power generation capability of 46.58 mJ per stroke. We used 30 cartridges. When a voltage of 3000 V is applied to a DE as a primary charge, the maximum energy generated is 1.4 J from 5 mm-stretched to relaxed states.

Differences in temperature, frequency, the amount of solar radiation input, the production of electricity and power generation efficiency were tested. The maximum

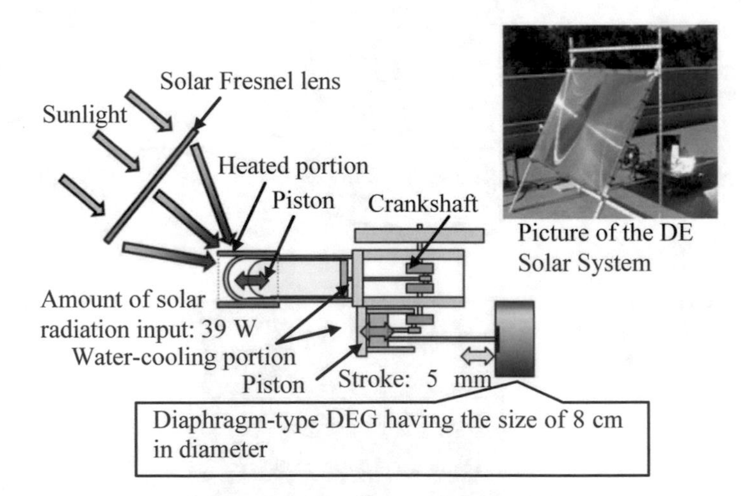

Fig. 28.11 DE Solar Heat Generator: Fresnel lens having the effective size of 1000 mm × 1350 mm and thickness of 3 mm. The capacity of the tube was about 44 cm^{3}. The piston stroke was 24 mm, and the diameter of the piston was 25 mm

production of electricity was 2.7 W at 12.2 Hz with a temperature difference of 674 K. The power generation efficiency, however, is very low.

Since it turns out that a thermal loss becomes large when the temperature of a heating head becomes high, to improve power generation efficiency, it is necessary to generate at low temperatures.

Substances evaporated below 40 °C that are used in binary power generation (geothermal power generation: e.g., power generation using Bentan: boiling point 36 °C., Ammonia: lower temperature) enables power generation at a considerably lower temperature. When such a device is attached to the front of a parked car in summer, the heat of sunlight is taken away to generate electricity that can keep the car interior cool.

Moreover, by using transparent DEs, there is no more discomfort and power can be generated by using the temperature difference between the human body/animal and the external temperature, enabling the recovery of some of the electricity used. Within the lab, we are developing flexible DE generators. It is becoming possible to make a flexible generator attached to clothes or tree leaves.

28.3.2 *Efficient Consumption of Energy Within Smart Cities*

Smart grids are not only efficient for the operation of power plants and power grids, but also for the installation of power generation facilities in places such as homes, offices and factories where power can meet the needs of those users. This means that a decentralized DE energy supply system not relying solely on conventional large power plants has to be built. The point is that it is important to use the device that stores power in the middle (EV battery, battery for cogeneration or electric double layer capacitor using DE) as efficiently as possible.

Figure 28.12 shows sites where power generation using DEs is possible and conceptual rendering of the generation systems. Those are divided into two types: (1) DE power generation systems in which the obtained electric power is consumed locally and (2) ultra large DE generation systems installed in the ocean.

(1) DE power generation system in which obitained electric power cosumed in locally:
   - (a) Wind Power Generators on tops of buildings (Chiba et al. 2007b; Chiba and Waki 2011)
   - (b) Water Mill Generators (Chiba 2007a)
   - (c) Waste energy Generators (Chiba and Waki 2011)
   - (d) Drain Generators (Chiba and Waki 2011)
   - (e) Wind Power Generators for Personal Houses (Chiba and Waki 2011)
   - (f) Solar Heat Generators (Chiba 2016)
   - (g) Wave Generators (Chiba et al. 2008, 2013, 2016, 2017a, b)
   - (h) Water Flow Generators (Chiba et al. 2017c)

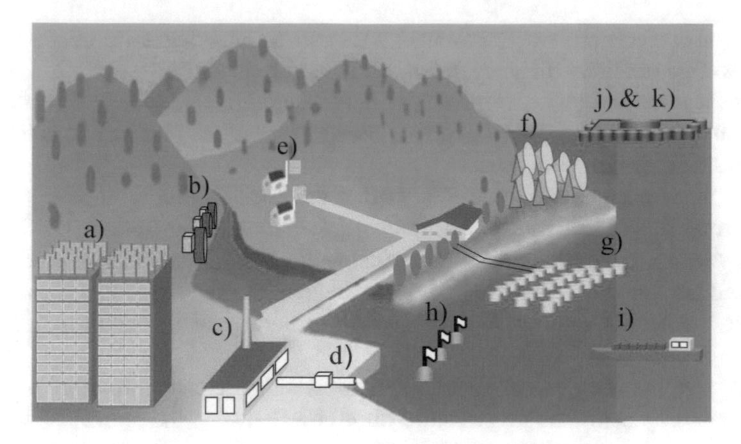

Fig. 28.12 Example sites where DE generator systems could be installed

(i) Tanker equipped with DE: It allows the ship to sway safely and always travel safely (Waki and Chiba 2017)

(2) Ultra large DE system which installed in the ocaen

(j) Wave Generators in Ocean (Chiba et al. 2008, 2017a)
(k) Hydrogen Production Plant (Chiba 2019c).

28.4 Promotion of Various Types of Green Technology

DE sensors are flexible, inexpensive, accurate, efficient and can be used anywhere, making them ideal. Since the sensor itself can generate electricity, batteries are unnecessary and it is possible to construct various kinds of global sensor systems.

In this section, we will briefly discuss the following items:

(1) The motors or engines of vehicles, trains, etc. could be replaced with DE motors. Existing factory robots could also be replaced with DE actuators.
(2) Global environmental measurement for accurate predictions using DE sensors
(3) DE sensors can also monitor the quality of water supplies and sewerage continuously, and can generate electricity with hydraulic power.

28.4.1 DE Motors

As mentioned in the section introducing the background of DEs, 0.15 g of DEs can lift a weight of up to 4 kg. Compared with the performance of dielectric elastomers, existing motors and actuators driven by air, dielectric elastomers have a very high output per unit weight, high driving speed, and very low power consumption (Waki

2017). In the near future, therefore, it is estimated to be at 400 horsepower at 100 g DE. In the case of current efficiency comparison of existing linear motor and DE linear motor (when comparing efficiency of 1 kg/cm (REN21 2018) at 1 g/cm (REN21 2018) of elastomer) DE linear motor is up to about 58 times more efficient when compared with a DC motor. When comparing the DE linear motor with the existing DC linear motor, the DE is about 123 times more efficient (Waki 2017). Even if the DE motor output is increased, the weight will eventually increase and will not change much. With DE vehicles using rubber (elastomers), it was no longer impossible to make them more efficient than existing car and train motors. In addition, when using an existing motor, if it moves quickly, heat builds up and movement is extremely slow, but the DE motor can be driven accurately regardless of the movement, and no heat is generated.

28.4.2 Global Environmental Measurement for Accurate Predictions Using dE Sensor

In recent years, global warming and accompanying abnormal weather patterns have begun to have an impact on our daily lives. To protect ourselves from the disasters brought about by abnormal weather, it is important to thoroughly understand the current situation, that is, how the global environment is changing. The monitoring of the global environment has been done by various countries individually, but to monitor environmental changes on a global scale, it will be necessary to build wide-ranging sensor networks. One of the major issues with that, however, is that there is no good method for obtaining electrical energy for running this system. Presently, many if not most of these sensor systems are powered by solar batteries. Yet in some locations and during some seasons, the daylight hours are extremely short, and in maritime and desert areas, salt and dust can dramatically reduce the electrical output. All this makes it difficult to maintain a stable sensor system. One way to solve this problem is a DE wave power generation system equipped with a DE sensor (Chiba 2019) (Fig. 28.13).

28.4.3 Monitoring the Quality of Water Supplies and Sewerage Using DE Sensors

Using DEG, electric energy can be obtained from the vibration caused by the flow of gas and water inside the pipe, and the electric power can be used to drive a DE sensor as well. Figure 28.11 shows a system in which a small DEG is attached to the inside of a water pipe to vibrate the DE by the flow of water to generate electricity (Chiba and Waki 2011). Although the power available from such a system is quite small, it

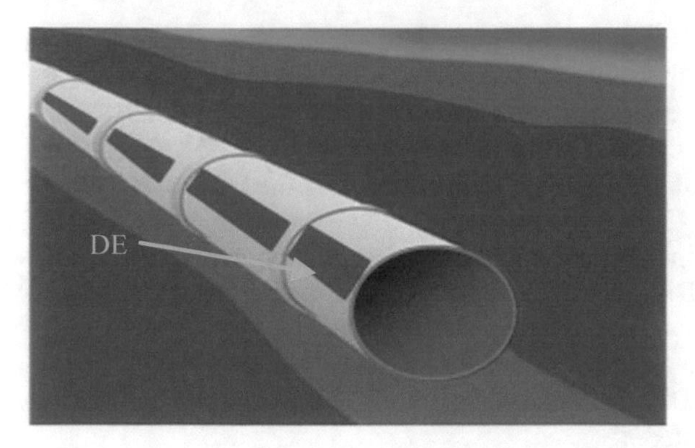

Fig. 28.13 Image of DE sensor and power generation system using water flow vibration (gallop movement): Inside the pipe, a DE sheet generator and sensors are installed (Chiba and Waki 2011)

is possible to obtain sufficient energy to operate the DE sensor, and it is possible to analyze water quality, flow, etc. even in places where power supply is difficult.

By using this system well, in the near future, energy can be recovered from the movement of the heart and blood, etc., and it can be used to drive artificial organs, internal sensors, micro analyzers and micro manufacturing equipment etc. (Chiba 2004).

28.5 Conclusion

In this paper, we described how to use DEs in order to build an eco-sustainable society by making good use of renewable energy and grids that distribute them well. A novel DE transducer could be applied to the two systems: (a) local generation of DC electric power and its networking and coupling with DE generation in the ocean, and (b) promotion of various types of green technology, such as DE motors for vehicles or trains, global environmental measurement, and monitoring of water quality and sewerage using DE sensors, etc.

We hope that those systems may help to counteract the issues associated with the various environmental challenges facing us today, such as global warming, over-population, or excess power consumption, to promote a better standard of living throughout the world.

References

Agency for National Resources and Energy in Japan, Energy report, March 26 2018

Anderson I et al (2012a) Multi-functional dielectric elastomer artificial muscles for soft and smart machines. J Appl Phys 112:04110

Anderson I, Gisby T, McKay T, O'Brienl B, Calius E (2012b) Multi-functional Dielectric Elastomer Artificial Muscles for Soft and Smart Machines. J Appl Phys 112(4):041101

Arena F et al (2018) Field experiments on dielectric elastomer generators integrated on A-OWC wave energy converter. In: Proceedings on OMAE 2018. https://doi.org/10.1115/OMAE2018-77830

Ashida K, Ichiki M, Tanaka M, Kitahara, T (2000) Power generation using Piezo element: energy conversion efficiency of Piezo Element. In: Proceedings of JAME annual meeting, pp 139–140

Bortot E, Gei M (2015) Harvesting Energy with Load-driven Dielectric Elastomer Annular Membranes Deforming Out-of-plane. Extreme Mech. Lett. 5:62–73

Brochu P, Yuan, W, Zhang H, Pei Q (2009) Dielectric Elastomers for Direct Wind-to-Electricity Power Generation, In: Proc. of ASME, Conference on Smart Materials, Adaptive Structures and Intelligent System 2009

Brouchu P. A., Li H, Niu X, Pei, Q. (2010) Factors Influencing the Performance of Dielectric Elastomer Energy Harvesters, In: Proc. SPIE, 7642, 7622 J

Chiba S et al (2004) Medical application of new electroactive polymer artificial muscles, Seikei-Kakou, vol.16, no.10

Chiba S et al (2007a) Extending applications of dielectric elastomer artificial muscle, In: Proceedings on SPIE, San Diego 18–22 Mar 2007

Chiba S et al (2007b) New opportunities in electric generation using Electroactive Polymer Artificial Muscles (EPAM). J Japan Instit Energy 86(9):743–747. https://doi.org/10.3775/jie.86.743

Chiba S et al (2008) Innovative power generators for energy harvesting using electroactive polymer artificial muscles, In: Bar Cohn Y (ed) Electroactive polymer actuators and devices (EAPAD) Proc. SPIE. Vol.6927, 692715 (1–9)

Chiba S et al (2013) Consistent ocean wave energy harvesting using electroactive polymer (dielectric elastomer) artificial muscle generators. Appl Energy vol 104, April 2013, Pages 497–502, ISSN 0306-2619

Chiba S et al (2016) Elastomer transducers. In: Advances in Science and Technology, Trans Tech Publication, Switzerland, vol. 97, pp 61–74. ISSN: 1662-0356. https://doi.org/10.4028/wwwsci enctific.net/AST.97.61

Chiba S et al (2017a) Simple and robust direct drive water power generation system using dielectric elastomers. J Mater Sci Eng B7(1–2), 39–47. https://doi.org/10.17265/2161-6213/2017.1-2.005

Chiba S et al (2017b) An experimental study on the motion of floating bodies arranged in series for wave power generation. J Mater Sci Eng A7(11–12): 281–289. https://doi.org/10.17265/2161-6213/2017.11-12.001

Chiba S, Hasegawa K, Waki M, Fujita K, Ohyama K, Zhu S (2017c) Innovative elastomer transducer driven by Karman Vortices in water flow. J Mater Sci Eng A7(5–6):121–135. https://doi.org/10.17265/2161-6213/2017.5-6.002

Chiba et al (2019a) Innovative power generations using dielectric elastomer artificial muscles Poster, Gren 2019, Dresden, Germany

Chiba S et al (2019b) Application to dielectric elastomer materials, power assist products, artificial muscle drive system. In: Next-generation polymer/polymer development, new application development and future prospects, Section 3, Chapter4, Technical Information Association, Japan, ISBN-10: 4861047382, ISBN-13: 978-4861047381

Chiba S et al (2019c) Recent progress on soft transducers for soft networks, In: A Hu et al (ed) Chapter 23, Technologies and eco-innovation towards sustainability II, Springer-Nature Singapore Pte Ltd, https//doi.org/https://doi.org/10.1007/978-981-13-1196-3_23

Chiba S, Waki M (2011) Power generation by micro/small vibration using dielectric elastomer. Functional Materials, 31(4):56–63, CMC, ISSN: 0286-4835

Chiba S, Waki M (2011b) Recent progress in dielectric elastomers (harvesting energy mode and high efficient actuation mode. Clean Tech, Nihon Kogyo Shuppan, Tokyo, Japan

Chiba S, Kornbluh R, Pelrine R, Waki M (2008) Low-cost Hydrogen Production from Electroactive Polymer Artificial Muscle Wave Power Generators, In: Proceedings of World Hydrogen Energy Conference 2008, Brisbane, Australia, 16–20 June 2008

Chiba S, Waki M, Asaka K, Suwa T, Wada T, Hirakawa Y (2012) Challenges for Loud Speakers Using Dielectric Elastomers (Invited), Abstract of IUMRS-ICEM 2012 C5-K125-002), C-5: Electroactive Polymer Actuators, Sensors and Energy Harvestors, Sept. 25-27, Yokokama, Japan

Huang J, Shian S (2013) Suo Z, Clarke D (2013) Maximizing the energy density of dielectric elastomer generators using Equi-Biaxial loading. Adv Func Mater 23:5056–5061

Jean-Mistral C. Basrour S. Chaillout J (2010) Comparison of electroactive polymer for energy scavenging applications, Smart Mater, Device, Model

Khan AMJ, Bhuyan G, Iqbal MT, Quaicoe JF (2009) Hydrokinetic energy conversion systems and assessment of horizontal and vertical axis turbines for river and tidal applications: a technology status review, Appl Energy 86:1823–35

Koh, S. J. A., Zhao, X., Suo, Z (2009) Maximal Energy That Can Be Converted by a Dielectric Elastomer Generator, Applied Physics Letters 2009: 94,26,262902-3.9

Kovacs GM (2018) Manufacturing polymer transducers: opportunities and challenges, In: Proc. of SPIE,10594–7

Krajacˇic´ G, Duic´ N, Carvalho MG (2011) How to achieve a 100% RES electricity supply for Portugal. Appl Energy 88:508–17

Lin, G., Chen, M., and Song, D. (2009) In: Proc. of the International Conference on Energy and Environment Technology (ICEET '09), 782–6

Liu AW, Lund H, Mathiesen BV, Zhang X (2011) Potential of renewable energy systems in China. Appl Energy 88:518–25

McKay T, O'Brien B, Calius E, Anderson I (2011) Soft Generators Using Dielectric Elastomers, Applied Physics Letters, 98 (14): 142903, 1–3

Moretti G, Fontana M, Vertechy R (2015) Parallelogram-shaped Dielectric Elastomer Generators: Analytical model and Experimental Validation. J Intell Mater Syst Struct 26(6):740–751

Moretti G, Pietro G, Papini R, Daniele l, Forehand D, Ingram D, Vertechy R, Fontana M (2019) Modeling and testing of a wave energy converter based on dielectric elastomer generators. InL Proceedinds on The Royal Society A, The Royal society publishing, https://doi.org/10.1098/rspa.2018.0566

Pei Q (2018) Dielectric Elastomers past, present and potential future, In: Proc. of SPIE,10594-4

Pelrine R, Chiba S (1992) Review of Artificial Muscle Approaches, In: Proceedings of third international symposium on micromachine and human science, Nagoya, Japan

Pelrine R, Kornbluh R, Chiba S et al. (1999) High-field defomation of elasomeric dielectrics for actuators, In: Proceedings of 6th SPIE Symposium on Smart Structure and Materials vol 3669, pp 149–161

REN21 (2018) Renewables 2018 Global Status Report, Paris, REN21 Secretariat 2018

Rourke F, Boyle F, Reynolds A (2009) Tidal energy update. Appl Energy 87:398–409

Segurado R, Krajacˇic' G, Duic´N, Alves L (2011), Increasing the penetration of renewable energy resources in S. Vicente, Cape Verde. Appl Energy 88:466–472

Van Kessel R, Wattez A, Bauer, P (2015) Analyses comparison of an energy harvesting system for dielectric elastomer generators using a passive harvesting concept: the voltage-clamped multi-phase system, In: Proc. SPIE Smart Structures and Materials + Nondestructive Evaluation and Health Monitoring, International Society for Optics and Photonics, 943006

Vertechy R, Papini Rosati GP, Fontana M (2015) Reduced Model and Application of Inflating Circular Diaphragm Dielectric Elastomer Generators for Wave Energy Harvesting. Journal of Vibration and Acoustics, Transactions of the ASME 137:11–16

Waki M (2017) Application of dielectric elastomer artificial muscle, In: Chapter 4, Development of soft actuator and application (Control Technologies for Practical Application), CMC. ISBN: 978-4-7813-1234-7

Waki M, Chiba S et al (2017) Development of Wave Generation Module for Small Ship Using Dielectric Elastomer. J Mater Sci Eng B-7(7-8):171–177. https://doi.org/10.17265-6221/2017.7-8.007

Yuan X, Changgeng S, Yan G, and Zhenghong Z (2016) Application review of dielectic electroactive polymers (DEAPs) and piezoelectric materials for vibration energy harvesting J. Physiscs: Conference series 744, 012077

Zhou J, Jiang L, and Khayat RE (2016) Dynamic analysis of a tunable viscoelastic dielectric elastomer oscillator under external excitation, Smart Material Structure, 25, pp 11, 025005

Zurkinden A, Campanile F, Martinelli L (2007) Wave energy converter through Piezoelectric Polymers, In: Proceedings on COMSOL User Conference 2007, Grenoble, France

Chapter 29
A Novel Approach to Artificial Energy-Loss Free Field Emitters: The Outstanding Cathodic Durability of High Crystallized Single-Walled Carbon Nanotubes

Norihiro Shimoi

Abstract We propose a novel approach to improve the durability and energy efficiency of field- emission (FE) devices with a simple structure employing high crystallized single-walled carbon nanotubes (SWCNTs) produced by high-temperature annealing as the cathode material. We succeeded in establishing the generality and efficacy underlying the electrical conductivity of a SWCNT by control of the crystallinity. We measured the FE durability and current fluctuation of the high crystallized SWCNTs, and found them to exhibit good stability for more than 1000 h with a high FE current density of 30 mA/cm^2 under an applied DC voltage. Moreover, we designed and successfully constructed elements of a planar electron emission source device composed of a cathode employing highly purified, crystalline SWCNTs dispersed homogeneously in organic solutions by a multi-step dispersion method. It was further confirmed that the field emitters with the high crystallized SWCNTs can be used in breakthrough applications with high loadings of over 3 A. Our new devices may have a significant impact as planar FE emission devices employing purified and high crystallized SWCNTs have a remarkable potential to indicate a new and excellent approach for establishing an artificial energy-loss free device.

Keywords High crystallization · Single-walled carbon nanotube · Low power consumption · Field emission · Planar lighting

29.1 Introduction

Since Rinzler et al. predicted the possibility of creating electrical devices using carbon nanotubes (CNTs) in a landmark paper (Rinzler et al. 1995), low dimensional carbon nano-materials—such as carbon nanofibers (CNFs) and CNTs—have been paid much attention for their introduction in electrical devices. Needle-shaped

---

N. Shimoi (✉)
Graduate School of Environmental Studies, Tohoku University, 6-6-20 Aoba, Aramaki, Aoba-ku, Sendai 980-8579, Japan
e-mail: norihiro.shimoi.c8@tohoku.ac.jp

Y. Kishita et al. (eds.), *EcoDesign and Sustainability I*, Sustainable Production, Life Cycle Engineering and Management, https://doi.org/10.1007/978-981-15-6779-7_29

CNTs especially possess excellent physicochemical properties, including high electrical and thermoconductivity, rigidness, and chemical stability. Single-walled carbon nanotubes (SWCNTs) can be grown to have metallic, semiconducting or semimetal characteristics by controlling a chirality of a tube rolling up a graphene sheet. The selection of the two types is available for developing and designing electronic devices (Hamada et al. 1992; Saito et al. 1992; Tanaka et al. 1992; Kim et al. 2002). SWCNTs, which express unique effects confined within low-dimensional structure, can be employed as quantum wires (Ebbesen et al. 1997; Wildoerm et al. 1998; Odom et al. 1998). Among the different types of CNTs, SWCNTs show particularly high Young's modulus (Treacy et al. 1996). SWCNTs have been used various applications, including electron sources as field emitters (Rinzler et al. 1995; Saito et al. 1998), probes of scanning probe microscopy (Dai et al. 1996), gas storage (Dillon et al. 1997), and electrically conductive materials for secondary batteries (Niu et al. 1997).

The crystal of SWCNTs as grown, including other CNTs, has some defects in their carbon networks, expressing their unstable physicochemical and electrical properties, and the crystal defects of CNTs have rendered their problem to use in electronic devices requiring stable reliability. The crystallization of untreated CNTs, mainly those grown by chemical vapor deposition, super growth, or laser ablation, has been paid no attention to develop in electronic devices; as it has proven that it was very difficult to handle CNTs artificially and manually. Also it was important that the control of their crystallinity and purity to obtain the good durability and stability in CNTs required for electron conductivity. Tohji et al. established purification methods of raw synthesized SWCNTs (Tohji et al. 1996) to obtain a perfect carbon network, and they improved the crystallinity of SWCNTs synthesized by arc discharge and succeeded in obtaining highly pure and crystalline SWCNTs by annealing them at over 1350 K and a low pressure under 10–5 Pa (Yamamoto et al. 2006). Together with Iwata et al., they succeeded to establish a method to analyze the crystallinity of highly crystalline SWCNTs by absorption of H_2 gas to their surface (Iwata et al. 2007). Based on these results, a process for synthesizing high crystallized CNTs, especially SWCNTs, has been gradually developed; however, a technical skill to employ high crystallized SWCNTs is yet unsatisfactory to use as assembly parts in an electrical device.

The application using high crystallized SWCNTs as field emitters is gradually promising and approaching practical utilization industrially. We have developed field emission (FE) devices employing high crystallized SWCNTs (Shimoi et al. 2013) and succeeded in employing FE electron sources with SWCNTs as the planar cathodic element of an artificial lighting device (Garrido et al. 2014). From these achievements, we believe that high crystallized SWCNTs will be necessary to decrease loading energy of FE devices, increasing their radioactive half-time independent of FE current density, and thereby correcting the FE electron emission homogeneity of planar artificial lighting devices (Garrido et al. 2014) with extremely low power consumption.

29.2 Experimental Implementation of SWCNTs

SWCNTs fabricated by arc discharge are suitable to synthesize high crystallized CNTs via an annealing process (Yamamoto et al. 2006). The SWCNTs were annealed at 1400 K in a low pressure under 10−5 Pa to obtain high crystallized and purified SWCNTs with technical conditions. The transmission electron microscopy (TEM; HR-3000, Hitachi High-Technologies Corporation) images of the SWCNT bundles (a) after and (b) before annealing as a reference, respectively, and the Raman-shift measurement (c) used to verify the high degree of crystallization of the SWCNTs are shown in Fig. 29.1. Un-annealed, merely purified SWCNTs were prepared previously as a comparative reference against the effects of crystallinity. A TEM image of individual SWCNTs in Fig. 29.1a is very sharp and clear with sidewalls of each tube shown as a straight line; however, a TEM image of individual SWCNTs with un-annealing process in Fig. 29.2b shows some crystal defects produced at the sidewalls of a SWCNTs bundle.

The crystallization of the treated SWCNTs were checked by Raman-shift measurements with a blue laser beam (Blue) at a 473 nm wavelength and 10 mW output power from a cobalt source, as shown in Fig. 29.1c. The ratio between the intensities of the G + band at 1590 cm−1 and the D band at 1351 cm−1 was 0.0096

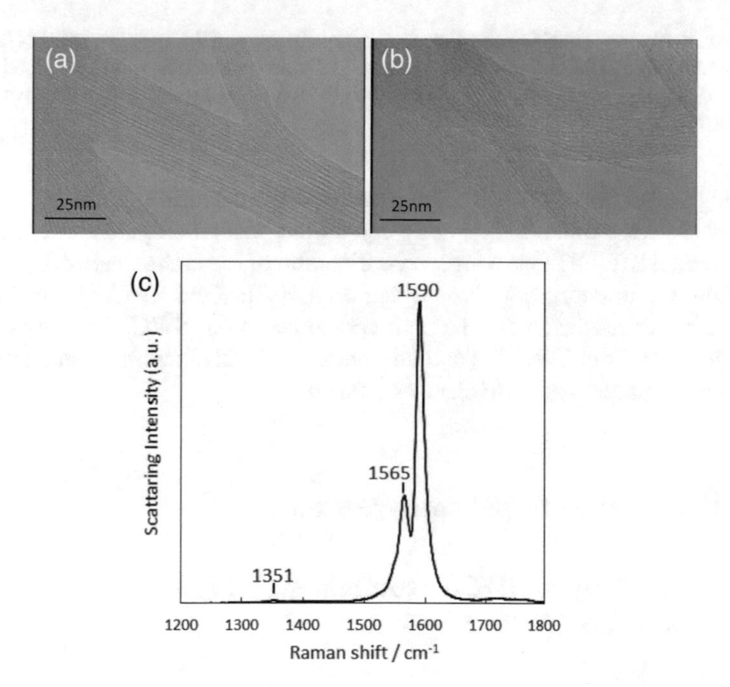

Fig. 29.1 The bundle images of treated and un-treated SWCNTs by TEM **a**, **b**, and a Raman spectrum of annealed SWCNTs **c**. **a** A bundle of highly crystalline SWCNTs with annealing treatment. **b** A bundle of SWCNTs having crystal defects. **c** Raman spectrum with G + and D band modes of high crystallized SWCNTs

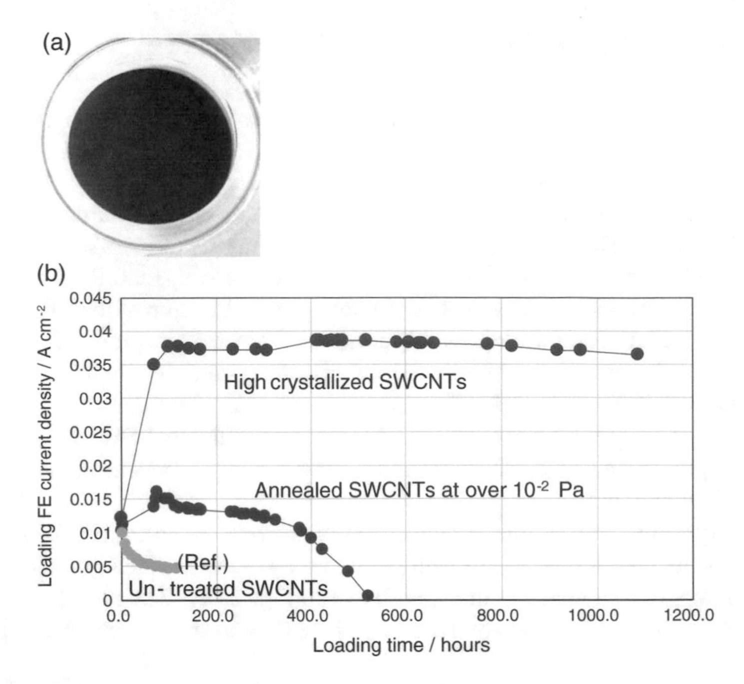

Fig. 29.2 **a** Buckypaper made by pressing a solution of dispersed SWCNTs or MWCNTs. **b** The radioactive half-time of FE current density up to 1000 h at an applied DC voltage. The loading time when employing high crystallized SWCNTs against the initial FE current density was over 1000 h without the attenuation of FE current

for the high crystallized SWCNTs after annealing studied in Fig. 29.1c. Although the SWCNTs were found to be highly crystalline, this result particularly confirms that the treated SWCNTs have high crystallization of the carbon network.

To show the importance of employing high crystallized SWCNTs in electrical devices, the increased effect of the high crystallization of SWCNTs was examined with high crystallized SWCNTs and un-annealed SWCNTs with crystal defects on their electrical properties as FE electron sources.

29.3 Measurements and compression

29.3.1 Durability of SWCNTs Loading Field Electron Emission Lifetime

The cathode electrodes prepared as Buckypaper made by pressing of solution-dispersed and filtered SWCNTs were used to measure FE lifetimes, as shown in Fig. 29.2. Fig. 29.2b shows the FE lifetimes up to 1000 h for high crystallized

SWCNTs by annealing at 1400 K and under 10–5 Pa, SWCNTs annealed at 1400 K and over 10–2 Pa, and un-annealed SWCNTs having crystal defects as a reference, respectively. The SWCNTs annealed at over 10–2 Pa, had more crystal defects in their carbon networks than SWCNTs annealed at 1400 K and under 10–5 Pa. After the treatment by annealing at a high temperature and vacuum, the SWCNTs synthesized by arcing had hardly crystal defects. These cathode electrodes were assembled into simple diode structures with conductive anodes and then a voltage of 580 V DC was applied, and they could obtain FE currents of 10 mA/ cm^2 as initial DC current in a vacuum of 10–4 Pa. The FE cathodes with the high crystallized SWCNTs and SWCNTs annealed at over 10–2 Pa were activated with FE loading, and FE current densities of each SWCNT increased to over 30 and 17 mA/ cm^2, respectively. The cause for this was assumed to arise from the cleaning of the surfaces of the SWCNTs' tips by flushing. This result expressed that the FE current density by the high crystallized SWCNTs was not attenuated at over 1000 h; otherwise, the radioactive half-time of SWCNTs annealed at over 10–2 Pa was 482 h, and the reference sample of un-annealed SWCNTs with crystal defects had a shorter emission lifetime of 105 h compared to those of the other SWCNTs. The control of the crystallinity of each SWCNT as field emitters is an effective improvement method for FE current durability.

Before verification of the FE lifetime, the IV characteristics for each SWCNT was initially measured, as shown in Fig. 29.3a. The turn-on field for the un-annealed SWCNTs was 2.01 V/μm. In contrast, the turn-on field for the high crystallized SWCNTs was 1.07 V/μm on the FE current density 0.01 mA/ cm^2. As shown in the lighting spot homogeneity of Fig. 29.3b, the probability densities of each SWCNT in the cathode films to induce FE resemble almost the SWCNTs with crystal defects and the highly crystalline SWCNTs.

The distribution of emission spots in the highly crystalline SWCNTs and un-treated SWCNTs are shown in Fig. 29.3b. The planar images on the right side in Fig. 29.3 show enlarged views indicated in the circled regions on the left zoom out images. We can find the FE curve properties in Fig. 29.3a depend on the crystallinity of each SWCNT. When the current density is uniform with the same supplied field, the densities of lighting spots having high- and low-brightness and the homogeneity of the planar lighting surface are almost similar between the two crystal types, i.e., highly crystalline SWCNTs and the un-treated SWCNTs with crystal defects. SWCNTs synthesized by arc discharge coexist with metallic and semiconductive properties from previous reports (Dresselhaus et al. 2005; Wu et al. 2011), and it is surmised that the high-brightness spots in Fig. 29.3b originate from the FE of semiconductive SWCNTs and the low-brightness spots originate from SWCNTs having metallic conduction. The chirality to roll up a graphene sheet comprising a SWCNT does not change, nor does the mixing ratio between SWCNTs with metallic properties and semiconductive SWCNTs with the increase of the crystallinity of SWCNTs (Wallace (1947); Saito et al. (1993); Dresselhaus et al. (1996); Spindt et al. (1976); Czerw et al. (2003)).

From the above-mentioned results, the SWCNTs' bundle size and the dispersion density of these bundles protruded from the cathode film were similar for the two

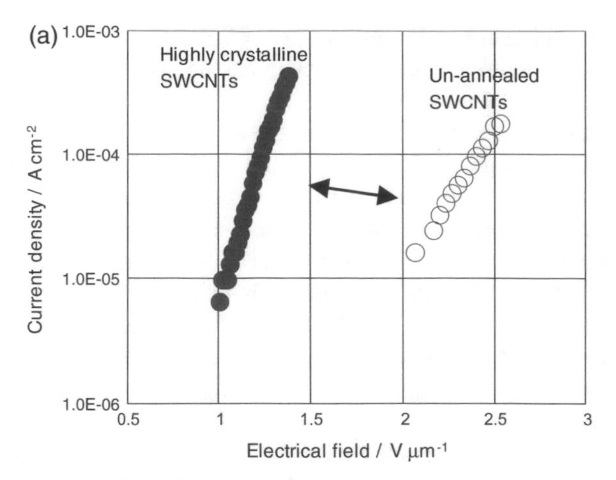

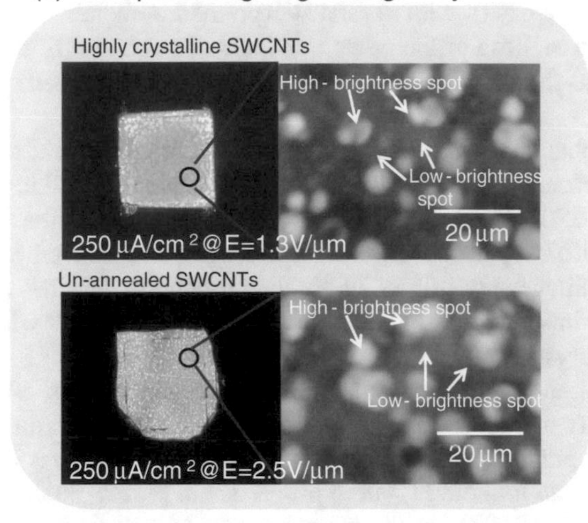

Figure 29.3 **a** FE characteristics of highly crystalline SWCNTs and SWCNTs with crystal defects. **b** Lighting spot homogeneity of highly crystalline SWCNTs and SWCNTs with crystal defects. The planar lighting images (left) and the distribution of lighting spots zoomed the left images (right)

types used as FE electron sources. However, the difference in their FE characteristics of the two types in Fig. 29.3a is impossible to explain by the morphology of SWCNTs dispersed in the cathode layer. We therefore guessed that the crystallinity of SWCNTs induces the difference in FE characteristics and electrical conductivity in SWCNTs. It was expressed that crystal defects in SWCNTs yield a local energy barrier that impedes the electrical conductivity of SWCNTs from impedance measurement by scanning tunneling microscopy (STM) (Czerw et al. 2003; Albrecht and Lyding 2003). Moreover, a conductive model of an electron passing through the inside of an SWCNT with crystal defects has been developed on the inelastic electron tunneling caused from energy barriers for the electrons existing in SWCNTs.

Spindt et al. have reported that the FE property follows the Fowler–Nordheim (F–N) tunneling model in quantum physics mechanism (Spindt et al. 1976). The enhancement of an electrical field concentrated on the top of a CNT tip depends on their shape. Electrons pass through a thin energy barrier by the electrical field enhanced on a CNT and they are emitted to the outside from the top of the CNTs. The FE equation obtained from the F–N tunneling model is following:

$$I = a * V^2 * \exp(-\frac{b}{V}) \tag{29.1}$$

where

$$a = \alpha * A * \beta^2 * \exp(\frac{B * 1.44E - 7}{\varphi^{0.5}})/(1.1 * \varphi) \tag{29.2}$$

$$b = \frac{0.95 * B * \varphi^{3/2}}{\beta} \tag{29.3}$$

$$A = 1.54E - 6, B = 6.87E + 7 \tag{29.4}$$

V is the applied voltage for driving FE current, and I is the FE current from the Eq. (29.1). Using the parameters of V and I, the electron emission site area α on the emitters, and the field enhancement factor β, which indicates the ratio between the intensity of the electrical field concentrated on the FE emitters (SWCNTs in this study) and the supplied electrical field, can be calculated. ϕ is substituted the work function of bulk carbon.

The FE current equation of Eq. (29.1) was originally modeled by elastic electron tunneling model. This model is valid only when electrons pass through in an FE emitter without the quantum–mechanical prevention, e.g. energy barriers. In this study, we constructed an FE model by combining the model of inelastic tunneling electrons passing through a SWCNT with crystal defects (Shimoi 2015).

The existence of ballistic electrons is speculated in a CNT having theoretical properties. However, when electrons pass through an SWCNT with crystal defects, the conductivity is impeded and energy barrier arising from crystal defects of a CNT is guessed to prevent the conduction of a CNT.

Therefore, the increased crystallinity of the SWCNTs can attribute to the improvement in the electrical properties observed above. A planar FE electron emitter using high crystallized SWCNTs will be expected to have high cathodic durability and high emission site homogeneity; it will also exhibit an FE long radioactive life-time enough to withstand practical use artificially as an FE device, as shown in Fig. 29.2.

29.3.2 Stability with High Current FE Property of SWCNTs

The utilization of high crystallized SWCNTs to construct an effective cathode element—with both a low FE fluctuation and stable FE current—relies on the ability to disperse them uniformly in liquid media. A solvent with In_2O_3-SnO_2 (tin-doped indium oxide; ITO) precursor as the conductive matrix material for the dispersion of high crystallized SWCNTs was used to construct a thin film containing the well-dispersed SWCNTs. The mixture of SWCNTs, the solvent including the ITO precursor and a non-ionic dispersant to disperse the SWCNTs was employed to provide an impact buffer product, following the stepwise dispersing and buffering of the large impact force generated by ultra-sonication using a homogenizer to receive the impact. After sintering a film coated with the mixture in a vacuum, the thin film was activated by scratching the film with a thin metal rod appropriately to obtain good FE properties. Figure 29.4 shows SEM images of a scratched ITO thin film containing the SWCNTs after sintering in a vacuum. This revealed the SWCNT bundles exposing on both sides of the edge in the nicks of the ITO film. SWCNT bundles protruding from the ITO film are indicated in white circles of Figs. 29.4b, c, and the SWCNT bundles in the ITO film were dispersed homogeneously and lay in random directions with protruding from the grooved face at even intervals.

Figure 29.5 expresses the FE current fluctuations and time-dependent thermal hysteresis on the both electrodes in a device with the cathode electrode employing high crystallized SWCNTs as field emitters with conductive anodes. The FE current from the planar diode applied a voltage of 2.75 kV DC and an FE current of 3.08 A was stable over 100 min, as shown in Fig. 29.5. In addition, the energy loss from both electrodes due to calorific heating under loading with a high FE current was measured by attaching thermocouples to each electrode, as shown in the inset of Fig. 29.5. The results indicated that the energy loss from the FE cathode with high crystallized SWCNTs was almost zero. On the other hand, the anode had a large heat loss; in fact, a method to reduce the energy loss from the anode electrode has not

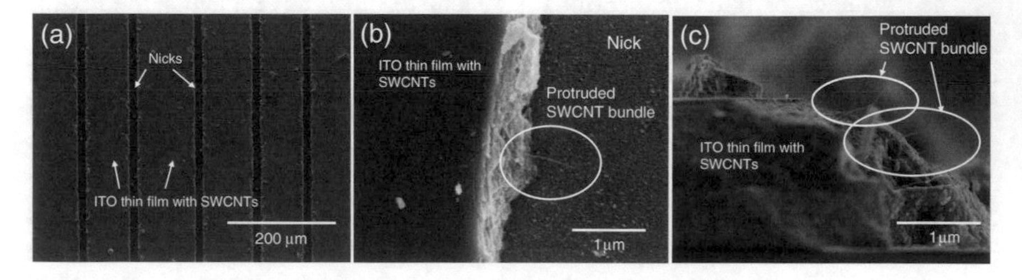

Fig. 29.4 SEM images of an ITO thin film with well-dispersed high crystallized SWCNTs. **a** Overview of a scratched ITO film. **b** Enlarged view of the grooved face of a scratched ITO film. A SWCNT bundle protruding from the wall of the ITO film is shown in a white circle. **c** Cross-sectional view of the edge of a scratched ITO film. The white circles indicate SWCNTs from a grooved face

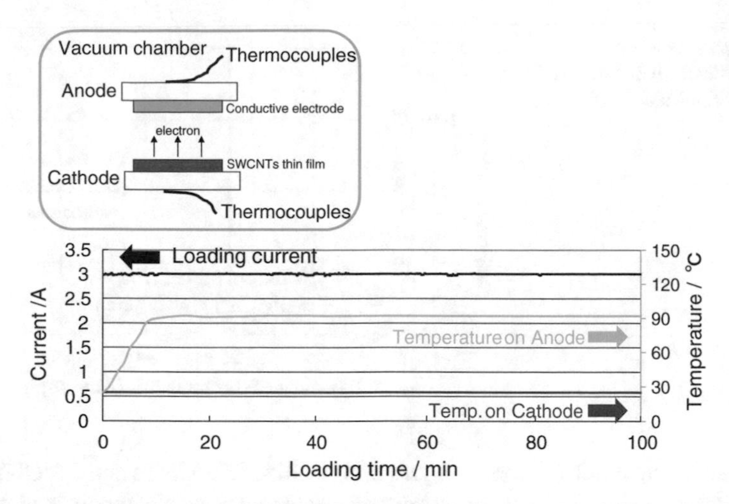

Fig. 29.5 Current fluctuations under 2.75 kV DC loading and time-dependent thermal changes over 100 min

yet been designed. The increased crystallinity of the SWCNTs prevented them from attenuating under a large FE current.

29.4 Conclusions and outlook

We have succeeded to apply high crystallized SWCNTs as planar FE sources for the first time ever. The homogeneous dispersion process with high crystallized SWCNTs is one of the essential elements required to fabricate electronic devices with a wet process. The developed thin films employing the SWCNTs as field emitters improve FE properties with low power consumption. However, conventional SWCNTs used in various electrical applications currently have carbon network defects that prevent the FE electron emission. From the above discussion for FE properties in this study, we examined the electric conductive in a SWCNT by using the inelastic electron tunneling model with energy barriers based on the crystal defects in a SWCNT.

Moreover, we have succeeded in obtaining a long FE radioactive life-time with a high loading FE current density by employing high crystallized SWCNTs. In this study, our FE devices did not even reach to their radioactive half-time during 1000 h with high FE current density, and we could ascertain the superior potential of high crystallized SWCNTs compared to other CNTs, as shown in Fig. 29.6 (Thuesen 2001; Bormashov et al. 2003; Xiang et al. 2005; Sheshin et al. 1999; Saito and Uemura 2000; Liu et al. 2007; Sung et al. 2008; Lee et al. 2008; Hu et al. 2010; Cho et al. 2007; Rao et al. 2000; Xu and Brandes 1999). Also, as the SWCNTs used as field emitters exhibited almost identical homogeneous dispersion states, their FE properties could be examined and directly compared in this study. We then

Fig. 29.6 Relationship between device half-life and loading FE current density

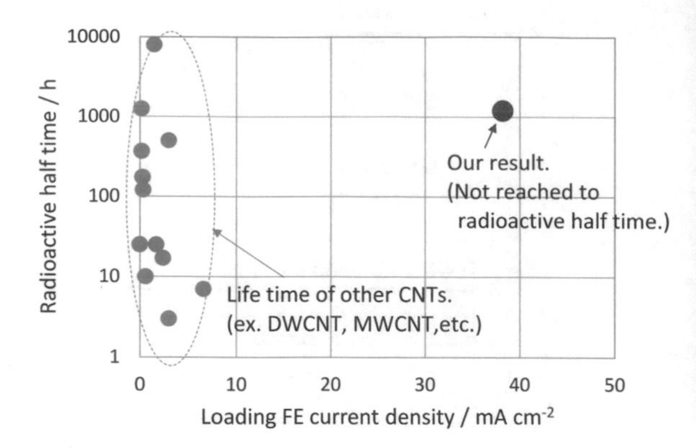

constructed field emitters employing high crystallized SWCNTs and SWCNTs with crystal defects using similar processing architectures for each cathode. A significant difference in the FE properties using each SWCNT depended on their crystallinity. We concluded that high crystallized SWCNTs can obtain a low turn-on and driving field for field emitters. It is speculated that the crystal defects in a SWCNT degrade the FE properties compared with those of high crystallized SWCNTs. The FE mechanism of SWCNTs with crystal defects can be explained by the inelastic tunneling model using energy barriers when electrons pass through the crystal defects in a SWCNT. In this study, we considered that a crystal defect in a SWCNT acts as a rectangular potential barrier to cause energy loss for FE (Shimoi 2015).

From the above-mentioned results, high crystallized SWCNTs is expected to exhibit almost zero energy loss from FE properties. We have given a brief explanation of the effect of the increased crystallinity of SWCNTs on their FE properties and then developed an electron flow model in a SWCNT. We anticipate that high crystallized SWCNTs will be utilized the artificial energy-loss free field emitters with large and stable FE current.

To prevent global warming through the stabilization of the concentration of CO_2 at a low level, a low-carbon society or carbon-positive state should be realized. We are responsible for contributing to global environmental conservation through the suppression of carbon use and energy consumption through the development of new electronic devices. All the electric equipment and electronic devices that support our lives consist of modular solid-state devices. A huge amount of energy is consumed for each solid-state device, from production to implementation to operation. To reduce all related energy from development to use in our daily lives and to develop a low-carbon society, we have been studying the development of devices that effectively use carbon-based nanomaterials. In this paper, we presented some of the process and device development technologies that maximally utilize CNTs. It is our hope that these technologies will promote the realization of a low-carbon society.

Acknowledgements This work was supported by JSPS Grant-in-Aid for Scientific Research(S) Grant Number 26220104 and partially by DOWA Holdings Co., Ltd. The authors gratefully appreciate the discussions and advice of co-researchers from DOWA.

References

Albrecht PM, Lyding JW (2003) Ultrhigh-vacuum scanning tunneling microscopy and spectroscopy of single-walled carbon nanotubes on hydrogen-passivated Si(100) surfaces. Appl Phys Lett 83:5029–5031

Bormashov VS, Nikolski KN, Baturin AS, Shesin EP (2003) Prediction of field emitter cathode lifetime based on measurement of I-V curves. Carbon 215:178–190

Cho Y, Lee S, An M, Kim D (2007) Transparent carbon nanotube field emission devices for display and lamp. Phys Stat Solid A 204:1804–1807

Czerw R, Webster S, Carroll DL, Vieira SMC, Birkett PR, Rego CA, Roth S (2003) Tunneling microscopy and spectroscopy of multiwalled boron nitride nanotubes. Appl Phys Lett 83:1617–1619

Dai H, Hafner JH, Rinzler AG, Colbert DT, Smalley RE (1996) Nanotubes as nanoprobes in scanning probe microscopy. Nature 384:147–150

Dillon AC, Jones KM, Bekkedahl TA, Kiang CH, Bethune DS, Haben MJ (1997) Storage of hydrogen in single-walled carbon nanotubes. Nature 386:377–379

Dresselhaus MS, Dresselhaus G, Eklund PC (1996) Science of Fullerenes and Carbon Nanotubes: Their Properties and Applications. Academic Press, Science of Fullerenes and Carbon Nanotubes

Dresselhaus MS, Dresselhaus G, Saito R, Jorio A (2005) Raman spectroscopy of carbon nanotubes. Phys Rep 409:47–99

Ebbesen TW, Lezec HJ, Hiura H, Bennett JW, Ghaemi HF, Thio T (1997) Electrical conductivity of individual carbon nanotubes. Nature 382:54–56

Garrido SB, Shimoi N, Abe D, Hojo T, Tanaka Y, Tohji K (2014) Plannar light source using a phosphor screen with single-walled carbon nanotubes as field emitters. Rev Sci Inst 85:104704

Hamada N, Sawada S, Oshiyama A (1992) New one-dimensional conductors: graphitic microtubules. Phys Rev Lett 68:1579–1581

Hu Y, Li Y, Zhu M, Hu Z, Yu L (2010) Field-emission of TiSi2 thin film deposited by an in situ chloride-generated route. J Vac Sci Technol B 28:1093

Iwata S, Sato Y, Nakai K, Ogura S, Okano T, Namura M, Kasuya A, Tohji K, Fukutani K (2007) Novel method to evaluate the carbon network of single-walled carbon nanotubes by hydrogen physisorption. J Phys Chem C Lett 111:14937–14941

Kim JY, Kim M, Kim HM, Joo J, Choi JH (2002) Electrical and optical studies of organic light emitting devices using SWCNTs-polymer nanocomposites. Opt Mater 21:147–151

Lee J-H, Yonathan P, Kim H-T, Yoon D-H, Kim J (2008) Optimization of field emission properties from multi-walled carbon nanotubes using ceramic fillers. Appl Phys A 93:511–516

Liu GY, Lv QN, Xia SH, Wang DA, Li TX, Li HY, Ding YG, Ju ZM (2007) An improvement on cold carbon field-electron emitter for commercial x-ray tube. In: 2007 IEEE 20th International vacuum nanoelectronics conference, pp 63–64.

Niu C, Sichel EK, Hoch R, Moy D, Tennent H (1997) High power electrochemical capacitors based on carbon nanotube electrodes. Appl Phys Lett 70:1480–1482

Odom TW, Huang JL, Kim P, Lieber CM (1998) Atomic structure and electronic properties of single-walled carbon nanotubes. Nature 391:62–64

Rao AM, Jacques D, Haddon RC, Zhu W, Bower C, Jin S (2000) In situ-grown carbon nanotube array with excellent field emission characteristics. Appl Phys Lett 76:3813

Rinzler AG, Hafner JH, Colbert DT, Smalley RE (1995) Field emission and growth of fullerene nanotubes. Mat Res Soc Symp Proc 359:61

Saito Y, Uemura S (2000) Field emission from carbon nanotubes and its application to electron sources. Carbon 38:169–182

Saito R, Fujita M, Dresselhaus G, Dresselhaus MS (1992) Electronic structure of chiral graphene tubules. Appl Phys Lett 60:2204–2206

Saito R, Dresselhaus G and Dresselhaus MS, (1993)"Electronic structure of double-layer graphene tubules. J Appl Phys 73:494–500

Saito Y, Uemura S, Hamaguchi K (1998) Cathode ray tube lighting elements with carbon nanotube field emitters. Jpn J Appl Phys 37:L346-348

Sheshin EP, Anashchenko AV, Kuzmenko SG (1999) Field emission characteristics research of some types of carbon fibres. Ultramicroscopy 79:109–114

Shimoi N (2015) Effect of increased crystallinity of single-walled carbon nanotubes used as field emitters on their electrical properties. J. Appl. Phys. 118:214304

Shimoi N, Adriana LE, Tanaka Y et al (2013) Properties of a field emission lighting plane employing highly crystalline single-walled carbon nanotubes fabricated by simple processes. Carbon 65:228–235

Spindt CA, Brodie I, Humphrey L, Westerberg ER (1976) Physical properties of thin-film field emission cathodes with molybdenum cones. J Appl Phys 47(12):5248–5263

Sung WY, Lee SM, Kim WJ, Ok JG, Lee HY, Kim YH (2008) New approach to enhance adhesions between carbon nanotube emitters and substrate by the combination of electrophoresis and successive electroplating. Diam Relat Mater 17:1003–1007

Tanaka K, Okahara K, Okada M, Yamabe T (1992) Electronic properties of bucky-tube model. Chem Phys Lett 191(5):469–472

Thuesen LH (2001) Effects of color phosphors on the lifetime of field emission carbon thin films. J. Vac. Sci. Tech. B 19(3):888–891

Tohji K, Goto T, Takahashi H et al (1996) Purifying single-walled nanotubes. Nature 383(6602):679

Treacy MMJ, Ebbesen TW, Gibson JM (1996) Exceptionally high Young's modulus observed for individual carbon nanotubes. Nature 381:678–680

Wallace RP (1947) The Band Theory of Graphite. Phys Rev 71:622–634

Wildoerm JWG, Venema LC, Rinzler AG, Smalley RE, Dekker C (1998) Electronic structure of atomically resolved carbon nanotubes. Nature 391:59–62

Wu B, Geng D, Liu Y (2011) Evaluation of metallic and semiconducting single-walled carbon nanotube characteristics. Nanoscale 3:2074

Xiang B, Wang QX, Wang Z, Zhang XZ, Liu LQ, Xu J, Yu DP (2005) Synthesis and field emission properties of $TiSi_2$ nanowires. Appl. Phys. Lett. 86:243103

Xu X, Brandes GR (1999) A method for fabricating large-area, patterned, carbon nanotube field emitters. Appl Phys Lett 74:2549

Yamamoto G, Sato Y, Takahashi T, Omori M, Hashida T, Okubo A, Tohji K (2006) Single-walled carbon nanotube-derived novel structural material. J Mater Res 21:1537–1542

Chapter 30
Development of Electrostatic Linear Motor for Insect-Type Microrobot

Genki Osada, Asuya Mizumoto, Satoshi Hirao, Yuichiro Hayakawa, Daisuke Noguchi, Minami Kaneko, Fumio Uchikoba, and Ken Saito

Abstract The authors aim to develop an insect-type microrobot system which can operate autonomously like as an insect by mounting a sensor, a power supply, a motor, and a controller in a millimeter-sized body. Realization of robots the same as the performance of organisms is difficult with current robotics. In particular, small size actuator with low power consumption, which can actuate by small-sized energy source are difficult to realize. In this paper, the authors will discuss an electrostatic linear motor to move the legs of the microrobot with low energy consumption using a small power source.

Keywords MEMS · Microrobot · Electrostatic motor

30.1 Introduction

In recent years, a microrobot of millimeter-scale has actively developed by using the micro electro mechanical systems (MEMS) technology. The microrobots are expected to be active in various fields that existing robots could not use. In the medical field, the microrobots expected in the examination of internal organs and blood vessels by taking the microrobots directly into the patient's body. Also, the microrobots are considered to be able to reduce the burden on patients by administering drugs directly to the affected area (Sitti et al. 2015; Li et al. 2018). In the sight of the disaster, the microrobots are expected to investigate narrow and dangerous spaces which are difficult for a human to enter (Li et al. 2016; Takeda 2011).

The authors are studying to realize a microrobot that can operate autonomously like insects. Small insects such as ants have excellent autonomous systems packaged to a small sized body. In modern robotics, there are difficulties in realizing a small and autonomously operating the microrobots. Miniaturization of a power supply,

G. Osada · A. Mizumoto · S. Hirao · Y. Hayakawa · D. Noguchi
Graduate School of Science and Technology, Nihon University, Tokyo, Japan

M. Kaneko · F. Uchikoba · K. Saito (✉)
College of Science and Technology, Nihon University, Chiba, Japan
e-mail: kensaito@eme.cst.nihon-u.ac.jp

Y. Kishita et al. (eds.), *EcoDesign and Sustainability I*, Sustainable Production, Life Cycle Engineering and Management, https://doi.org/10.1007/978-981-15-6779-7_30

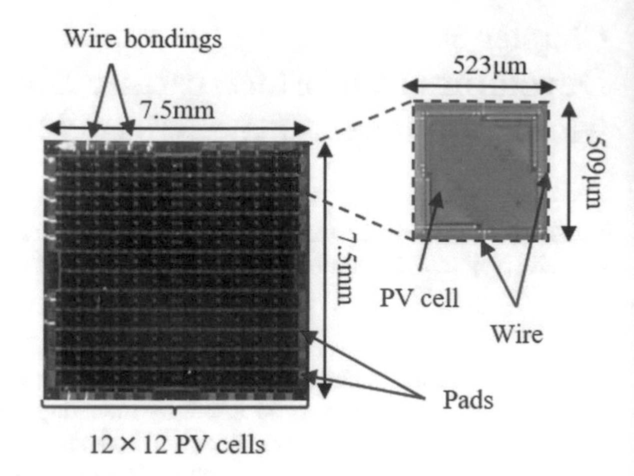

Fig. 30.1 High-voltage silicon PV cell array (Saito et al. 2017)

sensors, a control circuit, and motor are a difficult subject. In particular, small size motor with low power consumption, which can actuate by small-sized energy source is challenging to realize.

In the previous research, the authors proposed and demonstrated a microrobot that can replicate the tripod gait locomotion of insects (Tanaka e al. 2017). The microrobots legs actuated by a shape memory alloy (SMA) actuator. The small size components realized by MEMS technology. The SMA actuator provided a large deformation and force. However, the power consumed by actuating a single leg reached as high as 94 mW. Therefore, the microrobot has driven by the external power supply. The authors focused on electrostatic motors which can operate with low power supply (Saito et al. 2017). By achieving low power consumption as 1.3 mW, the microrobot can also be driven by a small-sized power supply. The electrostatic motor is suitable for miniaturization that electrostatic motors are operating contactless and the force of electrostatic motors is surface area dependent. Previously, the authors proposed the electrostatic motor. The electrostatic motor was made of silicon so that the microrobot can unify other silicon materials. The silicon materials integration can be expected to reduce costs and improve mass productivity. The electrostatic motor based on capacitive driven gap-closing actuators working in tandem to linearly displace a shuttle at a force output of over 1.5 mN without any static current.

However, the pullback motion by spring of electrostatic motors was not enough to operate the complete pullback motion. In this paper, the author designed an electrostatic linear motor using the electrostatic actuator for the pullback motion. Also, the rhombus-shaped spring designed to support the electrostatic linear motor.

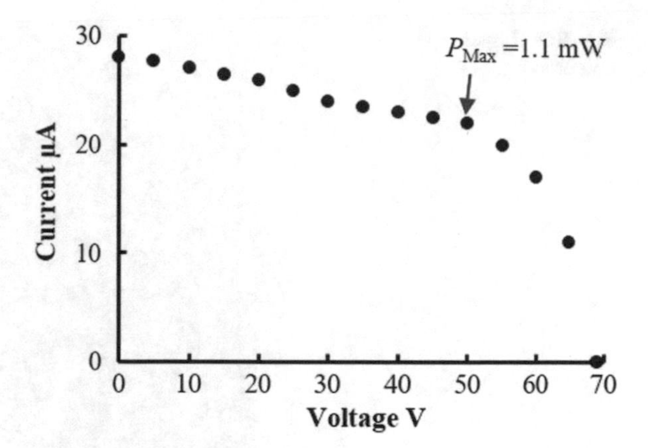

Fig. 30.2 I-V characterstics of the high-voltage Silicon PV cell array (Saito et al. 2017)

30.2 High-Voltage silicon PV cell

Figure 30.1 shows the high-voltage silicon PV cell array to drive the electrostatic motor provided by Prof. Yoshio Mita (Saito et al. 2017; Mori et al. 2014; Takeshiro et al. 2017). The PV cell array was a total area of approximately 7.5 mm × 7.5 mm. The PV cell array consists of 144 PV cells (509 mm × 523 mm) and connected in series. The array fabricated by CMOS post-process dry release and device isolating method.

Figure 30.2 shows the I-V characteristics of the high-voltage Silicon PV cell array. The output current of the PV cell array was measured using KEITHLEY 2000 MULTIMETER, and KIKUSUI PMC 500–0.1A used as a voltage source. The light source was a 54 LED array (6 W) with DC 6 V voltage source. When the voltage was 50 V, and the current was 22 μA, the maximum power (P_{Max}) was 1.1 mW.

30.3 Insect-Type microrobot

Figure 30.3 shows an insect-type microrobot driven by an SMA motor. Microrobot can replicate the tripod gait locomotion like an ant by moving each leg. The external dimensions are 4.6 mm × 9.0 mm × 6.4 mm. Microrobot is composed of a body made by MEMS technology and Integrated Circuit (IC) on the upper part of the body, and SMA in on the motor. An IC chip mounted on the robot can generate a gait pattern of the microrobot. The SMA motor drives the microrobot. In this paper, the motor changed to the electrostatic motor.

Figure 30.4 shows the leg of the microrobot. Figure 30.4a shows the parts of the leg. The leg parts made from a silicon wafer. The leg parts that are 100 μm thick 14 washers and 200 μm thick rod 1 to rod 6 develop using photolithography of MEMS technology. Leg parts made by Inductively Coupled Plasma (ICP) for high aspect ratio. Also, seven shafts are a diameter of 0.1 ± 0.002 μm and used as connecting

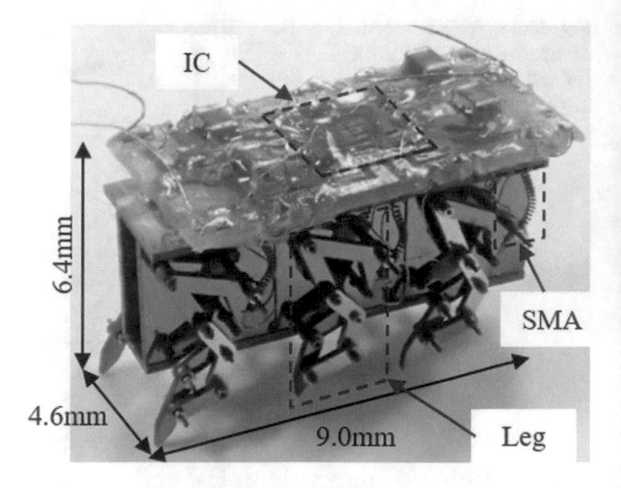

Fig. 30.3 Insect-type microrobot

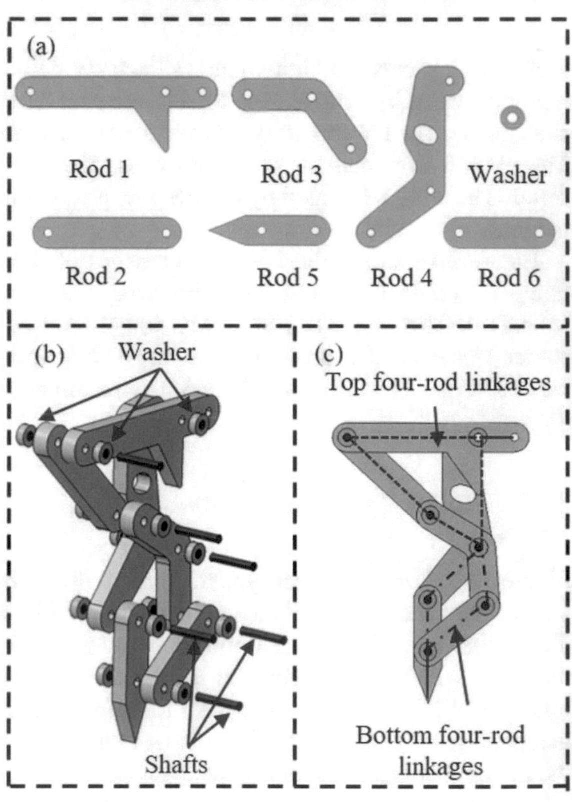

Fig. 30.4 Leg of the microrobot

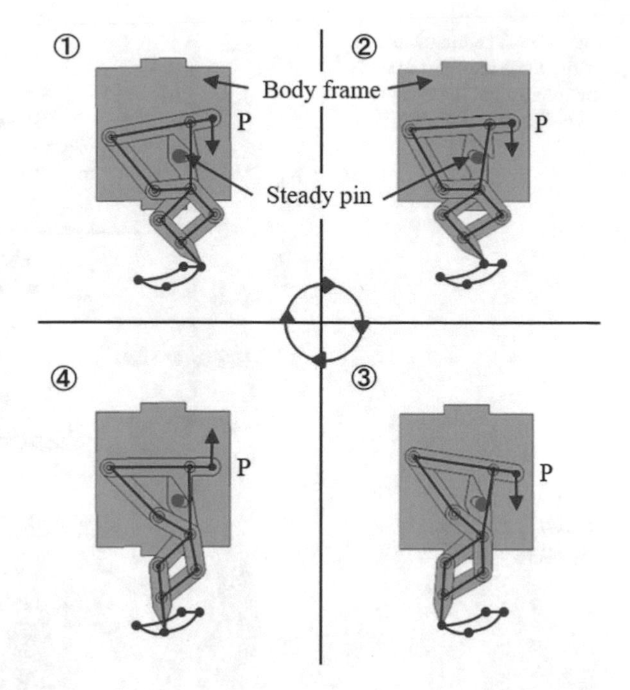

Fig. 30.5 Leg motion and trajectory of the leg

each part. The leg of the microrobot was assembled the leg parts shown in Fig. 30.4b by hand. The washer was the connected end of the shaft fixed using cyanoacrylate. Also, the authors design all leg parts to have a clearance of a 10 μm. Figure 30.4c shows the link mechanism of the microrobot leg. The microrobot leg has top four-rod linkages of composed of rod 1 to rod 4 and rod 3 to rod 6. The bottom four-rod linkages can convert linear motion from the motor into stepping motion. The design of each six legs of the microrobot is all the same except the order of assembling.

Figure 30.5 shows the leg motion and trajectory of the leg. Point P of Fig. 30.5 show the connection with the motor. We describe the operation of the microrobot leg. Point P moves in the order of 1, 2, and 3 by linear motion of the lower direction. When this move performed, steady pin generate the stepping motion because the trajectory of the leg bends by steady pin. Also, it works in order of 3, 4, and 1 by linear motion of the upper direction at point P. Also during this motion, leg returns to the origin position by steady pin because the trajectory of the leg bends by steady pin.

Figure 30.6 shows the characteristic of the force required to drive the microrobot leg against point P using micro force sensor. In Fig. 30.6, solid circle shows the force that point P of the leg displace of lower direction, solid triangle indicates the force that point P of the leg displace of upper direction. Also, a solid circle is called a pushing force, and a solid triangle is called a pullback force. According to Fig. 30.7, the maximum pushing force is about 0.25 mN, and the maximum pullback force is about 0.2 mN. From the figure, from the action limit of the leg drive. Therefore, the leg can be moved by a force of at least 0.25 mN and displacement of at least 250 μm.

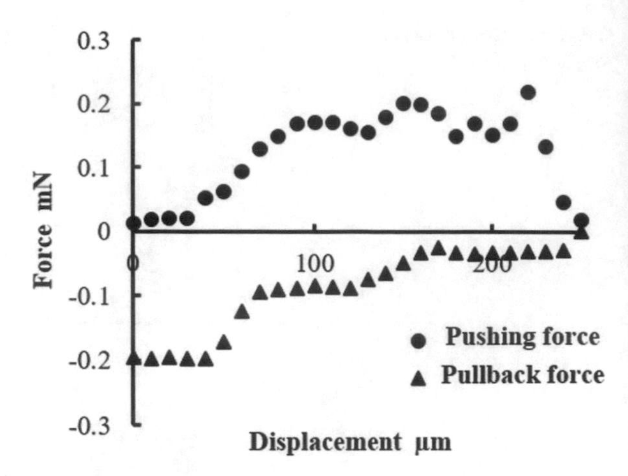

Fig. 30.6 The characteristic of the force required to drive the microrobot leg against point P (Saito et al. 2017)

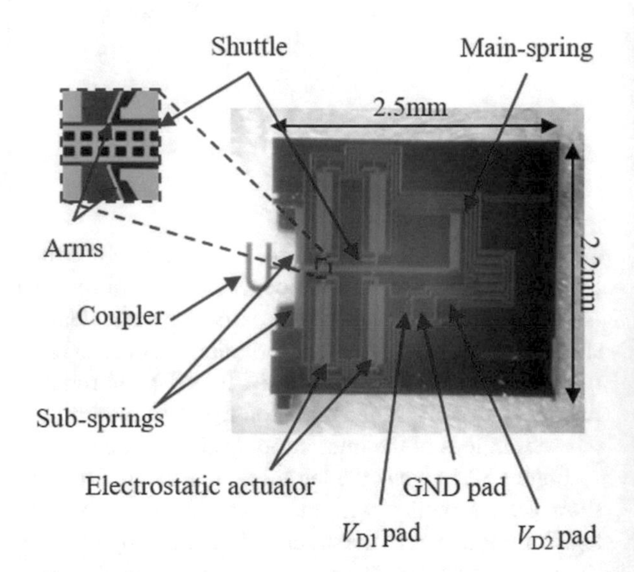

Fig. 30.7 Electrostatic motor (Saito et al. 2017)

30.4 Electrostatic Motor

30.4.1 Design Electrostatic Motor

The authors study the electrostatic motor that can drive the microrobot leg. As an alternative electrostatic motor can drive with low power consumption as compared to the SMA motor that has been mounted on the microrobot legs. The electrostatic motor is fabricated in a 3-mask silicon-on-insulator (SOI) process, which is a MEMS technology. The SOI wafers had a 40 μm device layer, 2 μm buried oxide, and 550 μm handle wafer. A layer of 100 nm-thick aluminum is deposited on the device-layer silicon to energize. The active part of the motor was floated by etching buried oxide.

Figure 30.7 shows the electrostatic motor. The electrostatic motor chiplet measures approximately 2.2 mm × 2.5 mm. The electrostatic motor is composed of a shuttle at the center, a coupler at the shuttle, four electrostatic actuators deployed two pairs for pushing the shuttle, arms deployed at the tip of the electrostatic actuators, main spring and sub spring that return to the original position, electrodes V_{D1}, V_{D2}, and GND. An electrostatic motor produces linear motion of the shuttle by energizing the electrodes.

30.4.2 Leg Drive Experiment by the Electrostatic Motor

Figure 30.8 shows the driving circuit of the electrostatic motor. The resistance value is $R_1 = R_2 = 2.2\ \text{M}\Omega$. The transistors were switched using a waveform generator to generate drive waveforms V_{D1} and V_{D2}. Figure 30.9a is the driving waveform using a voltage source. The driving waveform is a square wave with a pulse width 10 ms, pulse period 7.5 ms, pulse amplitude 60 V. Figure 30.9b is the driving waveform using the PV cell array. The driving waveform of Fig. 30.9b occurred a voltage drop

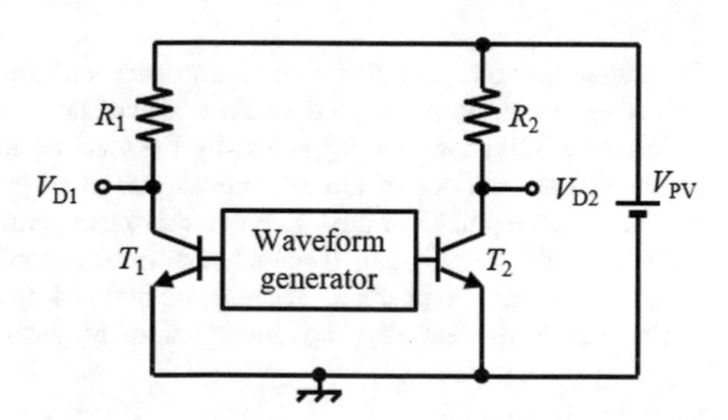

Fig. 30.8 Driving circuit of the electrostatic motor (Saito et al. 2017)

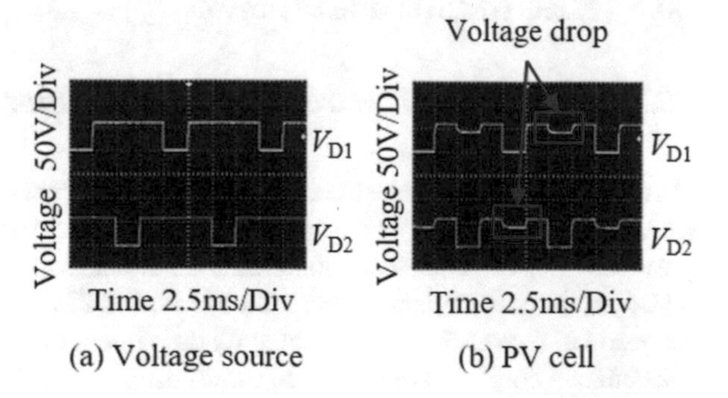

Fig. 30.9 Driving waveform of electrostatic motor (Saito et al. 2017)

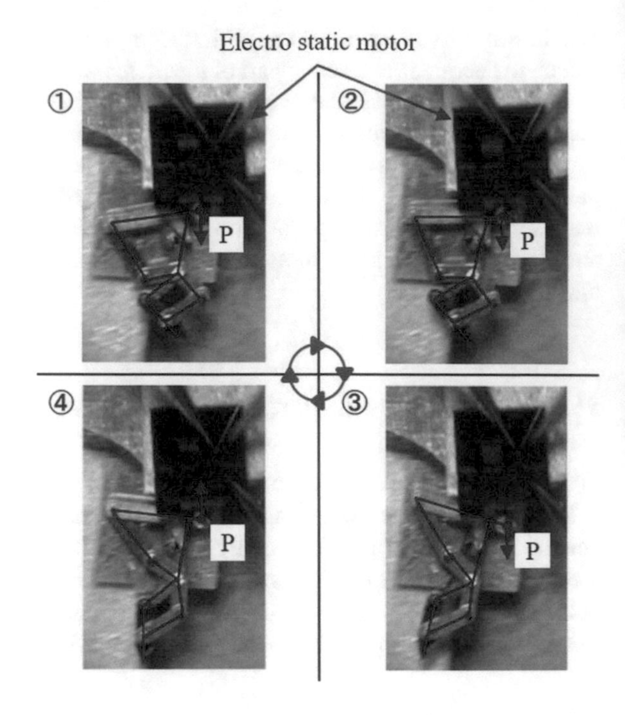

Fig. 30.10 Leg action by electrostatic motor (Saito et al. 2017)

because of the voltage of the PV cell array was insufficient. Even when using the PV cell array, the electrostatic motor could be driven normally.

Figure 30.10 shows the leg action by the electrostatic motor. The electrostatic motor and the one leg of the microrobot was connected through the axis to the coupler and the point P of the leg. When driving experiment by electrostatic motor, leg action of 4–1 in Fig. 30.10 cannot perform sufficiently, and leg position cannot return to the original position. Because the pullback force of main spring and sub spring was insufficient, stepping motion cannot be performed sufficiently.

30.5 Electrostatic linear motor

30.5.1 Design of Electrostatic Linear Motor

Figure 30.11 shows the electrostatic linear motor for the microrobot. The electrostatic linear motor fabricated in SOI process like the electrostatic motor described in the previous chapter. The electrostatic motor chiplet measures a 2.3 mm × 4.7 mm. In addition, the electrostatic linear motor is designed for the electrostatic motor to be disposed to face each other. The electrostatic linear motor is composed of a shuttle at the center, a coupler at the center of the shuttle, eight electrostatic actuators deployed two pairs for pushing the shuttle. Also, arms deployed at the tip of the electrostatic

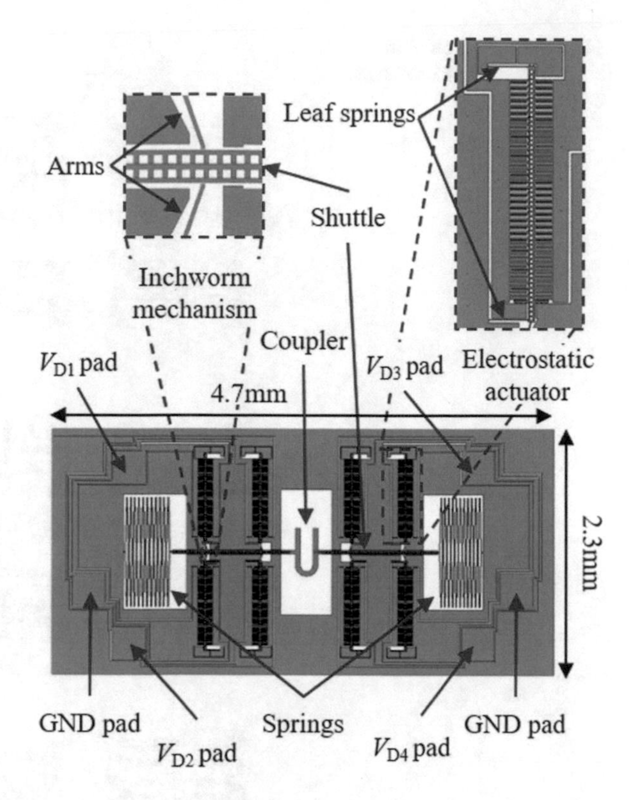

Fig. 30.11 Electrostatic linear motor

actuators, the rhombus-shaped spring to assist the driving of the electrostatic linear motor, leaf spring that returns driving part of the electrostatic actuator to the original position, electrodes V_{D1}, V_{D2}, V_{D3}, V_{D4}, and GND.

30.5.2 Motion of Electrostatic Linear Motor

Figure 30.12 shows the motion of the shuttle using the electrostatic actuator. As shown in Fig. 30.12, the electrostatic linear motor provides the delivering motion of the shuttle by driven the two pairs of electrostatic actuators at the same time. By repeating this delivering motion with four actuators, the shuttle is smoothly delivered to generate a large displacement. The electrostatic linear motor can be bi-directionally driving with high accuracy by designed the electrostatic motor on the left and right. In addition, the electrostatic linear motor is possible 300 μm in total at 150 μm on the one way of the left and right. The microrobot leg can take sufficiently the action, because the displacement by the linear drive is more than necessary displacement for the action of the microrobot leg.

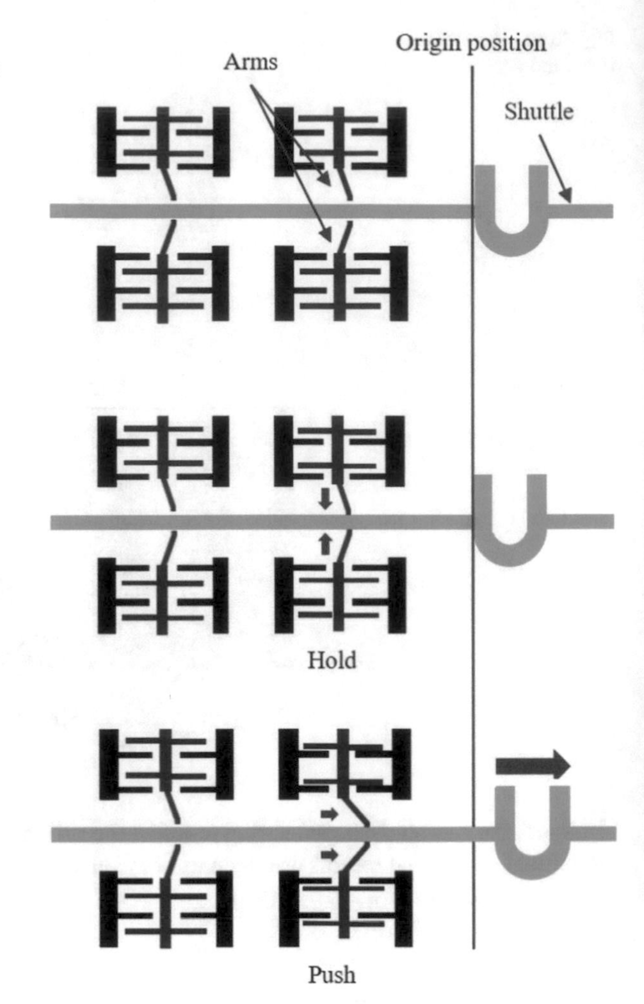

Fig. 30.12 Motion of the shuttle using the electrostatic actuator

30.5.3 Output of Electrostatic Linear Motor

Figure 30.13 shows the schematic view of the comb teeth of the electrostatic actuator. The electrostatic actuator can generate an electrostatic force between the comb teeth by applying a voltage. The generated force F_T of the electrostatic actuator can be determined from the dielectric constant ε, the comb teeth distance G_1, G_2, length L, thickness h, and the voltage V applied to the electrostatic actuator, as in Eq. (30.1).

$$F_T = \frac{\varepsilon h L}{2}\left(\frac{1}{{G_1}^2} - \frac{1}{{G_2}^2}\right)V^2 \tag{30.1}$$

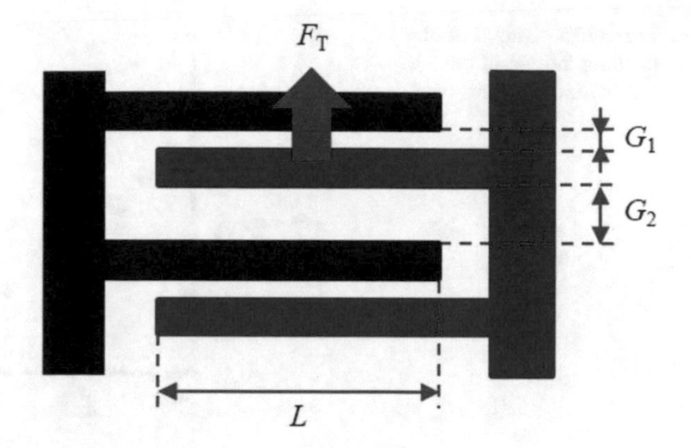

Fig. 30.13 Schematic view of the comb teeth of the electrostatic actuator

Since there are 70 comb teeth per the electrostatic actuator, and two electrostatic actuators deliver simultaneously, the generated force F_A of the electrostatic actuator is expressed by Eq. (30.2).

$$F_A = 2 \times 70 \times F_T \tag{30.2}$$

Figure 30.14 shows the transfer of force to the shuttle by the electrostatic actuator. As shown in Fig. 30.14, the arm collides with the shuttle and push the shuttle. Further movement causes the angle of the arm to change from 20° to 21°, pushing the shuttle 1 μm. The arm divides the output F_A of the electrostatic actuator into forces F_H in

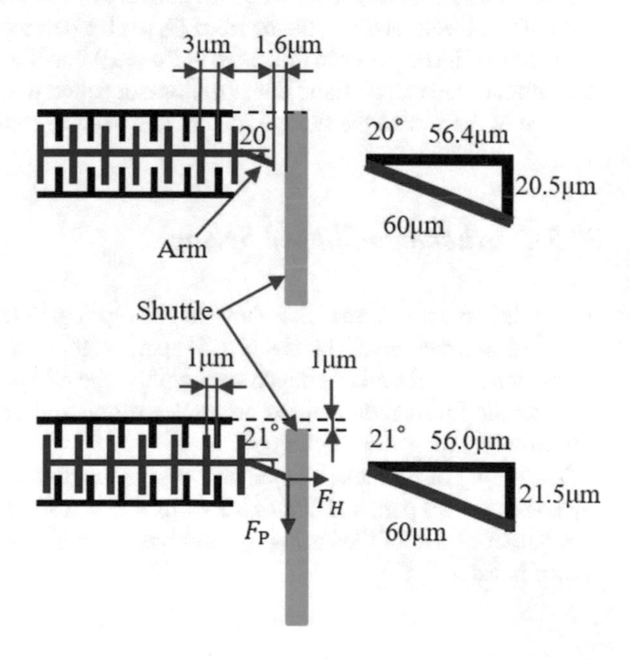

Fig. 30.14 Transfer of force to the shuttle by the electrostatic actuator

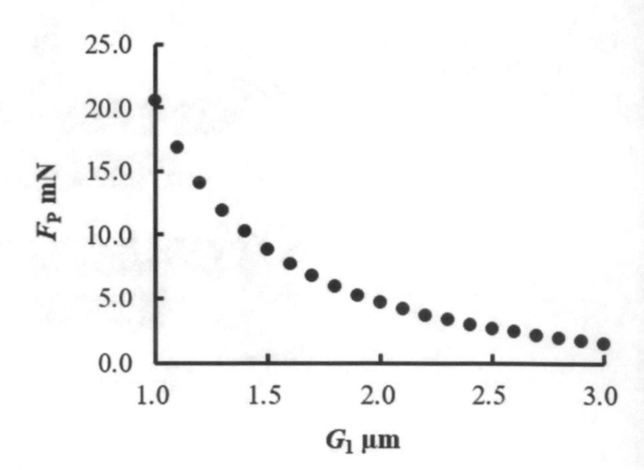

Fig. 30.15 Output of one pushing motion of the electrostatic linear motor

the direction of holding the shuttle and F_P in the direction of pushing the shuttle. From the above, the delivering force F_P of the electrostatic linear motor expressed by Eq. (30.3).

$$F_P = F_A \times \sin\theta \tag{30.3}$$

The electrostatic linear motor design $\varepsilon = 8.8 \times 10 - 12$ F/m, $h = 40$ μm, $G_1 + G_2 = 8$ μm. The output F_P of the electrostatic linear motor is derived from the Eqs. (30.2) and (30.3). Figure 30.15 shows the output F_P of one pushing motion of the electrostatic linear motor. The arm delivers the shuttle in the range of the comb teeth of the electrostatic actuator from $G_1 = 1.4$ μm to 1.0 μm. The output of the electrostatic linear motor in the range is $F_P = 10.2$ mN to 20.5 mN. From the above, the output of the electrostatic linear motor can action the microrobot leg sufficiently because of more than the force 0.25 mN that need to actuate the microrobot leg.

30.5.4 Rhombus-Shaped Spring

Figure 30.16 shows the rhombus-shaped spring. The rhombus-shaped spring designed a structure of 11 stacked 34 μm × 985 μm springs of the rhombus-shaped unit. The rhombus-shaped spring plays the role of returning the shuttle of the electrostatic linear motor to the original position and preventing the shuttle of the electrostatic motor from falling off.

The spring of rhombus-shaped unit design length $l = 500$ μm, width $b = 5$ μm, thickness $h = 40$ μm, and Young's modulus $E = 130$ GPa, Shuttle displacement x. The output of one of the spring of rhombus-shaped unit F_{S1} can be determined, as shown in Eq. (30.4).

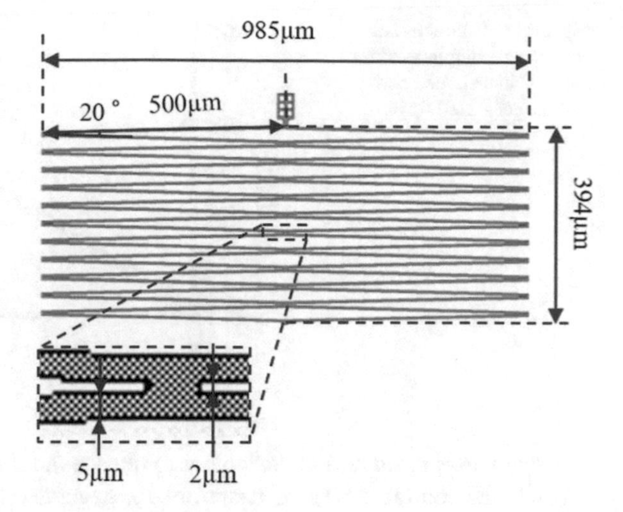

Fig. 30.16 Rhombus-shaped spring

$$F_{S1} = x \times k = x \times \frac{Ebh^3}{6l^3} \tag{30.4}$$

The rhombus-shaped spring has a structure in which 11 springs of the rhombus-shaped unit stack, as shown in Fig. 30.16. Figure 30.17 shows the schematic view of the rhombus-shaped spring motion. Since the shuttle of the electrostatic linear motor drive 150 μm to the left and right. Shuttle maximum displacement $x_{\mathrm{Max}} = 150$ μm.

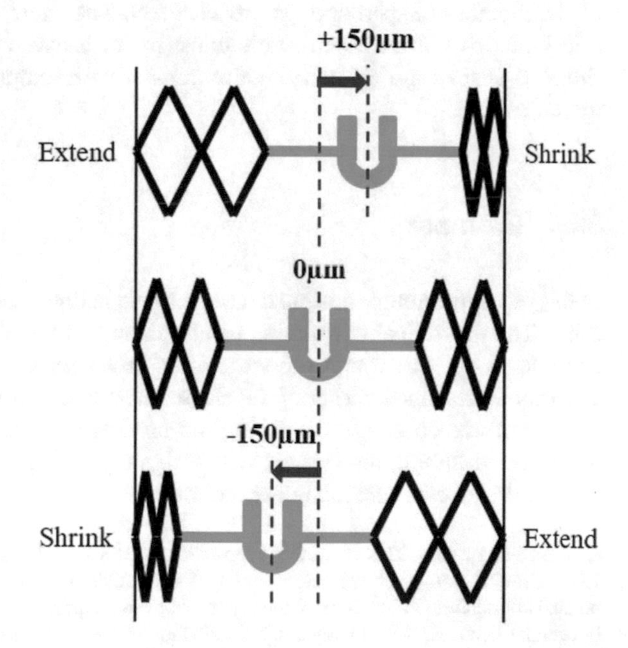

Fig. 30.17 Schematic view of the rhombus-shaped spring motion

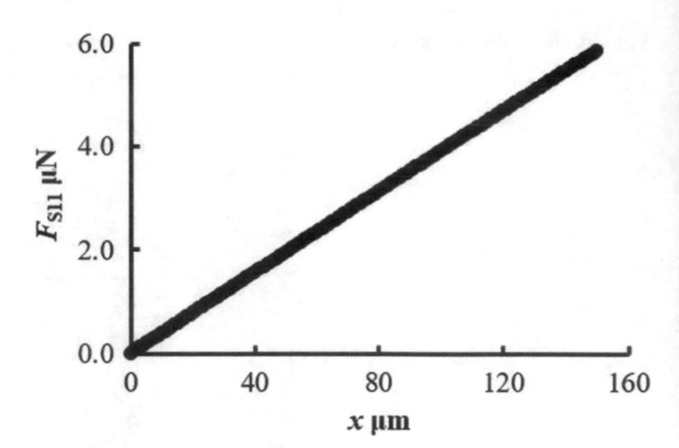

Fig. 30.18 Charactaristic between the displacement of the rhombus-shaped spring and the generated force

The rhombus-shaped spring design not to break even if the rhombus-shaped spring expands and contract 150 μm. Therefore, the generated force F_{S11} can be expressed by Eq. (30.5).

$$F_{S11} = \frac{F_{S1}}{11} = x_{Max} \times \frac{k}{11} \tag{30.5}$$

Figure 30.18 shows the characteristic between the displacement of the rhombus-shaped spring and the generated force. From Fig. 30.18, when the displacement is 150 μm, the generated force of the rhombus-shaped spring is 5.9 μN. Since two rhombus-shaped springs design on both sides of the shuttle, the generated force of the rhombus-shaped springs are 11.8 μN. The rhombus-shaped springs do not affect the drive of the electrostatic linear motor because the generated force of the rhombus-shaped spring is very small compared to the output of the electrostatic linear motor.

30.6 Summary

In this paper, the author designed the electrostatic linear motor for insect-type microrobot. The theoretical calculation results show that the electrostatic linear motor is possible to obtain enough output and displacement to drive the microrobot legs. The rhombus-shaped spring of the electrostatic linear motor design to have a small restoring force so as not disturbing the moving motion of the electrostatic linear motor. In the future, the authors will design the millimeter scale robot with silicon PV cell driven electrostatic linear motors.

Acknowledgements This work was supported by JSPS KAKENHI Grant Number JP18K04060. Also, part of this research was supported by Amano Institute of Technology Public Interest Incorporated Foundation. Fabrication of the microrobot was supported by the Research Center for Micro Functional Devices, Nihon University. Fabrication of the inchworm motors was supported by the

UC Berkeley Marvell Nanofabrication Laboratory. The authors would like to acknowledge the Berkeley Sensor and Actuator Center and the UC Berkeley Swarm Lab and Prof. Kristofer S. J. Pister for their continued support. VLSI Design and Education Center (VDEC), the University of Tokyo (UTokyo), and Phenitec Semiconductor are acknowledged for CMOS-SOI wafer fabrication. The Japanese Ministry of Education, Sports, Culture, Science and Technology (MEXT) is acknowledged for financial support through Nanotechnology Platform to UTokyo VDEC used for the PV cell postprocess. The authors would like to acknowledge the members of Mita Lab and Prof. Yoshio Mita for their continued support.

References

Li S, Jiang Q, Liu S, Zhang Y, Tian Y, Song C, Wang J, Zou Y, Anderson GJ, Han JY, Chang Y, Liu Y, Zhang C, Chen L, Zhou G, Nie G, Yan H, Ding B, Zhao Y (2018) A DNA nanorobot functions as a cancer therapeutic in response to a molecular trigger in vivo. Nature Biotechnol 258–264

Li H, Go G, Seong YK, Park JO, Park S (2016) Magnetic actuated pH-responsive hydrogelbased soft micro-robot for targeted drug delivery. Smart Mater Struct 25 1–9

Mori M, Lebrasseur KE, Mita Y (2014) Remote power feed and control of MEMS with 58 V silicon photovoltaic cell made by a CMOS post-process dry release and device isolation method. In: Symposium on Design, Test, Integration & Packaging of MEMS/MOEMS (DTIP2014), April 1–4, Cannes, France, 10.1109/DTIP.2014.7056670

Saito K, Contreras DS, Takeshiro Y, Okamoto Y, Hirao S, Nakata Y, Tanaka T, Kawamura S, Kaneko M, Uchikoba F, Mita Y, Pister KSJ (2019) Study on electrostatic inchworm motor device for a heterogeneous integrated microrobot system. Trans Japan Inst Electron Packa 12:E18–009–1-E18–009–7

Sitti M, Ceylan H, Hu W, Giktinan J, Turan M, Yim S, Diller E (2015) Biomedical applications of untethered mobile Milli/Microrobots. In: Proceedings of the IEEE, vol 103, No 2

Takeda M (2001) Applications of MEMS to industrial inspection" Technical Digest," MEMS 2001. In: 14th IEEE international conference on micro electro mechanical systems, pp 182–191

Takeshiro Y, Okamoto Y, Mita Y (2017) Mask-programmable on-chip photovoltaic cell array. In: Proceedings of the Power MEMS 2017, November 14–17, Kanazawa, Japan, pp 596–597

Tanaka D, Uchiumi Y, Kawamura S, Takato M, Saito K, Uchikoba F (2017) Four-leg independent mechanism for MEMS microrobot. Artificial Life and Robotics. September 2017, vol 22, Issue 3, pp 380–384, 10.1007/s10015–017–0365–2

Chapter 31
Investigation of the Behaviour of a New Nanogrid Concept and the Nanogrid Components

Ágnes Halász, Tamás Iváncsy, and Zoltán Ádám Tamus

Abstract In microgrids a distributed storage is efficient, but in the concept of microgrids the management of generation, storage and the flow between microgrids is needed. The economical situation has to be taken into account also. Nanogrids are smaller size of micrograds, but have the same main features. In an office environment one office could be one nanogrid. We simulated the behaviour of the different components of the office nanogrid to obtain information about the possible time necessary for charging and discharging a battery with given capacity. One possible extension to the nanogrid concept is the use of Information and Communication Technology (ICT) devices battery to extend the storage capacity. To optimize the charging of ICT devices the battery management system (BMS) has to be extended with a software solution to obtain information from the state of the batteries. Simulations are run to characterize the battery behaviour of ICT devices and its effect on the office nanogrid. With the users calendar entries the software has to predict when the fully charged state of the battery is expected, which also needs the charging times of the different devices batteries. Further investigation is needed to show if this prediction is also possible without the software part providing the battery state of charge.

Keywords Smart grid · Nanogrid · Grid storage extension · ICT devices · Energy management system

31.1 Introduction

The main difficulty of electric energy distribution is the balance between the energy production and consumption. Nowadays renewable energy sources have a big role in electric energy generation but the balancing in such an environment is challenging. Storage of electric energy is difficult, but it is a necessity. One possible way is to convert the electric energy to DC and use some sort of rechargable batteries.

Á. Halász · T. Iváncsy (✉) · Z. Á. Tamus
Department of Electric Power Engineering, Budapest University of Technology and Economics, Budapest, Hungary
e-mail: ivancsy.tamas@vet.bme.hu

Y. Kishita et al. (eds.), *EcoDesign and Sustainability I*, Sustainable Production, Life Cycle Engineering and Management, https://doi.org/10.1007/978-981-15-6779-7_31

31.1.1 Problems Emerging from Distributed Generation

The environmental friendly renewable energy sources and the electric energy generation from these are creating a new challenge for the electric energy distribution and so for the network. The power plants created with the new, renewable technologies are smaller in power capacity and so higher number of small plants provide the electric energy for the network. This changes the energy generation from the old centralised approach to a new decentralised one, which means that a different distribution network is needed for reliable operation than before. An other point is that the renewable energy sources are not present in the necessary amount all the time, and so the amount of electric energy produced by these plants are more difficult to predict and hold in balance with the consumption.

31.1.2 Possibilities of Smart Grids

The forecast difficulties of the renewable energy sources are leading to the necessity of a new type of distributed network, the so called smart grid. The goal of this new grid is a reliable, efficient and more secure operation despite the unpredictable nature of the energy sources.

The smaller parts of the smart grids are the so called microgrids, which are in a range of a few bigger buildings or a block of small buildings. These network parts are able to work independently or coupled together. The micrograds are low or middle voltage networks with energy generation and storage components beside the consumers and network parts. The local energy generation and storage makes the off grid operation also possible.

31.1.3 The Nanogrid

The nanogrid is a similar concept to the microgorid but in a smaller size, as the electric network of one building or more just a part of a building. In the case of the nanogrids the energy storage is not a necessity, but limits the off grid operation possibilities. The connection of more nanogrids together and share its resources is also possible with a more complex management system for the nanogrids. Usually these nanogrids connected together are also treated as a microgrid. To run nanogrids economically the sharing of the energy generating and storage resources is inevitable. (Díaz et al. 2015; Isikman et al. 2013; Burmester et al. 2017).

31.1.4 Nanogrid Control

Usually the nanogrid control is making decisions which source and which load is active. If an energy storage is present then in the decision making also the management of the storage is involved. The goal is to use the renewable source of the nanogrid and the sources outside the nanogrid only if there is no other possibility. If the outside sources have different costs, e.g. an other nanogrids renewable sources and the public distribution networks are also involved, then also the costs should be taken into account.

In all cases the control needs predictions about the amount of the generated power and the power demand of the loads in the nanogrid to balance the loads and sources and load the energy storage (Lujano-Rojas et al. 2012; Teleke et al. 2014). Some loads can be switched automatically on and off for shifting the demand in time and so lowering the costs for the consumed energy (Zhu et al. 2018).

The control has the following constraints to take into account by the decision making:

- power balance
- capacity
- output power
- battery operation
- load curtailment
- interruption duration
- number of interruptions
- interruption time interval
- shiftable load balance

For the control data from the weather conditions, solar irradiation and user behaviour should be present to be able to optimize the energy usage of the nanogrid and so create a close to optimal working state.

31.2 A New Nanogrid Concept

One possible nanogrid could be an office in an office building. The energy generation in this case could be wall and roof mounted photovoltaic cells. Each office has its own dedicated PV cells. The goal is to run the nanogrid so, that the generated energy from the PV cells runs the ICT devices in the office. Therefore energy storage is necessary to compensate the changes in the amount of generated energy due to the solar radiation changes.

31.2.1 Storage Possibilities in Nanogrids

It is worth to note that not only batteries can be used as energy storage systems for the nanogrids but other types of energy storage systems. Beside the well known battery types of lead-acid and lithium batteries one of the newest is the so called redox-flow battery. Also supercapacitors (or also called ultracapacitors) can be used, which also has several types like electrostatic double-layer capacitors (EDLCs), electrochemical pseudocapacitors and hybrid capacitors. The mechanical possibilities are flywheels, pumped-storages (hydroelectricity) and compressed air energy storages (Blaabjerg and Ionel 2017; Vilathgamuwa et al. 2011; Du et al. 2007; Zhang et al. 2012). The fact that one office as nanogrid has a relatively small energy storage demand makes it possible to use different storage types.

As our measurements shows for an office with two laptops with two external monitors and two phones a battery with a capacity of about 680 Wh and two 1.5 m^2 polycrystalline PV cell could be sufficient (Cserép et al. 2017). According to the measurements and calculations such a configuration can run the ICT devices trough the day even in winter when there is no direct sunlight on the PV panels.

31.2.2 Including ICT Device Batteries as Storage

Some of the ICT devices as laptops and mobile phones has a built in battery for their operation. The extension to the previous concept is to involve these batteries in the operation of the nanogrid. The mobility of these devices makes it necessary to have information if the user will use the devices out of office. The possible highest charge level should be provided for extended out of office usage. To provide this functionality the energy storage management system (ESMS) of the nanogrid should get information from the users calendar events to estimate the necessary time for charging the devices battery for the highest possible level. If the devices are used in the office the ESMS of the nanogrid can provide charging if sufficient energy is generated by the PV cells or switch off the power for the devices and force operation on the built in battery in case of low energy generation by the PV cells.

To the already presented constraints a further constraint is introduced which are the timetables of the users of ICT devices. These makes the control even more complex.

31.2.3 Battery Charging Estimations

To be able to estimate the charging process some information are necessary about the batteries from the devices and the energy storage of the nanogrid. The type of the energy storage, the capacity of the storage and the state of charge (SOC). The type and age of the energy storage lays down the state of health (SOH) and the state of life

(SOL). The SOH is a diagnostic indicator, which shows the capacity change of the energy storage originating from the ageing and usage. The SOL is also a diagnostic indicator which indicates the possibility of sever failures after a given ageing of the storage (Topan et al. 2016; Buchman and Lung 2018; Hou et al. 2017; Xia and Wei 2016). To get these information the nanogrids ESMS needs to communicate with the battery of the ICT devices, which means that a special software should be installed on the laptops and smart phones, however this is not possible on other devices e.g. DECT (Digital Enhanced Cordless Telecommunications) phones. If the installation of the special program is not possible then the inclusion of the device in the nanogrid is more difficult because the knowledge of the charging profile is necessary. If the battery voltage is known the charging SOC can be estimated and according this the constant current or constant voltage charging can be selected. The different charging types and the respective battery voltage, charging current and SOC are shown in Fig. 31.1. The third phase is the trickle charging which is also constant voltage charging with lower voltage level, closer to the battery voltage. This phase is not applicable to all battery types e.g. lithium-ion batteries. This charging type is also not selected by the nanogrids ESMS for any type of battery.

If the ESMS has the necessary informations it can make the decision if the battery can be charged and which battery will be charged.

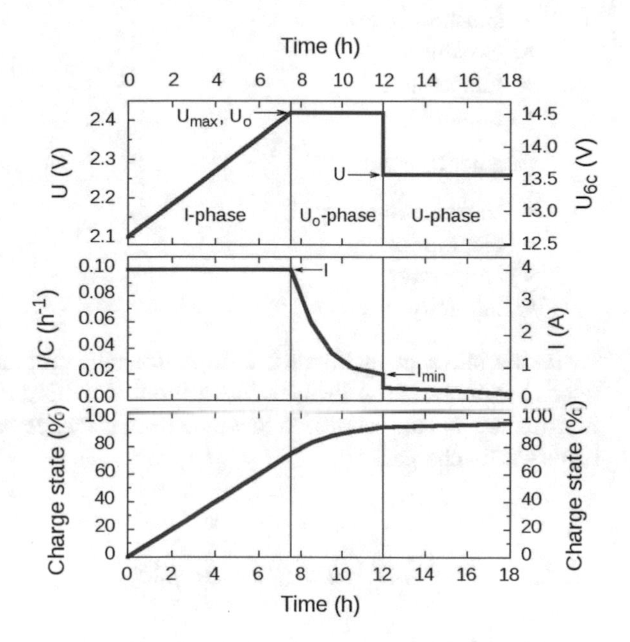

Fig. 31.1 The charging phases of a battery displayed in battery voltage, charging current and SOC

31.3 Working Principle for ESMS

31.3.1 Provided Information for ESMS

In our model the ESMS makes decision on the charging and discharging of the batteries upon a great number of collected information. These informations are as follows:

- technical parameters
  - charging characteristics of the devices
  - SOC of the devices
  - SOC of the energy storage in the nanogrid
  - capacity of the energy storages
  - power of the charger devices
- calendar events
- weather data
  - actual weather
    - sunshine intensity
    - cloudiness
    - temperature
    - humidity
  - weather forecast
    - sunshine intensity
    - cloudiness
    - temperature
    - humidity

Taking into account all these informations the program in the ESMS can decide which battery has to be charged. The difficulty in this decision is the high number of information and the estimation to which time is it necessary to have the ICT devices battery fully charged.

31.3.2 Decision Making Principle

The program for the decision making consists of two main task, one of them is the collecting of the necessary information and making the decision and the other is the estimation of charging rate upon the collected information.

If there is no event in the calendars for the day, in that case the actual solar irradiance on the PV cells and the SOC of the batteries are taking a role in the decision. If there is something scheduled for a given time there are more factors to take into account during the decision. The SOC of the devices battery and the remaining time to the event are the basic informations and an estimation should be made when to start to charge the battery to get a fully charged battery before the event starts.

According the provided informations the result of the decision can be the following:

- offering energy for other nanogrids
- charging the nanogrids battery
- charging the devices batteries
- charging the nanogrids and devices batteries
- claim energy from other nanogrids or the network

For the correct time calculations the controller needs an estimation on the power produced by the PV cells and the charging characteristics of the different batteries.

31.4 Battery Charging Measurements

31.4.1 Power Needs for Different Devices

A usual laptop power supply is 45 W to 65 W in power. The battery capacity is in the range of 55 Wh to 85 Wh usually. The battery of smartphones are in the range of 10 Wh to 15 Wh where the power supply is 5 W to 10 W of rated power.

The battery charging is done in the devices which also means that the devices can be used during charging. It means that the power necessary for charging is just one part of the rated power of the power supplies, the other part is the power consumption of the working device. The power consumption of the device is depending on the type of actual usage, e.g. using a word processor or running a simulation means a significantly different load for a laptop.

31.4.2 Charging Characteristics of Different Batteries

To get information about the batteries measurements of charging during device usage were performed. The measurements were done for different initial SOC of the batteries and for different usage of the devices during charging. On Fig. 31.2. two different measurement result can bee seen. The horizontal axis is time (hours and minutes), the vertical axis is the consumed power (watt).

We can observe the two different stages of charging, the constant current charging and the constant voltage charging. By the orange line (the 50% initial SOC) the change from the constant cuurent charging to the constant voltage charging is occured at

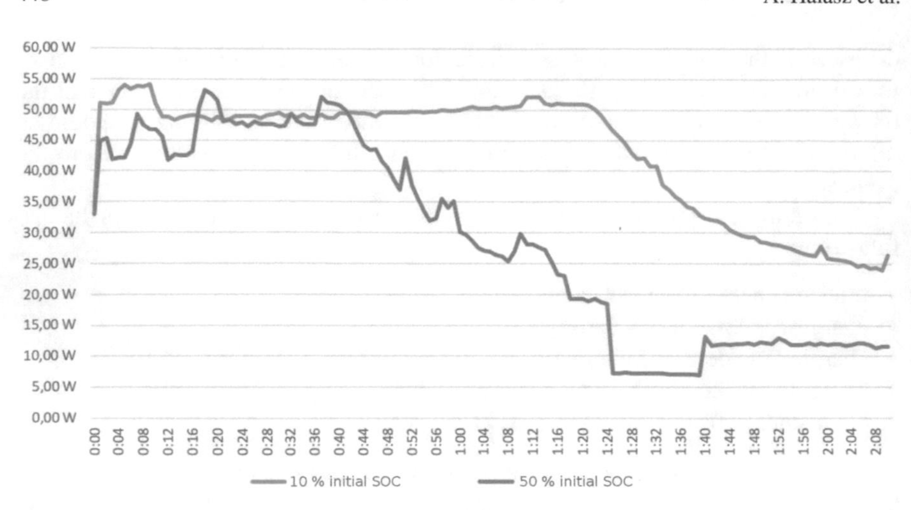

Fig. 31.2 The charging of laptop battery from different initial SOC values

about 40 min where for the blue line (10% initial SOC) this change occured at about 1 h 20 min. Till these points the power consumption is constant. The constant voltage charging is in both cases about 40 min long and has the same characteristics. The changes in the curves are due to the changing in the usage, so the different power consumption of the working device.

Similar charging characteristics can be observed for the charging of smartphones. The battery capacity in this case is smaller, also the charging power is smaller which means that the charging time is similar.

31.5 Modelling of the ESMS System

We created a simulation in Matlab 2018b from Mathworks to calculate the charging for the device batteries. Several simulations were run to see the effects on the irradiation and the initial SOC of the batteries. The simulations were run for two devices a laptop and a smartphone.

The simulations were run for different solar irradiation conditions and for different initial SOC for the batteries of the devices and the nanogrids own battery.

From the measured charging characteristics a SOC characteristic can be created. According to Fig. 31.1. the 80% SOC is the point where the constant current charging is changed to constant voltage charging. We assumed it for the batteries of both types of devices and for the nanogrid battery also. To the 80% value the SOC is increasing linearly and from there on the increase is $1\text{-}e^{-At}$ type till it reaches the 100% SOC value. We do not apply trickle charging, as most of the devices use lithium-ion batteries and this type of battery does not tolerate trickle charging.

To calculate the charging process the characteristics of the battery charging, the SOC of all batteries, the solar irradiance and the time of the calendar events are necessary. If all these information are present, we can simulate the charging of the devices.

One goal is to calculate if the batteries of the devices can be charged for the calendar event or if the calendar event is in time too close then the SOC of the device battery is the output. Also the SOC of the nanogrid battery is to be calculate.

To make the calculations a bit simpler we just used two type of solar irradiance, one for the shaded day (diffuse light) and one for total irradiance (the nominal power of the PV cells is reached).

We run the simulations for different cases. In Table 31.1 we see the results of the simulation. The time resolution were 5 min for the simulation. An event was scheduled for 60 min after the start of the simulation, so charging of the devices batteries was forced. We run two simulations with the same starting conditions of batteries and the same scheduled event but different solar irradiance.

We can see in Table 31.1 in the grey row, that for the scheduled event we get the phones battery charged to 44% and the laptop battery to 62% level. The remaining cells shows if the event is pushed further away for further 110 min then we can charge both the phones and laptops battery, whereas charging is forced in all cases.

The difference in irradiation is not seen in the Table 31.1 because the nanogrid battery was initially also at 10% SOC and the priority is on the charging of the devices battery.

If we examine the level of the nanogrids battery at the end of the charging (after 170 min) we can see that in the case one the battery SOC is 15% whereas in case two the SOC is only 9%. In case one we had overall higher irradiance than in case two as we can see on the SOC of the nanogrids battery.

31.6 Summary

A new nanogrid concept was proposed where the batteries of ICT devices in an office were also included in the nanogrids energy storage. This raises new possibilities and also new problems for the ESMS. The ESMS has more possibilities to store energy and select batteries to charge but the demand of the users of the devices should be taken into account. This means that the ESMS system should have knowledge about the activity of the users to be able to fully charge the laptops and phones batteries before the user leaves the office for an appointment.

The usual variability in renewable energy sources as solar irradiance and wind makes it hard to predict the necessary time for the charging. If the nanogrid has its own battery than the case can be easier as long as the battery's charging level is sufficient for charging the ICT devices batteries.

The simulations shows that the ESMS can calculate with the storage capacity of the ICT devices and it is also possible to calculate the necessary time for charging the ICT devices battery for the out of office usage.

Table 31.1 The charging of laptop and phone batteries

Time from start [min]	Phone battery SOC [%]	Laptop battery SOC [%]
0	11	11
5	13	15
10	16	19
15	19	24
20	22	28
25	25	32
30	27	36
35	30	41
40	33	45
45	36	49
50	39	54
55	41	58
60	44	62
65	47	66
70	50	71
75	53	75
80	55	79
85	58	82
90	61	85
95	64	87
100	67	90
105	69	92
110	72	94
115	75	95
120	78	97
125	81	99
130	83	100
135	85	100
140	88	100
145	90	100
150	92	100
155	94	100
160	96	100
165	98	100
170	99	100

According to the measured results a small office for two persons could be supplied by three PV panels and the usual battery could be extended with the batteries of the office devices.

References

Blaabjerg F, Ionel DM (2017) Batteries and ultracapacitors for electric power systems with renewable energy sources. In: renewable energy devices and systems with simulations in MATLAB® and ANSYS®, Boca Raton, FL, Taylor & Francis Group, 2017, pp. 320–334

Buchman A, Lung C (2018) State of charge and state of health estimation of lithium-ion batteries. in: 2018 ieee 24th international symposium for design and technology in electronic packaging (SIITME) pp. 382–385. 2018 https://doi.org/10.1109/siitme.2018.8599196

Burmester D, Rayudu R, Seah W, Akinyele D (2017) A review of nanogrid topologies and technologies. J Renew Sus Energy Rev 67:760–775. https://doi.org/10.1016/j.rser.2016.09.073

Cserép A, Halász Á, Iváncsy T, Tamus ZÁ (2017) A nanogrid concept for supplying ICT devices to improve the energy efficiency of small offices. In: Second international conference on DC microgrids (ICDCM), nuremburg, IEEE 2017, pp. 440–444. https://doi.org/10.1109/icdcm.2017.8001082

Díaz NL, Wu D, Dragičević T, Vásquez JC, Guerrero JM (2015) Stored energy balance for distributed PV-based active generators in an AC microgrid. In: 2015 IEEE power and energy society general meeting, Denver, CO, pp. 1–5. https://doi.org/10.1109/pesgm.2015.7286330

Du W, Chen Z, Wang HF, Dunn R (2007) Energy storage systems applied in power system stability control. In: 42nd international universities power engineering conference, Brighton, 2007, pp. 455–458. https://doi.org/10.1109/upec.2007.4468989

Hou Z-Y, Lou P-Y, Wang C (2017) State of charge, state of health, and state of function monitoring for EV BMS. In: 2017 IEEE international conference on consumer electronics (ICCE), Las Vegas, NV, pp. 310–311. https://doi.org/10.1109/icce.2017.7889332

Isikman AO, Yildirim SA, Altun C, Uludag S, Tavli B, Yldrm SA (2013) Optimized scheduling of power in an Islanded microgrid with renewables and stored energy. Globecom 2013 workshop—management of emerging networks and services (GC13 WS—MENS). Atlanta, USA, pp 861–866

Lujano-Rojas JM, Monteiro C, Dufo-López R, Bernal-Agustín JL (2012) Optimum load management strategy for wind/diesel/battery hybrid power systems. J Renew Energy 44:288–295. https://doi.org/10.1016/j.renene.2012.01.097

Teleke S, Oehlerking L, Hong M (2014) Nanogrids with energy storage for future electricity grids. In: IEEE PES T D conference and exposition. pp. 1–5 https://doi.org/10.1109/tdc.2014.6863235

Topan PA, Ramadan MN, Fathoni G, Cahyadi AI, Wahyunggoro O (2016) State of charge (SOC) and state of health (SOH) estimation on lithium polymer battery via Kalman filter. In: 2nd international conference on science and technology-computer (ICST), Yogyakarta, 2016, pp. 93–96. https://doi.org/10.1109/icstc.2016.7877354

Vilathgamuwa DM, Jayasinghe SDG, Lee FC, Madawala UK (2011) A unique battery/supercapacitor direct integration scheme for hybrid electric vehicles. In: IECON 2011–37th Annual conference of the IEEE Industrial Electronics Society, Melbourne, VIC, 2011, pp. 3020–3025. https://doi.org/10.1109/iecon.2011.6119791

Xia X, Wei Y (2016) Lithium-ion batteries State-of-charge estimation based on interactive multiple-model Extended Kalman filter. In: 22nd international conference on automation and computing (ICAC), Colchester, 2016, pp. 204–207. https://doi.org/10.1109/iconac.2016.7604919

Zhang X, Hao M, Liu F, Yu C, Zhao W (2012) Analysis and control of energy storage systems in microgrid. Second Int Conf Intell Sys Design Eng Appl, Sanya, Hainan 2012:1375–1379. https://doi.org/10.1109/ISdea.2012.529

Zhu L, Zhou X, Zhang X-P, Yan Z, Guo S, Tang L (2018) Integrated resources planning in microgrids considering interruptible loads and shiftable loads. J Mod Power Sys Clean Energy 802–815 https://doi.org/10.1007/s40565-017-0357-1

Part V
Sustainable Manufacturing

Chapter 32
Prediction of Width and Thickness of Injection Molded Parts Using Machine Learning Methods

Olga Ogorodnyk, Ole Vidar Lyngstad, Mats Larsen, and Kristian Martinsen

Abstract Injection molding is one of the major processes applied for production of thermoplastic products. Thermoplastic materials are used in manufacturing of dozens of products seen in everyday life, such as: car bumpers, children toys, bodies of electronic devices, etc. At the same time, plastic pollution is a well-known problem. One of the sources of this pollution is plastic scrap, which might appear because of using faulty process parameters during production process. To decrease amount of scrap, injection molding needs to include better process and quality control routines. Quality of a product can be defined in different ways and product dimensions can be one of criteria for accepting or declining a product. The following paper applies machine learning (ML) methods to predict width and thickness of the injection molded HDPE dogbone specimens with 4 mm thickness based on process parameter values used to produce the parts. Data used for creation of regression models with help of ML methods was acquired during an experiment, which included 160 machine runs during which 47 machine and process parameters were logged. Application of ML methods for training of product dimensions prediction models will increase overall intelligence level of injection molding machines and their compliance with Industry 4.0 standards. Beforehand prediction of product's dimensions will allow to decrease amount of scrap and energy consumption. This will contribute to more environmentally conscious use of thermoplastic materials and more sustainable design of manufacturing systems.

Keywords Machine learning · Artificial neural networks · k-nearest neighbors · Decision trees · Injection molding · Quality prediction

---

O. Ogorodnyk (✉) · K. Martinsen
Department of Manufacturing and Civil Engineering, Norwegian University of Science and Technology (NTNU), Gjøvik, Norway
e-mail: olga.ogorodnyk@ntnu.no

O. V. Lyngstad
Department of Materials Technology, SINTEF Manufacturing, Raufoss, Norway

M. Larsen
Department of Production Technology, SINTEF Manufacturing, Raufoss, Norway

Y. Kishita et al. (eds.), *EcoDesign and Sustainability I*, Sustainable Production, Life Cycle Engineering and Management, https://doi.org/10.1007/978-981-15-6779-7_32

32.1 Introduction

In the last thirty years, the popularity of the injection molding (IM) process had an increasing growth due to new applications in the fields of appliances, packaging and automotive industry (Fernandes 2018). As a result, today injection molding is one of the most frequently used processes for the production of thermoplastic parts in high volume and low cost.

Injection molding includes only four main phases: plasticization, injection, cooling and ejection, however, it is a rather complicated process due to the presence of non-linearities (Chen 2008). At the same time, due to its use for mass production, high process repeatability is extremely important (Ogorodnyk and Martinsen 2018) and keeping quality of the products as high and as similar as possible is a "must".

Depending on the application of the manufactured product, its quality can be defined in different ways. Usually, the good quality of the part means desired mechanical performance, dimensional consistency and proper appearance (Fernandes 2018). "*Dimensional consistency is a critical attribute for injection molded part quality and is highly dependent on various processing parameters*" (Panchal and Kazmer 2010). Product dimensions can be one of the criteria for accepting or declining a thermoplastic product.

The part quality, including dimensional accuracy, can be influenced by a significant number of factors such as general condition of the injection molding machine (IMM), mold that is in use, input material condition, drifting of the process parameter settings or the machine operator's fatigue (Kozjek et al. 2019). Some of the process parameters that may cause product quality variations are melt temperature, mold temperature, holding pressure, cooling time, etc. (Xu and Yang 2015). Because of the faulty setting of some of those parameters, such defects as warpage, sink mark, air traps, and weld lines might occur. As a result, prediction of final parts quality using certain process parameter values and "*optimal setting of injection molding process variables plays a very important role in controlling the quality of the injection molded products*" Mathivanan et al. 2010; Wortberg and Schiffers 2006).

The optimal process parameter settings were used to be determined by engineers and IMM operators based on their experience, intuition and trial-and-error (Guo 2012; Shi 2003). However, due to the increasing quality requirements new approaches for optimization of the plastics injection molding are being continuously developed. "*Researchers introduced the design of experiment (DOE), Taguchi orthogonal array and flow analysis software such as Moldflow Plastic Insight*" (Cheng et al. 2013).

Apart from statistical process control, design of experiments and Taguchi approach, machine learning methods show all of the necessary capabilities for the development of predictive models for the quality of the injection molded products (Cheng et al. 2013). Moreover, they have been proven to be better at dealing with non-linearities in comparison to conventional statistical methods such as linear regression (Ogorodnyk and Martinsen 2018).

Quality requirements are not the only challenge that the thermoplastics injection molding industry faces today. Another important issue is plastic pollution, which became a problem for the whole world. Keeping process parameters under control and beforehand prediction of quality of the plastic products can decrease amounts of produced plastic scrap and energy consumption, contributing to the more environmentally conscious use of the plastic materials.

This paper uses data from 160 machine runs during which 47 machine and process parameters were logged while producing dogbone specimens type A defined by ISO 527–2 (ISO 2012). REPTree (decision tree), random forest, k-nearest neighbors (kNN) and multilayered perceptron (MLP) machine learning algorithms were applied to create models for prediction of width and thickness of the focus parts. The results are discussed in terms of the ability to accurately predict the part's quality and interpretability of the chosen methods.

32.2 Literature Review

In recent years new methods for quality prediction and optimization of the injection molding process have been proposed and applied. Some of the approaches include the application of machine learning methods (Chen 2008; Manjunath and Krishna 2012; Kuo et al. 2007; Lotti et al. 2002), the use of simulation approaches (Berti and Monti 2013; Fu and Ma 2018) and the development of particular hardware solutions (Panchal and Kazmer 2010; Johnston 2015).

Machine learning methods

Machine learning methods (ANN, genetic algorithm, self-organizing maps, etc.) have been used for the development of prediction models of parts shrinkage, general parts quality, to find solutions for multi-objective optimization problem of the injection molding process parameters, etc. Different researchers have used numbers of samples ranging from 27 to 1000, some of them were collected through the simulation software, others during laboratory experiments.

For example, in (Chen 2008) the authors used a combination of a self-organizing map and a back-propagation neural network to create a dynamic quality predictor for the injection molding process. Nine process parameters were included in the model to predict the weight of the final part. To enhance the performance of the neural network, Taguchi's parameter design was also utilized. The dataset included 160 samples of experimental data. Manjunath and Krishna (Manjunath and Krishna 2012) applied forward and reverse mapping ANN to predict dimensional shrinkage of the produced part and appropriate set of process parameters to reach the required dimensional shrinkage correspondingly. The networks were trained using 1000 samples generated in the simulation software using equations reported by other researchers. In (Ogorodnyk et al. 2018) the multi-layered perceptron artificial neural network model and J48 decision trees algorithm is used to create models for prediction of injection molded parts quality. Data from 160 machine runs is utilized to train the models.

At the same time, Kuo, Su (2007) applied Taguchi quality method to establish the design of experiment for nine process parameters and analysis of variance to define the most important factors influencing the production process, while back-propagation neural network model was trained to tune the optimum conditions received from the Taguchi's quality method. The model was created using 180 data samples. In (Cheng et al. 2013) a methodology that includes variable complexity methods, constrained non-dominated sorted genetic algorithm, back-propagation neural network and MoldFlow analysis are applied to solve the multi-objective optimization problem of injection molding parameters. The data is generated using the MoldFlow analysis software.

In (Liu 2017) the orthogonal experiment with Taguchi method was performed, and ANOVA analysis was carried out to define which parameters are the most influential for the injection molding process. In total 81 samples were obtained and used to train models with the help of the back-propagation neural network and multi-class support vector machine methods. Finally, the multi-objective optimization was performed on the obtained models using the nondominated sorting genetic algorithm.

Lotti et al. (2002) used DOE and ANNs to create models for prediction of shrinkage of iPP plaques based on 30 data samples collected. The neural network model has shown better results in comparison to those calculated with the help of the MoldFlow™ software package. In (Gao 2018) injection process parameters optimization procedure is proposed. A combination of the response surface methodology, artificial neural networks, radial basis function, and Kriging surrogate is used.

Nagorny, Pillet (2017) collected thermal images of 204 rectangular specimens, as well as signals of pressure and temperature sensors situated in the mold and used the data to train models using support vector regression, random forest, k-nearest neighbors, stochastic gradient descent, bagging decision tree, Ada boosting decision tree, as well as convolutional neural network and long-short term memory network. Neural networks models show better results in comparison to other regression methods for prediction of parts quality. While in (Altan 2010) Taguchi method, design of experiments, analysis of variance were used to design the experiment and choose the most influential process parameters. In total 27 samples of data were collected and used in this study. ANN model was trained to predict the part's shrinkage in injection molding.

Simulation approaches

Another approach to optimization of the injection molding process is the use of numerical simulations and different sorts of simulation software, such as in (Berti and Monti 2013), for example. Here an approach based on a combination of numerical simulations, response surface methodology, and stochastic simulations is proposed to create a virtual prototyping environment to enable robust optimization of the IM process. At the same time, in (Fu and Ma 2018) authors propose a method to simulate the influence of the early part ejection on its final quality using the integration of MoldFlow™ and Ansys™ on the contrary to the stand-alone molding simulation.

Liau, Lee (2018), on the other hand, explain that simulation should not be based only on the historical data, but needs to include current process data to make a meaningful decision. Their solution is a framework for the development of a digital twin for injection molding that models the entire process and allows bidirectional control of the physical process. In (Guo 2012) fractional factorial design of experiments was used to screen some of the injection molding process parameters obtained from the process simulation in the MoldFlow™ software. Later the parameters were used to create a mathematical model with the help of central composite design and finite element simulation to predict warpage during the plastic injection molding.

Hardware solutions

Some other researchers use hardware development to enhance control and prediction of the outcomes of the IM process. For example, in (Panchal and Kazmer 2010) the authors propose a button cell type in-mold shrinkage sensor. The sensor's performance was validated and compared to traditional shrinkage prediction methods. "*The sensor signals acquired during each molding cycle were analyzed to validate the sensor performance in a design of experiments as a function of packing pressure, melt temperature, cooling time, and coolant temperature*" (Panchal and Kazmer 2010). In (Johnston 2015) an auxiliary process controller for online multivariate optimization of the injection molding process is proposed. The objective function includes terms related to process variation and energy control.

32.3 Experimental Setup

The research described in this paper includes the following steps:

Step 1: Process data collection. To collect the data design of experiment has been used, namely Latin Hypercube sampling technique (Seaholm et al. 1988). 32 combinations of *holding pressure, holding pressure time, backpressure, cooling time, injection speed, screw speed, barrel temperature, and mold temperature* process parameters were included in the DOE. Each of these combinations has been launched five times, resulting in 160 machine runs in total and 320 HDPE dogbone specimens, since during each run two of them are produced. This should be sufficient for the application of machine learning algorithms, as the minimum recommended number of samples according to (Scikit-Learn 2018) is 50.

The use of DOE resulted in the following ranges for width and thickness [mm] for the dogbone specimens 1 and 2: thickness1 $\in$ [3.3, 3.84]; thickness2 $\in$ [3.32, 3.84]; width1 $\in$ [9.02, 9.95]; width2 $\in$ [8.84, 9.96]. According to the CAD model of the target parts, the width is 10 and the thickness is 4 mm. However, 3–3.5% shrinkage rate is considered a regular shrinkage rate for the HDPE, thus resulting in a width of 9.65–9.7 mm and thickness of 3.86–3.88. It is possible to say that the use of the design of experiments allowed to vary the width and thickness of the parts quite significantly covering values close to the best possible ones, as well as those significantly lower.

During the experiment “ENGEL insert 130” vertical injection molding machine with CC300 control unit has been used and 47 machine and process parameters from the built-in by the machine manufacturer sensors were logged. The logging system has been developed using the Python programming language to be able to establish a connection with the machine and access all the necessary process parameter values.

Step 2: Data pre-processing. The logged experimental data included values of different machine and process parameters for every 0.5 s during every production cycle. However, some of the signals vary less than the others during one production cycle but change a lot from one cycle to another. As a result, the logged data needed certain “pre-processing” before any further use. Pre-processing resulted in one value of a process parameter per production cycle, some of the parameter values were averaged, some were taken the maximum or minimum value of, etc. An example of the structure of the data after the “pre-processing” is shown in Table 32.1. The top row includes names of the logged parameters and each following row corresponds to one data sample. The “pre-processing” has been done using scripts developed in the Python programming language.

Step 3: Quality data collection. After the production of the focus parts, the quality data needed to be obtained. Since dimensional consistency is one of the important criteria for the part’s acceptance, the produced dogbone specimens were measured using ZEISS DuraMax coordinate measuring machine (DuraMax 2019).

The measurements have been done according to the “*ISO 16012: Plastics—Determination of linear dimensions of the test specimens*” (ISO 2015), the accuracy of the machine in the temperature 18–22 °C is ±2.4 μm. The precision error doesn’t exceed ±0.02 mm for dimensions <10 (thickness) and ±0.1 (width) for ≥ 10. Figure 32.1 depicts one of the specimens being measured by the coordinate measuring machine.

Step 4: Important parameters selection. After obtaining both process and quality data, it was necessary to understand which parameters should be included in the quality prediction model. First, all parameters that had constant values during all 160 runs were excluded.

Secondly, six parameters (*machine time, shot counter, good parts counter, bad parts counter, parts counter and machine date*) were eliminated, because they do not contain any necessary information about the process.

Thirdly, feature selection algorithms were applied to define which parameters out of 25 that were left are the most influential. According to the Correlation-based feature selection algorithm with the Best First search algorithm only seven parameters should be included in the model: *cushion after holding pressure, plasticizing time, holding pressure time, cushion smallest value, injection work, holding pressure and*

Table 32.1 Experimental data

ID	Max screw speed	Pressure at switchover	Cushion size after holding pressure
1	239.18	1058.69	4.90
2	239.40	1059.35	4.92
3	239.52	1059.62	4.98

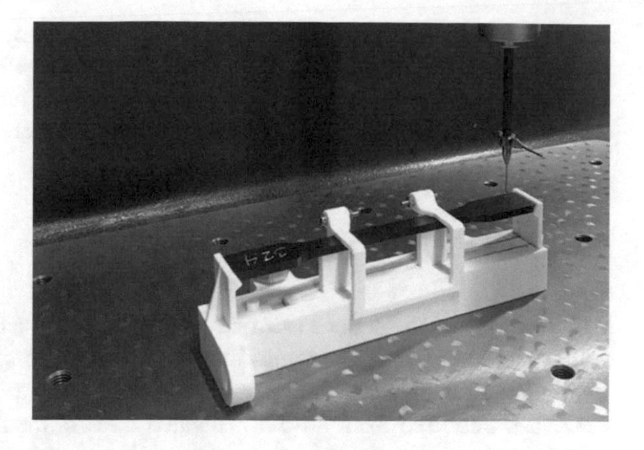

Fig. 32.1 Coordinate measurement of the specimens

tool temperature. If Greedy Stepwise search algorithm is used instead of the Best First one, then the same seven parameters were chosen, as well as *the last ejector position* parameter.

ReliefF algorithm, on the other hand, doesn't exclude unnecessary in its opinion parameters but ranks all the parameters. The higher the ReliefF score gets a parameter, the more important it is. If the ReliefF score is negative, it means that the parameter is unimportant. Parameters that scored lower than 0.02 were considered unimportant, thus resulting in 18 parameters that were to be included into the width and thickness prediction models (*cushion after holding pressure, tool temperature, holding pressure, backpressure, injection speed, cushion smallest value, cushion average value, cooling time, holding pressure time, barrel temperature, average temperature in the zone 2 of the nozzle, pressure at the switchover, maximum screw speed, screw speed, plasticizing time, last cooling time, flow number and injection work*).

Step 5: Training of the width and thickness prediction models. The last step includes training the predictive models for the width and thickness of the focus parts. Separate models for prediction of width and thickness are built for the first and the second dogbones produced during a single cycle. This results in four models per method (model for prediction of the thickness of dogbone 1, the width of dogbone 1, the thickness of dogbone 2 and width of dogbone 2) and the number of parameters included. Decision trees, random forest, k-nearest neighbors and artificial neural network algorithms are used to create the models.

32.4 Brief Description of the Used Machine Learning Methods

The ML algorithms used in this study (REPTree decision tree, random forest, k-nearest neighbors and multilayer perceptron) were induced and evaluated using WEKA – WAIKATO Environment for Knowledge Analysis (Waikato Environment for Knowledge Analysis 2019). 10-folds cross-validation was used to estimate the skill of the model on the new data. The correlation coefficient, mean absolute error and root mean squared error was used to estimate the difference between the measured and predicted values of width and thickness of the dogbone parts.

32.4.1 REPTree (Decision Tree)

REPTree stands for Reduced Error Pruning Tree. It is a decision tree algorithm that follows the regression tree logics through the creation of multiple trees in different iterations (Kalmegh 2015), information gain is used as a splitting criterion in this case. The best tree from all the generated ones is then chosen and considered as representative. Pruning of the tree is done using the mean square error on the predictions of the tree. The result of the algorithm application is a predictive model in the form of a decision tree or a set of if–then rules. In this study minimum number of instances in a leaf has been tested to see how it would influence the prediction quality of the obtained model.

32.4.2 Random Forest

Random forest is an ensemble learning algorithm that includes many separate learners (Breiman 2001). Each tree is created using a random sample of cases from the dataset. The collection of several tree predictors, in this case, is called a forest. In the case of regression, the model's output "*is the average of the responses over all the trees in the forest*" (Kalmegh 2015). In the experiment, different numbers of trees in the forest are tested.

32.4.3 K-Nearest Neighbors

The k-nearest neighbors algorithm uses parameter similarity to predict values of the new data points (Kononenko and Kukar 2007). The new point is assigned value depending on the values of instances closest to it. The number of closest instances is the k input parameter of the algorithm. In the case of regression, the average value of the closest data points is taken as the model's output. Euclidean distance is used in the experiment to calculate distances between the data points, a different number of nearest neighbors is also used to build the models.

32.4.4 Multilayer Perceptron

The last algorithm used in this experiment is the Multilayer Perceptron (MLP) ANN. It is one of the classic ANN models and is based on the sequence of layers of neurons interconnected between each other with weights. Every time a new sample of data is given to the network, those weights are adjusted accordingly. In addition to that, "*layer-to-layer mapping is activated with a non-linear function*" (Ogorodnyk et al. 2018). In this study, the sigmoid function is used as an activation function and the number of neurons in the hidden layer of the network is calculated as (*number_of_parameters* + *number_of_output_neurons*)/2. In this experiment, the MP was built using a different number of parameters as suggested by different feature selection algorithms (7, 8, 18 and 25).

32.5 Results

Predictive models for the thickness and width of the injection molded dogbone specimens have been built using different machine learning algorithms presented in the previous section. Depending on the algorithm, parameter configurations were varied to see how they affect the quality of the resulting model. Figure 32.2 shows the dependence of the correlation coefficient of the measured and predicted values of the part's dimensions and different algorithm parameters that were tested. The quality of the models was estimated using the 10-folds cross-validation procedure.

Table 32.2 shows the performance of tested algorithms using the correlation coefficient, mean absolute error (MAR) and root square mean error (RSME). The numbers are shown for the best algorithm parameter configurations. For the REPTree the best minimum number of instances in a leaf for thickness1, thickness2, and width2 is equal to 2, while for width1 it is 5. In the case of the random forest algorithm, the number of trees that gives better results for thickness1 and width1 is 100, for thickness2 and width2 it is 200. At the same time for the kNN method the best numbers of neighbors are 5 and 1, the first number shows better results for the thickness1 and width1, while the second number for thickness2 and width2. For the MLP, on the other hand, 8 features have shown the best result for the width2 model, 25 for thickness2 and 18 for thickness1 and width1 models.

32.5.1 Interpretation

The only algorithm out of those used in this study that is easy to interpret for a human being is REPTree decision tree algorithm. Figure 32.3 shows the structure of a decision tree with the minimum number of instances in a leaf equal to 2 that has shown the best results for prediction of the width of dogbone 1.

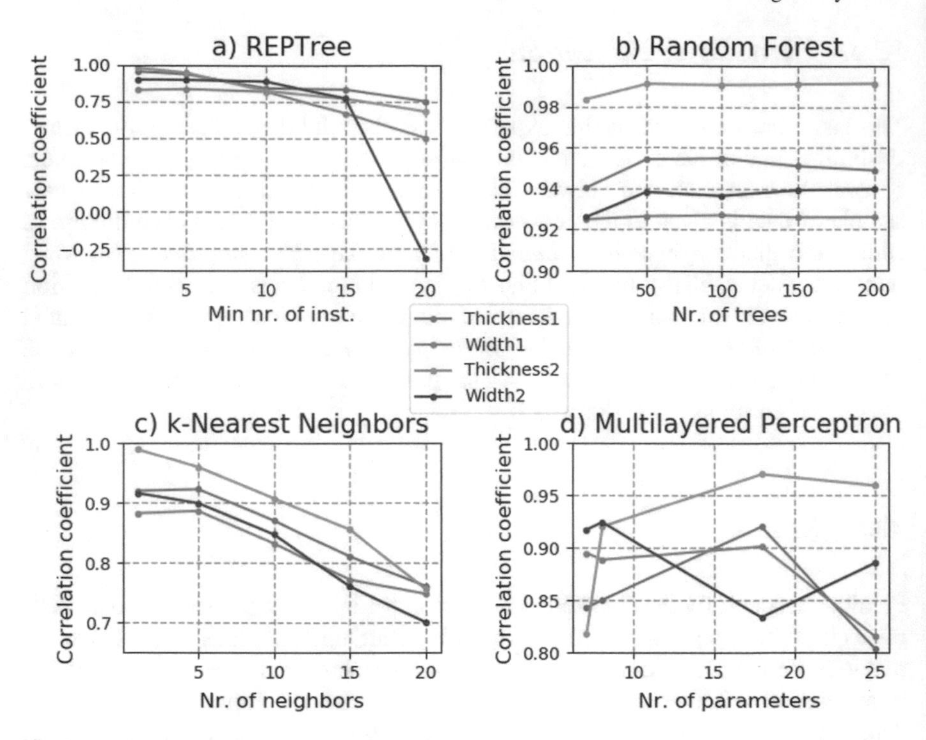

Fig. 32.2 Prediction quality of tested algorithms

Table 32.2 Performance of the tested algorithms

Perf. meas	REPTree	Random Forest	kNN	MLP
Thickness 1				
Corr. coef	0.9566	0.9547	0.9226	0.9203
MAR	0.0172	0.0171	0.0227	0.0243
RMSE	0.0355	0.037	0.047	0.0528
Width 1				
Corr. coef	0.8357	0.9271	0.8863	0.9014
MAR	0.0722	0.0499	0.0788	0.0644
RMSE	0.1172	0.0788	0.0976	0.1002
Thickness 2				
Corr. coef	0.9811	0.9913	0.9893	0.977
MAR	0.0144	0.0118	0.0108	0.0198
RMSE	0.0261	0.0189	0.0194	0.0388
Width 2				
Corr. coef	0.9017	0.9396	0.9159	0.9245
MAR	0.0697	0.0516	0.0621	0.0785
RMSE	0.1288	0.1012	0.1213	0.118

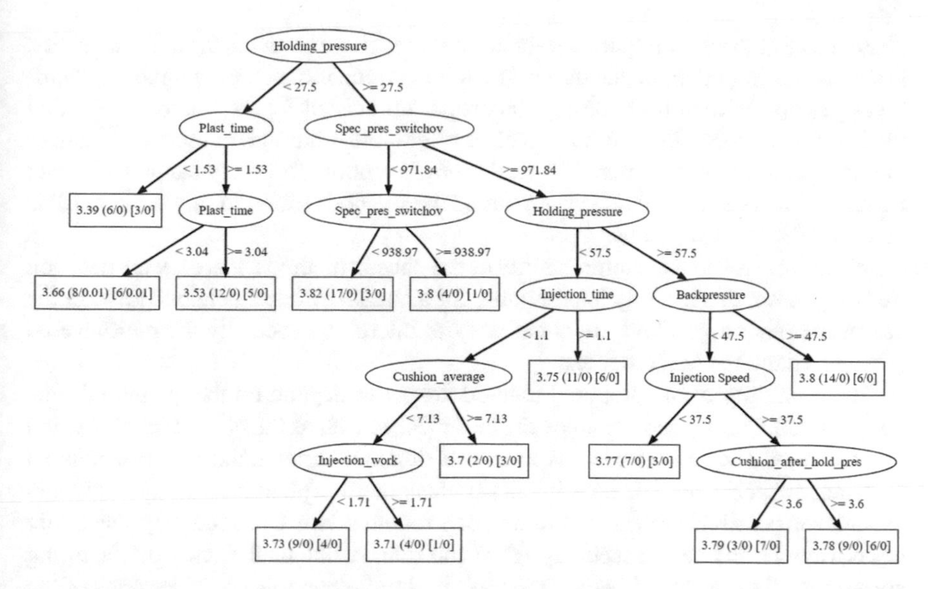

Fig. 32.3 Obtained structure of a decision tree for prediction of thickness1

Since pruning was enabled during the algorithm's work, not all the 25 parameters were included in the tree. The decision tree uses *holding pressure, plasticizing time, pressure at switchover, injection time, average cushion value, injection work, back-pressure, cushion after holding pressure and injection speed* as parameters to take the decision about the value of the thickness of the dogbone specimen.

For example, if looking at the tree from the very top, we can easily get the following: if the value of the holding pressure is less than 27.5, and value of plasticizing time is less than 1.53, then, on average, the value of the width of the first dogbone will be 3.39. Numbers in the brackets indicate the number of samples belonging to that leave and number of instances from the pruning.

32.5.2 Discussion

The results show that all the machine learning methods used in this study show the high capability of predicting the width and thickness of the focus parts. The best results in terms of the correlation coefficient are achieved using the random forest algorithm, while the second-best result is received through the application of the MLP. At the same time, REPTree decision tree performs a little bit better, than kNN. However, overall all four methods show quite high prediction capabilities on this data. Overfitting should not be present since the 10-folds cross-validation technique has been used in order to estimate skill of the models on the new data.

All these algorithms require certain tuning to receive a model of the highest quality. For kNN it is important to find the best matching number of nearest neighbors, multi-layer perceptron requires choosing the proper number of layers, neurons, selecting the learning rate, etc. In the case of decision trees, the minimum number of instances in a leaf is an important parameter, while for the random forest choosing the correct number of trees is crucial. However, out of the methods tested in this study MLP is probably the hardest to tune.

When it comes to the interpretability of the chosen methods, it is only the decision trees algorithm that is easily interpretable for a human. Models obtained through the use of the other algorithms are not as easy to interpret, especially the multilayered perceptron and the random forest.

The reliability of the proposed methodology can depend on the quality of data, computational tools used to induce the methods, as well as the correct interpretation of the results. The data used in this study was obtained through the above-mentioned laboratory experiment. The experiment was designed and planned in a way that would exclude the possibility of getting missing or erroneous data. Unfortunately, that might not always be the case, especially when the data is obtained from manufacturing companies. As mentioned before, the use of DOE helped to vary dimensions of the produced parts quite significantly. This allowed obtaining a dataset that includes dimension values close to the ideal ones, as well as those remarkably lower.

Use of different knowledge discovery platform or application of variation of any machine learning algorithm presented might lead to receiving results that are slightly different from the ones presented in this study. The ability to correctly interpret obtained results is also crucial, as a misunderstanding of certain model quality measures might lead to missing important aspects related to the model's performance capabilities.

The models presented in this study are capable of predicting dimensions of only the dogbone part since the dataset they are trained on doesn't include samples related to any other products. To create models for the prediction of dimensions of other products, the corresponding dataset needs to be obtained at first. At the same time, it is hard to say how accurate will the models' predictions be if the mold for production of the dogbones is changed. However, it is possible to assume that if mold has properties similar to the one used in the experiment, the models' predictions should be of similar quality. Such an approach could be a good starting point for manufacturing industries, including thermoplastics injection molding, to start analyzing and interpreting their data.

Amounts of data logged through manufacturing execution systems, enterprise resource planning systems, etc. continue to grow. There might be interesting and unrecognized at the moment patterns, that could potentially improve the quality of manufactured products, the overall flow of the production process, decrease amounts of produced scrap and energy consumption.

Future work includes obtaining more data, possibly for products of different geometry and material and testing machine learning methods on larger amounts of data. In addition to the real production or experimental data, simulated data can be used to pre-train the prediction models for further use of transfer -learning, as shown in

(Tercan 2018). This might help to increase the generalization abilities of the models so that they are able to predict dimensions and final quality not only for the dogbone specimens from HDPE but for a bigger variety of products and materials. It is also planned to do extended optimization of the model's parameters to possibly increase the prediction quality.

32.6 Conclusion

Injection molding is one of the major processes for the manufacturing of products from thermoplastic polymers. It is easily suited for the production of large amounts of plastic parts of different shapes and sizes (Mathivanan et al. 2010). However, the final quality of the plastic product depends on many factors, such as injection molding machine, mold in use and parameter settings (Liu 2017). Choosing the correct production parameters might be a complicated task leading to production of large quantities of scrap. That is why it is important to predict the final part quality based on the chosen process parameters.

This study presents an approach that involves the application of the machine learning methods, such as REPTree decision tree, random forest, k-nearest neighbors and multilayer perceptron ANN to create prediction models for the thickness and width of the HDPE dogbone specimens.

The model assessment is based on the application of 10-folds cross-validation, as well as correlation coefficient, mean absolute error and root square mean error. For all the algorithms some of their parameters were varied to see how it would influence the quality of the obtained prediction model.

The results of the study indicate that such an approach has the potential to facilitate the process of finding optimal production conditions and decrease amounts of produced plastic scrap in the field of thermoplastics injection molding.

References

Altan M (2010) Reducing shrinkage in injection moldings via the Taguchi, ANOVA and neural network methods. Mater Des 31(1):599–604

Berti G, Monti M (2013) A virtual prototyping environment for a robust design of an injection moulding process. Comput Chem Eng 54:159–169

Breiman L (2001) Random forests. Mach Learn 45(1):5–32

Chen W-C et al (2008) A neural network-based approach for dynamic quality prediction in a plastic injection molding process. Expert Syst Appl 35(3):843–849

Cheng J, Liu Z, Tan J (2013) Multiobjective optimization of injection molding parameters based on soft computing and variable complexity method. Int J Adv Manuf Technol 66(5–8):907–916

Fernandes C et al (2018) Modeling and optimization of the injection-molding process: a review. Adv Polym Technol 37(2):429–449

Fu J, Ma Y (2018) Computer-aided engineering analysis for early-ejected plastic part dimension prediction and quality assurance. Int J Adv Manuf Technol 98(9–12):2389–2399

Gao H et al (2018) Process parameters optimization using a novel classification model for plastic injection molding. Int J Adv Manuf Technol 94(1–4):357–370

Guo W et al (2012) Prediction of warpage in plastic injection molding based on design of experiments. J Mech Sci Technol 26(4):1133–1139

ISO (2012) ISO 527-2:2017 Plastics—Determination of tensile properties—Part 2: Test conditions for moulding and extrusion plastics. 2012 [cited 2019 31.01.2019]; Available from: https://www.iso.org/standard/56046.html

ISO (2015) ISO 16012:2015 Plastics—Determination of linear dimensions of test specimens. [cited 2019 25.06.2019]; Available from: https://www.iso.org/standard/63481.html

Johnston S et al (2015) On-line multivariate optimization of injection molding. Polym Eng Sci 55(12):2743–2750

Kalmegh S (2015) Analysis of weka data mining algorithm reptree, simple cart and randomtree for classification of indian news. Int J Innovative Sci Eng Technol 2(2):438–446

Kononenko I, Kukar M (2007) Machine learning and data mining: introduction to principles and algorithms. Horwood Publishing

Kozjek D et al (2019) Data mining for fault diagnostics: A case for plastic injection molding. In: 52nd CIRP conference on manufacturing systems (CMS), June 12–14, 2019. Ljubljana, Slovenia: Procedia CIRP

Kuo C-FJ, Su T-L, Li Y-C (2007) Construction and analysis in combining the Taguchi method and the back propagation neural network in the PEEK injection molding process. Poly-Plast Technol Eng 46(9):841–848

Liau Y, Lee H, Ryu K (2018) Digital twin concept for smart injection molding. IOP Conf Ser: Mater Sci Eng. IOP Publishing

Liu J et al (2017) Multiobjective optimization of injection molding process parameters for the precision manufacturing of plastic optical lens. Math Probl Eng 2017

Lotti C, Ueki M, Bretas R (2002) Prediction of the shrinkage of injection molded iPP plaques using artificial neural networks. J Injection Molding Technol 6(3):157

Manjunath PG, Krishna P (2012) Prediction and optimization of dimensional shrinkage variations in injection molded parts using forward and reverse mapping of artificial neural networks. Adv Mater Res. Trans Tech Publ

Mathivanan D, Nouby M, Vidhya R (2010) Minimization of sink mark defects in injection molding process–taguchi approach. Int J Eng Sci Technol 2(2):13–22

Nagorny P et al (2017) Quality prediction in injection molding. in 2017 IEEE international conference on computational intelligence and virtual environments for measurement systems and applications (CIVEMSA). IEEE

Ogorodnyk O et al (2018) Application of machine learning methods for prediction of parts quality in thermoplastics injection molding. Int Workshop Adv Manuf Autom Springer

Ogorodnyk O, Martinsen K (2018) Monitoring and control for thermoplastics injection molding a review. Procedia CIRP 67:380–385

Panchal RR, Kazmer DO (2010) In-situ shrinkage sensor for injection molding. J Manuf Sci Eng 132(6):064503

Scikit-Learn (2018) Choosing the right estimator. [cited 2018 13.05]; Available from: https://scikit-learn.org/stable/tutorial/machine_learning_map/index.html

Seaholm SK, Ackerman E, Wu S-C (1988) Latin hypercube sampling and the sensitivity analysis of a monte carlo epidemic model. 23(1–2):97–112

Shi F et al (2003) Optimisation of plastic injection moulding process with soft computing. Int J Adv Manuf Technol 21(9):656–661

Tercan H et al (2018) Transfer-learning: bridging the gap between real and simulation data for machine learning in injection molding. Procedia CIRP 72:185–190

WEKA—Waikato environment for knowledge analysis. [cited 2019 04.03.2019]; Available from: https://www.cs.waikato.ac.nz/ml/weka/

Wortberg J, Schiffers R (2006) Online quality prediction in injection molding processes (ICM 2006). In: 2006 IEEE International conference on mechatronics. IEEE

Xu G, Yang Z (2015) Multiobjective optimization of process parameters for plastic injection molding via soft computing and grey correlation analysis. Int J Adv Manuf Technol 78(1–4):525–536
ZEISS. ZEISS DuraMax. (2019) [cited 2019 25.06.2019]; Available from: https://www.zeiss.com/metrology/products/systems/coordinate-measuring-machines/production-cmms/duramax.html

Chapter 33
Analysis of Manufacturing Costs for Powder Metallurgy (PM) Gear Manufacturing Processes: A Case Study of a Helical Drive Gear

Babak Kianian and Carin Andersson

Abstract This paper presents a thorough manufacturing cost analysis for a powder metallurgy (PM) manufactured gear (using press and sintering technique—P/S), through a case study of a R&D 4th drive gear (a helical gearwheel) utilizing both real usage data from six companies and some estimated data from literatures. The chosen costs calculation method, performance part costing (PPC), is comprehensive and designed to follow the manufacturing process routes. The result is an excel based cost calculation tool for analysis of performance driven manufacturing costs for use by practitioners with the capability of serving as a decision support system (DSS) in production development issues. The impacts of three scenarios of (1) gear rolling process automation, (2) overall equipment effectiveness (OEE) parameters improvement and (3) increase in energy price on the total PM gear manufacturing costs are simulated, results are presented and discussed, sensitivity analyses are performed, and outcomes are further discussed.

Keywords Manufacturing cost · Powder metallurgy (PM) · Gear manufacturing · Decision support system (DSS) · Sustainable production · Sustainability

33.1 Background and Motivation

The automotive industry has sought for fuel economy and resource efficiency for many decades now. The selection decisions of alternative manufacturing techniques for a particular product (e.g. a gear) are essentially constructed with traditional industry cost structures (Kamps et al. 2018). However, negative sustainability effects of our production activities and their severe consequences globally has urged the manufacturing industry to adopt a more holistic view of technology adoption and deployment. There are many factors to consider when a manufacturer aims at comparing different production techniques e.g. profitability, sustainability, skills

B. Kianian (✉) · C. Andersson
Department of Mechanical Engineering, Division of Production and Materials Engineering, Lund University, Box 118, 221 00 Lund, Sweden
e-mail: babak.kianian@iprod.lth.se

Y. Kishita et al. (eds.), *EcoDesign and Sustainability I*, Sustainable Production, Life Cycle Engineering and Management, https://doi.org/10.1007/978-981-15-6779-7_33

supply and allocation, knowledge (know-how) and information on processes, and material utilization—to just name of few. Nevertheless, there is a lack of thorough methods and tools to assist companies' decision makers in their analysis of both cost and sustainability (Kianian and Andersson 2018; Kianian et al. 2019).

The electrification trend has begun to show its considerable effects on the transportation industry globally (in particular on the automotive industry). New drivetrains need to be designed and produced based on the ever increasing automakers' requirements (Kotthoff 2018) e.g. higher performance, better noise vibration harshness (NVH) behavior, and in the same time the cost is asked to be reduced to gain or sustain market competitiveness. The manufacture of motor vehicles accounts for 100 million per year globally and is uninterruptedly growing. Hence, components manufacturers constantly strive for more performance and cost efficient materials and manufacturing processes (Kruzhanov 2018). Nearly 75% of iron powder production globally is utilized in powder metallurgy (PM) processes, and the component manufacturers for automotive industry consume more than 80% of produced powder components worldwide (Kruzhanov 2018).

Therefore, in general PM technology, and in particular press and sintering (P/S) are rather well established in the manufacturing industry but their usage is limited in drivetrain applications e.g. transmission gears given its current shortcomings in e.g. material properties, technological limitations and performance gaps (Kianian 2019) (Kianian and Andersson 2017). To tackle these limitations, there has been many studies on P/S design and process simulation, performance data generation and also full system validation e.g. demonstrator cases illustrating P/S gear manufacturing technological readiness (Leupold et al. 2017). Nevertheless, beyond those cases showcasing PM production capabilities, studies showing P/S components manufacturing economic aspects e.g. P/S gear manufacturing cost analysis are rare or maybe not publicly available (Kianian 2019; Economic considerations and for powder metallurgy structural parts 2012; Steinert 2015).

In an ongoing research project, a cost and sustainability based decision support system (DSS) is intended to be developed for comparison of conventional machining processes and PM gear manufacturing techniques (Kianian and Andersson 2017). As a baseline, detailed knowledge regarding the manufacturing cost of machined gears has been created using a manufacturing cost model developed in the research project (Kianian and Andersson 2018). Afterward, the current state of the art of P/S gear manufacturing process routes were reviewed based on the final gear's performance and quality. A life cycle costing (LCC) comparative case study between conventional and P/S gear manufacturing techniques acquisition costs indicated that P/S acquisition costs were nearly three times higher than that of machining processes (Kianian 2019). Now as the next step, the main objective of this paper is to perform a detailed P/S gear manufacturing cost analysis with the intention of bridging the gap on lack of studies on P/S gears economic aspects as mentioned in the previous paragraph. The scenarios development and analytical capabilities of the selected manufacturing cost model is also illustrated.

33.2 Manufacturing Cost Analysis

To be able to compare and generate in depth knowledge, a cost model with representation of the important cost drivers that vary between different technologies are required. There are immense variations of methods and models to evaluate manufacturing cost, and literature reviews comparing them for both specific applications and holistically have been done previously (Jönsson 2012; Schultheiss et al. 2018; Windmark 2018), and also reported by this paper's authors (Kianian and Andersson 2018). These reviews show that the performance part costing (PPC) model is the most comprehensive one, and is therefore selected to be used in this study.

The selected PPC is developed to follow the manufacturing process routes. It is created for the discrete manufacturing processes based on batch size and it provides part's (e.g. gearwheel) cost per unit. It was originally developed to assess and compare alternative production systems to support decision makers in manufacturing firms with their production development agenda e.g. selection among different production location alternatives. It integrates technical performance parameters with economical parameters to determine the entire influence of production performance on cost in order to suggest improvement opportunities and find or create best cost efficiency (Jönsson 2012; Ståhl et al. 2007).

The design of the PPC model is based on the principle cost components, e.g. raw material, tool, equipment during operation, standby and downtime, personnel, maintenance and quality cost, which are indispensable for the complete manufacturing of a part from raw material to end-product. Equations (33.1), (33.2) and (33.3) show the cost model and Table 33.1 defines its cost components including their units.

$$\begin{aligned}\boldsymbol{k} = k_A + \frac{k_B}{Q} + \frac{k_{CP} \cdot t_0}{Q \cdot P \cdot 60} + \frac{k_{CS}}{60}\left(\frac{(1-\mathrm{A}) \cdot t_0}{Q \cdot P \cdot A} + \frac{T_{su}}{N_0} + \frac{T_b}{N_0} \cdot \frac{1-U_{IC}}{U_{IC}}\right) \\ + \frac{k_D}{60}\left(\frac{t_0}{Q \cdot P \cdot A} + \frac{T_{su}}{N_0} + \frac{T_b}{N_0} \cdot \frac{1-U_{IC}}{U_{IC}}\right) + \frac{k_M}{60} + \frac{k_{SUM}}{N_0}\end{aligned} \tag{33.1}$$

$$\boldsymbol{k_{CP}} = \frac{K_0 \cdot \frac{i(1+i)^n}{(1+i)^n - 1} + k_{ren} \cdot \frac{N_{ren}}{n} + Y \cdot k_Y + T_{plan} \cdot \left(\frac{k_{Mh}}{h_{P,M}} + k_{ph}\right)}{T_{plan}} \tag{33.2}$$

$$\boldsymbol{k_{CS}} = \frac{K_0 \cdot \frac{i(1+i)^n}{(1+i)^n - 1} + k_{ren} \cdot \frac{N_{ren}}{n} + Y \cdot k_Y}{T_{plan}} \tag{33.3}$$

The complete manufacturing cost per part (k) is calculated in an aggregated procedure where cost associated with each manufacturing step is added as the input cost to the next. The cost of raw material though is incorporated in the first manufacturing step and with doing that, the quality issue cost, which may occur along the manufacturing steps and cost distribution for each manufacturing step can be grasped and

Table 33.1 PPC input parameters (Ståhl et al. 2007)

Description	Symbol	Unit
Running/maintenance hours	$h_{P,M}$	h/h
Hours per year and shift	h_{year}	h/year
Interest rate	i	%
Annual cost of producing a product	K	SEK
Investment	K_0	SEK
Tool, lubricant and additive costs	k_A	SEK/unit
Material cost per part	k_B	SEK/unit
Personnel costs	k_D	SEK/h
Cost of Maintenance	K_M	SEK
Maintenance cost of equipment/hour	k_{Mh}	SEK/h
Eq. running costs (e.g. energy)	k_{ph}	SEK/h
Cost for renovation	k_{ren}	SEK
Additional costs	K_{SUM}	SEK
Annual cost of production area	k_y	SEK/m2
Estimated equipment lifetime	n	Year
Nominal batch size	N_0	Unit
Numbers of operators	n_{op}	-
Number of batches connected to a specific tool	n_{pA}	-
Number of renovation during the lifetime of the equipment	N_{ren}	SEK
Quality	Q	%
Performance	P	%
Availability	A	%
Nominal cycle time per part	t_0	Min
Production time for a batch	T_b	h
Required production time	T_p	h
Total paid and planned production time during a given period	T_{plan}	h
Set-up time	T_{SU}	Min
Machine utilization	U_{IC}	%
Area of production and area needed to facilitate production	Y	M2

evaluated. In the Eqs. (33.2) and (33.3), k_{cp} is the equipment cost during the operation (hourly), and k_{cs} is equipment cost during standby and downtime (hourly) (Ståhl et al. 2007).

33.3 Gear Manufacturing: Industrial Case Study

The chosen object for this study is a R&D 4th drive gear (a helical gearwheel) for a M32 six-speed manual transmission gearbox used in passenger cars such as Opel Insignia or SAAB 95 as shown in the Fig. 33.1 below. The optimized (incl. 12 weight reduction holes) P/S version of this gearwheel has not been mass-produced. The reason for which it is chosen for this case study is the fact that some limited data is either publically available or can be obtained through non-disclosure agreement (NDA) from interviewed companies (Kianian 2019). Based on interviews with PM part makers the batch size of 30,000 parts and the annual production of 400,000 parts is selected for this analysis to prototype a real production scenario.

For the manufacturing cost calculation, at first the manufacturing process routes for the P/S gear manufacturing were identified as illustrated in Fig. 33.2 below. Metal powders are porous material by nature and that consequently influences quality and tolerances of gears manufactured with metal powders due to residual porosity (Kianian 2019). Hence, it is a common practice to introduce a second densification process (e.g. gear rolling) to produce a final gear nearly without any porosity in the surface (Kianian 2019; Andersson and Flodin 2017). The actual production process routes for this study gear had not utilized gear rolling; however, it is included in this case study. Since one of the main objective of this research project is to identify and analyze different P/S process routes for gear production in order to compare their final performance (e.g. strength) and quality (e.g. tolerances) to conventional machined gears.

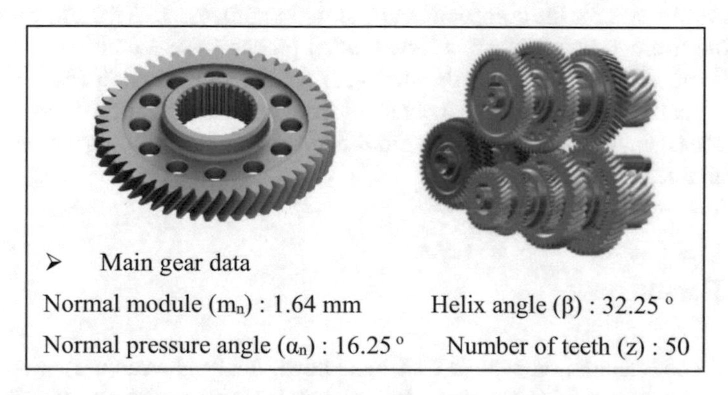

➢ Main gear data
Normal module (m_n) : 1.64 mm
Helix angle (β) : 32.25 °
Normal pressure angle (α_n) : 16.25 °
Number of teeth (z) : 50

Fig. 33.1 **a** CAD model of the gearbox layout (to the right). **b** redesigned 4th drive gear overview with weight reduction holes (to the left) (Larsson et al. 2014; Andersson and Flodin 2017)

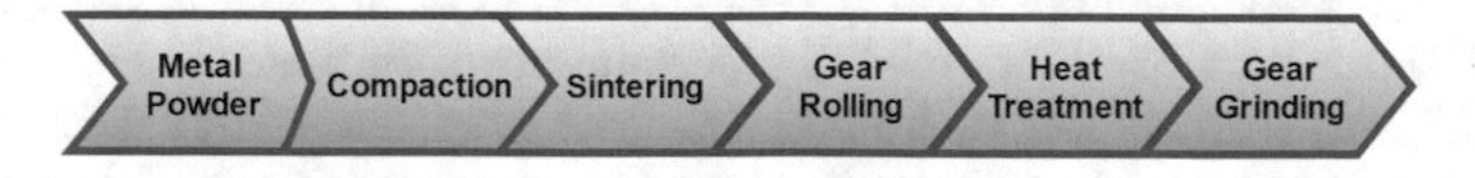

Fig. 33.2 PM process routings for the case study helical gear

As mentioned earlier (see Sect. 33.2) the selected cost analysis method (PPC model) is comprehensive and designed to follow, in this case, P/S gear manufacturing process routes, and that can make the input data gathering a challenge. This challenge has been previously recognized in other studies trying to assess machine tool manufacturing process costs too (Schlosser et al. 2011). In our case, finding a PM gear maker, which has a comprehensive overview and knowledge about the whole P/S gear manufacturing process routes both technical and economical aspects, and on top of that, willing to share this information was not possible.

Hence, drawings and data on both technical performance parameters and cost parameters for each manufacturing step were gathered by a combination of methods e.g. through both face-to-face and phone interviews of a powder metal producer, PM production equipment providers, a PM part maker and a conventional gear producer. The selected interviewees have different responsibilities e.g. production managers, process engineers, sale/marketing managers and high management (e.g. CFO). In total, 6 different companies and 11 people have been interviewed between the years 2016–2018. In addition, data has been collected also by visiting the companies' production facilities and from detailed conversation during different large-scale PM international congress and exhibitions (e.g. Euro PM) between the years 2017–2018. Therefore, the data retrieval challenge mentioned earlier for P/S gear manufacturing process routes has become an advantage to the authors since it motivated them to investigate a broader value chain, resulted to access, and create knowledge for more audiences.

Assumptions were made based on the findings from literatures (Kruzhanov and Arnhold 2012; Zapf and Dalal 1983) when there was a lack of real usage data, which was the case for the compaction and sintering processes of the PM gear in this study. The average values were selected when processes had multiple data points in the literatures. For the metal powder, gear rolling and grinding (incl. washing) processes real usage data could be obtained from the manufactures. Heat treatment (HT) process was outsourced for the production of this case study gear and its cost were available.

33.4 Result

An excel cost calculation tool (a cost breakdown analysis structure) was built as the main outcome of this case study. It is a modular tool to evaluate manufacturing cost with the capability of defining and simulating different scenarios, and it can be customized to follow and fit any company's facilities layout. The principle for the creation of the excel tool is to utilize the PPC model presented in Eqs. (33.1), (33.2) and (33.3) (see Sect. 33.2) for each P/S processing route illustrated in Fig. 33.2 (see Sect. 33.3). In the excel sheet, each processing route is represented by a sheet and then cost drivers (e.g. tool, labor) for each processing route are individually calculated, interconnected and summarized.

General technical and financial data are initially built in the excel tool. These include throughput time (days), annual volume and batch size (parts), cost of capital (%), facility rent (currency/m^2), energy price (SEK/kWh), number of shifts, work time (hour/year), labor cost (currency/hour) and currency exchange rates. For the majority of these general factors the same values as of the authors' previous case study (conventional wrought steel gear manufacturing cost analysis Kianian and Andersson 2018) are given for the comparability purpose. Since in authors' future research, PM and machining gear manufacturing costs will be compared, the hypothesis will be that the location (Sweden) and general costs associated with it for both manufacturing techniques are the same.

The major cost drivers for this case study are material cost (metal powder mixes), tool cost (e.g. compaction tooling cost, investment cost), equipment cost (e.g. press, belt furnace) during operation and downtime, OEE, personnel cost, and partially (due to lack of data) maintenance and quality inspection costs. The distribution of PM gear manufacturing cost, calculated with PPC model in the base case analysis, over its process routes is shown in Fig. 33.3.

The finishing process cost (gear grinding incl. washing of the final part) is the major manufacturing cost driver (35%), based on Fig. 33.3. The reason can be high investment cost, high complexity of the part and high labour cost, at least in this case study, as it accounts for 60% of the total grinding cost. Other studies have been suggested that costly finishing processes (e.g. gear grinding or honing) account for 20–25% of total PM manufacturing costs, and if the strength and quality of PM part after heat treatment (HT) process is sufficient, the finishing processes are not necessary.

Material (25%) and HT (17%) costs are the next significant cost drivers in this study. The interviewed PM part maker outsourced the HT process (carburizing, oil quenching, tempering); therefore, how its cost is calculated is not available to the authors. However, as Fig. 33.3 shows the combination of gear grinding (35%) and HT

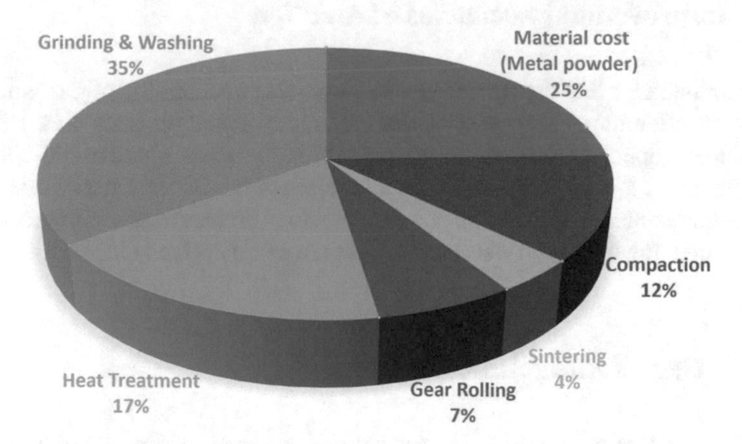

Fig. 33.3 PM gear manufacturing cost distributions over its manufacturing process routes—the base case analysis

(17%) costs account for more than half (52%) of the total PM gear manufacturing cost calculated in this study. Hence, any improvement opportunities in these two processes can benefit PM part makers both economically and environmentally to a large extent, since both metal powders production and HT process are prominent energy consumers. Discussion section (see Sects. 33.5.2 and 33.5.3 below) elaborates on metal powder and HT process more in detail. Optimization of these processes are dependent on the part's material, weight, size and complexity (Kruzhanov and Arnhold 2012).

Compaction cost (12%) is the next major cost driver for this study gearwheel, and the main reason is its very high tool investment cost (e.g. basic and part specific adoptions and helical gear drive system costs) generally and due to the complex helical gear in this case. The equipment (e.g. hydraulic press) cost is high as well, however because of the press high energy efficiency (Kruzhanov and Arnhold 2012) and competitive cycle time (7 stroke/min in this case), the compaction operational cost was only 25% of the total compaction cost in Fig. 33.3. Whereas its tool cost accounts for 65% of the total compaction cost.

Gear rolling (7%) and sintering (4%) processes are the minor cost drivers in this study. Rolling process tool (die and clamping accessories) cost accounts for 45% of the total gear rolling cost in Fig. 33.3, due to many die refurbishment and replacement times for the annual production volume of 400,000 units. Whereas, gear rolling operational cost accounts for 40% of the total gear rolling cost because of its competitive cycle time and very short setup time. Sintering process is considered as the most energy intensive process with rather low efficiency (less than 30%) due to different loses (e.g. radiation, ineffective loading) (Kruzhanov and Arnhold 2012; Zapf and Dalal 1983; Bocchini 1983). The sintering furnace utilized in this study is a belt furnace with conveyor belt width of 0.6 m and sintering capacity of 300 kg/h.

33.5 Improvement Scenarios Simulation

It is mentioned earlier that the PPC model provides capability of designing and simulating different production scenarios e.g. improvement opportunities. Different improvement opportunities regarding gear rolling process automation, and OEE improvement incl. sensitivity analyses were discussed with the interviewees at the different companies. Based on the discussions, two different improvement scenarios are presented, together with a scenario where the energy price is increased.

33.5.1 Gear Rolling Automation

One of the production improvement scenarios is automation of gear rolling process, which increases the equipment investment cost by nearly 27% as/in the base case analysis in this study. Based on data obtained from interviews, the gear rolling automation

Table 33.2 Gear rolling automation sensitivity analysis

Gear Rolling Process Automation Scenario	Investment cost	PM gear cost	Rolling process cost	Equ. cost during operation	Equ. cost during downtime
Lower bound value	20% ▲	0.2% ▼	3.4% ▼	19% ▲	51% ▲
Base case value	27% ▲	0.04% ▼	0.6% ▼	25% ▲	59% ▲
Upper bound value	33% ▲	0.08% ▲	12% ▲	30% ▲	67% ▲

decreases the equipment availability (A) by 5% and consequently decreases the OEE by 4.5% in this case. Results from this scenario simulation indicate that gear rolling automation has a marginal influence on both the total PM gear manufacturing and rolling process costs. These marginal cost reductions are due to eliminating labor cost. Thus, gear rolling cost share after process automation is still 7% of total per part PM gear manufacturing cost as it is illustrated in Fig. 33.3 above. The equipment costs during operation increased by 25% and during downtime or standby increased by 59% in this study.

The gear rolling process's equipment manufacturer and provider in this case study outsources its automation process. Despite the fact that the data, which is utilized in the base case analysis, is recent real in-use data from November 2018, a lower bound value of 25% decrease in automation cost, and an upper bound value of 25% increase in automation cost are selected for sensitive analyses purpose. These lower and upper bound values increase the equipment investment costs respectively by 20% and 33%, as it is shown in Table 33.2 above. The gear rolling cost share is decreased by 1–6% of the total PM gear manufacturing cost per part when the automation cost is reduced, in comparison to the base case analysis shown in Fig. 33.3 (see Sect. 33.4). However, when the automation cost increases, the gear rolling cost share stays the same as the base case analysis being 7%.

33.6 OEE Improvement

Overall equipment effectiveness (OEE) is an important part of the PPC model as it helps to evaluate and visualize the cost impact of reduced downtime and/or the increased quality yield. OEE is calculated based on measurement of three factors of availability (A), performance (P) and quality (Q) in manufacturing processes (Kianian and Andersson 2018).

In the Fig. 33.3 PM gear manufacturing cost is calculated based on the data obtained from interviews, suggesting that OEE for PM industry is in the neighborhood of 65–75%. The OEE used as the base case analysis is nearly 69% (A: 70%,

P: 100%, Q: 98%). The practical OEE can in worst cases be as low as 40–50%. For world-class manufacturing, OEE for non-process industry is reported to be in the range of 85–92% previously (Bicheno 2004). For sensitivity analyses, the lower bound value of 50% (A: 55%, P: 96%, Q: 95%) is selected for OEE, and in order to assess the effect of an improved OEE on the PM gear manufacturing cost, the world class OEE of 85% (A: 85%, P: 100%, Q: 99%) is selected as the upper bound value. These OEE parameters percentages are assumed to be the same for all processing routes in this case study's analysis.

The distributions of PM gear manufacturing costs, calculated with the PPC model with an OEE of 50% and an improved OEE of 85%, over PM process routes are illustrated respectively in Figs. 33.4 above and 33.5 below.

Based on this sensitivity analysis, the largest influences are on the grinding and sintering processes, as it is shown in Table 33.3 below. These process costs' fluctuations are due to increasing and decreasing equipment costs during downtime, impairing and improving parts' quality hence increasing and decreasing cost of poor quality and labour cost.

Material and HT costs were not affected despite the changes in OEE in this case study, since only their purchased prices were used in the excel calculation tool and they were not calculated by the PPC model unlike the rest of the cost drivers. Hence,

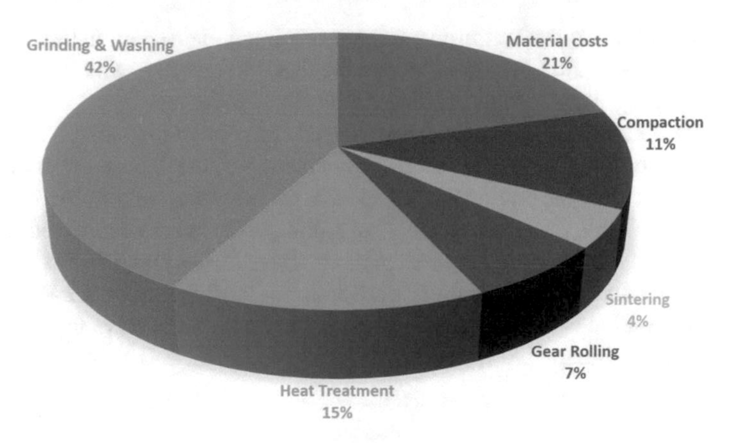

Fig. 33.4 PM gear manufacturing cost distributions over its manufacturing process routes with an OEE of 50%

Table 33.3 OEE sensitivity analysis

OEE improvement and sensitivity analysis	Total PM gear manufactur-ing cost	Grinding cost	Sintering cost	Gear Rolling cost	Compaction cost
OEE 50%	19% ▲	43% ▲	32.5% ▲	19% ▲	11% ▲
OEE 85%	9% ▼	19% ▼	16% ▼	10% ▼	7% ▼

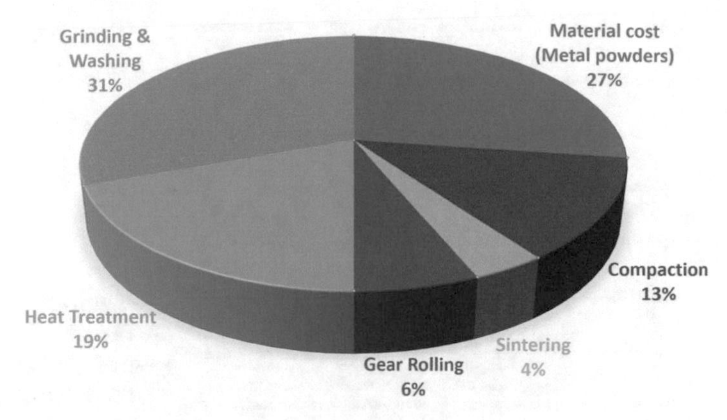

Fig. 33.5 PM gear manufacturing cost distributions over its manufacturing process routes with an improved OEE

their costs' shares increased in proportion to the rest of cost drivers in Figs. 33.4 and 33.5. This is further discussed in the next section—discussion (see Sect. 33.6.2).

33.6.1 Increase in Energy Price

There has been continuous growing demand for energy in the manufacturing sector and its high potential of scarcity has forced different industries to evaluate energy efficiency of their current operations for almost half a century (Kaufman 1980). Given global sustainability challenges. There is growing consensus worldwide to minimize greenhouse gas (GHS) emissions. Industry is accountable to 35% of all energy/process emissions (Azevedo et al. 2018). Thus, energy conservation is of vital importance for sustainability practices (e.g. sustainable production). The PM industry is considered as an energy and material efficient industry by many previous studies looking for alternative sustainable production processes (Kruzhanov 2018; Zapf and Dalal 1983; Bocchini 1983; Kaufman 1980; Azevedo et al. 2018). Therefore, the impact of an increase in energy price on PM gear manufacturing cost is evaluated.

Figure 33.6 illustrates the energy prices for industrial customers in (Sweden 2018). The EL price is illustrated for a period of 32 years (1986–2017). The average EL price during this period is 45 öre/kWh, which is approximately the same price as what interviewed companies paid in 2018. It can be noted from the Fig. 33.6 that EL price has been rather stable during the course of analysis, with an exception of between years of 2002 (36 öre/kWh)–2003 (71 öre/kWh), which it had been nearly doubled. In this case study, two scenarios for EL price increase are defined. The first scenario is similar to what had happened in reality between years of 2002–2003 (EL priced almost doubled), and the second scenario is a hypothetical case in which the EL price would be 4 times higher than the current price.

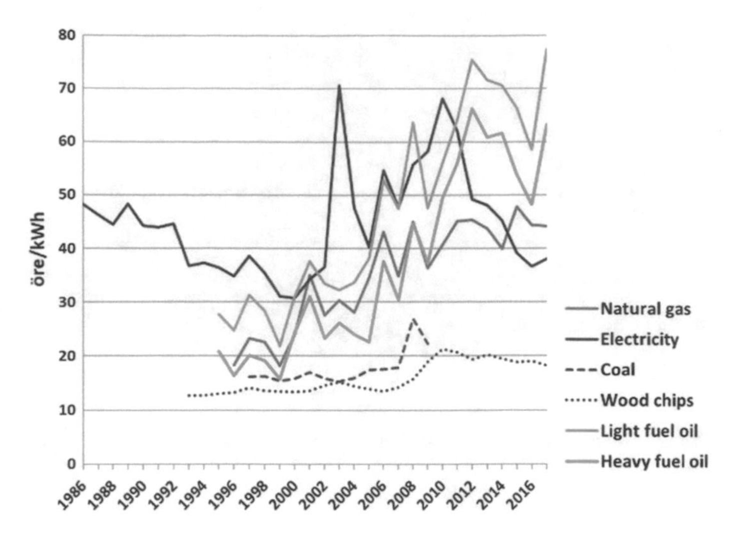

Fig. 33.6 Energy prices for industrial customers in Sweden from 1986 to 2017 (real prices), öre/kWh—courtesy of Swedish Energy Agency (Energimyndigheten—SCB) (Energy in Sweden Facts and Figures 2018)

The results of the simulations of the two scenarios doubling and quadrupling of EL prices, shows respectively 0.5% and 1.6% increases in the total PM gear manufacturing costs per part, and similar to the previous simulation case in Sect. 33.5.2 material and HT costs were not affected in these simulations. Nevertheless, in real practice both material and HT costs will increase as the energy price increases. However, this study has not quantified to what extent their costs would increase unlike the rest of the PM processes as follow. The largest influence is on the sintering process with 7% when doubling EL price, and 21% when quadrupling EL price, increases in process cost per part. As discussed earlier, PM furnace processes (HT, annealing and sintering) are the prominent energy consumers, and that explains the largest increases in sintering costs in comparison to other processes. And since these PM furnace processes have similar efficiencies, the HT process cost would probably increase similar to the sintering process.

Grinding process has an increase of 0.5% when doubling EL price, and 1.6% when quadrupling EL price in process costs per part, and compaction and gear rolling processes both have an increase of 0.3% when doubling EL price and 1% when quadrupling EL price in their costs per part.

33.7 Discussion

In this section, the outcomes of the three scenarios presented in Sects. 33.5.1 and 33.5.2 above are further discussed.

33.7.1 Gear Rolling Automation

The results in Table 33.2 (Sect. 33.5.1) show that gear rolling process's automation increases the equipment utilization and its operational cost. The equipment cost during downtime and idle are increased as well, which can be realized due to setup time, automation disturbances and preventive maintenances. However, the sensitivity analysis's outcomes indicate that higher automation cost, increases gear rolling cost considerably and the total PM gear manufacturing cost marginally, in comparison to the two other scenarios (see Table 33.2). Hence, the automation financially feasible options, in this case study, is between the lower bound value and base case analysis selected.

33.7.2 OEE Improvement

Figure 33.7 below illustrates the total PM gear manufacturing cost distributions over its manufacturing process routes' cost parameters with the base case analysis OEE of 69%. Comparisons among numbers in Fig. 33.8 below indicate that reducing and increasing OEE parameters are evidently impacted the total PM manufacturing costs respectively negatively and positively.

Since the costs's shares in these tables are relative, some explanations of how OEE impact costs are needed. E.g. equipment costs during downtime are increased with reduced OEE and decreased due to increased availability (A) and quality issues. Even if the availability is set to be 100%, there will be still some precentages of equipment downtime cost because of inclusion of setup time. Labour costs are also increased

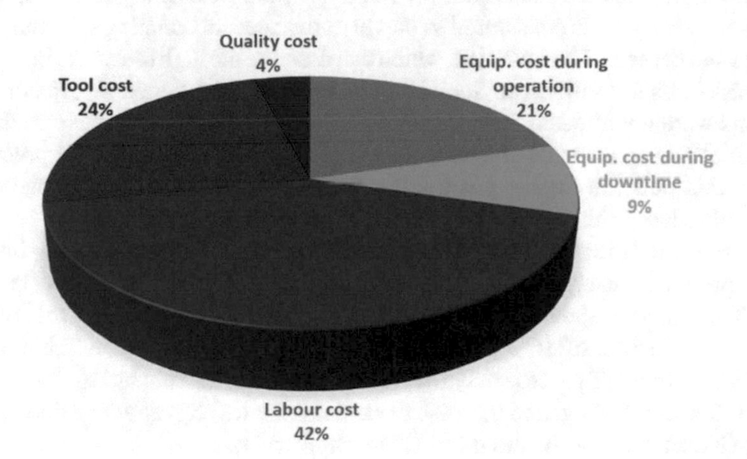

Fig. 33.7 PM gear manufacturing cost distributions over its manufacturing process routes' cost parameters (OEE 69%)

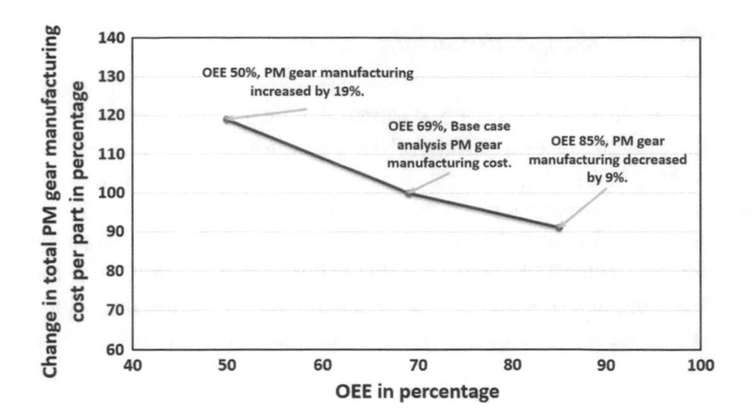

Fig. 33.8 Influences of OEE on the total PM gear manufacturing cost per part

with reduced OEE and decreased with improved OEE because operators are still occupied even if the equipment is on idle or subject to downtime.

Tool and equipment costs during operation are decreased with OEE of 50% and this mainly could be an outcome of the equipment poor availability. They both are increased after OEE improvement, as the availability of equipment increased and its standby mode and downtime decreased. A possible improvement opportunity to reduce the manual labour cost (particularly in a labour-intensive country like Sweden) is process automation as it is discussed previously above (see Sect. 33.5.1). High tool costs could be a result of insufficient or ineffective maintenance. The authors assume that the high tool cost is because of its very high investment cost, and many refurbishment and replacement times due to a high annual production volume (400,000 unit). This is in-line with the reality in the PM industry since the tool in this case study's gear is considered to be very advance and complex with number of punches and core rod (Flodin 2018), which can describe the high tool costs in Fig. 33.7 regardless of its utilisations. In Sect. 33.5.2 above, it is mentioned that material and HT costs were not affected despite the changes in the OEE. Nevertheless, the cost of material's scrap is included in calculation of quality cost for each PM processing route in this study. In these cases the fluctuations may be due to increased (lower OEE) and reduced (higher OEE) cost of poor quality and scrap rate.

Regarding HT efficiencies, PM furnace processes (HT, annealing and sintering) are the prominent energy consumers, and they have similar efficiencies (less than 40%). The sintering showed 32.5% increase in process cost with the OEE of 50%, and a cost reduction of 16% after OEE improvement per part (see Table 33.3 in Sect. 33.5.2), thus HT process may lose and gain similar benefits by e.g. less or more efficient furnace loading, and increase or decrease losses (e.g. at walls, doors, in the fumes) (Kruzhanov and Arnhold 2012; Bocchini 1983).

33.7.3 Increase in Energy Price

Given the increase importance of sustainability issues and future associated risks to manufacturing, it is important to view these results through the lens of possible future sustainability related challenges in order to ascertain if this case study can shed light on possible future outcomes. There is a clear link between energy issues with sustainability as the climate change challenge is inter linked with carbon heavy energy sources, which must be phased out, and replaced with greener alternatives. This means that the already volatile energy market is likely to become even more volatile in the years to come, so any simulations of the effects of energy costs are pertinent to those managing manufacturing sites.

It should be noted that, the cost increases calculated in this paper (see Sect. 33.5.3) due to inflated energy costs show only the direct energy costs associated with the manufacturing of the part in question. If energy prices increase in the manufacturing location and also in other areas of the parts value chain, then it is safe to assume that there will be knock on effects to the raw material (e.g. metal powders) costs as the production of the raw material is also from an energy intensive industry and the production companies are likely to increase the sales price of their raw material in order to absorb the costs associated with the energy increase. This effect could also be mirrored in each and every supplier in the value chain e.g. transportation costs, logistics costs, inventory and storage costs.

33.8 Conclusion and Outlook

This study addresses the gap in PM gear manufacturing cost assessment (Kianian 2019; Kianian and Andersson 2017; Kruzhanov and Arnhold 2012) with providing in-depth manufacturing cost analysis of a PM gear. Previous studies quantified the material and energy consumptions of PM processes but failed to detail the associated costs (Kruzhanov and Arnhold 2012; Zapf and Dalal 1983; Bocchini 1983; Kaufman 1980; Azevedo et al. 2018). This is due to the fact that, PM transmission gear manufacturing is not industrially adopted largely; hence, the data collection for the cost analysis was assembled from multiple industrial sources. This study shows the capability of the PPC model to analyse a complete set of cost drivers, showing the cost for each process step from powder to finished gear, see Fig. 33.3. The scenario capability is utilized to analyse two different improvement scenarios and a scenario of increased energy price. The conclusion drawn from these analyses are that automation can be cost efficient if the performance is not affected negatively and that OEE has a substantial impact on manufacturing cost, see Fig. 33.8.

As mentioned earlier (see Sect. 33.5.3) PM industry has been regarded as energy and material efficient by many studies, which have sought for alternative sustainable production processes (Zapf and Dalal 1983; Bocchini 1983; Kaufman 1980). This approbation is still verified by more recent studies (Kruzhanov 2018; Kruzhanov and Arnhold 2012; Azevedo et al. 2018). Although most of these studies acknowledged sustainability issues or at least the environmental aspects, neither comprehensive environmental assessment nor life cycle perspective have been conducted or considered in their investigations.

Steel PM emission share in industrial emission is currently small; however, this could change and become more notable due to the high pace advancement, adoption and deployment of emerging technologies like metal additive manufacturing (AM) and its new material development (Azevedo et al. 2018). In order to contribute to the field with the main intention of reducing the aforementioned gaps, the present authors will conduct comprehensive sustainability assessments e.g. environmental and social LCA, as their future work. In addition, their previous LCC studies (Kianian et al. 2019; Kianian 2019) will be further advanced to include environmental and social aspects as well. These will be performed by identifying the parameters, which drive the sustainability aspects and align them with cost drivers and its components in the PPC in order to create a DSS for companies in their pursue for alternative sustainable production systems.

Acknowledgements The authors would like to express their sincere thanks to the case study companies for their cooperation and continuous support of this research project. The financial support from Swedish Foundation for Strategic Research (SSF) through the 'Nanotechnology Enhanced Sintered Steel Processing' project (grant no. GMT14-0045) is also appreciated.

References

Andersson M, Flodin A (2017) Redesigning and prototyping a six speed manual automotive transmission of powder metal gears. In: Proceedings of the JSME international conference on motion and power transmissions 0:01–11

Azevedo JMC, Cabrera Serrenho A, Allwood JM (2018) Energy and material efficiency of steel powder metallurgy. Powder Technol 328:329–336

Bicheno J (2004) The new lean toolbox: towards fast. PICSIE Books, Moreton Press, Buckingham, Flexible Flow

Bocchini GF (1983) Energy requirements of structural components: powder metallurgy v. Other Prod Process Powder Metall 26(2):101–113

Economic Considerations for Powder Metallurgy Structural Parts (2012) In: Powder metallurgy review. https://www.pm-review.com/introduction-to-powder-metallurgy/economic-considerations-for-powder-metallurgy-structural-parts/

Energy in Sweden Facts and figures (2018) In: Swedish Energy Agency (Energimyndigheten—SCB). https://www.energimyndigheten.se/en/news/2018/energy-in-sweden---facts-and-figures-2018-available-now/

Flodin A (2018) Powder metal through the process steps. Gear Technology: 8

Jönsson M (2012) Cost-conscious manufacturing-models and methods for analysing present and future performance from a cost perspective. Lund University, Lund

Kamps T, Lutter-Guenther M, Seidel C, Gutowski T, Reinhart G (2018) Cost- and energy-efficient manufacture of gears by laser beam melting. CIRP J Manuf Sci Technol 21:47–60

Kaufman SM (1980) Energy consumption in the manufacture of precision metal parts from iron powder. SAE Tech Paper Ser 1:1311–1318

Kianian B (2019) Comparing acquisition and operation life cycle costs of powder metallurgy and conventional wrought steel gear manufacturing techniques. Procedia CIRP 81:1101–1106
Kianian B, Andersson C (2018) Analysis of manufacturing costs for conventional gear manufacturing processes: a case study of a spur engine gear. Proc Int Gear Conf 1:393–402
Kianian B, Andersson C (2017) Sustainability-conscious powder metallurgy gear manufacturing: an analysis of current manufacturing challenges. Proc VDI-Berichte 2294(2):1251–1264
Kianian B, Kurdve M, Andersson C (2019) Comparing life cycle costing and performance part costing in assessing acquisition and operational cost of new manufacturing technologies. Procedia CIRP 80:428–433
Kotthoff G (2018) NVH potential of PM gears for electrified drivetrains. Gear Technology: 4
Kruzhanov VS (2018) Modern manufacturing of powder-metallurgical products with high density and performance by press-sinter technology. Powder Metall Met Ceram 57(7–8):431–446
Kruzhanov V, Arnhold V (2012) Energy consumption in powder metallurgical manufacturing. Powder Metall 55(1):14–21
Larsson M, Andersson M, Rauch P (2014) Compaction of a helical PM transmission gear. In: Proceedings of World Congress PM
Leupold B, Janzen V, Kotthoff G, Eichholz D (2017) Validation approach of PM gears for eDrive application. Proc VDI-Berichte 2294(1):119–130
Schlosser R, Klocke F, Döbbeler B, Riemer B, Hameyer K, Herold T, Zimmermann W, Nuding BA, Schindler M, Niemczyk M (2011) Assessment of energy and resource consumption of processes and process chains within the automotive sector. In: Proceedings of the 18th CIRP International Conference on Life Cycle Engineering—Glocalized Solutions for Sustainability in Manufacturing 45–50
Schultheiss F, Windmark C, Sjöstrand S, Rasmusson M, Ståhl JE (2018) Machinability and manufacturing cost in low-lead brass. Int J Adv Manuf Technol 99(9–12):2101–2110
Ståhl JE, Andersson C, Jönsson M (2007) A basic economic model for judging production development. In: Proceedings of the 1st Swedish Production Symposium
Steinert J (2015) Technical and economical viability of surface rolled PM gears for manual car transmission. In: Proceedings of Euro PM Reims
Windmark C (2018) Performance-based costing as decision support for development of discrete part production: Linking performance, production costs and sustainability. Lund University, Lund
Zapf G, Dalal K (1983) Raw material and energy conservation in production of sintered PM parts. Powder Metall 26(4):207–216

Chapter 34 Cyber Physical System in Inverse Manufacturing

Shinsuke Kondoh, Hitoshi Komoto, and Mitsutaka Matsumoto

Abstract The vision and the concept of cyber physical system (CPS) has been gaining more and more interest in recent years. Physical artifacts become more and more closely intertwined with their corresponding models in cyberspace and can be designed and controlled as cyber physical artifacts that are to exhibit optimal performance throughout their entire life cycles by using these models. A connected car, which allows the car to share internet access with other devices to enable safe and optimal navigation, and manufacturing equipment that allows information sharing with other devices for optimizing its operation and maintenance are typical examples of such artifacts. For the success of cyber physical artifacts, an adequate description of use case scenarios as well as usage patterns and functions of systems to be developed is essential. Although many researches have been conducted to improve effectiveness and efficiency of production stages by using CPS, less attention is paid to those in products' recovery processes (e.g. reuse, repair, remanufacturing, and recycling). To demonstrate the potential of CPS adoption in product recovery processes, the paper designs and proposes CPS utilization scenarios in product recovery processes that can improve through life performance of artifacts.

Keywords Cyber physical system (CPS) · Product recovery · Industrial internet reference architecture (IIRA)

34.1 Introduction

The vision and the concept of cyber physical system (CPS) has been gaining more and more interest in recent years (Monostori et al. 2016; Park et al. 2012; Borgia, 2014; Törngren and Grogan 2018). Physical artifacts become more and more closely intertwined with their corresponding models in cyberspace and can be designed and controlled as cyber physical artifacts (CPA) that are to exhibit optimal performance throughout their entire life cycles by using these models. A connected car, which

S. Kondoh (✉) · H. Komoto · M. Matsumoto
National Institute of Advanced Industrial Science and Technology (AIST), Tsukuba, Japan
e-mail: kondou-shinsuke@aist.go.jp

Y. Kishita et al. (eds.), *EcoDesign and Sustainability I*, Sustainable Production, Life Cycle Engineering and Management, https://doi.org/10.1007/978-981-15-6779-7_34

allows the car to share internet access with other devices to enable safe and optimal navigation, and manufacturing equipment that allows information sharing with other devices for optimizing its operation and maintenance are typical examples of such artifacts.

For the success of cyber physical artifacts, an adequate description of use case scenarios as well as usage patterns and functions of systems to be developed is essential. In this context, a number of architecture as well as modelling methods [e.g. (Borgia 2014; Törngren and Grogan 2018; VDI/VDE 2015; Industrial Internet Consortium 2017; Lee et al. 2015; Benjamin et al. 2017; Tomiyama and Moyen 2018; Riel et al. 2018; Lanza et al. 2015; Tao et al. 2018; Braune 2018; International data spaces association (2019); Ghosh et al. 2019)] have been proposed and utilized in many case studies. Typical examples include Reference Architecture Model Industrie 4.0 (RAMI4.0) (VDI/VDE 2015) and Industrial Internet Reference Architecture (IIRA) (Industrial Internet Consortium 2017). Application areas of CPS are, for example, assembly planning [e.g. (Lanza et al. 2015)] and diagnostics of production equipment [e.g. (Tao et al. 2018)].

Although many researches have been conducted to improve effectiveness and efficiency of production stages by using CPS, less attention is paid to those in products' recovery processes (e.g. reuse, repair, remanufacturing, and recycling). To demonstrate the potential of CPS adoption in product recovery processes, the paper attempts to design and propose CPS utilization scenario in product recovery processes (i.e. customized disassembly) that can improve through life performance of artifacts.

Section 34.2 briefly explains IIRA that are used to model and design a CPS utilization scenario. Section 34.3 describes the customized disassembly scenario mainly focusing on business and usage views of the scenario. Section 34.4 discusses the result of design. Section 34.5 summarizes the paper and provides future outlook for the research.

34.2 Industrial Internet Reference Architecture

We utilize IIRA to model and design CPS application scenario in product recovery processes. According to Industrial Internet Consortium (Industrial Internet Consortium 2017), we briefly introduce basic concept of IIRA in this section. IIRA is a standards-based open architecture for Industrial Internet of Things (IIoT), which is proposed by Industrial Internet Consortium (IIC). It specifies IIoT system architecture comprising *viewpoints* and *concerns.* A viewpoint is used for framing the description and analysis of specific system concerns. Concerns refer to any topic of interest pertaining to the system to be developed. IIRA provides four viewpoints as an architectural template to define and clarify unique requirements for the IIoT system to be developed.

Business viewpoint

The business viewpoint addresses the concerns of the stakeholders involved. It defines values and objectives in establishing an IIoT system in its business and regulatory context. It also defines how the IIoT system achieves the objectives through identifying fundamental system capability. Typical stakeholders in the viewpoint are business decision-makers, regulatory bodies, and system engineers.

Usage viewpoint

The usage viewpoint addresses the concerns related to expected usage of an IIoT system. It is typically represented as sequences of *activities*. Activities are a coordination of more basic unit of work, which is called a *task*. A task is carried out by stakeholders assuming a *role* to deliver intended functionality or capabilities of the system. Here, a role is a set of capacities assumed by stakeholders to initiate and participate in the execution of some tasks or functions in an IIoT system. Note that assignment of roles to stakeholder is business specific. Typical stakeholders described in the viewpoint are any individuals who are responsible for the specification of the IIoT system (e.g. system engineer) and those who are responsible for execution of the activities (e.g. production manager, machine tool operator etc.).

Functional viewpoint

The functional viewpoint defines and clarifies the functional components in an IIoT system, their structure and interrelation, interactions and interfaces with their surrounding environment as well as other systems to support the overall activities described in the usage viewpoint.

Implementation viewpoint

The implementation viewpoint deals with the technologies utilized to implement functional components and communication scheme among them.

Among the four viewpoints, we use the business viewpoint and the usage viewpoint in the paper because we want to clarify and discuss the values provided by introducing IIoT system in product recovery processes.

34.3 Use Case: Customized Disassembly

34.3.1 Background and Motivation

Disassembly refers to the process of breaking down an End of Life (EOL) product into a separate part in order to recover value from its constituent components as well as materials. Because each individual product has been used in different conditions and maintenance policies, the conditions and the residual life of their components significantly differ from each other. Consequently, different products have different sets of usable (or profitable) components and materials to be recovered. This causes significant variation in disassembly processes. In addition, disassembly process itself differs from product to product because different design requires different disassembly means and tasks. Thus, customization of disassembly process to each individual product is quite promising for recovering maximum value from it so as to enhance material circulation in society. To support such customized disassembly, information sharing among all stakeholders involved in a whole life cycle of a given product is essential.

34.3.2 Business Viewpoint

The vision of customized disassembly is to achieve higher utilization of components and materials retrieved from EOL products so as to make the Earth more sustainable. The value focused in the use case is provision of essential information and services that helps disassembly firms to minimize disassembly cost as well as to maximize their profit by selling recovered components and materials. The other value perceived by original equipment manufacturers (OEM) will be obtaining cheaper (and sometimes more reliable) components from disassembly firms. This can also lead environmental load reduction, which can be perceived as value for the Earth.

Fundamental capabilities required in the use case are (i) information sharing among all stakeholder involved in a whole life cycle of a product and (ii) adequate selection of life cycle options (e.g. reuse, recycling, landfill, and so forth) for each individual component retrieved from EOL products based on the information from OEMs and users.

Figure 34.1 shows the business viewpoint of customized disassembly. We first identifies 13 stakeholders (including the Earth) as shown in the figure. Major stakeholder in the use case are summarized as follows,

Disassembly firm (DF):

The party who operates disassembly firm to retrieve usable components and materials from EOL products. DF is a user of customized disassembly support service and provides cheaper and sometimes more reliable components to OEM. DF is responsible for adequate reuse, recycling, and EOL treatment of discarded products.

Customized disassembly service provider (CDSP):

The party who provides customized disassembly support service with DF based on knowledge and information about individual EOL product.

Information provider (IP):

The party who provides useful information for customized disassembly such as product (component, and material) specification, disassembly means, technological information about disassembly equipment, residual lifetime of the components, legislation about and subsidy for reuse and recycling, acceptable conditions of reuse components and so forth. Some of these information can be directly obtained from

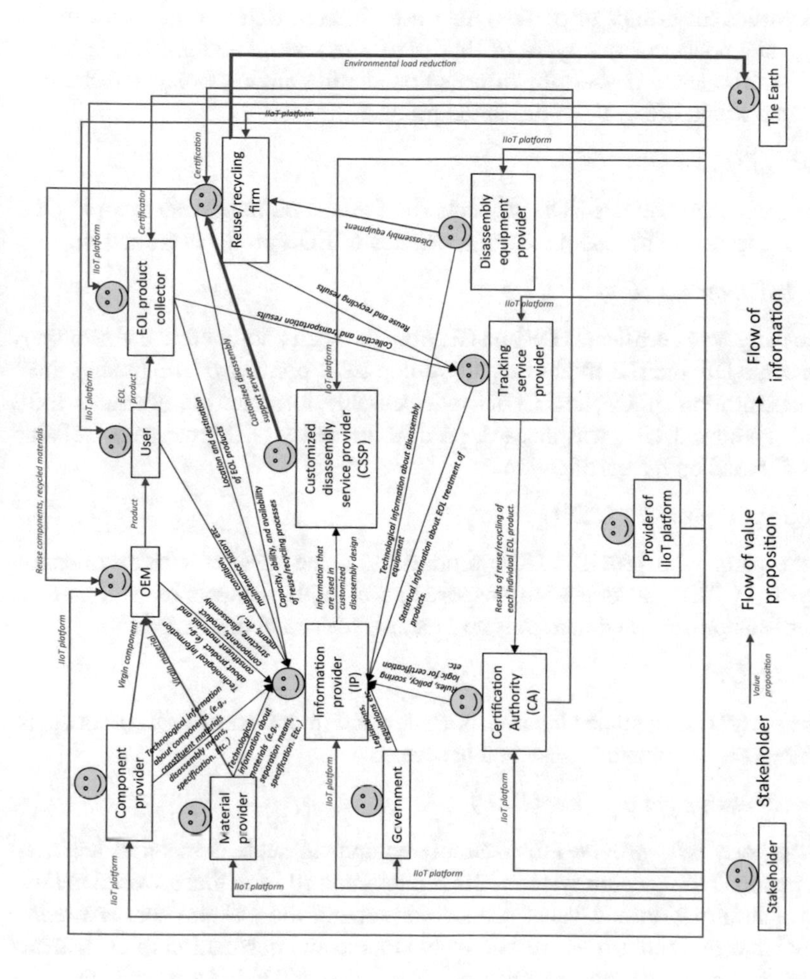

Fig. 34.1 Business viewpoint of the use case scenario: customized disassembly

OEM, disassembly equipment providers, government, and so forth, but other should be estimated by interpreting and/or analyzing publicly available information which are to be stored and shared in IIoT service platform.

Provider of IIoT platform (PoIIoT)

PoIIoT is the party who provides and operates IIoT platform that are used in the case for sharing information and knowledge among multiple stakeholders. Because different parties (or groups of parties) may use different IIoT service platform for their particular concerns, two types of IIoT platform, *private* and *public* platforms, might exist. The latter is used for information sharing among every stakeholders involved in a whole life cycle of discarded products.

Collector of EOL product (CE)

CE is the party who collects EOL products from users and transports them to DFs. CE is also responsible for adequate transportation of EOL products from user.

Certification authority (CA)

CA is the party who certificates DF and CE from the legal and/or ethical viewpoints. Because some DF and CE might illegally dump EOL product or might ship them abroad, certification of DF and CE is indispensable for ensuring adequate EOL treatment of products. User who discards products as well as OEMs can select reliable DF and CE based on the certification.

Tracking service provider (TSP)

TSP is the party who provides EOL products tracking service with certification authority (CA). TSP tracks every transportation of EOL products as well as their constituent components and materials and records their final destination.

Government (GV)

GV is the party who regulates the transportation and treatment of EOL products so as to achieve environmentally sustainable society.

Disassembly equipment provider (DEP)

DEP is the party who provides disassembly equipment such as material handling devices, manipulators, vision systems, 3D printers, and so forth that are used in DFs.

Flow of information and value proposition among the stakeholders are represented by black and red arrows, respectively. Source and destination of each arrow correspond to a provider and a receiver of information and value, respectively.

For example, IP collects technological information about product, disassembly equipment, materials, and components, as well as legislation and regulation information and usage data. Then, IP analyzes and interprets them and provides useful information with CDSP. CDSP determines optimal life cycle options for each individual product based on information provided by IP, generates bill of disassembly process (BODP) customized to each individual product, and provide them with disassembly firms. CDSP also provides information that ease execution and automatization of disassembly processes described in BODP.

In the value network, DF pays for customized disassembly support service, CDSP pays for information provided by IP, and every stakeholder using an IIoT service platform pays for usage for the platform.

34.3.3 Usage Viewpoint

Based on the business viewpoint described in Sect. 34.3.2, we have developed a usage viewpoint of customized disassembly as shown in Fig. 34.2. We have identified two roles (i.e. disassembly planner (DP) and disassembly engineer (DE)) in CDSP and three roles (i.e. EOL product acceptance staff (AS), plant manager (PM), and disassembly operator (DO)) in DF at first. Then, we have designed how IIoT system support them by identifying activities that are to be executed by them.

Major activities are summarized as follows;

Data collection stage

- IP periodically collects publicly available information regarding, (1) regulation and legislation about waste treatment and transportation, (2) technological information about disassembly equipment, (3) design information of products including a list of constituent materials, product and component structure, disassembly means, and so forth, (4) usage condition and maintenance history of each individual product, and (5) specification of the components and materials that are used in products in markets.

Product acceptance and inspection stage

- AS receives EOL products.
- AS inspects EOL products based on pre-defined instruction.
- AS uploads inspection results.
- AS accepts and stocks EOL products in stock yards.

Disassembly planning stage

- DP receives a list of accepted EOL products with their inspection results.
- DP determines life cycle options for each component included in the collected products considering profit by selling their recovered items as well as disassembly cost. When predicting profit, DP consults with IP. IP informs acceptable conditions of reuse components (and recycled materials) for OEM as well as their price. IP also informs usage data and maintenance history of each individual product as well as its design information (if possible), so that DP estimates the condition and residual life of each component included in the collected products.

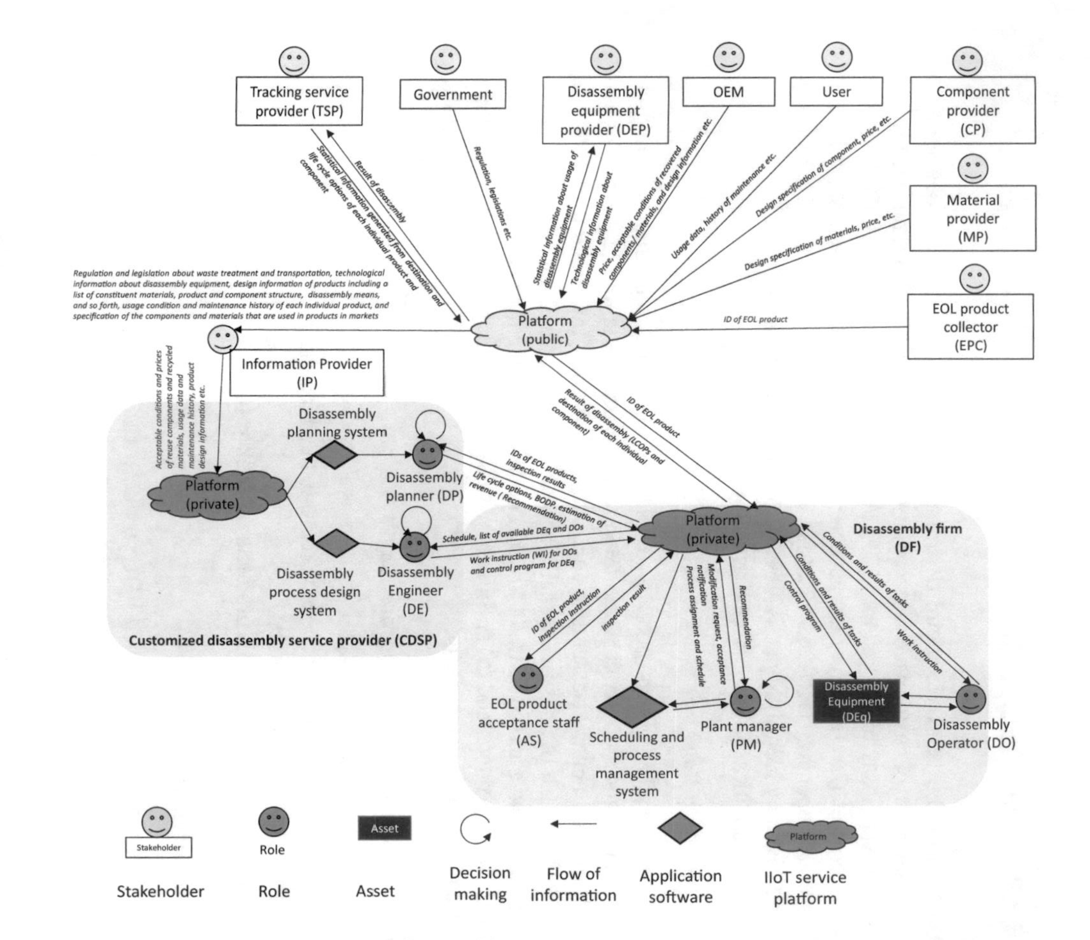

Fig. 34.2 Usage viewpoint of use case scenario: customized disassembly

- DP generates bill of disassembly process (BODP) based on life cycle options of each component considering cost and revenue associated with them. If necessary, DP consults with PM for estimating disassembly cost.
- DP sends recommending life cycle options and corresponding BODP for each EOL product by showing resulting revenue and cost.
- PM receives the life cycle options and BODP for each product and determines whether accept them or not. If PM does not accept them, PM sends modification request to DP.
- PM makes (or updates) disassembly schedule based on BODP and evaluate it considering the capacity and availability of work forces and disassembly equipment. If some part of BODP cannot be executed by currently available disassembly equipment and disassembly operators, PM also sends modification request to DP.
- After PM accepts life cycle options and BODP of each EOL product, DE generates work instruction and control program for disassembly equipment consulting with IP. IP informs technological information about disassembly equipment. DE also consults with PM when detailing work instruction for disassembly operator, if necessary.
- DE sends work instruction and control program for disassembly equipment to PM.

Disassembly execution stage

- DO executes disassembly tasks based on the instruction receiving from DP via PM in cooperation with disassembly equipment that are controlled by the program provided by DE.
- DO and disassembly equipment (DEq) reports the status and results of tasks when they terminates.
- Scheduling and process management system records the status of the tasks and updates and modifies future schedule of the system.

To support the activities and information exchange among roles and stakeholders involved, a public IIoT service platform and two private IIoT service platforms are to be implemented. The former is used for information exchange among all stakeholders involved. The latter two is used inside of two stakeholders (i.e. CDSP and DF) for their particular purposes (e.g. generation of BODP, control program for DEq, work instruction for DOs, and scheduling and management of disassembly tasks).

Because the tasks determining adequate life cycle options for each individual component, generation of BODP and corresponding work instruction and control program are time consuming tasks when customizing them to each individual product, we also identified two application software (i.e. disassembly planning system and disassembly process design system), which are to be implemented to support DP and DE.

34.4 Discussion

Through the modeling focusing on the business viewpoint, we have identified 13 stakeholders that are to be involved for realizing customized disassembly for each individual product. Among them, CA, TSP, and the Earth are those who are specific to product recovery process.

Discussions about IIoT service platform which enables communication and information sharing among multiple stakeholders mainly focus on those who are involved in production stage [e.g. (Monostori et al. 2016)]. However, the business viewpoint suggests importance of information sharing with CA and TSP. IIoT will be quite useful for material tracking by nature, implementation of such IIoT platform will be quite promising. The third stakeholder, the Earth, should also be taken into account when environmental aspects are included in concerns.

Through the design of the usage viewpoint, we have identified two fundamental roles (i.e. DP and DE) for supporting PM to customize product assembly process to each individual product. In the usage viewpoint shown in Fig. 34.2, we assume DEP, OEM, CP, MP, and user will provide technological and usage information. In order to make this happen, we have to design incentive for each stakeholder. Example of such incentive for DEP is receiving statistical usage information (e.g. frequency of usage of each part in the device) from the user (i.e. DF), which can be used to improve design and control software of its devices, in exchange of their technological information. In order to design such incentive for each stakeholder, secure data matching service such as international data spaces (International data spaces association 2019) will be useful.

34.5 Summary and Outlook

To demonstrate the potential of CPS adoption in product recovery processes, the paper designs and proposes CPS utilization scenario: customized disassembly focusing on the business and usage viewpoint of IIRA. As a result, we have identified recovery process specific stakeholders that are to share information as well as roles and application programs that has to be implemented in the use case.

This means current development effort for CPS is not sufficient for adopting CPS to product recovery process, although CPS has a significant potential for customizing (and individualizing) product recovery processes. Designing scenarios in product recovery processes as well as those in production and usage stages, thus, is quite useful for highlighting the future technological development needs.

To explore the potential and future development need for adopting CPS in product recovery processes, a variety of scenarios focusing on different tasks such as cleaning, maintenance, remanufacturing and so forth are to be explicitly described and shared among a variety of stakeholders with different concerns. The design of customized disassembly proposed in the paper provides a starting point for the exploration.

Future work includes following topics;

- Development of the functional viewpoint and the implementation viewpoint of the use case by detailing each activity and the contents of data and information exchanged among multiples roles in the usage viewpoint,
- Identification of essential functional components as well as their enabling technologies for realizing customized disassembly services,
- Serious discussion and evaluation on the designed use case with multiple experts including reuse/recycling plant managers, engineers, and operators as well as system engineers engaged in development of IIoT based service platform, and
- Quantitative evaluation of cost and profit of DFs after adopting customized disassembly support system so that CDSP and other stakeholders can evaluate their profitability for generating the service.

References

Benjamin S, Anwer N, Mathieu L, Wartzack S (2017) Shaping the digital twin for design and production engineering. CIRP Ann Manuf Technol 66(1):141–144

Borgia E (2014) The internet of things vision: key features, applications and open issues. Comput Commun 54:1–31

Braune A et al (2018) Usage viewpoint of application scenario value-based service, Platform INDUSTRIE 4.0

Ghosh A, Ullah A, Kubo A (2019) Hidden Markov model-based digital twin construction for futuristic manufacturing systems. Artif Intell Eng Des Anal Manuf 33(3):317–331

Industrial Internet Consortium (2017), The industrial internet of things volume G1: reference architecture. https://www.iiconsortium.org/IIC_PUB_G1_V1.80_2017-01-31.pdf (Accessed 16, Jan. 2019)

International data spaces association (2019) Reference architecture model version 3.0, https://www.internationaldataspaces.org/wp-content/uploads/2019/03/IDS-Reference-Architecture-Model-3.0.pdf

Lanza G, Benjamin Haefner B, Kraemer A (2015) Optimization of selective assembly and adaptive manufacturing by means of cyber-physical system based matching. CIRP Ann Manuf Technol 64(1):399–402

Lee J, Bagheri B, Kao HA (2015) A cyber-physical systems architecture for industry 4.0-based manufacturing systems. Manufacturing Letters 3:18–23

Monostori L, Kadar B, Bauernhansl T, Kondoh S, Kumara S, Reinhart G, Sauer O, Schuh G, Sihn W, Ueda K (2016) Cyber-physical system in manufacturing. CIRP Ann Manuf Technol 65(2):621–641

Park K-J, Zheng R, Liu X (2012) Cyber-physical systems: milestones and research challenges. Editorial Comp Commun 36(1):1–7

Riel A, Kreiner C, Messnarz R, Mch A (2018) An architectural approach to the integration of safety and security requirements in smart products and system design. CIRP Ann Manuf Technol 67(1):173–176

Tao F, Zhang M, Liu Y, Nee AYC (2018) Digital twin driven prognostics and health management for complex equipment. CIRP Ann Manuf Technol 67(1):169–172

Törngren M, Grogan PT (2018) How to deal with the complexity of future cyber-physical systems? Designs 2:40

Tomiyama T, Moyen F (2018) Resilient architecture for cyber-physical production systems. CIRP Ann Manuf Technol 67(1):161–164

VDI/VDE. ZVEI (2015) Reference architecture model industrie 4.0 (RAMI4.0). VDI/VDE, ZVEI, Düsseldorf, Germany, Status Report

Chapter 35
E-Catalogues of Equipment for Constructing an Injection Molding Digital Eco-Factory

Michiko Matsuda, Tomoaki Kondo, Wakahiro Kawai, Jun Hamanaka, Naohisa Matsushita, Shinichiro Chino, Susumu Fujii, and Fumihiko Kimura

Abstract A digital eco-factory has been proposed by the authors for the simultaneous simulation of environmental performance, productivity and manufacturability. A virtual production line is constructed as a multi-agent system by connecting virtual equipment implemented as software agents which are automatically generated from equipment models. These models are provided as equipment e-catalogues. An e-catalogue of equipment includes a dynamic behavior model and a static property model of manufacturing equipment. In the previous papers, the above-mentioned concept was applied to constructions and usages of a digital eco-factory for PCA (Printed Circuit Assembly). In this paper, the above concept is applied to the construction of a digital eco-factory for injection molding. A production line consists mainly of an injection molding machine. An injection mold, a molding extraction robot, and a mold temperature controller are connected to the injection molding machine. The connection is not a sequential connection. Therefore, in the virtual injection molding line, more complicated interaction and control among equipment agents are

M. Matsuda (✉)
Kanagawa Institute of Technology, Kanagawa, Japan
e-mail: matsuda@ic.kanagawa-it.ac.jp

T. Kondo
K.T.System Co., Ltd, Tokyo, Japan

W. Kawai
OMRON Corporation, Tokyo, Japan

J. Hamanaka
Hitachi, Ltd, Ibaraki, Japan

N. Matsushita
Institute of Industry Promotion-Kawasaki, Kanagawa, Japan

S. Chino
Mitsubishi Electric Corporation, Tokyo, Japan

S. Fujii
Kobe University, Hyogo, Japan

F. Kimura
The University of Tokyo, Tokyo, Japan

Y. Kishita et al. (eds.), *EcoDesign and Sustainability I*, Sustainable Production, Life Cycle Engineering and Management, https://doi.org/10.1007/978-981-15-6779-7_35

required. Therefore, e-cataloging of equipment models for injection molding lines is more complicated. Trial implementation of e-catalogues for the above equipment has been executed for construction of a virtual injection molding line. On this virtual injection molding line, an environmental performance simulation in terms of electrical energy consumption simulation can be performed. Because this virtual line is structured by equipment agents which are generated from selected equipment e-catalogues, it can simulate environmental performance with various views such as from an equipment level, a production line level and a factory level.

Keywords Equipment modeling · Behavior modeling · Cyber physical production system · Virtual equipment · Multi agent system

35.1 Introduction

For a forecast of productivity and manufacturability, simulation systems have been used. As a simulation execution field, a virtual production line which is called a digital factory is constructed (Freedman 1999; Bley and Franke 2004). A digital factory involves the modelling of an actual factory (Kuehn 2006; Gregor et al. 2009). A virtual production line in the digital factory is constructed by virtual equipment which is a set of models of actual equipment on the actual production line. A virtual equipment can be implemented as software such as a software agent. A virtual production line can be constructed as a multi-agent system (Monostori et al. 2006). A digital factory is the basis of CPPS (Cyber-Physical Production System) (Monostori 2014).

A digital eco-factory has been proposed by the authors for a simulation of environmental performance in addition to productivity and manufacturability (Matsuda et al. 2012; Matsuda and Kimura 2012, 2013, 2015). A digital eco-factory is constructed as a multi-agent system. An environmental performance simulation strongly requires to model dynamic behavior of the equipment. On the other hand, it is difficult for users of a production simulation system such as a production system designer and an operator to write a software program of agents. Thus, it was proposed that a user constructs a virtual production line by selecting an adequate equipment model from a repository. These equipment models are called e-catalogue of equipment. A preliminary study for a PCA (Printed Circuit Assembly) line had been done, according to the above concept. E-catalogues of equipment are implemented and a trial construction of a virtual PCA line and simulations of electrical energy consumption has been executed (Matsuda et al. 2015, 2016).

The above concept is newly applied to the construction of a digital eco-factory for injection molding. The connection of equipment to construct a virtual line is not a sequential connection like a PCA. In the virtual injection molding line, more complicated interaction and control among equipment agents are required. Therefore, e-cataloging of equipment models for injection molding lines is more complicated. Trial implementation of e-catalogues for the above equipment has been executed for construction of a virtual injection molding line. In this paper, this new application to injection molding is introduced.

35.2 A Digital Eco-Factory

35.2.1 Concept of a Digital Eco-Factory

Environmental performance of the planned production scenario can be examined in addition to productivity and manufacturability when the digital eco-factory is used. Simulation results on the digital eco-factory are used as references during production execution. There are three major use cases for the digital eco-factory as shown in Fig. 35.1 (Matsuda et al. 2015). The first use case is at the period of configuring the production line and/or the factory. In this use case, a configuration plan of a production line is determined and/or performances of newly introduced equipment is examined by constructing a virtual production line and simulation. The second use case is used determining a production plan. After process planning which is supported by the product design tool such as a CAD/CAM system, the production

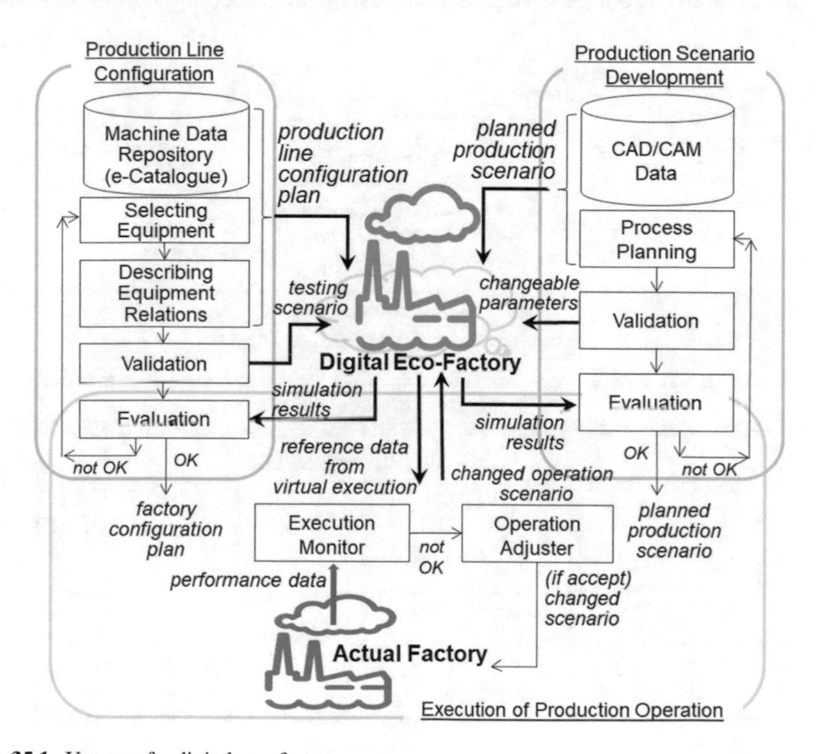

Fig. 35.1 Usages of a digital eco-factory

plan for the product is evaluated using simulation on the already constructed virtual production line. The evaluation is repeated by changing parameters and plans until satisfactory environmental efficiency and productivity data are obtained. The third use case is at the period of executing the production. In this use case, monitored data from the actual production line is compared with reference data from the digital eco-factory. If trouble is detected, a changed operation plan is generated and validated using a digital eco-factory.

35.2.2 Construction of a Digital Eco-Factory Using e-catalogues

To construct virtual production lines requires modeling an actual shop floor and its components, including their activities. Multi agent technologies are applied. All component equipment in the production line are configured as software agents. Figure 35.2 shows the procedure for the construction of a digital eco-factory. When using the digital eco-factory at first, a user constructs his/her own virtual production line on the computer. This constructing procedure starts from the step of selecting equipment models from e-catalogues, and a user provides configuration data which

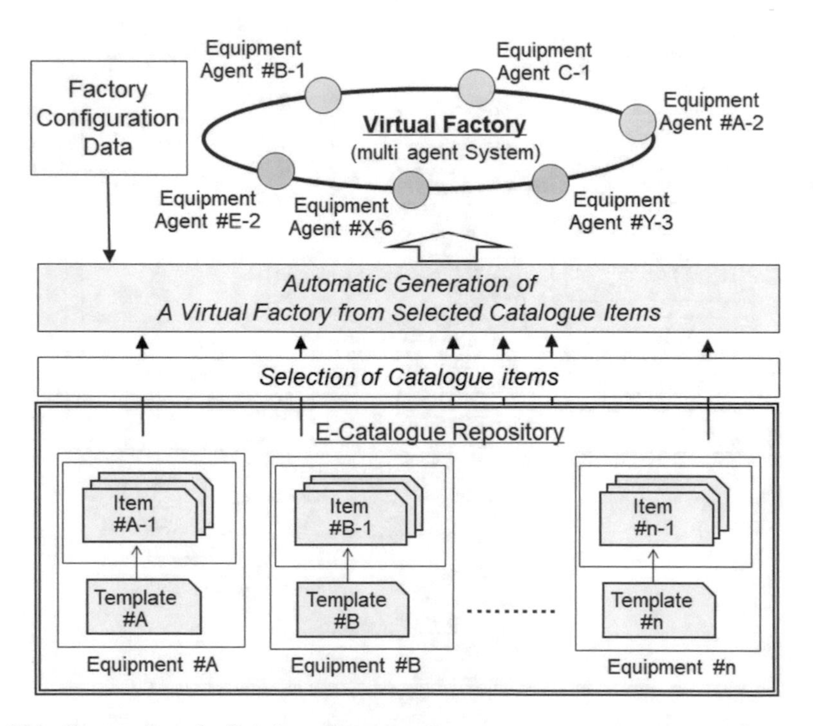

Fig. 35.2 Construction of a digital eco-factory

describe the connecting relationship between equipment, control policy and so on. Then, programs for machine agents and descriptions for the virtual production lines are generated automatically (Matsuda et al. 2016).

An e-catalogue of equipment is a group of one template and equipment items which are created using the template. The equipment template and/or equipment items are registered in a shared repository. An equipment template and an equipment item are required to include descriptions which specify properties, behavior resulting from equipment's activities, and communications with outside and/or other equipment. An equipment template is a schema representing a model for each equipment type including behavior. An equipment item is an instance of an equipment template filled with values. An equipment item is a model of a specific equipment. Template and item are implemented using a data description language such as XML or JSON.

35.3 Preliminary Study on Printed Circuit Assembly

35.3.1 A Digital Eco-Factory for Printed Circuit Assembly

A proposed multi-agent based construction of the digital eco-factory is applied to a PCA line and a trial digital eco-factory has been implemented as shown in Fig. 35.3. An e-catalogue creation system was prepared for supporting this trial. E-catalogues for a solder paste printer, an electronic part mounter and a reflow soldering oven were prepared using this system. When the simulation on the digital eco-factory is executed, input of a production scenario such as product data, schedule and operation data are required. These data affect the equipment behavior. The digital eco-factory for PCA is structured on a commercially available multi-agent simulator "artisoc" [13]. Figure 35.4 shows a screen shot of the simulation execution example using the generated virtual production of the PCA. In this example, there are two production lines. The graphs in the left column show conditions of line 1 and the right column shows conditions of line 2. The graphs in the second row show changes in the number of PCBs on the lines and change in the number of finished PCAs. The graphs in the third row shows changes in energy consumption of each machine. The graphs in the bottom row shows the total amount of energy consumption of each machine (Matsuda et al. 2015, 2016).

35.3.2 Summary of Preliminary Study

The trial digital eco-factory was used in a few industries as an experiment. As a result, the practical usefulness was confirmed especially about time qualitative changes of environmental performance. This trial shows how to construct the digital eco-factory with accuracy and high usability. This trial also shows prominent maneuverability at

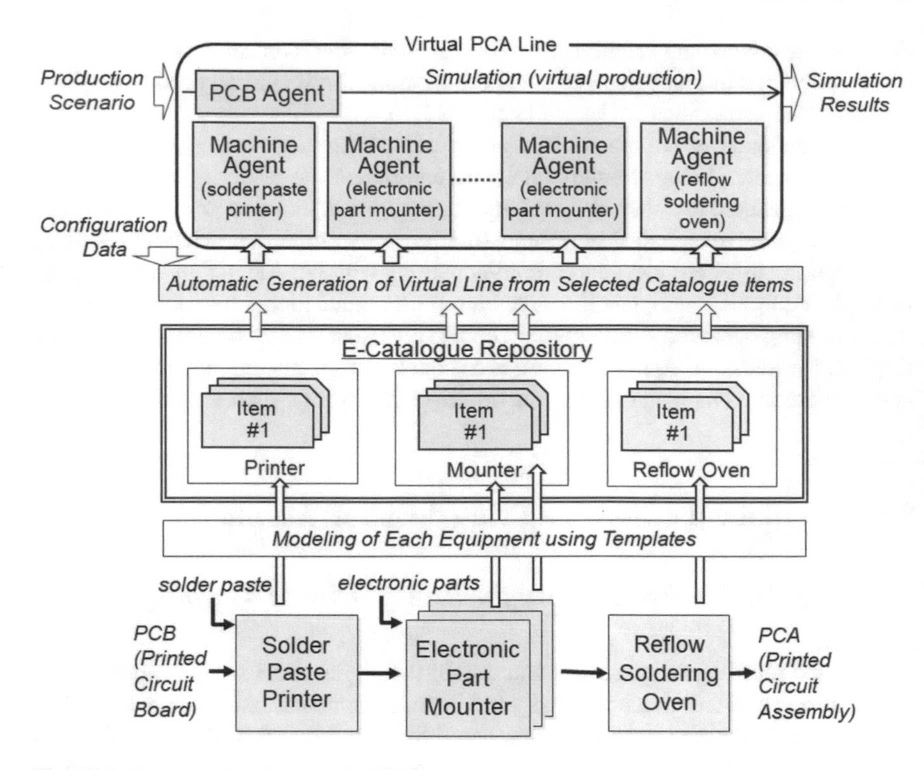

Fig. 35.3 Construction of a virtual PCA line

the usage stage. The user of this digital eco-factory can easily customize the configuration of the factory, target production scenario, granularity of simulation parameters, etc. This comes from modeling of each equipment including the performance simulation procedure which is equipment behavior. At this moment, an e-catalogue of a buffer had been introduced for more precise simulation of the virtual production (Matsumoto and Matsuda 2017).

This trial implementation on a PCA also shows how to make software agents of component equipment on the production line using an equipment model and how to construct a virtual production line as a multi-agent system. On a PCA line, interaction of equipment agents is simple. In other words, equipment agents are connected in sequence. The former positioned equipment finishes the job and then the next positioned equipment starts the job and finishes the job, and so on. However, in most of the production lines, there are more complicated interactions among equipment. Here is a requirement for the extension of an e-cataloging method.

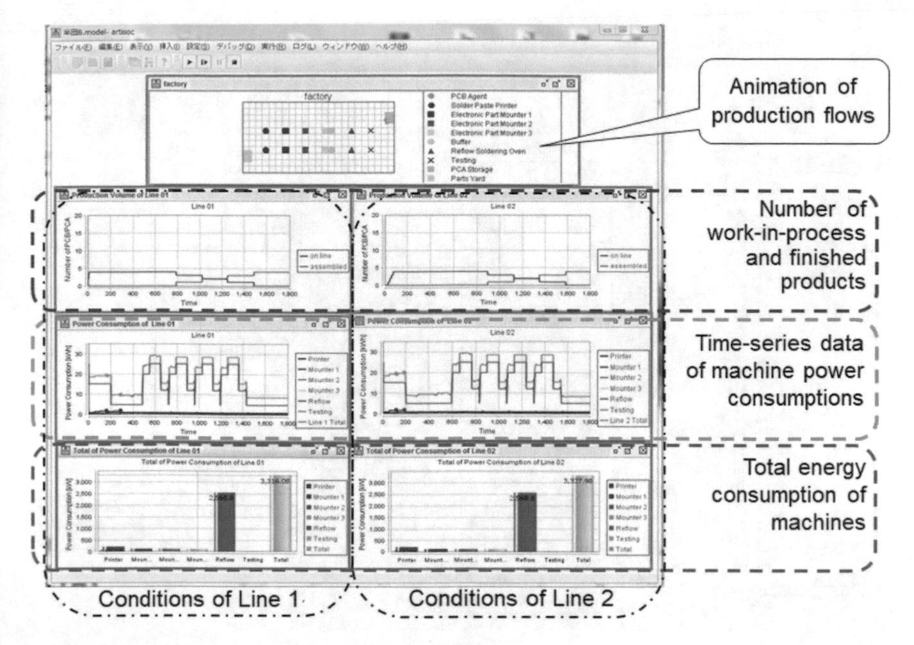

Fig. 35.4 Example simulation using a virtual PCA line

35.4 Application on Injectin Molding

35.4.1 Modeling of an Injection Molding Line

The proposed concept for a digital eco-factory has been applied to injection molding. The structure of an injection molding line is shown in Fig. 35.5. An injection molding line consists of an injection molding machine, a molding extraction robot, a mold temperature controller and a mold. A mold is mounted on an injection molding machine. a mold temperature controller and a molding extraction robot are connected with an injection molding machine. An injection molding machine controls starting

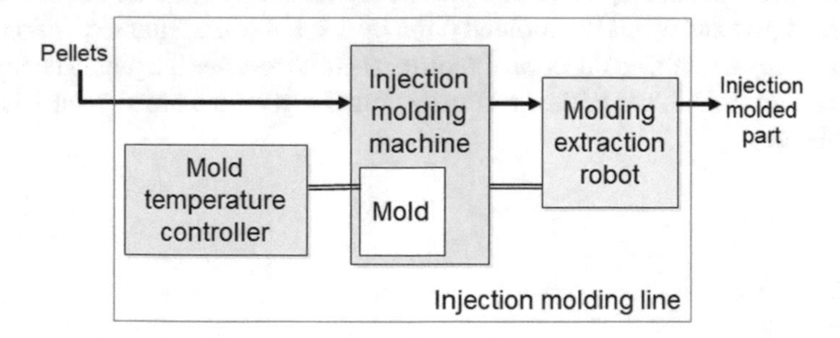

Fig. 35.5 Structure of an injection molding line

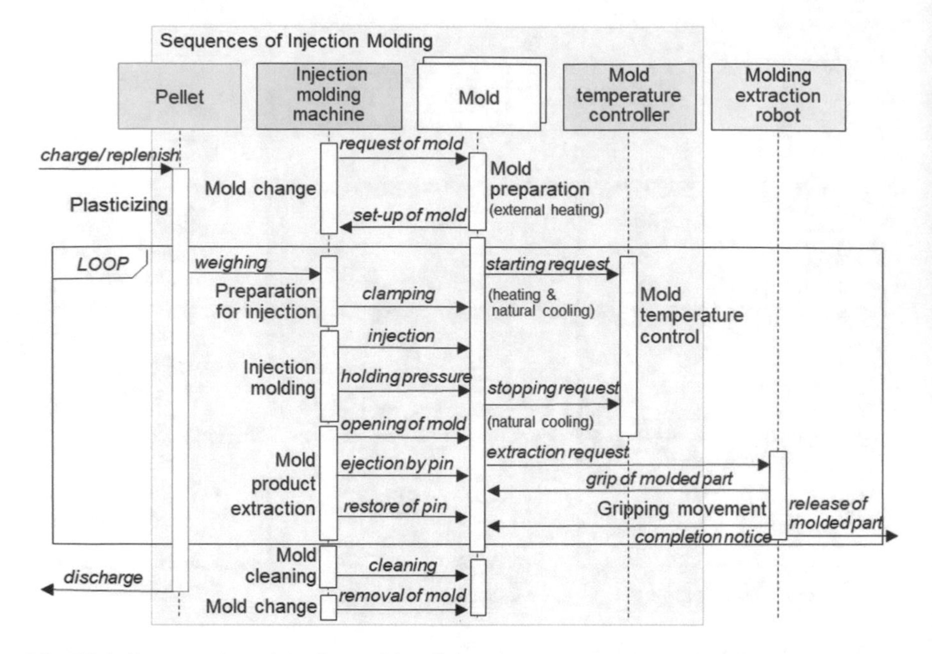

Fig. 35.6 Sequence in an injection molding line

and stopping the actions of a mold temperature controller and a molding extraction robot. Interactions of all equipment are shown in Fig. 35.6 which is a UML sequence diagram.

Sequence in an injection molding line is as follows. Pellets which are the raw material of the injection molded part are put into an injection molding machine, and are heated and melted. A mold is set on an injection molding machine. After these preparations, the injection molding process cycle starts. An injection molding machine requests the mold temperature controller to control the mold's temperature by heating or cooling. Melted pellets are weighed. A mold is clamped. An injection molding machine executes injection and holding pressure. An injection molding machine requests to stop the mold's temperature control. An injection molding machine makes an opening of mold, requests to a molding extraction robot to extract the molded part and ejects the molded parts by a set of pins. After extraction of the molded part by a robot, an injection molding machine restores the pins. The injection process returns to the start of the process cycle and repeats the process until the order is completed.

35.4.2 Virtualization of an Injection Molding Line

A virtual injection molding line is constructed as a multi-agent system as shown in Fig. 35.7. Component agents are an injection molding machine agent, a mold temperature controller agent, a molding extraction robot agent, a mold agent and a pellet agent. E-catalogues for an injection molding machine, a mold temperature controller and a molding extraction robot are prepared. An injection molding machine agent, a mold temperature controller agent and a molding extraction robot agent are generated corresponding to selected e-catalogue items by referring to operation data. Operation data includes the injection molding machine ID with a mold temperature controller ID and a molding extraction robot ID, a molded part ID, a number of the molded part, a mold ID, and so on. A mold agent is generated by referring to the operation data and the mold data. A pellet agent is generated according to the product data and the pellet data. The product data includes a type of a molded part, usable mold IDs, used pellet ID and so on. The pellet data include material, color, weight, required hopper temperature, required cylinder temperature, and required nozzle temperature.

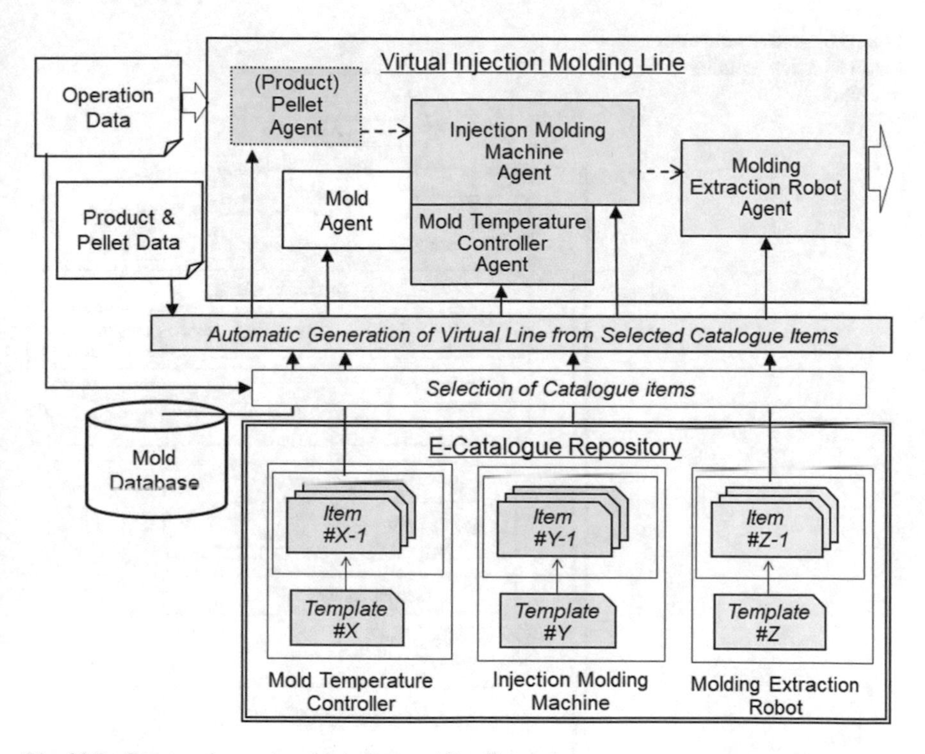

Fig. 35.7 Constructing a virtual injection molding line

35.5 E-catalogues of Equipment for Injection Molding

35.5.1 An Injection Molding Machine

The behavior of an injection molding machine is modeled as an activity flow shown in Fig. 35.8. When receiving an injection molding order in the idling state, the injection molding state starts. Before this, an adequate mold is prepared and changed. The injection molding process cycle is repeated until completion of the ordered amount. There are four processes in the process cycle: injection molding, opening of a mold, ejection by ejector pins and restoration of the ejector pins. At the beginning of injection molding process, the starting of the mold temperature control is requested.

At the end of the injection molding process, the stopping of the temperature control is requested for cooling of a mold. After cooling the mold, it is opened and a request to the molding extraction robot is sent out. Then the molded part is ejected by the ejector pins. After receiving acknowledgement of the completed extraction from the robot, the ejector pins are restored.

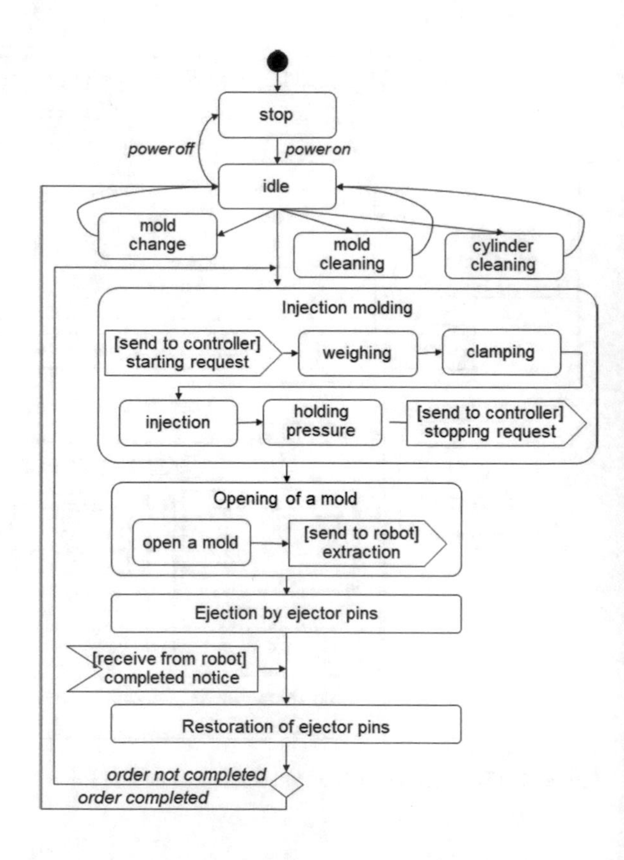

Fig. 35.8 Activity diagram for an injection molding machine

In the template for an injection molding machine, the abovementioned activity flows are described. The following parameters are described in the property part of the template. Properties are referred to by activity descriptions.

- machine ID
- associated mold temperature controller ID
- associated molding extraction robot ID
- action time (start-up, mold change, cleaning, weighing, clamping, etc.)
- electric energy consumption (start-up, cleaning, heating, weighing, ejecting, etc.)
- calculation formula for energy consumption (injection, holding pressure, cooling)
- calculation formula for cumulative energy consumption, etc.

A part of activity descriptions in the template for an injection molding machine using XML is shown in Fig. 35.9. The modelling of a specific equipment is to create the equipment item by filling up the template by value. Some values of parameter or variable in the calculation formula can be provided by operation data or product data.

```
<Equipment-Behavior-Catalogue>
  <equipment type="InjectionMoldingMachine">
    ......
    <activity>
      <status name= "Injection"> ← start description of "Injection molding" behavior
        <source status= "Idle"> ← previous state "Idel"
          <condition ope="==">
            <arg> <param name="InjectionInstruction" ref="operation"/> </arg>                  condition
            <arg> <param name="@True" type="Boolean"/> </arg> </condition> </source>         for starting
        ......
        <pre>
          <dynamic name= "CurrentStatus" ref="variables">
            <param name="@Injection" type="String"/> </dynamic>                                description
          ......                                                                                for pre-
          <dynamic name= "CurrentPowerConsumption" ref="variables">                             processing
            <param name= "PowerConsumption-Injection" ref="statics"/> </dynamic>

          ......
          <dynamic name= "Send-StartingRequestToTempareture Controller" ref="operation">
            <param name="@True" type="Boolean"/> </dynamic> </pre>
        <do>
          ......                                                                                description
          <dynamic name="CumulativePowerConsumption_kWh-Injection" ref="variables">             for main
            <param name="CumulativeInjectionTime" ref="variables"/>                             processing
            <param name="CurrentPowerConsumption" ref="variables" ope="*"/>
            <param name="@3600" type="Decimal" ope="/"/> </dynamic> </do>
        <post>
          <dynamic name="Send-StoppingRequestToTempareture Controller" ref="operation">          description
            <param name="@True" type="Boolean"/> </dynamic>                                     for post
          ......                                                                                processing
          <dynamic name="InjectionInstructio" ref="operation">
            <param name="@False" type="Boolean"/> </dynamic> </post> </status>
      <status name= "OpenMold"> ← start description of "Injection molding" behavior
        <source status="Injection">
        ......
    </activity> </equipment> </Equipment-Behavior-Catalogue>
```

Fig. 35.9 A part of template for an injection molding machine described by XML

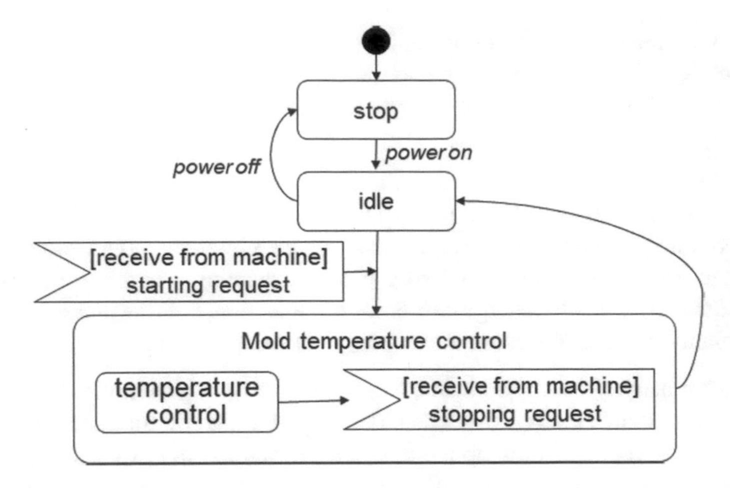

Fig. 35.10 Activity diagram for a mold temperature controller

35.5.2 A Mold Temperature Controller

The behavior of a mold temperature controller is modeled as an activity flow shown in Fig. 35.10. When receiving the request for starting the temperature control from a injection molding machine, the controlling of the mold temperature for keeping the constant mold temperature is started. Constancy of mold temperature is kept by repeating heating and cooling/not heating. When receiving the request for stopping temperature control from an injection molding machine, the temperature controlling ends.

In the template for a mold temperature controller, the abovementioned activity flows are described. The following parameters are described in the property part of the template.

- controller ID
- action time (start-up, heating, etc.)
- heating capacity
- electric energy consumption (start-up, idling, heating)
- calculation formula for energy consumption (heat retention)
- calculation formula for cumulative energy consumption, etc.

35.5.3 A Molding Extraction Robot

The behavior of a molding extraction robot is modeled as an activity flow as shown in Fig. 35.11. When receiving the request for extraction of a molded part from an injection molding machine, the extraction process is started. To extract the molded part, a robot moves to the gripping position, grips the molded part, moves back to

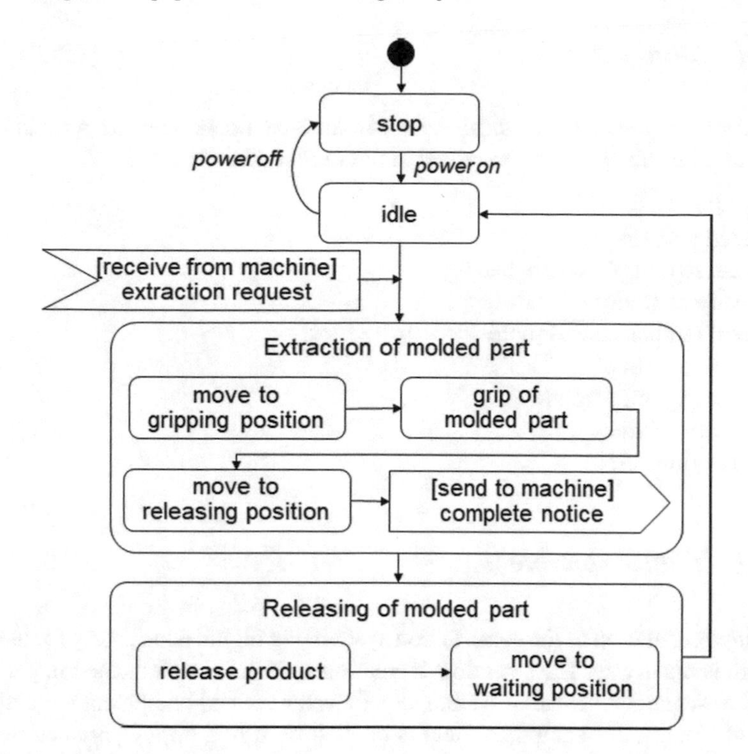

Fig. 35.11 Activity diagram for a molding extraction robot

the releasing position, and sends out a completion notice to an injection molding machine. At the releasing position, the runner is cut and the molded part is released. Finally, a molding extraction robot moves back to the waiting position.

In the template for a molding extraction robot, the abovementioned activity flows are described. The following parameters are described in the property part of the template.

- robot ID
- waiting position (x, y, z)
- action time (moving to gripping position, gripping of molded part, moving to releasing position, releasing of molded part, moving to waiting position)
- electric energy consumption (moving to gripping position, moving to releasing position, moving to waiting position)
- calculation formula for cumulative energy consumption, etc.

35.5.4 Mold Data

Mold data are prepared in a database not in an e-catalogue because a mold has no behavior. The following are described in the mold data.

- mold ID
- molded part ID
- mold size (length, width, height)
- capacity of cavity
- current temperature of mold
- mold set-up time
- pre-heating temperature
- pre-heating time
- cooling time, etc.

35.5.5 Trial Experiment

Trial implementation of the construction system of a digital eco-factory for injection molding is underway. The e-catalogue creation system is implemented in parallel.

In the e-catalogue creation system, the abovementioned equipment templates are installed already. An equipment item is created by filling up the parameters of the corresponding template with concrete values, calculation formula of electric energy consumption, etc. At the time when a virtual production line is constructed for simulation, some of parameters of the template are retained as variables. Their values are provided from production scenarios, operation data and/or product data. Equipment items in e-catalogues can be provided by equipment vendors, and also by production engineers using practical data from their factory.

A virtual injection molding line is constructed as a multi-agent system of equipment agents generated from the selected equipment items in the above e-catalogue. This virtual injection molding line is a core of the digital eco-factory for injection molding. Simulation using this digital eco-factory for injection molding will provide time series changes of electrical energy consumption from various views such as an equipment level, a production line level and a factory level. This trial digital eco-factory for injection molding is structured on a free multi-agent programmable modeling environment "NetLogo" [15]. Figure 35.12 shows a part of screen shot of the simulation execution example using the generated virtual injection molding line. Changes of electric energy consumption for the injection molding machine and the molding extraction robot are shown. This screen shot was taken when one process cycle finished.

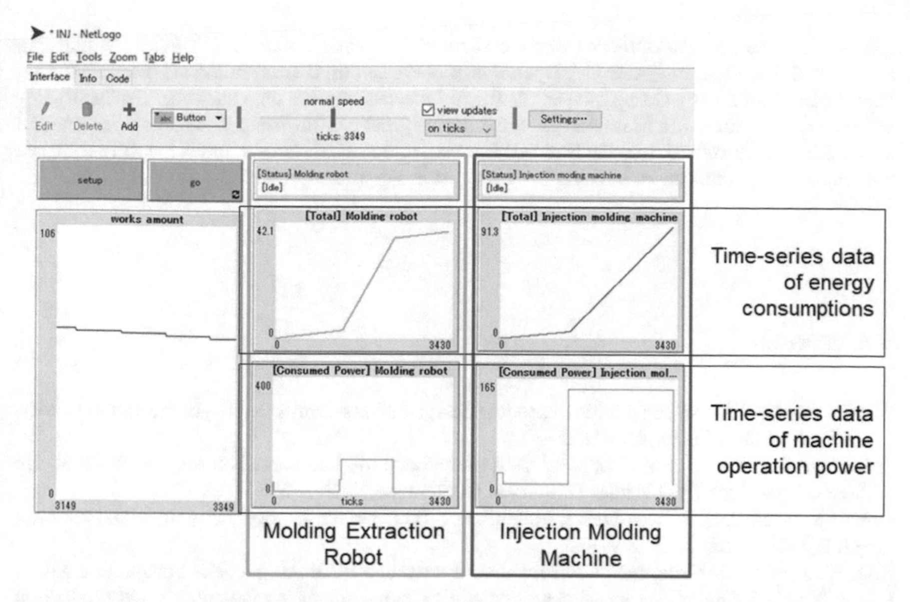

Fig. 35.12 Example display during the simulation

35.6 Conclusions

A virtual production line is a core of the digital eco factory. Environmental performance simulation on a virtual production system provides useful results for planning, operation and execution of sustainable manufacturing. The proposed virtual production line construction method using equipment catalogues has made possible a model-based construction.

A virtual production line is constructed as a multi-agent system. Its components are equipment agents which are derived from equipment catalogues. In other words, an equipment catalogue is a static description of an equipment model including its behavior. And an equipment agent is an executable model which is derived from the static description. As a result, a production line is modeled as a multi-agent system on which simulation is executable.

In this paper, the method for e-cataloging of equipment has been extended for constructing a digital eco-factory for injection molding with more complicated interactions among equipment. This extension indicates that the digital eco-factory with proposed e-cataloging method can be applied to more complicated production systems.

As a future work, by using implemented trial systems, experimental usage of a digital eco-factory is planned in industries. The result of the trial usage will show the feasibility of the proposed method and will lead to further practical developments in a digital eco-factory.

Acknowledgements The authors thank members of Technical Committee "DEcoF (Digital Eco Factory)" (2012. Oct.) by FAOP (FA Open Systems Promotion Forum) in MSTC (Manufacturing Science and Technology Center), Japan for fruitful discussions and their supports. Especially, the authors are thankful to Mr. Masayuki Kitajima for his great contributions. The authors are grateful to Dr. Udo Graefe, retired from the National Research Council of Canada for his helpful assistance with the writing of this paper in English. This work is supported by JSPS KAKENHI KIBAN (A) 18H03826.

References

Bley H, Franke C (2004) Integration of product design and assembly planning in the digital factory. CIRP Ann Manuf Technol 53(1):25–30

Freedman S (1999) An overview of fully integrated digital manufacturing technology. In: Proceedings of the 1999 Winter Simulation Conference, pp 281–285

Gregor M, Medvecký Š, Matuszek J, Štefánik A (2009) Digital factory. J Autom Mob Rob Intel Syst 3(3):123–132

Kozo Keikaku Engineering Inc., Users manual of artisoc, mas.kke.co.jp/cabinet/manual-en.pdf

Kuehn W (2006) Digital factory—integration of simulation enhancing the product and production process towards operative control and optimization. I J Simul 7(7):28–39

Matsuda M, Kasiwase K, Sudo Y (2012) Agent oriented construction of a digital factory for validation of a production scenario. Procedia CIRP 3:115–120

Matsuda M, Kimura F (2012) Configuration of the digital eco-factory for green production. Int J Autom Technol 6(3):289–295

Matsuda M, Kimura F (2013) Digital eco-factory as an IT support tool for sustainable manufacturing, digital product and process development systems, IFIP advances in information and communication technology vol 411, Springer, pp 330–342

Matsuda M, Kimura F (2015) Usage of a digital eco-factory for sustainable manufacturing. CIRP J Manuf Sci Technol 9:97–106

Matsuda M, Sudo Y, Kimura F (2015) A multi-agent based construction of the digital eco-factory for a printed-circuit assembly line. Procedia CIRP 41:218–223

Matsuda M, Matsumoto S, Noyama N, Sudo Y, Kimura F (2016) E-catalogue library of machines for constructing virtual printed-circuit assembly lines. Procedia CIRP 57:562–567

Matsumoto S, Matsuda M (2017) Construction of a virtual production line including a buffer for simulation of electric power consumption. In: Procedia CIRP vol 63, pp 465–470

Monostori L (2014) Cyber-physical production systems. Roots, Expect R&D Challenges: Procedia CIRP 17:9–13

Monostori L, Váncza J, Kumara S (2006) Agent-based systems for manufacturing. CIRP Ann Manuf Technol 55(2):697–720

NetLogo https://ccl.northwestern.edu/netlogo

Chapter 36
Additive Manufacturing for Circular Manufacturing: Trends and Challenges—A Survey in Japan, Norway, and India

Mitsutaka Matsumoto, Shingo Hirose, Kristian Martinsen, Suryakumar Simhambhatla, Venkata Reddy, and Sverre Guldbrandsen-Dahl

Abstract Circular manufacturing such as product remanufacturing, refurbishment, and repair is a key element in promoting a circular economy, whereas enhancing its resource-efficiency-increasing-effects is often dictated by the critical process of material surface restoration. One of the key technologies that enable effective material surface and geometry restoration is additive manufacturing (AM). This study presents, firstly, a survey of the existing industrial usage of AM in circular manufacturing, which includes applications of re-coating, cladding, and thermal spray. The survey was mainly conducted in three countries, namely, Japan, Norway and India. Secondly, the study presents a review of research and development of the applications of metal 3D printing—an advanced AM. Challenges hindering the promotion of AM applications in circular manufacturing are laid on a process-to-process basis. In general, such challenges include advancement of process automations, design for restoration, quality enhancement of metal 3D printing-based restoration, and cost reduction of the process applications.

Keywords Circular manufacturing · Remanufacturing · Material surface restoration · Additive manufacturing

M. Matsumoto (✉) · S. Hirose
National Institute of Advanced Industrial Science and Technology (AIST), Tsukuba, Japan
e-mail: matsumoto-mi@aist.go.jp

K. Martinsen
Norwegian University of Science and Technology (NTNU), Gjøvik, Norway

S. Simhambhatla · V. Reddy
Indian Institute of Technology Hyderabad (IITH), Hyderabad, India

S. Guldbrandsen-Dahl
SFI Manufacturing, Gjøvik, Norway

Y. Kishita et al. (eds.), *EcoDesign and Sustainability I*, Sustainable Production, Life Cycle Engineering and Management, https://doi.org/10.1007/978-981-15-6779-7_36

36.1 Introduction

Increasing the resource efficiency of economies is one of the keys in promoting a sustainable society. To increase the resource efficiency, promotion of material saving, material recycling, and prolonging the product/component lifetimes are indispensable. In recent years, activities prolonging the product/component usage time have become increasingly important. Such activities specifically include product remanufacturing, refurbishment, repair, and direct reuse (RRRDR), and are alternatively called value-retention processes (VRPs) (UNEP-IRP 2018). In this paper, we call these activities wholly as circular manufacturing. They attract attentions because they offer not only the opportunity to achieve significant environmental impact reduction, but also to create economic opportunities for cost reduction, value-retention, and employment opportunities (UNEP-IRP 2018).

Promoting circular manufacturing is quite challenging in developing process technologies, system technologies, and business management and policy arrangement (GIS RIT 2017; Matsumoto et al. 2016). This study focuses on the process technologies for RRRDR specifically for remanufacturing which enables high-value retentions. The process technologies indispensable for remanufacturing include techniques for disassembly/assembly, cleaning, inspection, material restoration, reliability assessment, and so on (GIS RIT 2017; Matsumoto et al. 2016). Among these technologies, material restoration technologies are the highlight of this study, with specific focus on additive manufacturing (AM). The material restoration critically influences the remanufacturing's effects on prolonging product/component lifetimes and thus, the effects of enhancing resource efficiency. The surface of the materials of used products/components can contain fracture, fatigue, wear, corrosion, etc., which need restoration for remanufacturing. The term AM is used, in some cases, to specifically indicate 3D printing, but in general, AM also includes techniques as welding, thermal spray, cladding, laser engineered net shape, cold spray, and so on. This study uses the term AM in the latter sense. At the same time, the study distinguishes 3D printing from other AM techniques.

This study conducted a survey of AM techniques that are used and can be potentially used in remanufacturing. It also explored the challenging issues in the techniques. The first survey was on the AM techniques used in remanufacturing today, based on visits and interviews with industries mainly in Japan, Norway, and India. The subsequent survey was focused on metal 3D printing that has been researched intensively in the last decade. The industrial applications today are mostly limited to new product/component manufacturing, however, the technique has already started to be applied in remanufacturing, and demonstrates a great potential that field in the future (GIS RIT 2017; Matsumoto et al. 2016). This study surveyed the applications of the metal 3D printing largely based on literature review. It further conducts an exploration of the challenges on the techniques.

The rest of this article is organized as follows. The next section outlines the challenges in process technologies of remanufacturing, and specifies the importance of material restoration technologies in the field. Section 36.3 summarizes the AM techniques used in today's remanufacturing, and outlines the challenges in the application of these techniques. Section 36.4 presents a review of the applications of metal 3D printing in remanufacturing, along with a summary of the challenges on the techniques. The final section summarizes the main points of the article.

36.2 Material Restoration in Remanufacturing

RRRDR prolong the usage duration of products and components, and thus contribute to saving materials and energy in manufacturing. In RRRDR, remanufacturing can retain the highest value of products and components; its process typically consists of: disassembly, cleaning, component inspections, restoration of components and materials, reassembly, and testing (Fig. 36.1). More importantly, remanufacturing restores the products (used products) to the original, as-new condition and performance or better (UNEP-IRP 2018).

Remanufacturing is employed in various product areas, the main areas being aerospace, heavy-duty and off-road equipment (HDOR), automobile parts, machinery, electronics and IT, and medical devices (Table 36.1).

Continuous research and development (R&D) enhance remanufacturing. It is specifically needed to assure the quality of remanufactured products and to control the costs at a reasonable level. Each of the processes shown in Fig. 36.1 requires technological development; nevertheless, while the techniques for these processes are

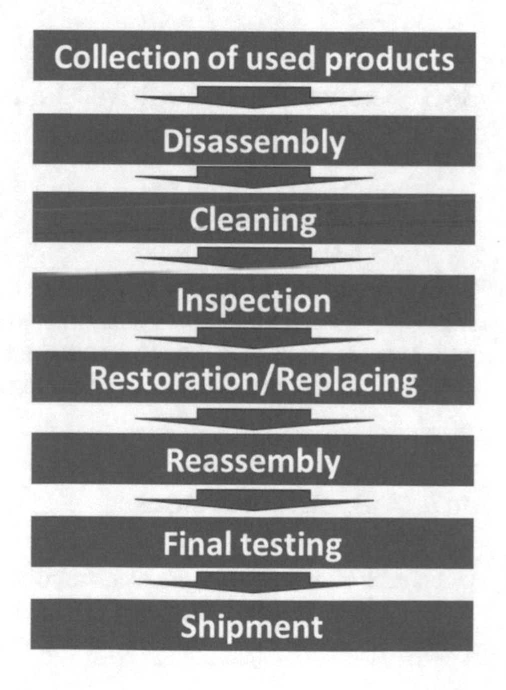

Fig. 36.1 Remanufacturing process

Table 36.1 Product areas of remanufacturing and market scales in the US and Europe

Product area	Production in US International Trade Commission (USITC) (2012) (Billions USD)	Production in Europe Eurpoean Remanufacturing Network (ERN) (2015) (Billions Euro)
Aerospace	13.0	12.4
HDOR	7.8	4.1
Auto parts	6.2	7.4
Machinery	5.8	1.0
Electronics and IT	2.7	3.1
Medical devices	1.5	1.0
Retreaded tires	1.4	–
Consumer products	0.7	–
Rail	–	0.3

important, technologies for inspection and material restoration are especially important for products for which material deterioration significantly affects the product performance and reliabilities. Such products include those in aerospace, HDOR, machinery, and automobile parts sectors.

Hence, this study investigated material restoration processes in remanufacturing, focusing on the conduct of a survey of AM techniques in these processes. Surveys in the industry were mainly targeted at the sectors of aerospace, HDOR, and automobile parts.

36.3 Conventional Additive Manufacturing Techniques Used in Remanufacturing

36.3.1 Survey Method

To investigate the material restoration processes in present remanufacturing, an author visited and interviewed nine firms to survey the process. Most of the firms were Japan-based companies. The product areas included: automobile parts, HDOR, aerospace, machinery, and power-generating facilities.

36.3.2 Application Cases

Oftentimes after usage of a product, the components incur surface damages. In some cases, the damages can be restored by scraping off the damaged part, a method categorized under subtractive manufacturing. Nonetheless, the technique is applicable when the depth of the layers to scrape off is within the range of tolerance. In other cases, the damages are restored by making up the part, under what is known as AM. AM techniques used in present-day remanufacturing include welding, cladding, and thermal spray.

The automobile parts sector is one of the sectors in which remanufacturing is observed actively. AM is used in the sector. The sector, however, is oftentimes associated with a relatively strong cost constraint, originating from the principle that the price of the original products in this sector is relatively lower than those of other sectors such as aerospace and HDOR. Figure 36.2 shows a case of AM being applied in automotive parts remanufacturing. Here, the company is a third-party remanufacturer of engines. The figure shows the process of restoration of damage on the surface of an engine crankcase. The damaged part is, first polished, then cladded by arc-welding, and then is finished by polishing. Cladding by welding is also used to restore component breakage. Figure 36.3 shows a case of another company. The drive shaft in the figure had broken splines, restored by cladding. These AM processes are a key in remanufacturing; however, they rely heavily on manual work, and thus, as

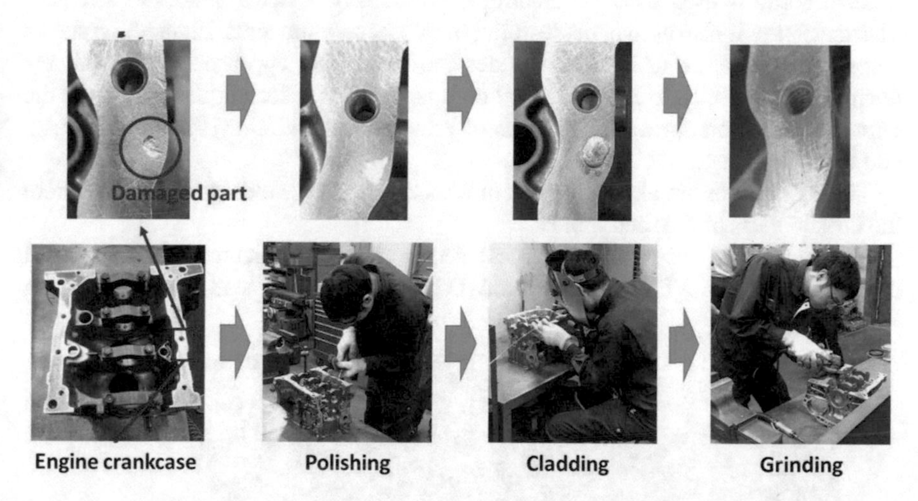

Fig. 36.2 Restoration of the surface of crankcase in automobile engine remanufacturing

Fig. 36.3 Restoration of splines of automobile driveshaft by cladding

labour wage increases in these countries, these processes become less economically feasible. This is the reason why component restoration is less done today.

HDOR is another sector in which remanufacturing is actively conducted. Typical examples of HDOR equipment include excavators, bulldozers, backhoes, asphalt pavers, and trucks. Remanufactured components in the sector include engines, starters, alternators, turbochargers, transmission axles, hydraulic cylinders, hydraulic pumps, reduction gear equipment, and so on. Of these components, material surface restoration of hydraulic cylinder is a critical remanufacturing task. Figure 36.4 shows a case of damages (scratch and corrosion) on the inner surface of a hydraulic cylinder tube. Damage on the cylinder surface may cause leakage of lubricant oil, resulting in the functional degradation of a product. As such, restoration of the damage is important in remanufacturing. The remanufacturing process for the tube starts with an inspection of the damage. If such damage is assessed as restorable, the cylinder (tube and/or rod) is electroplated, and then the plated surface is planarized by honing. In some cases, thermal spray is used instead of electroplating. Generally, thermal spray is costlier, but because an electroplating facility needs a large initial investment, thermal spray is used when an electroplating facility is not available in a region. Thermal spray is also used in the restoration of other components. Figure 36.5 shows a case of restoration of a shaft. Besides the mentioned approaches, various AM techniques are used and researched by companies. These techniques vary from the wire-arc spray, bore spray, high-velocity oxy-fuel spray, cold spray, thin film coatings, and laser cladding.

Dies and molds are also remanufactured. Cladding by welding is used to restore the broken parts of dies and molds.

The aerospace sector is the largest sector of remanufacturing in the global market (International Trade Commission (USITC) 2012; Eurpoean Remanufacturing

Fig. 36.4 Scratch and corrosion on hydraulic cylinder tube

Fig. 36.5 Thermal spray for surface restoration of a shaft

Network (ERN) 2015). Because the components are generally expensive, remanufacturing is conducted for various components; at the same time, the quality requirements for the components are governed by high and stringent remanufacturing standards. Customarily, remanufacturing is driven in large part by the mandatory aircraft safety inspections prescribed by national aircraft certification authorities (International Trade Commission (USITC) 2012). Cladding by welding is also widely applied for remanufacturing in the sector; cladding is used to restore some broken parts of such components as compressors, burners, casing, and disks. Nevertheless, the increasingly high requirements for the performance of the components make remanufacturing in this sector difficult. Turbine blades (Fig. 36.6), for example, were previously commonly repaired, but in recent years they have not been repaired that much. In turbine blade remanufacturing, the top coating layer is first removed, and then a new layer is coated. Today, turbine blades are used to tolerate increasingly high temperatures, to increase engine fuel efficiencies. To assure the high-temperature mechanical property of the blades, a new type of additive substances is added to the substrates. However, these new materials cause chemical reactions during the re-coating process in remanufacturing that degrade the blades' mechanical properties, thereby making remanufacturing a difficult task.

Fig. 36.6 Aircraft turbine blade

36.3.3 Trends and Challenges

AM has been widely used in remanufacturing, among which cladding by welding and re-coating are the most popular. The trends and challenges in these techniques are as follows. Firstly, the cost is critical in remanufacturing. Thus, cost reduction of the AM techniques is important. As restoration processes generally rely on manual work, as described in the automobile parts remanufacturing cases, automation or semi-automation of the inspection and these restoration processes should be advanced. Cost reduction is also important because a survey in the United States revealed that AM techniques are used mostly by large companies, not much by small-and-medium sized enterprises (GIS RIT 2017). To further activate circular manufacturing, development and commercialization of affordable AM techniques is necessary.

One approach to reduce the cost of re-coating is to develop local re-coating processes. In re-coating, generally the coated layer is removed from the entire surface and then a new layer is coated on the substrate. However, when only the damaged part of the layer is removed and re-coated, the total cost be reduced.

Another challenge is increasing the value and the quality of restoration. Conditions of damages in products vary depending on the customers' usage of the products. As such, remanufacturers can customize the restoration depending on the usage pattern of the customers, which can lead to an increase in the value of remanufacturing and the effects of prolonging the product lifetimes.

Technological development of material restorations for new types of materials is also needed. As with the case of aircraft turbine blade repair, new types of materials can make remanufacturing more challenging. In this case, re-coating techniques that do not harm the substrate by re-coating should be developed. Product design for remanufacturing (DfReman), or material design for restoration, is equally important.

36.3.4 Application in Surface Embedded Sensors

Another potential application of AM in circular manufacturing is the development of intelligent tooling based on surface-embedded sensors. Feasibly, AM can be employed to realize surface-embedded sensors; machine tools can become smarter by perceiving their own states and the state of the surrounding environment. The key enablers for this capability include smart sensors embedded into machine tools, a system that is useful in evaluating the tool's health conditions and predicting its remaining lifetime, as well as a possible breakdown or malfunction before that (Yang et al. 2018). The system is also useful for monitoring of the manufacturing process within the tools. Martinsen et al. conducted R&D of surface-embedded sensors used in injection molding dies (Martinsen et al. 2017). In principle, pressures and temperature distributions within the mold cavities are critical parameters for injection molding of thermoplastics components, whereas in-mold sensors are highly useful for improving the manufacturing process controls. In the research, the sensors were

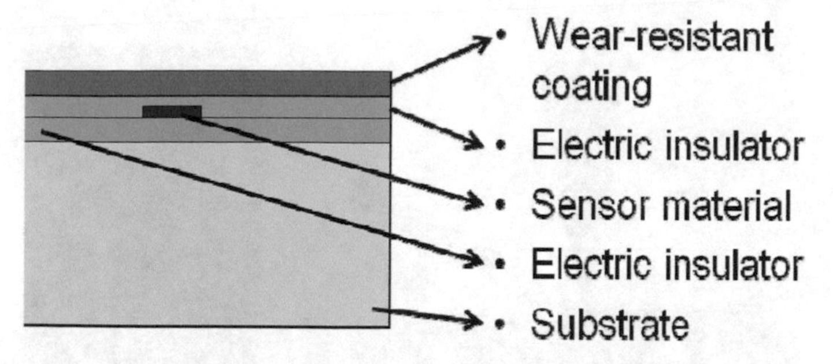

Fig. 36.7 Sensors embedded in coating layers (Martinsen et al. 2017)

fabricated on the surface of inserts on a bridge tool for injection molding using direct write thermal spray (Martinsen et al. 2017). Figure 36.7 shows the embedded sensors that were applied between layers of electric insulators. The data from the sensors were used to calibrate the process simulation of conformal cooling channels, which can enhance the analyses capabilities of the tool, process, and product behaviors, such as thermal stress, friction properties, and inlet issues (Martinsen et al. 2017).

36.4 Metal 3D Printing Applications in Remanufacturing

36.4.1 Metal 3D Printing

In a broad sense, AM refers to the process of creating an object by building something up. It is also often used as a synonym of 3D printing, which is actually a type of AM in the former sense. 3D printing is defined as the process of fabricating 3D parts through layer-by-layer deposition of materials. Metal 3D printing was first used to develop prototypes in the 1980s. As the technique improved by the early 2000s, it was employed to create functional products. More recently, companies have begun using metal 3D printing as integral part of their business processes.

3D printing has the potential to provide sustainability advantages, such as the generation of less waste during manufacturing due to it being an additive process, the capability to optimize geometries and create lightweight components that reduce material consumption in manufacturing and energy consumption in use, the increase in opportunities to repair, restore, and remanufacture due to the ability to rebuild 3D structures, the subsequent reduction in transportation in the supply chain, and inventory waste reduction due to the ability to create spare parts on-demand (Ahn 2016; Ford and Despeisse 2016). Of such advantages, the focus of this study is on the potential of the technique in increasing the opportunities for repair, restoration, and remanufacturing.

A number of different 3D printing methods exist currently. The American Society for Testing and Materials grouped these methods into seven categories including

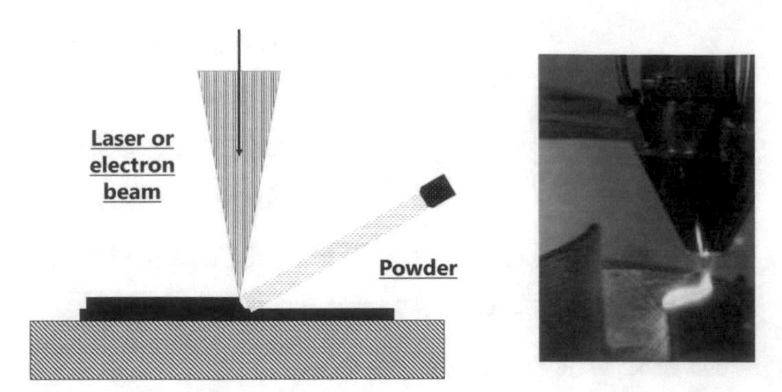

Fig. 36.8 Directed energy deposition (DED) process

binder jetting, powder bed fusion, sheet lamination, and directed energy deposition (DED). Of these, DED is well suited for remanufacturing as it can create a desired geometry along an arbitrary trajectory (Matsumoto et al. 2016; Yang et al. 2018). DED uses a focused thermal energy beam to melt and fuse materials. The typical DED process is illustrated in Fig. 36.8. Here the thermal energy beam is usually laser or electron beam. The materials added are typically in the form of the powder injected into the fusing zone with a flow of powder and inert gas through a nozzle.

36.4.2 Applications of Metal 3D Printing in Remanufacturing: A Review

Metal 3D printing applications in remanufacturing are still in an immature stage; most are in the R&D stage.

Research has been undertaken most actively in the aerospace sector. The companies and research organizations have applied DED to the repair of various aircraft components, including turbine blades, blade disks (blisks), compressor drums, compressor seals, mobiles, drive shafts, drive couplers, titanium components, etc. (Ahn 2016; Mudge and Wald 2007). Mudge and Wald, for example, reported the repair of the worn bearing housing of a gas turbine engine, cost of which was about 50% of a newly manufactured product (Mudge and Wald 2007). They also reported the repair of a drive shaft and a comparison of the repair by DED and thermal spray. The hardness and the corrosion resistance of the repaired region by DED were greater than those of the repaired region by thermal spray (Mudge and Wald 2007). Additionally, the spalling phenomenon was not observed in the region repaired by DED, whereas severe spalling appeared in the region repaired by thermal spray (Mudge and Wald 2007). Such a case, the repair cost of DED was approximately 50% that of the thermal spray (Mudge and Wald 2007).

Similarly, water turbine remanufacturing by DED has been researched. In the remanufacturing, after measurement of the amount of wear, the additive machine added the material (Stellite) directly on the worn surface, consequently prolonging the expected lifetime of the turbine (Matsumoto et al. 2016).

Remanufacturing of dies and molds is another example of DED applications. Figure 36.9 illustrates a process of remanufacturing worn-out dies. The region containing cracks was processed by pocket machining, and was restored by laser metal deposition (a DED type), and was finished by post machining (Fig. 36.9). In a case where a company applied DED to the repair of a die, the service life of the repaired die was extended by 250% (Ahn 2016). Moreover, DED-based remanufacturing can lead to extended product lifetime, shorter lead time, cost and energy savings (Ahn 2016). AM process has been applied not only to the repair of worn-out dies, but also to the remanufacture of dies for new products and processes (Matsumoto et al. 2016). There is also a study on remanufacturing dies with the hard-faced layer consisting of a wear-resistant material (Ahn 2016).

There were also application cases of other metal 3D printing techniques besides DED. Siemens Power Service, for instance, applied the powder bed fusion process for the repair of damaged gas turbine burner tips wherein the tips were machined before they were placed into the powder bed of a selective laser melting machine, and before new tips were built onto the machined surface (Fig. 36.10) (Matsumoto et al. 2016).

There are explanatory studies on assessment of energy use and environmental impacts of AM processes (Kellens et al. 2017). The merits of AM techniques, ultimately, should be judged based on not only their technical and economic advantages, but also their environmental effects.

In India, metal 3D printing is mostly looked at in the context of repairing, refurbishing, and remanufacturing old components. Examples of the regional R&D include refurbishment of railway components with laser-based DED; and repair of forging dies with the use of wire-arc AM.

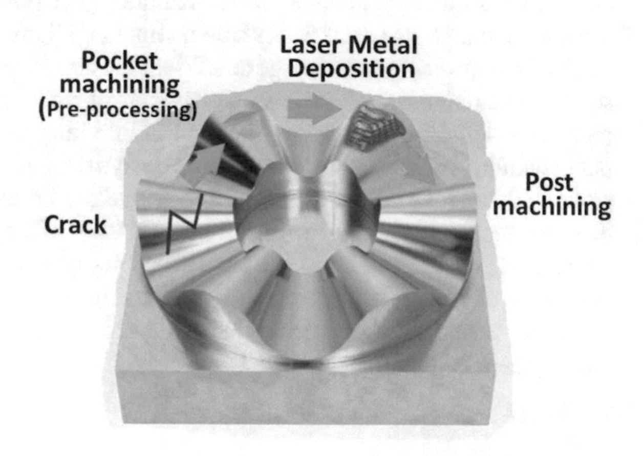

Fig. 36.9 Illustration of die remanufacturing by DED . (*Source* Okuma corporation)

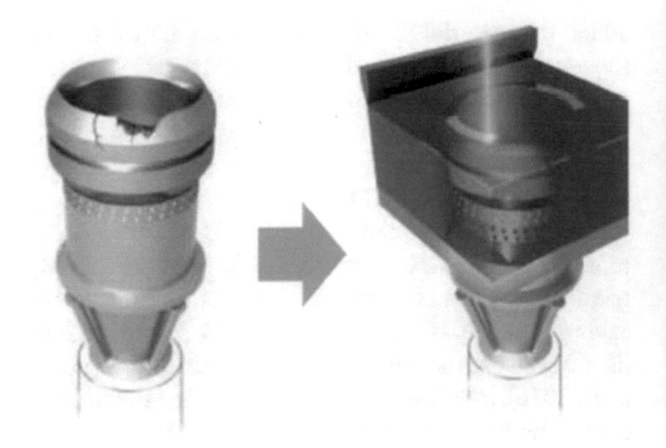

Fig. 36.10 Gas turbine burner tip repair by powder bed fusion (Matsumoto et al. 2016)

36.4.3 R&D Challenges

As earlier mentioned, the application of metal 3D printing in remanufacturing is still in the early stage of development. Nevertheless, the technology can be seen as having a huge potential in the future. Realizing such potential requires the achievement of two goals, namely, (a) improving the quality of the restored region and the interface (the border between the original part and added region), and (b) reducing the process cost.

These goals can be achieved through advancing the technological development of each of the processes for metal 3D printing-based remanufacturing. Figure 36.11 provides an outline for these processes, which include (1) condition assessment and digitization, (2) restoration process, and (3) post machining and inspection.

For the initial stage, as the objects are deformed, torn, and have other defects, it is essential to acquire the geometrical data of the object, practically through various 3D scanning methods as laser triangulation, structured light, and probing methods. The material composition and damage conditions are also inspected at this stage. The next stage is restoration, which mainly involves path planning, material definition, process parameter definition, preprocessing, and 3D printing processing (AM process). In path planning, in case the original digital design data of the product is available, the scanned data of the object are compared with the original design data using reverse engineering software, and the information is used to create the tool path plan. The restoration process is followed by post machining and inspection process, mainly consisting of tool path planning, post-processing, and post-inspection.

These processes require technological development. The major challenges of such development include:

- Material composition analysis technique

DED can deposit a wide range of materials including: tool steels, stainless steels, titanium and titanium alloys, aluminum alloys, nickel-based alloys, cobalt-chromium

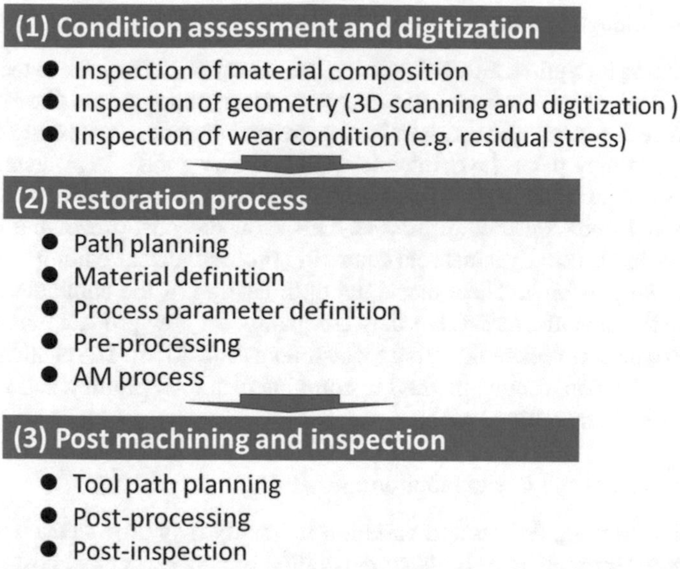

(1) Condition assessment and digitization

- Inspection of material composition
- Inspection of geometry (3D scanning and digitization)
- Inspection of wear condition (e.g. residual stress)

(2) Restoration process

- Path planning
- Material definition
- Process parameter definition
- Pre-processing
- AM process

(3) Post machining and inspection

- Tool path planning
- Post-processing
- Post-inspection

Fig. 36.11 Metal 3D printing-based remanufacturing process

alloys, copper alloys, gold and silver (Matsumoto et al. 2016). As the material composition of the region to repair is often unclear, a method for quickly analyzing the elemental composition of the materials would be essential. Developing analysis techniques such as advanced X-ray fluorescence is a challenge.

- Computer-aided manufacturing (CAM) to create flexible machining path

Users of the 3D printing device expect such to create any type of geometry once the numerical data of the geometry (computer-aided design data) are input to the device. Actually, the conditions under which a certain geometry can be created are limited due to differences in shape and the materials. The processing conditions are device-specific know-how, and are accumulated as "process recipe." In order to advance 3D printing-based remanufacturing, it is necessary to develop flexible CAM software and build a system that enables improving and upgrading the process recipe.

- Material and structural property analyses

To enhance the reliability of remanufacturing by 3D printing, further analyses and understanding of the material and structural properties are necessary. The strength of the restored region and interface is affected by structural properties of, for instance, crystallinity and crystal anisotropy. Defects such as porosity also affect the strength. However, local failure mechanisms are still not completely understood. Uncertainty in the material properties is currently a challenge for the establishment of reliable remanufacturing processes using metal 3D printing.

- Post machining planning

Post machining is required after 3D printing the damaged part to achieve the desirable dimensional tolerance, hardness and surface roughness and to ensure like-new performance. There are techniques to minimize the surface roughness, for example, through optimizing building paths, layer thickness, and powder grain size (Matsumoto et al. 2016). However, a better surface from the 3D printing process usually means a more expensive and time-consuming process. This is where post-processing comes in. Post-processing is usually a material-removing process such as cutting and abrasive processes like polishing. There exists the optimization of the combination of AM and subtractive manufacturing, but they are mainly for new product manufacturing and not designed for remanufacturing (Matsumoto et al. 2016). The challenge on this respect is to develop a sound method to combine metal 3D printing and subtractive manufacturing in an optimal way.

- Post-inspection and in-situ monitoring

In metal 3D printing, defects and variation in quality may arise in each product as it is fabricated through layer-by-layer deposition of materials; thus, post-inspection is important. An X-ray computerized tomography scan technique can be used to inspect the fabricated product; however, it is expensive and comes with a problem, wherein detection becomes difficult as the defect size becomes smaller. This issue can be resolved via development of highly precise and low-cost quality inspection techniques and establishment of an appropriate inspection standard. In addition, if the defect occurrence during the process can be detected and further reflected in the process parameters immediately, the quality of restoration can improve significantly. Development of such in-situ monitoring technique is also a challenge.

- Cost reduction of the process

Cost reduction of the process is important to expand the use of metal 3D printing in remanufacturing. Here, price reduction of the device is a challenge. The high price of powder materials today is another hindrance factor for the expansion of the applications; thus, reducing the powder cost is also a challenge.

36.5 Conclusion

The study presented a survey of existing industrial usage of conventional AM in remanufacturing and a review of R&D of metal 3D printing applications in remanufacturing. Conventional AM is widely used in remanufacturing, of which cladding by welding and re-coating are the most popular. Further promoting the usage of conventional AM in circular manufacturing essentially incorporates cost reduction and quality enhancement of AM, and associates with the challenges of advancing the automation of the restoration processes, developing partial re-coating processes, and material design for restoration.

On the other hand, metal 3D printing has been intensively researched in the last decade, with it being used initially in prototype manufacturing. The technique has great usage potential in remanufacturing in the future. Nevertheless, realizing such potential would necessitate an increase in the quality (strength) of the restored region by 3D printing and again, minimal cost. The challenges for this objective include the development of quick material composition analysis technique, CAM to create optimal machining path, increase of material and structural property understanding, development of post machining, post inspection, and in-situ monitoring techniques.

Furthermore, material surface restoration and geometry restoration processes are critical in enhancing the resource-efficiency-increasing-effects of circular manufacturing. In order for advanced AM to have a significant impact on the industry and reduction of the environmental impacts of manufacturing, R&D of process technologies outlined in this study should be realized. In addition, a product design for remanufacturing should be adopted, and a production system for smart remanufacturing, a supply chain for AM-based remanufacturing, and a business model for remanufacturing must be established. Future work includes identification of country-specific priorities of R&D in AM-supported circular manufacturing. The priority of R&D differs in different industries and product areas, and thus, the priority differs in different economies. An appropriate application of AM techniques is one of the keys in realizing circular manufacturing.

Acknowledgements This study is partially supported by INMAN project which is funded by Norwegian INTPART program.

References

Ahn DG (2016) Direct metal additive manufacturing processes and their sustainable applications for green technology: a review. Int J Pr Eng Man-GT 3(4):381–395

European Remanufacturing Network (ERN) (2015) Remanufacturing market study. European Commission

Ford S, Despeisse M (2016) Additive manufacturing and sustainability: an exploratory study of the advantages and challenges. J Clean Prod 137:1573–1587

GIS RIT (2017) Technology roadmap for remanufacturing in the circular economy

Kellens K, Baumers M, Gutowski TG, Flanagan W, Lifset R, Duflou JR (2017) Environmental dimensions of additive manufacturing. J Ind Ecol 21(S1):S49–S68

Martinsen K, Gellein LT, Boivie KM (2017) Sensors embedded in surface coatings in injection moulding dies. Procedia CIRP 62:386–390

Matsumoto M, Yang SS, Martinsen K, Kainuma Y (2016) Trends and research challenges in remanufacturing. Int J Precision Eng Manuf—Green Technol 3(1):129–142

Mudge RP, Wald N (2007) Laser engineered net shaping advances additive manufacturing and repair. Welding J 86(1):58–63

U.S. International Trade Commission (USITC) (2012) Remanufactured goods: an overview of the U.S. and global industries, markets, and trade. USITC Publication 4356, Investigation No. 332–525

UNEP-IRP (2018) Re-defining value—The manufacturing revolution. Remanufacturing, refurbishment, repair and direct reuse in the circular economy

Yang SS, Raghavendra AMR, Kaminski J, Pepin H (2018) Opportunities for Industry 4.0 to support remanufacturing. Appl Sci 8(1177):1–11

Printed in the United States
by Baker & Taylor Publisher Services